The Elements in the Medieval World – Interdisciplinary Perspectives: Water

Elements, Nature, Environment

MULTIDISCIPLINARY PERSPECTIVES FROM THE ANCIENT TO THE EARLY MODERN WORLD

VOLUME 1

The titles published in this series are listed at *brill.com/ene*

The Elements in the Medieval World

Interdisciplinary Perspectives

Water

Edited by

Marilina Cesario
Hugh Magennis
Elisa Ramazzina

BRILL

LEIDEN | BOSTON

Cover illustration: Landscape study of water, sky and pine trees by Albrecht Dürer, ca. 1497. Watercolour and bodycolour (no. 125889001) ©The Trustees of the British Museum.

The Library of Congress Cataloging-in-Publication Data is available online at https://catalog.loc.gov
LC record available at https://lccn.loc.gov/2024027478

Typeface for the Latin, Greek, and Cyrillic scripts: "Brill". See and download: brill.com/brill-typeface.

ISSN 2950-0079
ISBN 978-90-04-54842-8 (hardback)
ISBN 978-90-04-69650-1 (e-book)
DOI 10.1163/9789004696501

This book is printed on acid-free paper and produced in a sustainable manner.

Contents

PART 3
Water in Thought and Practice

PART 4
Romance, Religion, Myth

Foreword: Water Power

David N. Livingstone

In 1943 the distinguished American anthropologist, Ruth Benedict, paused in her obituary of one of America's 'great scientists', the celebrated ethnologist and linguist Franz Boas, to comment on his transmutation from geophysicist to cultural anthropologist. She attributed this conversion to his experience of preparing his doctoral dissertation at the University of Kiel on 'The Nature of the Color of Sea Water' combined with his field expedition to Baffinland in 1883. Boas's time there, according to Benedict, taught him that 'if we are ever to understand human behavior we must know as much about the eye that sees as about the object seen'. It was among the Canadian Inuit, she claimed, that Boas came to understand 'once and for all that the eye that sees is not a mere physical organ but a means of perception conditioned by the tradition in which its possessor has been reared'.[1] Here, no doubt, the lines of historical causation tracking the sea-change in Boas's intellectual outlook to changes in the colour of sea water are rather too starkly drawn. As he moved from physical materialist to cultural historicist many forces were evidently at work. And yet it was his experiments centring on the absorption of light by a range of distilled water samples that led him to the psychophysics then under development by the physicist and psychologist Gustav Fechner. Reflecting on these manoeuvres, George Stocking observed that 'On the basis of sensory experiments in which he was both subject and experimenter, Boas argued that there were *situational* factors (e.g., the mental state of the subject) which affect perception'. This realisation encouraged him to occupy a conceptual space somewhere between what has been called 'the hairy philosophy of monistic materialism' and the 'historicist spirit of romantic idealism'.[2] What further cemented this entire line of thinking were Boas's connections with the tradition of German historical geography which impelled him towards a more holistic and affective appreciation of humanity's entanglements with the natural order. In this philosophical niche he found it possible to synthesise an intellectual interest in the physics of nature with an intensely emotional response to the physical world. As Simon Schaffer has powerfully reminded us, however superannuated some of Boas's

1 Benedict R., "Obituary. Franz Boas", *Science* 97.2507 (1943) 60–62.

2 Stocking G.V., "From Physics to Ethnology: Franz Boas's Arctic Expedition as a Problem in the Historiography of the Behavioral Sciences", in Stocking G.W. (ed.), *Race, Culture and Evolution: Essays in the History of Anthropology* (Chicago: 1982) 142, 140. Here Stocking resorted to the language of the Princeton historian of science, Charles Gillespie.

ideas plainly are, his efforts to keep in tandem what he 'famously called "the logics of science" and "the logics of life"' continue to have resonance.[3]

Why do I begin these remarks on a suite of chapters dwelling on the myths and meanings of water throughout the length and breadth of medieval culture with a tale of a late nineteenth-century physicist transmuting into a cultural anthropologist? The story of course centres on the power of water to induce conceptual awakening. As one contributor to this collection of essays notes, the cultural history of water can well be seen as emblematic of broader changing attitudes towards the world in general, while another laments the way in which different conceptions of water as a physical entity have largely been ignored by historians. But Boas's transformation also displays something of the relevance of historical geography to inquiring into the nature and meaning of water as *both* a physical phenomenon *and* a cultural construct. As a geographer-historian, my own academic centre of gravity lies in rather later periods of history than those of the experts whose scholarship is on display in these pages. Nevertheless, given the salience of Boas's experiences, I have been encouraged to accept the editors' gracious invitation to provide some prefatory reflections on this collection in hopes of furthering the conversation between medieval scholars and historical geographers, as well as between past and present.

What further reinforced this aspiration was the realisation that contributors to this collection had found inspiration in a variety of cultural geography projects. The work of old friends, such as Anne Buttimer on water symbols and the quest for wholeness, and Paul Cloke, along with Owain Jones on the culture of trees, informs investigation into the narrative power of water in twelfth-century Byzantine texts. Perhaps not surprisingly the exposition of medieval oceanic imaginaries draws on the writings of John Allen on early North American exploration, John Short on the role of cartography in indigenous encounters in the New World, Lesley Cormack on the evolution of the science of geography in the early modern period, and Matthew Edney on cartographic knowledge in the early Atlantic World. Elsewhere the rich resources available in the monumental project on the history of cartography spearheaded by David Woodward and Brian Harley, are evident in chapters exploring what are called the 'demonic seas', and early Arabic conceptions of the hydrological cycle. Alongside these, reference is to be found to the earlier interventions of Eva Taylor, the first woman to hold an academic chair of geography in the United Kingdom, on the history of navigational practices, as well as the more recent, and controversial,

3 Schaffer S., *From Physics to Anthropology – And Back Again* (Cambridge: 1994) 42.

social history of the Mercator projection by Mark Monmonier. Likewise, an inquiry into the polyvalent meanings of water in the Gothic Bible finds value in work on the geography of religion by practitioners such as David Sopher, Lily Kong and Chris Park. My list could certainly be extended. But multiplying testimonials is not my aim. It is rather to report that links between geographical research and medieval scholarship turn out to be both more expansive and more engaging than I had hitherto imagined.

For my own part, the remarkably rich diversity of themes that wend their way through this collection of essays incorporates subjects whose relevance has continued to reverberate down through the centuries that followed, and in some cases right up to the present day. They have never been served an eviction notice even if they have undergone architectural remodelling. As one contributor to this collection notes, there are residual elements of the medieval mindset that continue to shape human relations with the natural environment in our own time. Understanding these enduring attitudes is a real desideratum. And this is all the more urgent in the light of the accelerating problems of the Anthropocene era with its unprecedented human-induced climate change. Unearthing the roots of the impulse to indulge in a too-easy separation of humans from the natural environment, and uncovering resources for mitigating it, makes the interrogation of past texts a critically significant undertaking. Indeed it is the environmental crisis that currently faces humanity that animated Dipesh Chakrabarty to call for a radical rethinking of the entire humanist tradition in the light of the revelation of *Homo Sapiens* as an aggressive and destructive environmental agent.[4]

Mapping, even in minimal detail, the outlines of the rich legacy we have inherited from the medieval ontologies of water, is obviously beyond both the scope of these reflections, and indeed my capacity to take the true measure of that inheritance. Instead I have chosen to dwell on a few relatively discrete, through frequently inter-related, themes that seem to me to have proven to be resilient in the face of changing intellectual fashions. At least in reading through the variety of ways in which water was conceptualised and mobilised in different arenas during medieval times – medical, mythological, hagiographical, theological, legal and so on – the following presented themselves to me as conspicuous landmarks on this volume's intellectual horizon.

4 Chakrabarty D., "The Climate of History: Four Theses", *Critical Enquiry* 35.2 (2009) 197–222. There is certainly much to this claim, though I have highlighted some reservations in Livingstone D.N., *The Empire of Climate: A History of an Idea* (Princeton, NJ: 2024).

1 Water, Weather and Wellbeing

The manifold ways in which water and wellbeing were associated throughout the Middle Ages is a striking *Leitmotif* in the essays that follow. Hydrological theories of health occupied a prominent place in the medical arts of the time. Sometimes this took the form of the role of water-drinking in dietary regimens; sometimes the prescribing of bathing therapies for dealing with various diseases came to the fore. We learn that in medieval Irish culture, for example, bathing the sick was an integral part of medical treatment. These practices were frequently connected with reports of miraculous cures involving immersion in water as well as folk remedies invoking the curative powers of holy water. During the later medieval period, therapeutic balneology flourished. Here bathing procedures were intimately connected with Hippocratic humoralism and were therefore governed by guidance on the appropriate seasons of the year for patients of different temperaments to subject themselves to water-cure treatments. The reason was simply because different humours were believed to become dominant at different points in the seasonal cycle.

In one way or another, associations between water, weather and wellbeing persisted over the centuries that followed. During the mid-nineteenth century, for example, hydropathy underwent a conspicuous revival largely courtesy of the endeavours of a Silesian farmer Vinzenz Priessnitz (1799–1851) and Sebastian Kneipp (1821–1897), a Dominican monk. Its popularity spread quickly.[5] Among the literati who availed themselves of water-cure facilities was Charles Darwin, who underwent hydropathic treatment at Dr James Gully's facility in the Malvern Hills.[6] And even though many were sceptical about its medicinal value, Gully made a fortune out of the practice. As for Darwin himself, he responded well to the treatment. Seen in its wider setting, the Victorian middle- and upper-class enthusiasm for hydropathy can be interpreted as a return to ancient natural healing as 'a drugless system of therapeutics intended to tranquilise and stimulate the nervous system', and thus as a point of resistance to the prevailing medical and social assumptions of the time.[7]

5 See, for example Price R., "Hydropathy in England 1840–70", *Medical History* 25 (1981) 269–280; Bradley J. – Dupree M.W., "A Shadow of Orthodoxy? An Epistemology of British Hydropathy 1840–1858", *Medical History* 47.2 (2003) 173–194; Legan M.S., "Hydropathy in America: A Nineteenth Century Panacea", *Bulletin of the History of Medicine* 45.3 (1971) 267–280. For the practice of hydrotherapy in colonial contexts see Jennings T., *Curing the Colonizers: Hydrotherapy, Climatology, and French Colonial Spas* (Durham, NC: 2006).

6 See Desmond A. – Moore J.R., *Darwin* (London: 1992) 363–374. Gully was the author of *The Water Cure in Chronic Disease*.

7 Price, "Hydropathy in England" 270.

Perhaps more conspicuously, the latest revitalisation of the Hippocratic *Airs, Waters and Places* medical tradition in our own time speaks to the urgent health risks facing the globe on account of global warming, environmental change and extreme weather events. In various locations, severe flooding has brought about a marked escalation in the occurrence of cholera, gastroenteritis and typhoid. And it is predicted that higher ambient temperatures and flood-related run-off will increase the incidence of various types of cancer. Significantly, the 2003 World Health Organisation report, *Climate Change and Human Health,* staged the contemporary climate-health crisis as 'an old story writ large'. In substantiating the claim that the current climate-change challenge recalls the 'ancient struggle' between climate and health, the report turned to Hippocrates as its first port of call. The recognition of how human wellbeing could be affected by anthropogenic climate change, it noted, was 'a recent development'. But it insisted nonetheless that the fundamental realisation that human health and disease were closely linked to airs, waters and places had been heralded by 'the Greek physician Hippocrates' who drew attention to the causal connections between seasonal variation and the outbreak of epidemics, and called on medical practitioners to attend to ecological conditions when establishing settlements.[8] These self-conscious references to the Hippocratic tradition might seem merely ornamental, but as the eminent historian of medicine, Charles Rosenberg, noted, there remain 'striking if schematic continuities between the holistic world of Hippocratic medicine and today's medical understandings'. Perhaps most significantly, he continued, is the perpetuation of the emphasis that the tradition of *Airs, Waters, Places* placed 'on the body as always situated and always in process, always interacting with the environment that sustains and threatens it'.[9]

The links between water, weather and wellbeing, moreover, have long extended beyond the physical realm into matters of mental health. As this collection indicates, in his *Etymologies*, Isidore of Seville insisted that psychological harmony could be re-established through balneological practices since water had the power to banish sorrow and exorcise anxiety. In early Christian culture, healing encompassed both body and soul, and a variety of biblical incidents in both Old and New Testaments stage water as a central element in the restoration of psycho-physical wholeness. In more recent times, that holistic promise continued to attract many seekers in search of wellbeing, both

---

8 McMichael A.J., "Global Climate Change and Health. An Old Story Writ Large", in McMichael A.J. – Campbell-Lendrum D.H. – Corvalán C.F. – Ebi K.L. – Githeko A.K. – Scheraga J.D. – Woodward A. (eds.), *Climate Change and Human Health* (Geneva: 2003) 17.

9 Rosenberg C.E., "Epilogue: *Airs, Waters, Places*. A Status Report", *Bulletin of the History of Medicine* 86.4 (2021) 661–670, 668.

psychological and physiological, to spa waters and thalassotherapy centres as health-promoting getaway sites. In the nineteenth century, what has been dubbed the 'wellness craze' gripped many members of the rich urban leisure classes who flocked to both hot– and cold-spring resorts for mental relaxation and respite from nervous exhaustion and pathological anxiety.[10]

Matters of mind and meteorology have intertwined in other ways too. In this collection one contributor, drawing on contemporary extended mind theory in tandem with Latour's actor-network proposals, argues that human minds are continually shaped and reshaped by engagement with the external landscape as well as by internal neurological chemistry. In a related key, other scholars have long argued for the role of water and other environmental conditions in moulding the mental capacities necessary for the growth of civilisation. Karl August Wittfogel's theory of hydraulic civilisations, and its reanimation in Harold Dorn's ecological reading of the development of science and technology, are cases in point.[11] But perhaps the most conspicuous manifestation of the mind-meteorology nexus has been the notable concern about the influence of climate change on mental health and mass psychology. Feelings of sadness, anger and low self-esteem are reported to be more frequent on rainy days, while acute and chronic mood swings, sometimes resulting in suicide, have been chronicled in the wake of hurricane-induced flood disasters. These and a host of comparable weather-related disorders have been the subject of extensive study in recent decades and have generated a myriad of websites devoted to providing sufferers from climate sensitivity with medical and psychological advice.[12]

2 Hydrotheology

Bringing human mental states within the compass of the agency of water, points to another prominent theme in the chapters that follow – the spiritual significance of water throughout the medieval epoch. The diverse ways in which water and the spirit surfaced in the period are both extensive and

10 Smith C., *City Water, City Life* (Chicago: 2013).

11 Wittfogel K., *Oriental Despotism: A Comparative Study of Total Power* (New Haven: 1957); Dorn H., *The Geography of Science* (Baltimore, MD: 1991).

12 For the merest sampling see Clayton S. – Krygsman K. – Speiser M., "Mental Health and Our Changing Climate. Impacts, Implications, and Guidance", *American Psychological Association* (2017); Overy C. – Tansey E.M. (eds.), *The Recent History of Seasonal Affective Disorder* (SAD) (London: 2014); Klimstra T.A. – Denissen J.A. – Koot H.M. – van Lier P.A.C. – Frijns T. – Keijsers L. – Quinten L. – Raaijmakers A.W. – van Aken M.A.G. – Meeus W.H.J. "Come Rain or Come Shine: Individual Differences in How Weather Affects Mood", *Emotion* 11.6 (2011) 1495–1499.

remarkable. In the light of our reflections above, it is not surprising that physical *and* spiritual ills were often elided in the medical regimens of the time. Here we find too that in the narratives of the lives of saints, water was a dominant feature in accounts of miracles, cleansing and healing. This serves to recall the biblical image that those who drink the water of life will never thirst again. Whether in the context of baptism, foot-washing, ritual cleansing or curing leprosy, the symbolic significance of water is conspicuous. At the same time, an inquiry here into the semantic domain of water in the Gothic Bible discloses how the trope of 'living water' shared its ontological habitat with the perception of water as a threat, a space of chaos, a site of peril. Whether in narratives of stilling the tempest, surviving shipwreck, or needing a heavenly pilot to secure safe passage through life's stormy seas, water served as an archetypal theological and moral symbol.

In the biblical and medieval theology of water, the River Jordan occupies a privileged ecological niche. It is the border of the Promised Land, the setting for the healing of Naaman and the location of the baptism of Christ. And, as we discover in this collection, it likewise appears as a potent symbol of spiritual advancement in the medieval hagiographical literature. Perhaps not surprisingly it was the subject of a range of projects in the western artistic tradition. Because Christ's baptism was of such iconic moment in Christian theology, the River Jordan was understandably the subject of major artistic endeavour. Depicting the colourless character of this watery substance on canvas was of huge interest, but, as this volume discloses, presented technical problems of a particularly taxing sort as practitioners struggled to resolve the representational challenge of how to depict a transparent medium by a variety of devices – cocoons, mounds, ribbons.

Out of these spiritualising impulses, supplemented by images of the deep as monstrous – as in the story of Jonah and the whale, the portrayal of Job's Leviathan, the seven-headed beast of the Apocalypse arising out of the sea, and the like – a whole theology of the sea as a space of sin as well as salvation was sculpted. We learn that water, at once a medium of cleansing and a site of demonic forces, occupied the theologies of the likes of Augustine, Gregory, Basil and Chrysostom. Much more specifically, attempts to explain the mysteries of the hydrological cycle dating back to Aristotle's meteorology, commanded the attention of Christian and Islamic scholars alike. A variety of more or less naturalistic accounts resorting to thermal or mechanical explanations of this phenomenon were developed alongside, and sometimes in tandem with, providential readings of the natural order. In this volume's analysis of the water cycle in Arabic meteorology, we discover that such problems as the salinity of sea water and the natural philosophy of hail could be given a teleological rendering even if the mechanics of their operation were entirely natural. In this way divine action could be seen as thoroughly compatible with natural causation, and the hydrological cycle could thus be viewed as a manifestation of the wisdom of God.

In one form or another, these strains of thought continued to reverberate over the following generations. For me, three later manifestations in particular stand out. First, the hydrological cycle continued to play a critical role in the elaboration of a scientised natural theology. The thought that the water cycle bore witness to the wisdom of God captured the minds of numerous natural philosophers and theologians, sometimes being presented as the *locus classicus* of the teleological argument. In the late seventeenth century the mathematician John Keill could detect the hand of Providence in explaining the abundance of the earth's water as a means of supplying sufficient vapours for the needs of the whole world. In the years that followed, Richard Bentley told the readers of his Boyle lecture, *A Confutation of Atheism, from the Origin and Frame of the World* (1693), that he could readily perceive divine action in the globe's water-cycle, as did William Derham in his *Physico-Theology* of 1713 and William Wollaston in his 1722 *Religion of Nature Delineated*. Soon the College of Hamburg philosopher-theologian Johannes Albert Fabricius could bring out a theological synthesis under the title *Hydrotheologie* (a companion to his *Pyrotheologie*) in his mission to synthesise spiritual and scientific learning.[13] In one form or another, hydrological natural theology persisted throughout the nineteenth century and in some cases well into the twentieth. The United States naval officer and pioneer of oceanography Matthew Fontaine Maury (1806–73), for instance, found in the mechanistic operations of marine and atmospheric circulation systems and of energy transfers between land, sea and air, traces of the divine architect, while the Swiss-born geologist-geographer Arnold Guyot (1807–1884) at Princeton viewed the oceans, springs and rivers as indications of God's superintendence of nature.[14]

Second, the theologisation of the hydrological cycle resonated with the ongoing inclination to engage in what might be called 'moral meteorology', namely subjecting various climatic zones to moral judgment. Nowhere perhaps was this more conspicuously on view than in the tropical world, or torrid zone, as it was frequently termed, during the colonial period. At least since the eighteenth century, tropical climates, while sometimes associated with ideas

---

13 See Tuan Yi-Fu, *The Hydrologic Cycle and the Wisdom of God. A Theme in Geoteleology* (Toronto: 1968); Wright J.K., *The Geographical Lore of the Time of the Crusades. A Study in the History of Medieval Science and Tradition in Western Europe* (London: 2018); Glacken J., *Traces on the Rhodian Shore* (Berkeley, CA: 1976).

14 I have discussed their contributions in Livingstone D.N., *The Geographical Tradition: Episodes in the History of a Contested Enterprise* (Oxford: 1992).

of paradise, were typically regarded as pestilential spaces of disease, degeneration and debauchery. European colonists in particular were believed to be placed in great moral peril as they sought to settle in the equatorial zone. A whole range of treatises on tropical medicine incorporated sermonic advice on the moral jeopardy Europeans faced in colonial situations, and on a variety of measures to be taken to safeguard their own and their children's moral integrity. These moralistic homilies were routinely accompanied by a suite of racial denigrations that attributed the depravity and lethargy of native peoples to the dictates of their hazardous climatic regimes. Later, as colonial ties weakened and decolonisation began to tighten its grip, various politicians and administrators sought to find ways to exploit the rich untapped resources of tropical nature by resorting to theories of supervision from the metropolitan north, colonial stewardship, and dual-mandate rule. Throughout, morality and meteorology were intimately intertwined.

Third, and rather less formally, the sense of spirituality that many attached to water during the medieval era has continued to resurface in different guises. In this connection it is not perhaps surprising to find an assessment of the semantic field of water, cognisant of the fact that 'living water' embraces both spiritual and literal dimensions, calling attention to the significance of contemporary work on ecotheology and symbolic biogeography. These allusions speak to what has been called 'geopiety'. The term was coined in the mid-twentieth century by the historian and geographer John Kirtland Wright to convey the sense of 'emotional piety aroused by awareness of terrestrial diversity' and was later reenergised by Yi-Fu Tuan in his work on topophilia.[15] Wright traced the sentiment across the American intellectual landscape of the eighteenth and nineteenth centuries. But its legacy stretched well beyond. Echoes can clearly be heard in the preservationist insistence that wilderness is sacred space, in the proclamations of the deep ecologists, and in the deliberations of ecotheology.

3 Aqueous Bi-polarities

Snaking its way through many of these chapters is the sense of profound ambiguity that attends the medieval understanding of water. Frequently this takes the form of a sequence of bi-polarities that steer human sensibility in

15 Wright J.K., "Notes on Early American Geopiety", *Human Nature in Geography: Fourteen Papers, 1925–1965* (Cambridge: 1966) 251; Tuan Yi-Fu, *Topophilia: A Study of Environmental Perception, Attitudes and Values* (New York: 1990) and "Geopiety: A Theme in Man's Attachment to Nature and Place", *GeoJournal* 25 (1991) 387–392.

diametrically opposing directions. Among these dissonances, the contrary motions of hope and despair, life and death, sin and salvation, darkness and light, and their like are clearly audible. We learn, for example, that the oceans were believed to be zones where devils linger and demons dwell. And yet they could also be the very spaces where the human spirit might soar, freed from the constraints of terrestrial life. In some contexts the sea was a site of danger and darkness, yet also a sphere of promise for fishing, hunting and other ways of making profit. Whether in Viking mythology or Muslim scholarship, the sea was a dangerous place where risks and riches were to be encountered in equal measure. Mariners could find in the oceans a domain of both fortune and strife. In biblical contexts, the portrayal of the seas as evil stands in tension with the realisation that the worlds of water nonetheless remain under the power of God. The ruin of the world through Noah's flood, the destruction of the Egyptians in the Red Sea, the comparison of the wicked to troubled waters, and the rising of the Antichrist from the ocean depths all conspire to reinforce the image of the watery world as treacherous, unpredictable and menacing. At the same time, Hebrew waters brought ritual cleansing, priestly ablution, Christian baptism, and were the domain where miracles like Christ walking on the water or calming the sea-storm demonstrated his power over nature. Something similar attended Old Norse poetry and other romantic narratives. Instances of poetic portrayals of flowing water as an overwhelmingly perilous force obliterating human life and material culture are on display here. So too in twelfth-century Byzantine novels, the different 'shapes' that water assumed, to use the vocabulary of the analysis below, could allude to death, despair and the dark abyss, while at the same time advertising its beneficial features. As a metaphor for the struggles of a frail humanity cast upon life's billows and tempest-tossed – to use an image from Victorian hymnody – the waters spoke of both bane and blessing, hazard and hope.

In other ways too, water's power was double-edged. In medical writings, therapeutic bathing could be healthful or harmful depending on such circumstances as the season of the year and thus the dominant humour. Medical meteorology came to dominate the medical landscape in later centuries as it dwelt on place, pathology and the weather. Local climates and ecologies were thus swept into numerous treatises dealing with the geographical distribution of health and disease many of which provided lists of pathological spaces to be avoided and prophylactic sites – frequently spa towns – which patients were recommended to visit and 'take the waters'. With the flourishing of neo-Hippocratic ways of thinking during the first two-thirds of the nineteenth century, medical climatology and 'medical geography, broadly conceived, reigned as the queen of the medical sciences'.[16] In medieval romance literature the seas were frequently staged as boundaries or pathways, enclosures or conduits. The waters were central to narratives of parting, death and separation, not least of lovers, and were often associated with either the severing or renewing of ties with distant loved ones. In Norse cosmology the boundary effects of water are such that they not only divide the globe into discrete landmasses, but, as the analysis of water in Old Norse myth reveals, separate the realm of the gods from the world of giants. At once, waterways both constrain and enable movement from place to place, from realm to realm, from cosmos to cosmos.

In our own day, confronted with the vision of melting ice caps and rising sea levels, the horrors of over-abundant water, flooding and land degradation have taken on apocalyptic dimensions as ecological eschatologists deliver terrifying prophecies of climate disaster, human devastation and mass migration. At the same time, the leisure possibilities afforded by lakes and seas continue to attract tourists to sites where water sports and beach-bathing abound. In other venues, it is the scarcity of water that brings death and destruction. James Lovelock, polymathic scientist and inventor, insisted that drought was likely to ignite civil and international wars, while Sir Crispin Tickell, UK diplomat and environmentalist, was long convinced that water and war have always been tightly associated. Indeed an entire academic industry has flourished in recent decades devoted to specifying the connections between climate change and conflict and to demonstrating how drought, water pollution, flooding and the like, trigger civil unrest, insurrection, revolution and international strife.[17] Not surprisingly, climate change has entered national security agendas as one of the greatest challenges to the very survival of the human race. Plainly now, no less than in the medieval period, water in one form or another, as well as in one place or another, is both baneful and beneficial, sinister and salubrious.

4 Artistry, Authority, Agency

Finally, in perusing the chapters that follow, three further points of analysis awakened me to the persistence of other themes that have continued to echo

16 Numbers R.L., "Medical Science before Scientific Medicine: Reflections on the History of Medical Geography", *Medical History Supplement* 44.20 (2000) 217–220, 217.

17 See, for instance, Dyer G., *Climate Wars: The Fight for Survival as the World Overheats* (London: 2010); Welzer H., *Climate Wars: Why People Will Be Killed in the Twenty-First Century* (Cambridge: 2012).

down the centuries since the medieval period. Ever since the 'unnavigable mind' of Isidore of Seville, to use the words of Kelly Grovier's poem in this volume, subjected an unwieldly world to the alchemy of T-and-O reduction, the cartographic impulse has been a fertile site of strategic artistry. It is on display in Opicinus de Canistris's fourteenth-century drawings where the seas are visualised as monsters with grotesque features, the Mediterranean as devilish, and the continents take human form. These 'body-worlds', as they are termed here, continued to find cartographic expression in many ways. The Ditchley portrait of Queen Elizabeth I standing on a map of English counties by Christopher Saxton published in 1579, for example, was intended to confirm her imperial power over her realm. In a different key, the stream network of Kent was symbolised by a half-kneeling man in a map by Christopher Packe in 1736 where the drainage pattern was integrated with the human vascular system.[18] The discussion of Opicinus's portrayal of the figure of Africa as a woman in a corrupt world recalled to me what I have elsewhere termed the racial cartographics of a range of scientific writers,[19] as well as Frederick Rose's sequence of political cartoon maps of Europe around 1900 where the Russian empire featured as an octopus with tentacles strangling Poland and Persia.[20] Such endeavours reveal that maps act as rhetorical devices of persuasion in different cultural arenas. That they have informed, and continue to inform, the geographical imaginations of society seems beyond doubt.

The image of Elizabeth I exercising dominion over her realm is a reminder that issues of sovereignty and jurisdiction have long engaged lawmakers in legislating for who may exert authority over land and water. The ways in which Germanic and Roman legal principles were brought to bear on public and private waters, and on matters centring on rights of use and ownership, are explored in one of the chapters below. How public waters should be regulated for private use, whether domestic, agricultural, or recreational, was a stubborn problem requiring careful law-making, and is brought into clear light here. The medieval legislation governing marine waters is another focus of analysis as laws evolved in different venues in response to different needs and desires. Those regulating northern waters faced different challenges from their southern counterparts, arising from different trading practices, patterns of investment, international alliances, and the prevailing relationships among

18 This latter is discussed in Tuan, *The Hydrologic Cycle* 29–30.

19 Livingstone D.N., "Cultural Politics and the Racial Cartographics of Human Origins", *Transactions of the Institute of British Geographers* 35.2 (2010) 204–221.

20 This map underwent numerous transformations. They can be viewed at: https://illustrationchronicles.com/Mapping-Chaos-Fred-W-Rose-s-Serio-Comic-War-Map. Last accessed May 14, 2024.

ship-owners, crews and merchants. For me, this analysis brought two matters to mind. First, today's law of the sea, together with legislation about riparian rights, downstream pollution, territorial waters and fishing rights, remains rooted in how the lawmakers of the Middle Ages structured the discourse on inland water and maritime regulation. Second, the recognition that water legislation took different shapes in different regional, political and economic contexts resonates with Montesquieu's claim in *De L'Esprit des Lois*, dubbed heretical at the time, that law-making bore the stamp of the clime in which it came to birth. The moral relativism that Montesquieu's geography of law seemed to introduce into legislative philosophy was deeply disturbing to the French religious establishment of his day, and the work was placed on the *Index of Prohibited Books* in 1751.

If these legal deliberations highlight the effort to impose human authority on nature's waterways, a theme running in the opposite direction works to decentre the human species from its natural habitat and to disable the anthropocentric stance that sustains it. Attributing agency to natural processes is to widen our vision to a more-than-human world. Water is one of those natural entities to which agency is now often attributed, and its 'agentive faculty', as one contributor calls it, is being welcomed into the arena of historical explanation as a powerful actant (to use Bruno Latour's term[21]). Agency, to put it another way, is not a human-only trait; it is rather dispersed into objects, forces and other beings. In this volume, what commands attention is the mobilisation of water's potent agency in narratological and poetic contexts. Major plot events are determined by water, transitions between narrative sequences are frequently conveyed through the image of sea voyaging, and, in Old Norse Eddic poetry, water features as a sublime yet terrible power. In our own time there can surely be little doubt that the future story of humanity will be shaped in forceful ways by the agency of water. Whether through drought or deluge, humanity's wellbeing and wealth are bound up with water. Famines and floods alike are transforming human life across the globe. At the same time, attributing too much influence to the watery realm and according it a greater agency than is warranted, can result in blaming the weather for things that are much more intimately bound up with governance, greed and political ambition. Water may indeed cause wars, but water has yet to pull a trigger.

21 Actant is one of the dominant concepts in Latour's actor-network-theory. See Latour B., *Science in Action: How to Follow Scientists and Engineers through Society* (Cambridge, MA: 1988) and *Reassembling the Social: An Introduction to Actor-Network-Theory* (Oxford: 2007).

If the observations stimulated by these essays are in the right neighbourhood, then I am convinced that this volume is not an exercise in exegetical autopsy but rather is an expedition into the intellectual and cultural archaeology of the present. By this I don't mean to advocate what is often referred to as presentism, or Whig history – an anachronistic reading of the past as leading inevitably to a later orthodoxy. Rather it is a kind of history of the present in which my purpose has been to try to discern how present-day tribulations and triumphs alike have been shaped by how generations before us imagined, interrogated and inhabited the natural world.[22] That at least has been my aim in these reflections.

Bibliography

Benedict R. "Obituary. Franz Boas", *Science* 97.2507 (1943) 60–62.

Bradley J. – Dupree M., "A Shadow of Orthodoxy? An Epistemology of British Hydropathy, 1840–1858", *Medical History* 47 (2003) 173–194.

Chakrabarty D., "The Climate of History: Four Theses", *Critical Inquiry* 35 (2009) 197–222.

Clayton S. – Manning C. – Krygsman K. – Speiser M., *Mental Health and Our Changing Climate: Impacts, Implications, and Guidance* (Washington, DC: 2017).

Desmond A. – Moore J., *Darwin* (London: 1991).

Dorn H., *The Geography of Science* (Baltimore, MD: 1991).

Dyer G., *Climate Wars: The Fight for Survival as the World Overheats* (Oxford: 2010).

Garland D., "What is a 'History of the Present'? On Foucault's Genealogies and their Critical Preconditions", *Punishment and Society* 16 (2014) 365–384.

Glacken C.J., *Traces on the Rhodian Shore: Nature and Culture in Western Thought from Ancient Times to the End of the Eighteenth Century* (Berkeley, CA: 1967).

Gully J., *The Water Cure in Chronic Disease* (London: 1846).

Jennings E.T., *Curing the Colonizers: Hydrotherapy, Climatology, and French Colonial Spas* (Durham, NC: 2006).

Klimstra T.A. – Denissen J.J.A. – Koot H.M – van Lier P.A.C. – Frijns T. – Keijsers L. – Quinten L. – Raaijmakers A.W. – van Aken M.A.G. – Meeus W.H.J., "Come Rain or Come Shine: Individual Differences in How Weather Affects Mood", *Emotion* 11 (2011) 1495–1499.

Latour B., *Science in Action: How to Follow Scientists and Engineers through Society* (Cambridge, MA: 1987).

22 Here I am not using the concept of a 'history of the present' in any explicitly Foucaultian way. See Garland D., "What is a 'History of the Present'?" *Punishment and Society* 16.4 (2014) 365–384.

Latour B., *Reassembling the Social: An Introduction to Actor-Network-Theory* (Oxford: 2007).

Legan, L.M.S., "Hydropathy in America: A Nineteenth Century Panacea", *Bulletin of the History of Medicine* 45 (1971) 267–280.

Livingstone D.N., *The Geographical Tradition: Episodes in the History of a Contested Enterprise* (Oxford: 1992).

Livingstone D.N., "Cultural Politics and the Racial Cartographics of Human Origins", *Transactions of the Institute of British Geographers* 35 (2010) 204–221.

Livingstone D.N., *The Empire of Climate: A History of an Idea* (Princeton, NJ: 2024).

McMichael A.J., "Global Climate Change and Health: An Old Story Writ Large", in McMichael A.J. – Campbell C. – Lendrum D.H. – Corvalán C.F. – Ebi K.L.– Githeko A.K.– Scheraga J.D. – Woodward A. (eds.), *Climate Change and Human Health* (Geneva: 2003) 1–17.

Numbers R.L., "Medical Science before Scientific Medicine: Reflections on the History of Medical Geography", in Rupke N.A. (ed.), *Medical Geography in Historical Perspective* (London: 2000) 217–220.

Overy C. – Tansey E.M. (eds.), *The Recent History of Seasonal Affective Disorder* (SAD), Wellcome Witnesses to Contemporary Medicine 51 (London: 2014).

Price R., "Hydropathy in England 1840–70", *Medical History* 25 (1981) 269–280.

Rosenberg C.E., "Epilogue: *Airs, Waters, Places.* A Status Report", *Bulletin of the History of Medicine* 86 (2012) 661–670.

Schaffer S., *From Physics to Anthropology – and Back Again* (Cambridge: 1994).

Smith C., *City Water, City Life: Water and the Infrastructure of Ideas in Urbanizing Philadelphia, Boston, and Chicago* (Chicago: 2013).

Stocking G.W., "From Physics to Ethnology", in Stocking G.W. (ed.), *Race, Culture, and Evolution: Essays in the History of Anthropology* (Chicago: 1982) 133–160.

Tuan Yi-Fu, *The Hydrologic Cycle and the Wisdom of God: A Theme in Geoteleology* (Toronto: 1968).

Tuan Yi-Fu, *Topophilia: A Study of Environmental Perception, Attitudes and Values* (Englewood Cliffs, NJ: 1974).

Tuan Yi-Fu, "Geopiety: A Theme in Man's Attachment to Nature and Place", in Lowenthal D. – Bowden M. (eds.), *Geographies of the Mind: Essays in Historical Geosophy in Honor of John Kirtland Wright* (New York: 1976) 11–39.

Welzer H., *Climate Wars: Why People Will Be Killed in the Twenty-First Century*, trans. P. Camiller (Cambridge: 2012).

Wittfogel K., *Oriental Despotism: A Comparative Study of Total Power* (New Haven, CT: 1957).

Wright J.K., "Notes on Early American Geopiety", Chapter 14 of Wright J.K, *Human Nature in Geography* (Cambridge, MA: 1966) 250–285.

Wright J.K., *The Geographical Lore of the Time of the Crusades* (New York: 1965).

Figures

Notes on Contributors

Michael Bintley
is Associate Professor in Medieval English Literature at the University of Southampton. His research is transdisciplinary, focusing on the intersection of early medieval textual and material culture in relation to landscapes and environment, trees and plant-life, buildings and settlements, and minds and matter. He is author of *Trees in the Religions of Early Medieval England* (2015), *Settlements and Strongholds in Early Medieval England: Texts, Landscapes, and Material Culture* (2020), and co-author, with Dr Kate Franklin, of *Landscapes and Environments of the Middle Ages* (2023). With Professor Richard North he edited and translated *Andreas: An Edition* (2016), and he has co-edited several interdisciplinary collections of essays.

Tom Birkett
is Senior Lecturer in Old English and Old Norse at University College Cork, Ireland. His main research interests include the poetic traditions of early medieval England and medieval Scandinavia, the reception of medieval literature, Norse mythology and runology. His publications include the monograph *Reading the Runes in Old English and Old Norse Poetry* (2017) and the co-edited collections *Translating Early Medieval Poetry* (2017) and *The Vikings Reimagined* (2020), as well as articles on the *Poetic Edda* and the reception of Norse myth. His popular retelling of the *Norse Myths* was published in 2018 and translated into several languages, and his illustrated children's *Legends of Norse Mythology* was published in 2023.

Laura Borghetti
is currently a PhD candidate and associate member of the DFG-funded Training Research Group 1876 'Early Concepts of Humans and Nature' at Mainz University, completing a doctorate on the literary depictions of winds in Byzantine texts from the ninth to the twelfth century. She is a former research associate at Mainz University (2016–2019). Her main research interests revolve around questions of narratology, metaphor theories, agency and ecocriticism within Middle-Byzantine literature.

Rafał Borysławski
is an Associate Professor in the Institute of Literary Studies, University of Silesia, Poland, and co-founder of its Centre for Old Norse and Old English Studies. His research focuses chiefly on Old English culture and literature, the uses of

apophatic philosophy in the early European Middle Ages, medieval emotionology (with particular emphasis on fear and its derivatives), and the questions of medieval culture associated with the field of social history and femininity. He is the author of *The Old English Riddles and the Riddlic Elements of Old English Poetry* (2004) and he has co-edited seven volumes related to medieval studies and the intersections of history, historiography and philology, including *Emotions as Engines of History* (2021, with A. Bemben), *Aspects of Royal Power in Medieval Scandinavia* (2018, with J. Morawiec), *HistoRisus. Histories of Laughter and Laughter in History* (2016) and *Cryptohistories* (2015).

Marilina Cesario
is Reader in the Earliest Writings in English at Queen's University Belfast. She is an expert in early medieval history and literature and Anglo-Latin connections. She has published extensively in the fields of early medieval weather and astronomy, prognostication, reception of classical mythology in the early Middle Ages and on manuscript studies. She is the editor with Hugh Magennis of *Aspects of Knowledge: Preserving and Reinventing Traditions of Learning in the Middle Ages* (2018).

Marusca Francini
has been Associate Professor of Germanic Philology at the University of Pavia's Department of Humanities since 2002. Her areas of interest in research include the Gothic Bible, biblical translations in Germanic medieval traditions, Old Norse Skaldic sagas, the Riddarasögur and Lombard onomastics. She is the author of *Saga di Björn, campione dei valligiani di Hítardalr* (2004), *Edizione sinottica del Vangelo di Giovanni in gotico del Codex Argenteus* (2009) and *Il Vangelo di Giovanni in Anglosassone* (2019). She has also co-edited the special issue of the journal *Il Confronto Letterario* entitled *Eroi di carta e di celluloide – Il Medioevo Germanico nelle forme espressive moderne* (2002, with M.G. Saibene).

Kelly Grovier
is a writer and independent scholar. He has taught creative writing at Cambridge University, Oxford University and Aberystwyth University. He is the author of four collections of poetry and has published widely on British Romanticism and on the history of visual cultures. He is co-founder of the journal *European Romantic Review* and is author of *The Art of Colour: The History of Art in 39 Pigments* (2023), *On the Line: Conversations with Sean Scully* (2021), *A New Way of Seeing: The History of Art in 57 Works* (2018), *Art Since 1989* (World of Art series) (2015), *100 Works of Art that Will Define Our Age* (2013) and *The Gaol: The Story of Newgate, London's Most Notorious Prison* (2008).

Deborah Hayden
is Associate Professor in the Department of Early Irish at Maynooth University and previously held research fellowships at Christ Church, University of Oxford, and the School of Celtic Studies, Dublin Institute for Advanced Studies. She has published and taught widely on aspects of the intellectual and literary culture of medieval Ireland and the Insular world in its European context, and has particular interests in the history of linguistic thought and education and classical and medieval tradition; premodern Irish medical writing; early Irish lexicography; medieval Irish and Welsh legal tradition; and translation literature. She is Lead Editor of the peer-reviewed journal *Language & History*.

Simon Karstens
currently teaches early modern history as an interim professor at the University of Trier, Germany. Previously, he held a position as senior lecturer at the same institution. His fields of expertise include the history of the Habsburg Monarchy and the Southern Netherlands, the age of early transatlantic encounters and seafaring and public history. He has published three books on these topics. The first, titled *Lehrer-Schriftsteller-Staatsreformer. Die Karriere des Joseph von Sonnenfels 1733–1817* was published in 2011 and focused on the career of an Austrian court official. The other two examined failed attempts to colonise the Americas. *Gescheiterte Kolonien - Erträumte Imperien* (2021) was an in-depth analysis of primary sources. It engaged extensively with contemporary depictions of the ocean and the risks of seafaring in the context of early colonial projects. This was followed by a general study aimed at a broader audience titled *Untergegangene Kolonialreiche - Gescheiterte Utopien in Amerika* (2022). Simon Karsten's current research explores the history of seafaring women in the early modern era.

Andreas Lammer
is Associate Professor of History of Philosophy at Radboud University Nijmegen in the Netherlands. His primary research interests are in Greek, Arabic and Latin natural philosophy in both the Aristotelian and the Avicennian traditions, and more broadly in the transmission of philosophical and scientific literature from Greek into Arabic and from Arabic into Latin. He obtained his PhD in philosophy and Arabic studies from LMU Munich and published several papers on the notions of time, creation and nature in ancient, late ancient and Arabic philosophy as well as a monograph on the history of Greek and Arabic natural philosophy entitled *The Elements of Avicenna's Physics: Greek Sources and Arabic Innovations* (2018).

David N. Livingstone

a Fellow of the British Academy and Member of the Royal Irish Academy, is Emeritus Professor of Geography and Intellectual History at Queen's University Belfast. He works on the histories of geographical knowledge, evolutionary theory and science and religion. He is the author of several books including *Nathaniel Southgate Shaler and the Culture of American Science* (1987), *Darwin's Forgotten Defenders* (1987), *The Geographical Tradition* (1992), *Putting Science in its Place* (2003), *Adam's Ancestors* (2008) and *Dealing with Darwin* (2014). His latest book, *The Empire of Climate: A History of an Idea*, has just been published (2024).

Luca Loschiavo

is Professor of Medieval and Modern Legal History at Teramo University, Italy, and President of the 'Istituto Emilio Betti di Teoria e Storia del Diritto'. His main research interests include law and institutions (especially legal procedures) in late Antiquity and early Middle Ages; early Christian legal institutions; Roman-barbarian law codes (especially Lombard law); glossators and renaissance of Roman Law in the Middle Ages; Justinian's Novels in western legal tradition. Among the books he has published are *'Summa Codicis Berolinensis', Studio ed edizione di una composizione 'a mosaico'* (1996); *Figure di testimoni e modelli processuali tra antichità e primo medioevo* (2004); *Temi, personaggi e ambienti della scuola della glossa* (2008); *L'età del passaggio. All'alba del diritto comune europeo (secoli III–VII)* (2nd ed. 2019). As co-editor he has published *Forme stragiudiziali e straordinarie di risoluzione delle controversie nel diritto comune e nel diritto canonico* (2008, with P.A. Bonnet); *Novellae Constitutiones. L'ultima legislazione di Giustiniano tra oriente e occidente da Triboniano a Savigny* (2011, with G. Mancini and C. Vano); and *Civitas, Iura, Arma. Organizzazioni militari, istituzioni giuridiche e strutture sociali alle origini dell'Europa (secc. III–VIII)* (2015, with F. Botta).

Hugh Magennis

is Professor Emeritus at Queen's University Belfast, where he had previously been head of the School of English and director of the Institute of Theology, among other positions held. He has published widely on Old English and related literature, specialising particularly in saints' lives, translation and poetic tradition. Among his publications are *The Cambridge Introduction to Anglo-Saxon Literature*, *Translating Beowulf* (both 2011) and, most recently, the Dumbarton Oaks Medieval Library edition and translation *Anonymous Old English Lives of Saints* (2020) (with J. Kramer and R. Norris). Hugh Magennis is a Member of the Royal Irish Academy and a Fellow of the English Association.

Colin Fitzpatrick Murtha
is a PhD candidate at Radboud University, Nijmegen. His primary interest is in medieval Arabic philosophy, and his research deals mostly with value theory and natural philosophy. He is currently working on a dissertation on how Aristotle's *Meteorologica* was received in the medieval Islamic world, particularly in Avicenna's writings (d. 1037).

François Quiviger
is a Fellow of the Warburg Institute, University of London, where he previously worked as a librarian, curator of digital resources, researcher and teacher.

The main theme of his research is the history of cognition and sensation in so far as it pertains to the making and reception of images and to the relationship of humans to nature. With these questions in mind he has written on early modern art and art theories, academies, music, wine and banqueting. Recent books include *Leonardo da Vinci: Self, Art and Nature* (2019) and *The Sensory World of Italian Renaissance Art* (2010).

Elisa Ramazzina
is a Research Associate at Queen's University Belfast. Prior to this, she was a Marie Skłodowska-Curie Research Fellow at the same university for two years, where she worked on a project titled *Water and Baptism in Old English Poetry*. She also worked as a research assistant at the University of Oxford on the ERC-funded project CLASP: A Consolidated Library of Anglo-Saxon Poetry. Her research interests span the Germanic philological tradition, including Old English literature, as well as Middle English, Old High German, Middle High German and Anglo-Latin literatures. Her interests also include monster studies, medieval cartography, medieval science, medicine and cosmology, and border studies. Currently, she is working on editing a collection on medieval borders and the environment as well as authoring her first monograph on the Old English versions of the *Wonders of the East*.

Karl Whittington
is Associate Professor of History of Art at The Ohio State University. His research focuses on the intersection of medieval art with histories of science, gender and sexuality. He is the author of two books: *Body-Worlds: Opicinus de Canistris and the Medieval Cartographic Imagination* (2014) and *Trecento Pictoriality: Diagrammatic Painting in Late Medieval Italy* (2023), as well as a volume of essays co-edited with B. Keene, *New Horizons in Trecento Italian Art* (2021).

Introduction

Marilina Cesario, Hugh Magennis and Elisa Ramazzina

The present four-volume edited collection *The Elements in the Medieval World: Interdisciplinary Perspectives* is the first multidisciplinary and cross-disciplinary examination of the elements in intellectual and artistic endeavours as well as in daily life, covering the period from Late Antiquity to roughly 1500. Starting with *Water* and *Earth*, before moving on to *Air*, and concluding with *Fire*, the series devotes a volume to each element.

As explained more fully in the Introduction to the *Earth* volume, the idea of 'the elements', which are essential to all creation, whether animate or inanimate, has long been at the heart of human understanding of the universe.[1] Since ancient times, the structure and complexity of matter have been explained through the elements, which were also thought to constitute the universe and enable life. Based on Greek philosophy, mainly on the works of Plato and Aristotle, the four elements – earth, water, air, and fire – were fundamental to medieval conceptions of nature.

In addition to their intellectual and religious value, they were seen to have practical uses, impacting the human body, affecting the weather and constituting the cosmos, as evidenced in the writings of European philosophers such as Isidore of Seville (6th–7th century) and Byrhtferth of Ramsey (11th century), as well as by Arabic scholar including Sergius of Rēš ʿAynā (d. 536) and Avicenna (d. 1037).

In a similar vein to the other three books that make up the series *The Elements in the Medieval World: Interdisciplinary Perspectives*, the introduction to the present volume will present a broad survey of medieval conceptions of the 'Water' element, before focusing specifically on the contents of this volume. The first part of the introduction is comprised of four subsections, the titles of which predominantly reflect the four thematic strands which make up the whole collection. Reflective of this book's coverage, the introduction particularly foregrounds western intellectual traditions, conscious too, however, of the influence of such traditions on Islamic thought and of non-Western perspectives.

1 For an introduction on elemental theories throughout the medieval period see the "Medieval Ideas on the Four Elements" section of the "Introduction" in Cesario M. – Magennis H. – Ramazzina E. (eds.), *The Elements in the Medieval World: Interdisciplinary Perspectives. Earth* (Leiden – Boston, MA: forthcoming).

 | DOI:10.1163/9789004696501_002

1 Water in Natural Philosophy

Thales of Miletus was one of the first to theorise the importance of water for living beings, promoting it as the principle of everything, since it provides life-giving moisture and generates the heat necessary for life'.[2] Aristotle explains that heat is vital for all things and that their 'seeds' have a moist nature, and he describes the relationship between water and air, which are inextricably linked through evaporation and condensation in that 'water is potentially air, while air is potentially water, though in another way'.[3] His *Meteorology*[4] categorises the interrelation of air and water, naming exhalation of water 'vapour', air condensing into water 'cloud', and what remains from the reverse (clouds liquefying into water) 'mist'. 'Dew' is vapour condensing into water when it is not warm enough for it to dry or cold enough to freeze, while 'hoarfrost' is when vapour freezes before condensing back into water.[5]

Isidore of Seville distinguishes two types of water: *aqua* ('water'), associated with *aequalis* ('equal'), is *stativa* ('motionless'), its surface being even, and *unda* ('wave'), which is *eminens liquor* ('rising water'), always in motion.[6] He associates water with fire, defining both as the most powerful elements concerning human life. Water is cold and moist, prevailing over the other elements, tempering the sky and making the earth fertile. Evaporation occurs, *aerem exhalationibus suis incorporantis* ('giving body to the air with its exhalations'),[7] when the sunrays heat the water, transforming it from a liquid into a gas. With condensation, water takes hold of the sky as vapour, rising where temperatures are low, where it cools down and returns to its liquid state, forming clouds, which fall to the ground as precipitation, providing sustenance to man and drenching all living creatures.

Bede's *De natura rerum* (Chap. 32) describes water's cycle similarly, comparing the tempering of elements into water droplets and their condensation to a brewing process, with infusion transferring flavours from one drink to

---

2 Aristotle, *Metaphysics*, ed. and trans. W.D. Ross (Oxford: 1924) I:III.

3 Aristotle, *Physics*, trans. R.P. Hardie – R.K. Gaye (1994–2009) IV:V:7. Internet Classics Archive. Accessed May 14, 2024. http://classics.mit.edu/Aristotle/physics.html.

4 Aristotle, *Meteorology*, trans. E.W. Webster (1994–2009) I:ix. Internet Classics Archive. Accessed May 14, 2024. http://classics.mit.edu/Aristotle/meteorology.html.

5 Ibidem I:X.

6 Isidore, *Etymologiarum sive originum*, ed. W.M. Lindsay (Oxford: 1911) XIII:XX.2. Digitised Edition University of Chicago, 2010. Accessed May 14, 2024. https://penelope.uchicago.edu/Thayer/E/Roman/Texts/Isidore/home.html.

7 *The Etymologies of Isidore of Seville*, trans. S. Barney – W.J. Lewis – J.A. Beach – O. Berghof – M. Hall (Cambridge: 2006) XIII:XII.3, 276.

another. Different condensation points do in fact depend on the element that the droplets are interacting with: they get 'decocted' when exposed to the sun's fire and heat, but if air currents cause them to change, they become fresh. He explains that this happens as *marinam aquam humo, vel dulcem marinis herbis infundentes, in contrarium solemus transferre saporem* ('even as we, by infusing sea water into the soil or fresh water into sea plants, are accustomed to transform [fresh or salty] savor into its opposite').[8] Indeed, if heated by the fire of solar rays, they are 'decocted',[9] but when altered by air currents, they become fresh. Raindrops frozen by wind and cold form ice crystals, generating snow (Chap. 35) and hail (Chap. 34).

Bede recognises water's dual nature – *aquarum duplici natura formantur et reformantur omnia quae cernuntur in terra* ('all things that are seen on earth are formed and reformed by the dual nature of the waters' (Ch. 38)[10] – noting that salty water nourishes and quenches all creatures living in the sea, while fresh water is essential for all beings breathing air. He concludes that both types of water are natural if they can be transformed back into one another. Interestingly, Isidore describes marine waters as salty and as bitter, due to the etymology of *mare* ('sea') from *amarus* ('bitter').[11] His explanation is possibly indirectly reminiscent of Aristotle's *Meteorology*, which also depicts seawater as bitter and salty due to its mixture with earthy components, and compares it to residual urine in the bladder, observing how, despite only sweet foods having been digested, this is salty and bitter.[12] Curiously, both Aristotle and Isidore, and consequently also Bede and his later countryman Ælfric of Eynsham[13] argue that evaporated seawater is turned into fresh water by the sun's heat, and having lost its salinity, it falls to the ground as rain.[14]

8 Bede, *De natura rerum* Chap. 32. Translations are from Kendall C. – Wallis F., *Bede: On the Nature of Things and On Times* (Liverpool: 2010) 93.

9 Bede, *De natura rerum* Chap. 32: '[…] *qua vel solis igne decocti vel aeris itinere mutati dulcescant*' ('[…] in this way, either decocted by the fire of the sun or altered by the current of the air, the vapours become fresh', Kendall – Wallis, *Bede: On the Nature of Things* 93).

10 Bede, *De natura rerum* Chap. 38 (Kendall – Wallis, *On the Nature of Things* 94).

11 Isidore, *Etymologiarum* XIII:XXIV:1. For a narratological analysis of Marie de France's *lais* according to Isidore's etymology of the sea, see Borysławski R., "Marie de France's Writing on Water" 344–371.

12 Aristotle, *Meteorology* II:II–III.

13 It is well known that Isidore's *De natura rerum* is the main source of Bede's *De natura rerum*; Ælfric, in his preface to his *De temporibus anni*, states that he has based his treatise on Bede's work. See Blake M. (ed. and trans.), *Ælfric's De temporibus anni* (Cambridge: 2009) 75–76.

14 Isidore, *De natura rerum* Chap. 22; Bede *De natura rerum* Chap. 22; Blake, *Ælfric's De temporibus anni* 93–94.

Around the thirteenth century reflections on water begin to proliferate. Robert Grosseteste, one of Aristotle's commentators, for instance, writes *De impressionibus elementorum* explaining the transformation of the elements through heat and fire, including how ice becomes water and water vapour. He describes water as a transparent body like air, icicles and glass,[15] and elucidates how the condensation[16] of light rays can heat water to such an extent that it is transformed into air.[17] This alteration occurs through the formation of *ampullae* ('bubbles') coming out of water and rising above its surface. Each bubble contains all four elements: earth, water, air and fire, and when too many form, their moist nature turns them into vapour or steam, forming clouds. The bubbles' composition can vary: those in which earth prevails producing dry smoke, those where air abounds create dense vapour, and those in which water predominates give humid vapour.[18] Grosseteste also explains mist and dew: the former is obtained when bubbles do not completely detach from the ground, continuing to drift when it is not warm enough. The latter appears when bubbles of mist lose their heat, falling to the ground.[19]

Thomas Aquinas discusses the four elements, identifying fire as the most noble and most similar to the heavenly bodies. In *De mixtione elementorum* he distinguishes between primary and secondary elementary qualities, with the former intrinsic in the element and the latter deriving from its proximity to another element. The element 'water' has two active qualities, coldness and wetness: the former belongs exclusively to water as such. It is active, as *frigidum sit activum patet etiam per hoc, quia congregat omnia* ('it is clear that cold is active as it gathers everything together'), including things having different

15 Sparavigna A.C., "Robert Grosseteste and the Four Elements", *International Journal of Sciences* 2.12 (2013) 42–45, 43.

16 Grosseteste explains the difference between the four elements through the mechanisms of condensation and rarefaction. In his *Hexameron*, he explains how, on the third day of Creation, the water of the firmament was condensed out of vapourous water and became a dense liquid. This water was then rarefied until it became air, which was in turn rarefied and transformed into fire. This demonstrates how water and air are more versatile elements than earth and fire, as the former can be further condensed and rarefied while the latter cannot. See Kedar Y., "The Nomological Image of Nature: Explaining the Tide in the Thirteenth Century", *Annals of Science* 73.1 (2016) 68–88, 72.

17 Grosseteste, Robert, *De impressionibus elementorum*, in Baur L. (ed.), *Die Philosophischen Werke des Robert Grosseteste* (Münster: 1912) 87–88. Documenta Catholica Omnia. Accessed May 14, 2024. http://www.documentacatholicaomnia.eu/04z/z_1168-1253__Grosseteste_Robert__De_Impressionibus_Elementorum__LT.doc.html.

18 Ibidem.

19 Sparavigna, "Robert Grosseteste" 43–44.

natures.[20] For example, when water freezes, it gathers not only ice crystals but also different kinds of debris. Wetness is a passive quality due to its proximity to air, allowing both elements to take on the shape of the container they are poured or blown into. Consequently, water can be defined as a simple body (like all elements) primarily cold and secondarily wet[21] due to its contact with air and humidity.

Aquinas summarises his account of water:

> [It is] wet, it is transparent (diaphanous), it is heavy, it has a downward falling motion, it has the capability to become warm (indeed hot), the capability to become cool (indeed cold), the capability to be either a liquid, or a solid (as in ice and snow), or a gas (as in steam).[22]

Also in the thirteenth century, Bartholomew the Englishman reaffirmed water's importance for living creatures in *De proprietatibus rebus*, dedicating an entire book (XIII) of his encyclopaedic treatise to the description of the nature of wells, pools, lakes, biblical rivers and the ocean. Like Isidore and Bede, he also ascribes a double function to water, as being vital both for creatures of the air and land, and for marine fish (XI). Echoing Thales, Hesiod and Pliny, he designates water as the principle of all things, and the most ancient and powerful of the elements[23] as it can devour the earth, extinguish fire, ascend to the sky and fall back to the ground as rain, becoming the source of sustenance for all forms of life. Bartholomew, recalling Isidore's etymology, describes the 'even' (*equa*) surface of water as motionless, thinner than the earth, clean and bright. He also notes its passive quality, also mentioned by Aquinas, since it takes on

---

20 Conway P. – Larcher R.F. (trans.), *Exposition of Aristotle's Treatise on Generation and Corruption, Book I, cc. 1–5* (Columbus, OH: 1964) 2. St. Isidore e-book library. Accessed May 14, 2024. https://isidore.co/aquinas/GenCorrup.htm#2 (bilingual).

21 Bobik J. (ed. and trans.), *Aquinas on Matter and Form and the Elements: A Translation and Interpretation of 'De Principiis Naturae' and 'De Mixtione Elementorum' of St. Thomas Aquinas* (Notre Dame, IN: 1998) 169–173; '*aquae vero proprie et secundum se competit ei frigidum, quod est secunda qualitas active, quasi privative se habens calidum, competit autem ei humidum secundario secundum propinquitatem ad aerem*' ('To water, properly and because of itself, belongs being cold, which is the second active quality, since it is privatively related to being hot. And being wet belongs to water secondarily, because of its closeness to air').

22 Ibidem 182.

23 Bartholomew the Englishman, XI:VI. *De proprietatibus rerum: Batman vppon Bartholome his booke De proprietatibus rerum, newly corrected, enlarged and amended*, ed. T. Batman, trans. J. Trevisa (London: 1582) – (Ann Arbor, MI: 2011). Accessed May 14, 2024. https://quod.lib.umich.edu/cgi/t/text/text-idx?c=eebo;idno=A05237.0001.001.

the contours of the containers into which it is poured.[24] Water has no qualities, flavours or colours of its own, deriving them from the places it flows and from the substances it touches. It has a changeable character as it can be sweet or salty, clear or cloudy, thin or thick and paradoxically, the clearer the water, the murkier it appears when the sun's rays hit it.

Bartholomew relates water to two other elements, as it comes partly from air, as in the case of rainwater, and partly from the earth, as in the case of springs and streams. To Bartholomew, rainwater is the best type of water, being clear, bright, thin and savoury, and abundant, especially when accompanied by thunder, which contributes to the formation of vapours and to making water clear and clean. However, water mixed with snow or melted snow are noxious, as they hurt the stomach and teeth, and cause coughing.[25]

Bartholomew also explains the optical phenomena of reflection and refraction of solar rays in water. He describes reflection comparing the water surface to a mirror capturing the sun's rays and sending them back towards the sky and reflecting other objects. Refraction is how water amplifies the rays that strike any submerged object, making it appear larger.[26]

Bartholomew's contemporary Roger Bacon argues in his *Communia naturalium*[27] that one of the universe's ruling powers is universal nature, which is in charge of the continuity, uniformity and regulation of natural bodies. While bodies go to their natural location (heavy bodies down and light ones up) due to their nature, in certain circumstances they are forced to obey the higher law of universal nature, when its demands outweigh their own natures' tendencies. The most important characteristic of matter's universal nature is its inclination to sustain and perpetuate material continuity.[28] Starting from these

24 Bartholomew the Englishman, *De proprietatibus rerum* XIII:1.

25 Ibidem.

26 Ibidem.

27 This treatise addresses the main themes in Bacon's philosophy, such as form, matter, the relationship between natural and individual causes, motion, and its attributes, i.e., emptiness, time, and space. See Crisciani C. – Pereira M., "Introduction to the Philosopher's Laboratory", in Bernardini P. – Rodolfi A. (eds.), *Roger Bacon's "Communia Naturalium:" A 13th Century Philosopher's Workshop* (Florence: 2014) 3–25, 5–6.

28 Bacon's theory on material continuity expanded Aristotelian concepts of matter. Its basis was provided by Bacon's concept of a universal nature, which governs the interactions between elements, phenomena and species. The fundamental feature of universal nature is its relentless urge to preserve material continuity, or, to put it another way, its repulsion of vacuum. Water seemed to demonstrate this particularly effectively, prompting Bacon to experiment with it to find out more about its properties. See also Kedar Y. – Giora H., "Roger Bacon (c. 1220–1292) and his System of Laws of Nature: Classification, Hierarchy and Significance", *Perspectives on Science* 25.6 (2017) 719–745.

assumptions, Bacon explains the Aristotelian concept of nature rejecting the void and matter occupying empty spaces. His 'hourglass experiment' (I.II.3–6) demonstrates the principle of universal nature through water, in that it prevents the liquid from flowing out and the formation of a vacuum, preserving matter's order and continuity. The experiment involves submerging an hourglass-shaped vessel with holes at the base and at the top in water until the air is replaced by the liquid entering through the openings. When the container is removed from the water, the liquid contained in the hourglass is not expelled from the hole at its bottom, as it would be expected according to water gravity.[29]

Bacon's study of the 'sphere' of water is a significant example of Aristotelian elemental theory, considering the four elements as concentric spheres with earth at the centre and fire in the outmost sublunar portion of the universe. Bacon's *Opus majus* (IV.IV.11) elaborates his complex law of water gravity, which states that water always flows to the lowest point, with its surface being equally distant from the centre of the universe.[30] Bacon explains that the water's surface forms a sphere around the centre of the world, as any lines drawn starting from the earth's centre and going towards any point of the water surface, have equal length. In this way, he also provides a geometrical explanation to the Isidorian etymology of *aqua*. Indeed, while explaining the theory of water gravity, he also elucidates the phenomenon of communicating vessels through an experimental approach, resorting to the well-known 'experiment of the two cups', described in the *Opus majus*.[31]

29 Ibidem 729–730.

30 Kedar – Hon, "Roger Bacon" 721–722.

31 This experiment involves placing two identical containers in different locations (one higher and one lower) and filling them with water. According to the law of gravity, the water surface in the upper cup is a portion of the sphere which is farther from the earth's centre, and, therefore, is larger than the water surface in the lower cup. The diameter of each cup forms a chord that is equal to each sphere, but the proportion between the chords and the spheres is different in that the upper cup's chord constitutes a smaller part of the larger sphere. Conversely, the lower cup's chord is a larger part of the smaller sphere. The water surface of the two bowls is therefore different: the smaller sphere of the lower bowl has a greater surface, while the surface of the upper bowl is smaller. In fact, the lower cup can hold more water than the upper one because the same amount of water occupies less space, i.e., more water can be poured into one lower place than in the same container that is in a higher location. As Kedar and Hon note, Bacon's explanation is not Aristotelian, but uses geometric proportions to explain water's behaviour. See also Kedar – Hon, "Roger Bacon" 723.

2 Images of Seas and Rivers

Isidore defines the sea (*mare*), as *aquarum generalis collectio* ('a general gathering of waters'),[32] be it fresh or salty, noting its bitterness (*aquae eius amarae*)[33] and its 'even surface' (*aequor*),[34] which, despite rippling up during storms, returns flat with clear weather. Bede explains that the evaporation caused by the sun makes the sea denser and more bitter, leaving everything heavy and earthy in the water (*De natura rerum* Chap. 41).

Both Isidore and Bede[35] suggest that the sea, despite the continuous flow of fresh waters into it, does not increase in size, either due to its waters consuming fresh ones, or due to the water cycle, which evaporates excess water due to clouds, the sun's heat, or winds. Alternatively, water *per occulta quaedam terrae foramina percolatus, et ad caput amnium fontesque revolutus recurrat* ('is percolated through certain hidden openings in the earth, and runs back again to the source of springs and fountains').[36]

The cosmology of the thirteenth-century scholar Johannes de Sacrobosco connects the sea's shape to that of individual waterdrops. As Sacrobosco elucidates in *De sphaera mundi*,[37] the sea surface is spherical (*rotunda*) and its water has a bulge (*tumorem*). He points out that this is easily demonstrated empirically by observing a ship leaving port with signals being set up on the seacoast. A person standing at the mast's bottom cannot see the signals; whereas, if the same person climbs onto the mast and the ship is stopped, the signals will be detectable. To explain this phenomenon, Sacrobosco suggests that the sea has a spherical shape, water is *corpus homogeneum* ('a homogeneous body'), and its individual parts tend to behave like the whole, with even the smallest waterdrops assuming a round shape.[38]

At the end of the sixteenth century Diego Pérez de Mesa, a Spanish humanist, who produced treatises on mathematics, geometry and astronomy, and trained navigators and sailors, wrote *Comentarios de sphera*[39] based on

32 Isidore, *Etymologiarum* XIII:14:1, trans. Barney – Lewis – Beach – Berghof, *The Etymologies* 277.

33 Isidore, *Etymologiarum* XIII:XIV:1.

34 Ibidem XIII:XIV:2.

35 Isidore, *Etymologiarum* XIII:XIV:2; Bede, *De natura rerum* Chap. 40.

36 Isidore, *Etymologiarum* XIII:XIV:2, trans. Barney – Lewis – Beach – Berghof, *The Etymologies* 277.

37 de Sacrobosco, *De sphaera mundi* 8.

38 Ibidem.

39 A tract describing the cosmos and entailing different scientific disciplines, including natural philosophy, elemental theory, astronomy, geography, meteorology and navigation.

Sacrobosco's work, further exploring the theory of water's sphericity, and speculating on the circularity of oceanic motion. Indeed, according to his cosmology, water, unlike earth, which moves in straight lines, drifts circularly on the terrestrial surface. To prove this, he reports the experiences of many sailors, who observed how the sea moves in circles, providing copious examples ranging from the Pacific to the Indian Oceans.[40]

Isidore is the scholar who proposed the broadest array of theses on tides. In *De natura rerum* (Chap. 40), echoing the ancient geographer Solinus, he offers an organicistic worldview, comparing sea mechanics to human respiration, placing cavities in the seabed to act as *mundi nares* ('nostrils of the world') and allowing the passage of *ventorum spiritus* ('winds' breath'), which, by blowing or exhaling through those passages, cause high and low tides. Relying on Ambrose, he also mentions the moon as a possible influence, and, drawing on Aristotle's *Meteorology*, he proposes the sun and the evaporation of seawater as plausible causes for the tides. Isidore's treatment of tides is inconclusive, admitting to not knowing which theory is correct and deferring to divine providence. In the *Etymologies*, however, he defines tides (*aestus*) as *maris accessus vel recessus, id est inquietudo* ('a rising or falling of the sea, that is, a restlessness'),[41] causing the ocean to swell and deflate. Isidore attributes human emotions to the swirling ebb and tide, noting how the sea surface looks restless.

Bede illustrates the relationship between moon and tides in his *De natura rerum* (Chap. 39), expanding his ideas further in *De temporum ratione* (Chap. 29). He connects tides and lunar phases, noting that the peak of high tide corresponds to full moon, and that both phenomena occur cyclically. He observes that the tides' level varies during the month, defining as *malinae* the rising tides and as *ledones* the decreasing ones and noting that these alternate, and that their timing varies according to location.[42]

Grosseteste's *De fluxu et refluxu maris* explains tides as water's movement through two fundamental mechanisms: condensation and rarefaction. Heaven is the efficient cause for water's movement,[43] and the moon's action over the

---

See Crowther K., "Sacrobosco's *Sphaera* in Spain and Portugal", in Valleriani M. (ed.), *De sphaera of Johannes de Sacrobosco in the Early Modern Period* (Berlin: 2020) 161–184, 179–180.

40 Ibidem.

41 Isidore, *Etymologies* XIII.XXVIII.1, trans. Barney – Lewis – Beach – Berghof – Hall, *The Etymologies* 278.

42 Jones C. (ed.), *Bedae opera de temporibus* (Cambridge, MA: 1943) 232–235.

43 Drawing on Aristotle, Grosseteste argues that change cannot take place without a transmission process, and that everything in the universe occurs due to the everlasting

sea is an example of the efficient causality governing the elements through rays and light. Indeed, the moon controls all that is cold and wet. Through its light, the moon sets the water in motion: its rays are incorporated into the water, intersecting and reflecting, thus creating heat. In this way, vapours are created in the sea and winds agitated through rarefaction. Grosseteste's explanation of high tide refers to water's physical characteristics when, as the sea swells, it is too viscous and heavy to release the vapours trapped inside it.[44]

Perceptions of the relation between sea and sky led to the development of navigation techniques, such as the those of the Vikings, whose well-known navigation skills led some groups to settle in Newfoundland.[45] Norse navigation techniques were based on astronomical instruments, which were hand-held and functional, as demonstrated by archaeological finds such as the eleventh-century Uunartoq dial fragment found in Greenland.[46] Among such finds was the horizon board, used to interpret directional information by combining seasonal azimuth angles of the rising and setting sun with a given latitude. Other navigation tools whose use is attested in Old Norse sources were the sunstone, the solar compass and the twilight compass, which exploited the position of the sun to locate the geographic north.[47]

Although the sea and ocean inspired a plethora of studies during the Middle Ages, freshwater courses also received some attention. Isidore proposes a taxonomy in the *Etymologies*,[48] distinguishing between different types of natural watercourses, such as *flumen* ('river') and *torrens* ('torrent') and artificial ones, such as *rivi* ('canals') and *puteus* ('well') for irrigation and extraction of drinkable water, respectively.[49]

---

revolution of celestial spheres. Consequently, each element cannot move neither on its own nor due to the other elements, but only due the heavens or to the stars. See Kedar, "The Nomological Image" 72.

44 Ibidem 73.

45 Kuitems M. – Wallace B. – Lindsay C. – Scifo A. – Doeve P. – Jenkins K. – Lindauer S. – Erdil P. – Ledger P. – Forbes V. – Vermeeren C. – Friedrich R. – Dee M., "Evidence for European Presence in the Americas in AD 1021", *Nature* 601 (2022) 388–391.

46 Balázs B. – Farkas A. – Száz D. – Blahó M. – Egri A. – Barta A. – Åkesson S. – Horváth G., "How Could the Viking Sun Compass be Used with Sunstones before and after Sunset? Twilight Board as a New Interpretation of the Uunartoq Artefact Fragment", *Proceedings of the Royal Society A* 470.2166 (2014) 1–18, 2.

47 Ibidem 3–4.

48 Isidore, *Etymologiarum* XIII:XIX.1.

49 Ibidem XIII:XIX.1; XIII:21.1–4; *De natura rerum*, Chaps. 43–44.

Bede dedicates only Chap. 43 in *De natura rerum* to freshwater,[50] focusing on a single river, the Nile, which is noteworthy for replacing heavy rains, which are absent in Egypt due to hot weather. Echoing Isidore,[51] he highlights the interaction of air and earth with the river's water to cope with drought and reports how in May the winds form sand dunes that block the Nile's mouth, making it swell and overflow, thus irrigating the surrounding plains.

Among later scholars, Bartholomew the Englishman, drawing on Isidore and Bede, argues that rivers originate from the sea through hidden passages at the spring's origin, and go back to the sea again, and that river circulation is made up of two specular flows, one taking place on the earth's surface and one below it (*De proprietatibus rerum* XIII:II).[52]

3 Water in Theory and Practice

Throughout history the need to supply water for drinking and for agricultural and urban planning purposes has led to landscape transformations, whether aiming to contain floods or to address water scarcity. Among these, the Byzantine Empire's hydraulic engineering works of the third century are remarkable, replacing and upgrading aqueducts, dams and reservoirs built or upgraded by the Romans with cisterns for rainwater collection.[53] They are the endpoints of masonry aqueducts, covered in waterproof mortar, supplying water to urban environments,[54] mostly built underground, such as the Basilica Cistern, which brought water from the Marmara springs to Constantinople, with only a small proportion in the open air, which was constructed in Constantinople to provide water to Valens's and Justinian I's imperial palaces.[55]

In ancient and medieval Islam, knowledge on the identification of water sources, excavation of wells and construction of irrigation-systems is a distinct discipline (*Ilm al-Riyāfah*, 'hydronymy'). The *qinqin* ('hydrologist') is a

50 Bede also discusses rivers and springs in his Commentary on Genesis (*On Genesis* 2:5–6, 2:10).

51 See also Isidore, *De natura rerum* Chap. 43.

52 Sarti L., "Dante and the Water Cycle in the *Commedia*: God, Physics, and Flowing Waters", *Dante Studies* 137 (2019) 1–22, 12.

53 Haut B. – Viviers D., "Water Supply in the Middle East during Roman and Byzantine Periods", in Angelakis A. – Mays A.L. – Koutsoyiannis D. – Mamassis N. (eds.), *Evolution of Water Supply Throughout the Millennia* (London: 2012) 319–350, 319–321.

54 Ibidem 346.

55 Mays – Sklivaniotis – Angelakis, "Water" 33.

professional figure[56] who uses different methods for identifying underground water sources, ascertaining their depth and facilitating the extraction of water. For example, through botany, the hydrologist observes the length and depth of the roots of some plants such as purslane and papyrus.[57] Geology and mineralogy help classify rocks and water is easily found if the color of the stones is green and black, but not where it is red and yellow. Topographical knowledge is essential as mountain ranges abound in water due to snow accumulation. Water is also found in semi-arid mountainous areas, but decreases as one moves away. Meteorology is also employed to detect aquifers: according to al-Karajī, author of a tenth-century treatise on *The Extraction of Hidden Waters*, if certain plants are covered in dew, there is a near-surface aquifer, and the same goes for foggy areas with black clay or earth.[58] Magic can be also used: the water-diviner combines scientific observations, the scrutiny of birds' behaviour, and obscure practices, such as holding an iron rod in their hands, which tilts downwards if it senses water.[59] Moreover, hydronymy is so important in medieval Islam that Ibn al-ʿArabī arranges about fifty aquifers alphabetically in his ninth-century *Kitāb al-biʾr* ('Book of the Well')[60] describing their characteristics such as width, depth, water-flow and taste.[61]

The importance of landscape transformation is notable in the works carried out by Germanic populations, such as the permanent diversion of some Icelandic rivers mentioned in *Sturlunga saga*,[62] such as the Öxara, in the Þingvellir area, which is diverted to supply water to the *Alþing*. Artificial canals were built to increase and decrease water-levels so as to minimise the accumulation of detrimental sediments and to enrich the soil with mineral nutrients in the wetlands.[63] Scandinavian populations also intervened in watercourses

---

56 Hassan A., "Arab Contributions to Hydronymy: The Detecting of Underground Water that Led to the Construction of Water Wells", in Long C. – Amarillo S.C. (eds.), *Islamic Heritage Architecture and Art III* (Southampton: 2020) 13–25, 13.

57 Ibidem 14–18.

58 Ibidem 14–15.

59 Savage-Smith E., "Introduction", in Savage-Smith E. (ed.), *Magic and Divination in Early Islam* (London: 2017) 11–36, 22.

60 One of the two terms for 'well' in the *Quran* is the singular noun *biʾr* (22:45), along with *jubb* (12:10).

61 Hassan, "Arab Contributions" 15.

62 These texts are part of the *Íslendingasögur*, narrating Icelandic history between 1117 and 1264 and transcribed in the thirteenth and fourteenth centuries.

63 Sigurðardóttir R. – Newton A. – Hicks M. – Dugmore A. – Hreinsson V. – Ogilvie A.E.J. – Júlíusson A. – Einarsson A. – Hartman S – Simpson I.A. – Vésteinsson O. – McGovern T., "Trolls, Water, Time, and Community: Resource Management in the Mývatn District of

to compensate for the rapid decline in sea-level[64] to improve access to ports and canals. The thirteenth-century deviation of some watercourses (such as the Finnish river Kumo and some ports in the Gulf of Bothnia) favoured settlements, agriculture, travel and trade. Some marine islets were turned into islands in freshwater lakes to promote transit, such as Stockholm, which became the largest commercial port in the region.[65]

Striking examples of interest in the mechanics of water and hydrotechnology are to be found in the works of the great *quattrocento* artist and polymath Leonardo da Vinci – and in the actual implementation of Leonardo's ideas in real life. Leonardo interprets waterflow through the lenses of a still distinctly medieval organicistic worldview: as illustrated in drawings and notes in his notebook *Codex Leicester* (fol. 34r), he sees the earth as a living system capable of self-regulation, with blood representing water;[66] cardiac membranes the ocean; inhalation, exhalation and heartbeat representing ebb and flow;[67] and water flowing from the mountains the blood that keeps them alive (*Manuscript H*, fol. 77r).[68]

Leonardo is fascinated by water, vortexes and turbulences, depicting them in drawings and annotations,[69] and around 1490, starting a treatise on the topic – unfortunately lost – as reported in *Manuscript A* (fol. 55v). His notes on water, however, were collected in a seventeenth-century anthology entitled *Del moto e misura dell'acqua*, published in 1828; *Codex Leicester* is mainly dedicated to water's motions and characteristics.[70]

Northeast Iceland", in Lozny L. – McGovern T. (eds.), *Global Perspectives on Long Term Community Resource Management* (Berlin: 2019) 77–101, 91.

64 Except for Denmark, where the opposite phenomenon has occurred, a sizeable area of Scandinavia has seen a rapid decline in relative sea level since Ice Age, with the northern region of the Gulf of Bothnia being the most affected and where the level is currently falling at a rate of about 80 cm per century. As a result of this phenomenon, dry ground regions have gradually grown, increasing the riches of coastal lands. See Sawyer B. *Medieval Scandinavia: From Conversion to Reformation, Circa 800–1500* (Minneapolis, MN – London: 1993) 44.

65 Ibidem.

66 Seneca's imagery of river circulation as corresponding to the human circulatory system was widely used in the late Middle Ages, such as in Saint Albert the Great's *De meteoris* (II:III) and Dante's *Purgatorio*, where Matelda's speech (XXVIII:121–126) explains how rivers originate in the mountains from an underground system of veins that transform vapour into water. See Sarti, "Dante" 6–10.

67 Giovannetti G., *Indagine su Leonardo: Pavia, Vigevano, il Ticino e l'università* (Cremona: 2015) 38.

68 Ibidem 40.

69 See for example *Codex Leicester* fol. 34r. Ibidem 28–29.

70 Ibidem 32.

Leonardo's observation of water's movements is at the basis of his attempt to understand the mechanisms of the creation of natural forms. He also identifies two main forces acting on water motion: gravity and internal friction or viscosity, describing the phenomena generated by their interaction. He also argues that water is not compressible: although it takes different forms, it always keeps its mass.[71]

The *Atlantic Codex*[72] (fol. 730r) argues that cities are living beings, with water having the same vital function as blood in the human body and lymph in plants.[73] It compares the relation between bodily humours and elements in the environment to the interaction between urban planning and ecological building, arguing that in the same way that the sick need doctors to determine how health works to maintain humoral balance, so an architect-doctor is needed to understand construction and the rules that hold buildings together. Thus, in the plan of his ideal city, waterways play a fundamental role as a vehicle of public hygiene preventing the spread of pestilence.[74] Leonardo designs a multifunctional 'urban organism' divided into two levels with the upper one devoted to residential area, parks, pedestrian transit and wealthy classes, and the lower one reserved to warehouses, shops, streets, watercourses, transport of people and goods and drains (*Manuscript B* fol. 16r). He also outlines a double water-system with inlet and sewage channels, inspired by his observation of the human vascular system. Finally, in *Codex Arundel* (fol. 270v) he draws a series of canals, which converge into a single central river, with turbid waters discharged through ditches and secondary watercourses and river waters kept at a clean and consistent level.

Leonardo was instrumental in the agricultural economy of the Duchy of Milan thanks to Ludovico il Moro's entrepreneurial foresight in exploiting the polymath's hydraulic engineering inventions.[75] Having been appointed hydraulic superintendent for the Duchy. Leonardo reorganised all watercourses to extend rice cultivation,[76] thus contributing to completing the reclamation of marshes with water-meadows and fountains started by Cistercian monks in the eleventh century. Leonardo believed in the necessity to confine

71 Ibidem 32–35.

72 The *Atlantic Codex*, preserved at the Biblioteca Ambrosiana in Milan, is the largest extant collection of writings and drawings by Leonardo.

73 In *Manuscript A* (fol. 26r) and in *Codex Arundel* (fol. 56v), Leonardo notes that, unlike blood, water is transformed into vapour and mists and then falls back as rain when small clouds' parts condense.

74 Ibidem 40–44.

75 Tinarelli A., *Il riso in 10 lezioni* (Vercelli: 2010) 12.

76 Giovannetti, *Indagine* 47, 55.

water inside artificial basins to ensure the durability of resources (*Manuscript K*, fols. 108v-109r). Therefore, he designed a water-network of *navili* ('canals' for irrigation and navigation) and artificial small-flow ditches for agriculture and to feed mills to ensure the durability of practical resources. Leonardo also contributed to the Duchy's hydrographic reconfiguration with drawings of a dry-dredger, lock-works, and stilt system (*Atlantic Codex* fol. 3r and fol. 935r)[77] and he oversaw other projects elsewhere.[78]

Leonardo's observation of hydrodynamics was essential also for his physiological researches, especially concerning the heart, circulatory system and fetal membranes. Galen's *De motibus liquidis*, in Mark of Toledo's translation from Arabic, is mentioned in *Codex Windsor* (f. 19102r) as foundational to his embryological studies.[79] In *Anatomical Manuscript A* (f. 19003r), he contrasts the veins with the sea, as the latter receives all the rivers into itself after vapour, risen to the air, falls back to the ground as rain, supplying watercourses. In contrast, blood, a metaphorical sea, is the origin of all veins as it supplies them, not *vice versa*.[80]

4 Romance, Religion, and Myth

Water is a transversal, fluid and changeable element that has been used metaphorically since ancient times. It hides innumerable dangers, but also offers the possibility of reunion and salvation. It is a universal symbol in literature, as 'it changes in response to shifts of sensibility'.[81] In the Arthurian cycle, for example, the sea symbolises Tristan's separation from Isolt and triggers memories

77 Ibidem 12–13.

78 Leonardo was fascinated by the practical uses of water, such as mills, bridges, cataracts and river trades. He depicted sandmen, gold prospectors, boats and digging machines in *Codex Windsor* (fol. 12644r). At Villa Sforzesca, the first prototype of villa-farmhouse in the Duchy of Milan, in the outskirts of Vigevano, he studied water as a propelling force for agriculture, improving the 'water stairs' of the Scala Mill to regulate its flow, maximising the use of gravity by directing the watercourse in the same direction as the wheel (*Manuscript H*, fols. 30v, 63r). He also planned improvements to baths and ornamental water basins, such as heating the water in Isabella of Aragon's stove and embellishing Pavia Castle's park with a fishpond (*Atlantic Codex* fol. 289r and fol. 1036v). See Pisani F., "I percorsi tematici in un museo d'agricoltura", *Acta Museorum Italicorum Agricolturae* 16 (1996–1997) 169–176, 170 and Giovannetti, *Indagine* 66–69, 121.

79 Ibidem 148.

80 Ibidem 224.

81 Sobecki S., *The Sea and Medieval English Literature* (Woodbridge: 2008) 4.

of their love.[82] Water is central in the formation of myths,[83] as for instance the river Tiber is entrusted with the basket containing Romulus and Remus and the Nile with Moses, saving the children and preparing them for a great future.

In the Middle Ages, the biblical Flood is interpreted as a myth with a twofold interpretation: on the one hand, water's destructiveness, guided by divine wrath, punishes sin; on the other, it purifies and gives hope of rebirth for all believers.[84] This cleansing and renewing power is reflected in the Christian Easter liturgy, during which baptism is administered.[85] Flood waters typologically represent those of the baptismal font,[86] destroying sin while giving new life to the Christian faithful.

Bede, for example, interprets the Flood episode in Book II (VI:13–14a) of the exegetical treatise *In Genesin*, linking God's punishment with Judgment Day and comparing the ark's construction to that of the Church. He associates the Flood with baptismal water, cleansing both the Church and its members, and with Christ's resurrection. He argues that the Flood can typologically represent both the baptismal font and the waves of temptation that test the world,[87]as water is a metaphorical container in which the faithful must decide whether to choose salvation or damnation.

Baptismal water was an important part of medieval life, essential to Christian practice.[88] As Augustine specifies, baptismal water and the divine word are indissoluble, without which the sacrament would not exist.[89]

Baptismal liturgy and rites have changed over time, but water's role remains fundamental. The Roman rite of baptism, for example, involves the bishop blessing the font, mixing water with chrism, and sprinkling it on bystanders, who collect some water from the font to bless their land and their homes. After the blessing, baptism is administered to children and adult catechumens by immersing their heads three times in the water, and the top of their heads

82 Ibidem 5.

83 Ibidem 6.

84 Anlezark D., *Water and Fire: The Myth of the Flood in Anglo-Saxon England* (Manchester: 2006) 1.

85 Ibidem 6.

86 Ibidem 38.

87 Kendall C. (trans.), *On Genesis: Bede* (Liverpool: 2008) 172–174.

88 Keefe S., *Water and the Word: Baptism and the Education of the Clergy in the Carolingian Empire Volume 1* (Notre Dame, IN: 2002) 2.

89 Augustine, *In Joannis evangelium tractatus* CXXIV, XV:4, 'Quid est baptismus Christi? Lavacrum aquae in verbo. Tolle aquam, non est baptismus: tolle verbum, non est baptismus' ('What is Christ's baptism? The water bath in the Word. Take away water, it is not baptism; take away the Word, it is not baptism'; our translation).

is anointed before confirmation.[90] The triple immersion of the catechumen, symbolising the death of sin and man's rebirth through the baptismal font, represents the Trinity. It also emblematises a new degree of purity, typified by the white robes catechumens wear, which represent their joy for rebirth, angelic beauty and chastity.[91]

Water is also central to Eastern Orthodox and Byzantine Catholic liturgies, where it plays a key role in baptism, exorcism and other blessing prayers. For example, in the Syriac version of the prayer for the waters' blessing of the *Holy Theophany*, God is firstly addressed as 'creator of the waters' and only secondly as 'maker of the universe', thus highlighting water's primacy in the invocation. Through prayer, the faithful ask God to bless the liquid, giving it power against any evil force, guaranteeing the purification and healing of soul and body to anyone touching it, whether by washing, sprinkling or drinking.[92]

Water's destructive nature is represented on world maps and in literature by terrible sea monsters, such as the giant Aspidochelone, which, in bestiaries, looks like an island, deceiving sailors, who disembark on it, lighting fires on its back, but sinks into the Ocean drowning them. To give just a couple of further examples of the association of seas and destructiveness, seas are presented as home to enormous creatures such as orcas, whales, and giant squid, which evoke both fascination and fear.[93] More like monsters than fish, they communicate a sense of threat and their monstrously huge bodies, hinting at the beastly excess of guilt, remind people of their own sins.[94] Some sea monsters even attack humans out of the water, as in *Königin Deudalinda mit dem meerwunder*, a German poem from 1562, about Queen Teudelinda's rape on the shores of Luneria by a marine monster with bat legs, hawk eyes, and hairy like a bear. From their union a monstrous child is born with an uncontrollable violence and an abominable cannibalistic appetite, raping young women, and devouring them after intercourse.[95]

---

90 Ibidem 46.

91 Ibidem 82.

92 Ruggieri V., "The Water in the Basin (embates, loutron) and Byzantine Euchology", *Studi sull'Oriente Cristiano* 21.2 (2017) 303–321, 304–305.

93 Palmer M., "Great Fishes and Monstrous Men (Shoreline)", in Tomaini T. – Mittman A.S. (eds.), *Sea Monsters* (Goleta, CA: 2017) 9–17.

94 Ibidem 16–17.

95 Zironi A., "Teodelinda e il mostro marino: tracce storiche e motivi letterari", in Ruggerini M.E. – Szöke V. (eds.), *Dee, profetesse, regine e altre figure femminili nel Medioevo Germanico* (Cagliari: 2015) 168–169.

5 The Present Volume

The thirteen chapters in this volume reflect the fundamental importance of water across the whole breadth of medieval endeavour and understanding. Ranging geographically from Ireland to the Arab world and from Iceland to Byzantium and chronologically from the fourth century CE to the sixteenth, the following chapters explore perceptions and theories of water from many points of view. They demonstrate that in the Middle Ages, not only – as throughout history (and indeed pre-history) – was water seen to be essential to all life and the supply and practical management of it to be a matter of life and death, but its mysteriousness was an object of scholarly fascination and its manifestations were the source of rich symbolism and imaginings. At the end of this book Kelly Grovier picks up on the mysteriousness of water and its contribution to human endeavour in a poem rich in medieval allusion. Here water is portrayed as both wondrous and mundane, active and passive, elusive but instrumental in human ritual and industry: the personified element exclaims,

> I shape the shift from spirit
> to engine, shoulder the weightless
> glister of mill wheels as they spin
> revolution to revolution.

Modern science has confirmed that water is indispensable to all life. As we have seen earlier in this introduction, the unique place of water among the elements was also recognised by medieval authorities. Isidore of Seville, the most influential of them in the early medieval west, highlights its special properties and qualities in its various forms: in his *Etymologies*, he declares, 'The element of water rules over all the rest, for water tempers the sky, makes the earth fertile, gives body to the air with its exhalation, ascends to the heights, and claims the sky for itself'.[96] In the east, the ninth-century Arab polymath Abū Yaʿqūb ibn Isḥāq al-Kindī saw the various life-giving forms of water as relating to each other in a never-ending cycle: as quoted below in the chapter by Colin Murtha and Andreas Lammer, for al-Kindī, 'All that which flows into the seas is little by little dissolved [into] the atmosphere through the revolution of the Sun and the planets, and so it becomes vapour, condenses as clouds, and becomes rain, snow, and hail, which return to the Earth, flow into the seas, ever in this cycle eternally for as long as the world remains'.[97]

---

96 *Etymologies* XIII: XII: 3; trans. Barney, *The Etymologies* 276.

97 al-Kindī, *Risāla fī l-ʿilla al-fāʿila li-l-madd wa-l-ǧazr* 111.2–4. See below 31.

The following chapters present contrasting approaches to such medieval understandings and perceptions, and the material studied is strikingly varied, though there are also strong interconnections across the volume. The influence of holy texts and venerable authorities, for example; the symbolism of water; perceptions of water in art and imagination; practical applications of water in 'real life'; sea and land: these are some of the cross-currents that can be traced in the pages below. The chapters consider water as an element, as a danger and as a resource.

Interconnections are apparent across the volume but, structurally, we have thought it appropriate, as in the other volumes in the series, to order the contributions in four broad thematic sections, each focusing on a distinct aspect of medieval treatments of water. Thus, the first section, 'Water in Natural Philosophy and Medicine', highlights health-giving properties and uses of water, as developed in learned writings but also in traditional practice; the second, 'Images of Seas and Rivers', studies imaginings of water as seen in artistic representations and as prompted by (changing) human experience of the sea; the third section, 'Water in Theory and Practice', considers water in philosophical thought and through the lens of modern theory, while also addressing the symbolic significance of words for water and the history of medieval written legislation concerning water, with reference to practical issues such as the ownership of water and the use of water; finally, section 4, 'Romance, Religion, Myth', is concerned with elaborations of water themes in examples of creative literature, both religious and secular, and with the mythic significance of the element, as evident in a particular culture (Norse). The volume ends with the poem by Kelly Grovier, 'Effusion on Affusion', quoted from earlier in this Introduction.

6 Chapters and Connections

The 'Water in Natural Science and Medicine' section consists of three contributions that range from high philosophical speculations to medieval medicine. In the first chapter Colin Murtha and Andreas Lammer focus on Islamic scientific understandings of the water cycle – i.e. water dissolved from the sea as rising vapour, returning to the earth in precipitation and flowing into the sea again – in a detailed account of a tradition of scholarship indebted ultimately to Aristotle's *Meteorology* but supplemented and adapted by pre-Avicenna Arabic thinkers (thus up to about the eleventh century CE). The authors demonstrate that scholars of the medieval Islamic tradition produced masterly theories that account for many of the puzzling features of water (such as the salinity of the sea), achieving in their writings a sophisticated balance between the complexity and simplicity of this vital element.

Deborah Hayden's chapter examines the rich corpus of medieval Hiberno-Latin and Irish-language literature that portrays water as a key element in the healing of both physical and spiritual ills. Hayden discusses the evidence provided by a range of texts, including hagiographical (there are interconnections here with Hugh Magennis's chapter), penitential and narrative, for the association between water and medicine in medieval Irish tradition. As well as drawing upon established scholarly editions of early Irish writings, Hayden makes use of material in unpublished manuscripts to explore the symbolism and representation of water as a curative element, arguing that these offer important additional insight into the development of the literary motif in question.

The other chapter in the first section is a survey by Elisa Ramazzina of developments in bathing practices in medieval Europe, thus taking forward the focus on health-giving water featured in Hayden's contribution. Ramazzina's chapter draws attention to the importance of the southern Italian Salernitan school (which flourished in the eleventh to thirteenth centuries), whose Arab-influenced treatises are shown to represent a turning-point in the treatment of baths, giving them a vital role in iatric practice. Along with stressing the complex relationship between late medieval balneology and elemental theory and examining the readership of balneotherapic treatises, Ramazzina also explores the significance of mineralogy and the influence of alchemy on the evolution of late medieval balneology.

The book's second section, 'Images of Seas and Rivers', consists of three chapters, the first of which, by Karl Whittington, explores the drawings and writings of the fourteenth-century Italian priest Opicinus de Canistris. In dozens of bizarre illustrations Opicinus created embodied maps of the Mediterranean world in which the landforms of Europe and North Africa and the surrounding seas were turned into human bodies; the Mediterranean itself is portrayed as a startling figure of the devil. Opicinus's drawings were based on accurate sea-charts but reimagine these in grotesque form, with the sea portrayed as a space of danger, sin and pollution, its waters turned into monstrous beings. Placing Opicinus's work in the context of Christian theological interpretations of the sea in symbolic terms, Whittington uses his examination of Opicinus's strange representations, and also of his extensive written texts, to explore perceptions and understandings in fourteenth-century Italy of the imagined dangers, threats and possibilities of the sea. The symbolism of water can be seen to be a recurrent theme across the present volume.

After Whittington's close focus on an individual artist, François Quiviger broadens the perspective by addressing an aspect of art history viewed over the course of a millennium. Quiviger considers depictions of the River Jordan in western art from late Antiquity to the early modern period, using the theme of the baptism of Christ in the Jordan to outline developments in the representation of water across the centuries, developments that reflect changing understandings of the element. The chapter traces technical changes, fluctuations and continuities over time exemplified in a wide range of artistic productions (decorated baptistries and baptismal fonts, mosaics, frescos, altarpieces, prayer books and other works), as practitioners strove to present recognisable figurations of an element that has neither stable shape nor colour. The solutions they came up with are seen to display an increasing awareness of water materiality and dynamics to the point that by the sixteenth century aquatic attributes such as fluidity, movement and flow eventually rose as aesthetic categories. Here we might also think of the eloquent image in Grovier's poem at the end of the book, of

> the alchemy
> of monks conjuring the surge
> from pulverised lazurite
>
> in Lindisfarne and Durrow.

In the final chapter in this section Simon Karstens explores the mindsets of Europeans in the fifteenth and sixteenth centuries in relation to the Atlantic Ocean, identifying contrasting and indeed contradictory perceptions concerning it. Contextualising his analysis by noting inherited medieval ideas about the ocean as a dark unknown place of danger and mystery (as reflected also in the contribution by Whittington), and placing such imaginings against the lived experience of seafarers, Karstens goes on to identify different late-medieval/early-modern meanings and understandings circulating in various social groups in Europe in the age of the first regular transatlantic voyages. He highlights colonial, commercial (including slavery), religious, navigational and military viewpoints, pointing out that contradictory perceptions of the Atlantic were an important source of conflict at both intellectual and, as countries and other groups such as privateers vied for control of the seaways, physical levels.

The third section of our volume is entitled 'Water in Theory and Practice'. Its three chapters range in content from the application of modern theory to the legal underpinnings of the use and control of water in everyday practice, underpinnings that are themselves connected to elemental theory.

In the first chapter Michael Bintley draws upon cognitive theory as he applies insights from Extended Mind Theory and Material Engagement Theory, with their key concept of exograms (objects in the material environment that interact with the brain), to the famous collection of Old Norse poems the *Poetic Edda* preserved in the thirteenth-century Codex Regius. A 'mind', according to these theories, is constituted not only by the brain, the central nervous system and the body, but also by the objects and environment encountered and engaged with by the human organism. Bintley regards the poems of the Codex Regius, which preserve the mythology and wisdom of earlier medieval Iceland, as constituting an exogram, affecting the interactions of its readers with the material world. Assessing the representation of water in various contexts – as the sea, as bodies of water on land and as an element of meteorological phenomena, and also in relation to fluids more directly related to human bodies – he argues that texts like those in the Codex Regius can be viewed as active components of the extended minds of humans.

The theory that features prominently in the next chapter is that of translation. Here Marusca Francini examines the translation, from Greek, of terms from the semantic field of water in Bishop Wulfila's fourth-century Gothic Bible, from which large parts of the New Testament survive. Intersecting with other chapters in the volume, Francini's contribution incorporates a study of the symbolic meanings of water in the gospels, highlighting especially 'living water' and water as a force of chaos. Making use of philological evidence, etymology and semantics, it investigates Wulfila's linguistic and stylistic strategies as he devised a new terminology in Gothic that managed to convey the metaphorical overtones of 'water' and related terms in the prestigious original.

This section concludes with Luca Loschiavo's sweeping history of legislation concerning water in the west, but with reference also to the Byzantine east, from Antiquity to the Renaissance. Stressing the importance of water as taught by Isidore and other authorities, Loschiavo traces the development of regulations for its practical use and control in its many manifestations, on land and at sea. Adopting a chronological approach, he discusses in the first half of the chapter topics such as laws pertaining to the provision and maintenance of water supply, public and private waters, water in Christian ritual, rights to water, ownership of water and Germanic (as opposed to Roman) legal principles. The second half of the chapter moves on to consider another complex aspect of water legislation, that concerning marine waters. Here Loschiavo chronicles the changing fortunes over the centuries of major centres of maritime activity and outlines the evolution of compilations on the 'law of the sea' that governed their affairs, from Greco-Roman beginnings to the regulatory texts of the late Middle Ages and beyond, these compilations reflecting the social and political history of their times.

The four chapters of the final section, 'Romance, Religion, Myth', take us to water in the imagined worlds of romance and hagiography and to the place of water in the Old Norse mythical complex as preserved in Icelandic literature. In the opening chapter Laura Borghetti examines the role and symbolism of water in the twelfth-century Byzantine romance *Rodanthe and Dosikles*, arguing that this work displays a blending of ancient Greek literature and Christian doctrine in which water, which features ubiquitously in the narrative, is presented as both harmful and beneficial, representing body and soul, darkness and light, sin and salvation. In its fluidity, instability and metamorphosis, water is seen as constituting an evocative metaphor of human life-experience, whose contradictions can be overcome by trusting in God's forgiveness and final salvation. Borghetti focuses mainly on the waters of the sea, as the *dramatis personae* tirelessly travel across it, but she also has some pertinent discussion of bathing, thus linking up with chapters in our first section.

Next comes Rafał Borysławski's chapter on the *lais* of the twelfth-century Anglo-Norman poet Marie de France, in which water, and in particular the water of the sea (though baths figure here too), is again seen to play an important narrative role in scenes of parting, distance and death and in the idea of the voyage and its capacity to both sever lovers and reconnect those that have been separated. Distinctive of Borysławski's approach is the contention that Isidorean etymologising is operative in the *lais*, notably in their play on perceived connections between *mare*, *mors*, *amarus* and *amor* (Old French *mer*, *mort*, *amer*, *amur*). Borysławski proposes that such etymologising may supplement the ambiguity of the narrative elements of Marie's narratives, suggesting that not only was reading through etymologies familiar to and expected among the more learned audiences of the *lais*, but that it may have contributed to and invited their multiple, and not infrequently contradictory, interpretations.

In the third chapter of 'Romance, Religion, Myth' Hugh Magennis uses the relatively extensive corpus of saints' lives in Old English prose and verse as a sample to offer an analysis of how water functions in hagiographical narrative. The study identifies four main strands to its use: in miracles, travel, cleansing (including baptism) and asceticism. As a rule, other than with reference to baptism, water is presented primarily in prose narrative as a literal entity rather than functioning symbolically. An exception to this rule is the *Life of Saint Mary of Egypt*, in which water represents boundaries and spiritual progression. Old English verse saints' lives make use of the same strands but they enrich their treatment by drawing upon the resources of traditional Old English poetry, most notably adapting a theme identified here as that of the journey of the hero by sea. In verse saints' lives water can also have more of a symbolic dimension than in prose, as demonstrated particularly in the case of the Vercelli Book poem *Andreas*.

The final chapter, by Tom Birkett, surveys the multifarious significance of water in Old Norse mythology, complementing Bintley's earlier chapter on Norse material. The first part of the chapter presents a consideration of the place of water in the creation (and destruction) of the Norse cosmos, particularly as described by Snorri Sturluson in his *Prose Edda*. It finds that Snorri combines, albeit somewhat uneasily, aspects of learned elemental theory, in which the elements combine to bring about creation, with ideas from pre-Christian sources, in which water occupies the primordial place. The chapter goes on to explain the central role that water plays in the structuring of the Norse cosmos, encircling, circumscribing and dividing the world and creating barriers and conduits between realms. Then in the second part of the chapter Birkett examines personifications of water as divine and supernatural beings, both positive and negative, before moving on to an account of Thor's frequent adventures concerning water, which are taken to be particularly revealing of the element's positioning within the cosmic struggle between gods and giants and indicative of male strength prevailing over a feminised element.

Birkett's wide-ranging chapter synthesises with reference to a specific culture major concerns about the nature and meaning of water that animate many of the contributions across our collection. Water, the indispensable element, is approached from many points of view in this book, philosophical/theological, medicinal, meteorological/scientific, economic, artistic and symbolic, in chapters that study how medieval people sought to understand this element in their various endeavours and explorations, both physical and mental, as they encountered its dangers, threats and possibilities. Water is ubiquitous in medieval experience and thought. 'All things that are seen on earth', writes Bede, 'are formed and reformed by the dual nature of water [either salt or fresh]', echoing Isidore's sentiment, as quoted above, 'The element of water rules over all the rest'.[98] Consciousness of the primacy of water among the elements and of its relation to the other elements will be widely exemplified in the following chapters.

Bibliography

Manuscripts

da Vinci, Leonardo. 1487–1489. *Manuscript B*, Paris, Institut de France.
da Vinci, Leonardo. 1478–1518. *Codex Arundel*, London, British Library.
da Vinci, Leonardo. 1478–1518. *Codex Atlanticus*, Milan, Biblioteca Ambrosiana.
da Vinci, Leonardo. 1478–1518. *Codex Windsor*, Windsor, Windsor Castle.
da Vinci, Leonardo. 1490–1492. *Manuscript A*, Paris, Institut de France.
da Vinci, Leonardo. 1493–1494. *Manuscript H*, Paris, Institut de France.
da Vinci, Leonardo. 1503–1507. *Manuscript K*, Paris, Institut de France.
da Vinci, Leonardo. 1509–1510. *Anatomical Manuscript A*, Windsor, Windsor Castle.

98 Bede, *De Natura Rerum* Chap. 38; trans. Kendall – Wallis, *On the Nature of Things* 94; for Isidore, see above 18.

da Vinci, Leonardo. 1510. *Codex Leicester.* DaVinci Codex: Leicester's Guide – Developed by WebUp S.p.a., 2016. Accessed May 14, 2024. https://hammercodex.com/.

Primary Sources

Aristotle, *Metaphysics,* trans. W.D. Ross (1994–2009). Internet Classics Archive. Accessed May 14, 2024. http://classics.mit.edu/Aristotle/metaphysics.html.

Aristotle, *Physics,* trans. R.P. Hardie – R.K. Gaye (1994–2009). Internet Classics Archive. Accessed May 14, 2024. http://classics.mit.edu/Aristotle/physics.html.

Augustine, *In Joannis evangelium tractatus CXXIV,* ed. J.P. Migne (Petit–Montrouge: 1841). Corpus Corporum University of Zurich. Accessed May 14, 2024. https://www.mlat.uzh.ch/browser?path=/38/900/765//7323.

Barney S. – Lewis W.J. – Beach J.A. – Berghof O. – Hall M. (eds. and trans.), *The Etymologies of Isidore of Seville* (Cambridge: 2006).

Bartholomew the Englishman. *De Proprietatibus Rerum: Batman uppon Bartholome his Booke De Proprietatibus Rerum, Newly Corrected, Enlarged and Amended,* trans. T. Batman, ed. T. Trevisa (London: 1582; reprint Ann Arbor: 2011) Accessed May 14, 2024. https://quod.lib.umich.edu/cgi/t/text/text-idx?c=eebo;idno=A05237.0001.001.

Bede, *De natura rerum,* ed. J.P. Migne (Petit–Montrouge: 1850), Corpus Corporum University of Zurich. Accessed May 14, 2024. https://www.mlat.uzh.ch/?c=2&w=Beda.DeNaRe2&path=Beda.DeNaRe2/8442.

Blake M. (ed. and trans.), *Ælfric's De Temporibus Anni* (Cambridge: 2009).

Conway P. – Larcher R.F. (trans.), *Exposition of Aristotle's Treatise on Generation and Corruption,* Book I, cc. 1–5 (Columbus, OH: 1964). St. Isidore e-book library. Accessed May 14, 2024. https://isidore.co/aquinas/GenCorrup.htm#2 (bilingual).

de Sacrobosco, Johannes, *De sphaera mundi* (Venice: 1501), Open Library. Accessed May 14, 2024. https://openlibrary.org/books/OL14636177M/Sphaera_mundi.

Grosseteste R., *De impressionibus elementorum,* in Baur L. (ed.), *Die Philosophishen Werke des Robert Grosseteste* (Münster: 1912) 87–89, Documenta Catholica Omnia. Accessed May 14, 2024. http://www.documentacatholicaomnia.eu/04z/z_1168-1253__Grosseteste_Robert__De_Impressionibus_Elementorum__LT.doc.html.

Isidore, *Etymologiarum sive Originum,* ed. W.M. Lindsay (Oxford: 1911), Digitised Edition University of Chicago, 2010. Accessed May 14, 2024. https://penelope.uchicago.edu/Thayer/E/Roman/Texts/Isidore/home.html.

Kendall C. (trans.), *On Genesis: Bede* (Liverpool: 2008).

Kendall C. – Wallis F. (trans.), *Bede: On the Nature of Things and on Times* (Liverpool: 2010).

Kendall C. – Wallis F. (trans.), *Isidore of Seville: On the Nature of Things* (Liverpool: 2016).

Jones C. (ed.), *Bedae Opera de Temporibus* (Cambridge, MA: 1943).

Wallis F. (trans.), *Bede: The Reckoning of Time* (Liverpool: 1988).

Secondary Sources

Anlezark D., *Water and Fire: The Myth of the Flood in Anglo-Saxon England* (Manchester: 2006).

Balázs B. – Farkas A. – Száz D. – Blahó M. – Egri A. – Barta A. – Åkesson S. – Gábor Horváth G., "How Could the Viking Sun Compass be Used with Sunstones before and after Sunset? Twilight Board as a New Interpretation of the Uunartoq Artefact Fragment", *Proceedings of the Royal Society A* 470.2166 (2014) 1–18.

Bobik J., *Aquinas on Matter and Form and the Elements: A Translation and Interpretation of "De Principiis Naturae" and "De Mixtione Elementorum" of St. Thomas Aquinas* (Notre Dame, IN: 1998).

Cesario M. – Magennis H – Ramazzina E., "Introduction", in Cesario M. – Magennis H. – Ramazzina E. (eds.), *The Elements in the Medieval World: Interdisciplinary Perspectives. Earth.* (Leiden – Boston MA: forthcoming).

Crisciani C. – Pereira M, "Introduction to the Philosopher's Laboratory", in Bernardini P. – Rodolfi A. (eds.), *Roger Bacon's "Communia Naturalium:" A 13th Century Philosopher's Workshop* (Florence: 2014) 3–25.

Crowther K., "Sacrobosco's *Sphaera* in Spain and Portugal", in Valleriani M. (ed.), *De sphaera of Johannes de Sacrobosco in the Early Modern Period* (Berlin: 2020) 161–184.

Hassan A., "Arab Contributions to Hydronymy: The Detecting of Underground Water that Led to the Construction of Water Wells", in Long J.C. – Dominguez Amarillo S. (eds.), *Islamic Heritage Architecture and Art III* (Southampton: 2020) 13–25.

Haut B. – Viviers D., "Water Supply in the Middle East during Roman and Byzantine Periods", in Angelakis A. – Mays L. – Koutsoyiannis D. – Mamassis N. (eds.), *Evolution of Water Supply Throughout the Millennia* (London: 2012) 319–350.

Kedar Y., "The Nomological Image of Nature: Explaining the Tide in the Thirteenth Century", *Annals of Science* 73.1 (2016) 68–88.

Kedar Y. – Hon G., "Roger Bacon (c. 1220–1292) and his System of Laws of Nature: Classification, Hierarchy and Significance", *Perspectives on Science* 25.6 (2017) 719–745.

Keefe S., *Water and the Word: Baptism and the Education of the Clergy in the Carolingian Empire Volume 1* (Notre Dame, IN: 2002).

Giovannetti G., *Indagine su Leonardo: Pavia, Vigevano, il Ticino e l'università* (Cremona: 2015).

Kuitems M. – Wallace B. – Lindsay C. – Scifo A. – Doeve P – Jenkins K. – Lindauer S. – Erdil P.– Ledger P. – Forbes V. – Vermeeren C. – Friedrich R. – Dee M., "Evidence for European Presence in the Americas in AD 1021", *Nature* 601 (2022) 388–391.

Mays L. – Sklivaniotis M.– Angelakis A., "Water for Human Consumption through History", in Angelakis A. – Mays L. – Koutsoyiannis D. – Mamassis N. (eds.), *Evolution of Water Supply Throughout the Millennia* (London: 2012) 19–42.

Palmer M., "Great Fishes and Monstrous Men (Shoreline)", in Tomaini T. – Mittman A.S. (eds.), *Sea Monsters* (Goleta, CA: 2017) 9–17.

Pisani F., "I percorsi tematici in un museo d'agricoltura", *Acta Museorum Italicorum Agricolturae* 16 (1996–1997) 169–176.

Ruggieri V., "The Water in the Basin (embates, loutron) and Byzantine Euchology", *Studi sull'Oriente Cristiano* 21.2 (2017) 303–321.

Sarti L., "Dante and the Water Cycle in the *Commedia*: God, Physics, and Flowing Waters", *Dante Studies* 137 (2019) 1–22.

Sawyer B., *Medieval Scandinavia: From Conversion to Reformation, Circa 800–1500* (Minneapolis, MN – London: 1993).

Savage-Smith E., "Introduction", in Savage-Smith E. (ed.), *Magic and Divination in Early Islam* (London: 2017) 11–36.

Sigurðardóttir R. – Newton A. – Hicks M. – Dugmore A. – Hreinsson V. – Ogilvie A.E.J. -Júlíusson A. – Einarsson A. – Hartman S. – Simpson I.A. – Vésteinsson O. – McGovern T., "Trolls, Water, Time, and Community: Resource Management in the Mývatn District of Northeast Iceland", in Lozny L. – McGovern T. (eds.), *Global Perspectives on Long Term Community Resource Management* (Berlin: 2019) 77–101.

Sobecki S., *The Sea and Medieval English Literature* (Woodbridge: 2008).

Sparavigna A.C., "Robert Grosseteste and the Four Elements", *International Journal of Sciences* 2.12 (2013) 42–45.

Tinarelli A., *Il riso in 10 lezioni* (Vercelli: 2010).

Zironi A., "Teodelinda e il mostro marino: tracce storiche e motivi letterari", in Ruggerini M.E. – Szöke V. (eds.), *Dee, profetesse, regine e altre figure femminili nel medioevo germanico* (Cagliari: 2015) 167–189.

PART 1

Water in Natural Philosophy and Medicine

CHAPTER 1

What Goes Up Must Come Down: The Water Cycle in Early Arabic Meteorology

Colin Fitzpatrick Murtha and Andreas Lammer

Introduction

> All that which flows into the seas is little by little dissolved [into] the atmosphere through the revolution of the Sun and the planets, and so it becomes vapour, condenses as clouds, and becomes rain, snow, and hail, which return to the Earth, flow into the seas, ever in this cycle eternally for as long as the world remains.[1]

These lines appear at the early stages of medieval Islamic philosophy, penned by the famous 'philosopher of the Arabs', Abū Yaʿqūb ibn Isḥāq al-Kindī (d. c. 256/870), who invites his reader to consider the cycle of water in the world we inhabit. The water cycle is depicted as a circular process and is likened to the revolution of the heavenly spheres.[2] Despite the constancy of the circular motion of the heavens and the omnipresence of the water cycle in everyday life, each stage of the water cycle when considered closely brings with it its own theoretical puzzles. Why does water vapour rise when all other forms of water fall and, ultimately, tend towards depth? How could the sheer variety of forms of precipitation be explained? How is it that rivers continue to flow even if precipitation is lacking, and why on Earth is the sea salty? Producing a scientific account in answer to these and other questions, often based on common observations, was indispensable for natural philosophers of medieval Islam and, almost invariably, theories about water at different stages of the water cycle fell under the purview of the discipline called 'meteorology'.

1 al-Kindī, *Risāla fī l-ʿilla al-fāʿila li-l-madd wa-l-ǧazr*, in *Rasāʾil al-Kindī l-falsafiyya*, ed. M.ʿA. al-H. Abū Rīda (Cairo: 1950–1953), vol. 1 111.2–4. The authors have greatly profited from Hans Daiber's pioneering contributions and Paul Lettinck's voluminous monograph *Aristotle's Meteorology*. All translations, unless noted otherwise, are our own.

2 See Aristotle, *Meteorology*, trans. E.W. Webster, in Barnes J. (ed.), *The Complete Works of Aristotle: The Revised Oxford Translation*, Bollingen Series 71.2 (Princeton, NJ: 6th ed. 1995) II.9, 346b35–347a8.

 | DOI:10.1163/9789004696501_003

In the philosophical curriculum, the science of meteorology was significantly placed between the physical sciences and the life sciences, promising both the concrete application of theoretical physical concepts and prerequisite knowledge for the study of organic life. The most important source for meteorological theories was Aristotle's *Meteorology*, a work which had been rendered into Arabic by Yaḥyā ibn al-Biṭrīq (d. c. 215/830), but in a highly paraphrastic and incomplete manner, omitting large portions of the original and occasionally deviating from it even on fundamental points. Although the commentaries on Aristotle's *Meteorology* of Alexander of Aphrodisias (fl. c. 200) and Olympiodorus (d. after 565) were also translated into Arabic according to the bio-bibliographer Ibn al-Nadīm (d. c. 385/995), they are no longer known to be extant in Arabic.[3] Despite its paraphrastic character, Ibn al-Biṭrīq's translation of Aristotle's *Meteorology* circulated in Islamic lands as early as al-Kindī (who after all employed Ibn al-Biṭrīq as a translator), and was known as late as the early thirteenth century and as far west as al-Andalus.[4] Additionally, and depending on how the testimony of Ibn al-Nadīm is read, there may have been another Arabic translation of Aristotle's *Meteorology* by the hands of Yaḥyā ibn ʿAdī (d. 363/974).

Clearly, then, Aristotle's theories were disseminated by multiple channels throughout the Islamic world, and medieval scientists and philosophers had much to contribute as they provided supplementary explanations for meteorological occurrences, first-hand observations, new analogies, and a refinement of Aristotle's theories. Many thinkers of the period, both well-known and lesser-known, were united in the effort to understand the natural world we inhabit and its plurality of meteorological occurrences, not least of all the water cycle – but these same thinkers were also often divided over just how to understand such natural facts, as we shall see.[5] Investigating a number of key texts that were available in Arabic up to the beginning of the eleventh century, the present contribution focuses on the water cycle before Avicenna (d. 428/1037)

3 See Peters F.E., *Aristoteles Arabus: The Oriental Translations and Commentaries on the Aristotelian Corpus*, Monographs on Mediterranean Antiquity 2 (Leiden – Boston, MA: 1968) 39–40.

4 It is quoted extensively by Averroes (d. 595/1198) in his *Middle Commentary on the 'Meteorology'* (*Talḫīṣ Kitāb al-Āṯār al-ʿulwiyya*) and was the basis of the first translation of an Aristotelian text into Hebrew by Samuel ibn Tibbon (d. c. 627/1230); see R. Fontaine's introduction in Aristotle, *Otot Ha-Shamayim: Samuel ibn Tibbon's Hebrew Version of Aristotle's* Meteorology, ed. and trans. R. Fontaine, Aristoteles Semitico-Latinus 8 (Leiden – New York – Cologne: 1995).

5 For lists of Islamic thinkers who wrote on topics on meteorology, see Sezgin F., *Geschichte des arabischen Schrifttums* vols. 8 (Leiden – Boston, MA: 1967–1984) vol. 7, esp. 233–305.

entered the stage, composed his own *Meteorology*, and transformed the tradition.[6] In order to assess the prevalent ideas on the hydrological circle, we shall examine (1) rising water, (2) falling water, (3) flowing water, and (4) deep water.

1 Rising Water

Imagine yourself sitting by a lake. There has been a downpour of rain from storm clouds overhead and the resultant swelling of streams that flow downward across the ground and into the lake have it brimming with fresh water. In considering this downpour, the downward course of water flowing into the lake, and the deep waters of the lake itself, one could perceive water as an element which takes a natural course toward depth. After all, as Aristotle reminds us, 'water flows into the deepest place'.[7] Echoing this, Ḥunayn ibn Isḥāq (d. 260/873) states in his compendium on meteorology that 'water is such as to (*min šaʾnihī*) seek depth', and in a similar way, Ps.-Olympiodorus, the as-yet unidentified author of an exposition of Olympiodorus's commentary on Aristotle's *Meteorology*, argues that river water, in flowing to the seas, 'yearns for' (*tatašawwaqu ilā*) its deep natural end.[8]

Now, imagine that some time has passed, as has the storm, and the Sun is shining upon the lake. As sunlight touches the surface of the water of the lake, you observe vapour rising into the air. Consequently, it seems that a survey of theories on those states of water which seek depth (falling water and flowing water) as well as water that has achieved depth (deep water) must also include an account of the water which rises as vapour. In rising, vapour seems to counteract the natural downward course of all other occurrences of water and takes what appears to be an equally natural upward course as if seeking height.

6 Due to the highly complex nature of that work, we exclude Avicenna from our analysis and focus, instead, on the state of the art leading up to the time in which he flourished. The details of his theories on mineralogy and meteorology still await investigation, much as his text remains to be translated. Both projects have been begun by the present authors.

7 Aristotle, *Meteorology*. II.2, 355b17, trans. Webster.

8 *Ǧawāmiʿ Ḥunayn ibn Isḥāq* 5, 41.120–121; *Tafsīr Alumfīdūrūs li-Kitāb Arisṭūṭālīs fī l-Āṯār al-ʿulwiyya*, in ʿAbd al-Raḥmān Badawī (ed.), *Šurūḥ ʿalā Arisṭū mafqūda fī l-yūnāniyya wa-rasāʾil uḫrā: Commentaires sur Aristote perdus en grec et autres épitres*, Recherches publiées sous la direction de l'Institut de Lettres Orientales de Beyrouth, Nouvelle série A: Langue arabe et pensée islamique 1 (Beirut: 1971) 106.24–107.1. Both Ḥunayn's and Ps.-Olympiodorus's statements stem from a discussion of the sea, as its saline waters are particularly 'such as to seek depth', as will be discussed below, 63.

The resounding and common answer as to how vapour comes to rise invokes the heat of the celestial bodies and, as is often pointed out, it is the Sun especially that imparts heat to terrestrial water, causing it to rise upward. That is certainly not to say that there was unanimity among philosophers about exactly *how* the celestial bodies impart heat to the sublunary world, but only *that* the heat of the Sun especially causes vapour to rise from earth-bound bodies of water.[9] There is also no absolute uniformity among texts surveyed here in the use of terminology intended to describe how the Sun acts on these waters. Ibn al-Biṭrīq's version of Aristotle's *Meteorology* states that the Sun 'kindles' (*alhaba*) the Earth and its waters, giving rise to dry and wet exhalations: smoke and vapour.[10] Ḥunayn and the Iḫwān al-Ṣafāʾ (the so-called 'Brethren of Purity' active in the fourth/tenth century) describe the Sun's heat as 'provoking' (*aṯāra*) the exhalations into rising, and Abū ʿAlī Aḥmad Miskawayh (d. 421/1030), a century later, has it that the water, once heated, 'is such as to be drawn toward' (*min šaʾnihī* [...] *an yanǧaḏiba ilā*) the source that heats the water: the Sun.[11] Despite these differences in phrasing and technical terminology, the idea remains the same, namely, that the celestial bodies impart heat to water, causing it to rise upward as vapour.

In Aristotelian meteorology, vapour is not just another meteorological occurrence to be studied alongside the variety of atmospheric weather phenomena. In fact, the vapour you observe rising into the air is an *essential* ingredient of Aristotle's *Meteorology* and no less than a 'principle' (*archê*), explicitly so-called.[12] In Arabic meteorological texts adopting Aristotle's theories, the rising vapour constitutes the material basis for all watery occurrences in the atmosphere, including clouds, rain, snow, hail, dew, frost, fog, and mist. Alongside the wet vapour, Aristotle introduces another meteorological principle,

9 Potential disagreement could arise from questions concerning the material composition of the celestial bodies, especially whether they consist of fire or not. If they, as Aristotle argued, do not consist of fire, then other explanations need to be explored, such as heat through friction or radiation – the former being Aristotle's and the latter being al-Kindī's answer.

10 Ibn al-Biṭrīq, *Kitāb Arisṭūṭālīs fī l-Āṯār al-ʿulwiyya* I.VI, 27.211. The closest language to this idea of kindling or provoking in Aristotle is at *Meteorology* 1.3, 341a17–18, where Aristotle states that 'motion is able to dissolve and inflame the air'. Otherwise, Aristotle mainly emphasises that the Sun 'warms' the Earth and that moisture 'is made to evaporate [...] and rises' (see, for instance, *Meteorology* 1.4, 341b6–7; 1.9, 346b24–26).

11 *Ǧawāmiʿ Ḥunayn ibn Isḥāq* 2, 33.46; *Rasāʾil Iḫwān al-Ṣafāʾ* XVIII.VII. 209.8; Miskawayh, in ʿAlī Abū Ḥayyān al-Tawḥīdī – Abū ʿAlī Aḥmad Miskawayh, *The Philosopher Responds: An Intellectual Correspondence from the Tenth Century*, ed. B. Orfali – M.A. Pomerantz, trans. S. Vasalou – J.E. Montgomery, Library of Arabic Literature (New York: 2019) vol. 2, 276.17.

12 Aristotle, *Meteorology* 1.4, 341b5–10.

which is hot and dry and rises from the Earth itself: smoke. Together, smoke and vapour constitute the required *material cause* for all meteorological phenomena, sometimes in interaction, sometimes alone. Yet, it is clear that watery occurrences in the atmosphere are predominately to be explained through the rising vapour, which is subject to different processes and conditions. Now, turning our gaze upward with the rising vapour, let us consider the changes it begins to undergo.

1.1 *Reaching the Apex*

Two main factors eventually bring the rising motion of vapours to an end, one thermic and the other mechanic. The former is the explanation found in Aristotle, for whom the higher regions of the Earth's atmosphere are colder than the air below, often explained by lesser influence from reflected sunlight, which is more dispersed the further it travels away from Earth after reflecting from its surface. This dispersal of reflected sunlight in the upper regions of the atmosphere leads to the vapour losing heat, and thus cooling down.[13] Heat, however, is the central quality that initiated and prolonged the ascension. Thus, as the vapour cools down, it also slows down and, ultimately, stops rising entirely at a certain height.[14] Apparently, then, it is due to temperature – or, more precisely, the loss of heat – that the rising vapour is prevented from reaching even higher regions.

The second, mechanical explanation is a Theophrastean adaptation and is usually discussed by referring to a form of pressure.[15] As we have just read,

---

13 See Aristotle, *Meteorogy* I.9, 346b26–35; see also al-Kindī, *Risāla fī l-ʿilla allatī lahā yabrudu aʿlā al-ǧaww*, 97.2–3, trans. P. Adamson – P.E. Portmann, *The Philosophical Works of al-Kindī*, Studies in Islamic Philosophy (Oxford – New York: 2012) 213.

14 See Aristotle, *Meteorology* I.9, 346b26–33, 347a8–10; I.11, 347b18–19; II.4, 359b34–360a3.

15 Olympiodorus, *In Aristotelis* Meteora *commentaria*, ed. W. Stüve, Commentaria in Aristotelem graeca XII.2 (Berlin: 1900) 80.30–33 explicitly states that Theophrastus added the account of *pilêsis* ('pressure, compression') to Aristotle's 'single' (*monon*) explanation by *psyxis* ('cooling'); see Theophrastus Arabus, *al-Qawl fī l-Āṯār al-ʿulwiyya* 7, 236.5–6; see also Steinmetz' comments in Theophrastus, *Der syrische Auszug der Meteorologie des Theophrast*, ed. and trans. E. Wagner, with annotations by P. Steinmetz, Akademie der Wissenschaften und der Literatur: Abhandlungen der Geistes- und Sozialwissenschaftlichen Klasse 1 (Wiesbaden: 1964) 51 and in Steinmetz P., *Die Physik des Theophrastos von Eresos*, Palingenesia: Schriftenreihe für klassische Altertumswissenschaft 1 (Bad Homburg v.d. Höhe: 1964) 217–221, where Theophrastus's emphasis of the role of mountains in cloud formation is noted, as well as the formation of precipitation in *De ventis* §§4.30–5.36. Ps.-Olympiodorus later utilises this mechanical explanation to account for the occurrence of rain in typically *hot* lands, that is, where Aristotle's thermal explanation may not suffice. Referring to the lands of Abyssinia, for instance, Ps.-Olympiodorus states that due to the abundance of mountains in these lands, clouds that are pushed by winds

cooled-down vapour accumulates at a certain height. Yet, more and more vapour still keeps rising from the waters on Earth. In consequence, the accumulation increases and thickens, thus generating pressure: the already cold vapour presses from above, preventing still warm vapour to rise further, and the warm vapour presses from below trying to rise. The amount of pressure may further be increased through topographical conditions, especially mountains, perhaps in combination with winds.[16]

Taking both explanations together, vapour is prevented from rising as it cools down, thereby effectively losing its impetus to rise further, but even if it had not cooled down, or had cooled down more slowly, it could not rise further, because the cold vapour which already accumulated above hinders it so that the vapour thickens and condenses before it could rise any further. Consequently, cooling down (as in Aristotle) and pressure (as additionally in Theophrastus) lead to condensation, i.e., cloud formation.[17] It should be noted that in his explanation of rain, Theophrastus refers particularly to mountain walls against which the vapours are pressed by winds, illustrating this process by referring to the ceiling of bathhouses. What we read in Theophrastus, then, are not one but two separate cases of pressure-induced condensation, one cause being whenever wind presses the vapours together on a horizontal axis against mountains ('wall'), the other being the constant rise of vapours which on a vertical axis keeps pressing against the stationary, cooled-down vapours above ('ceiling'). Both cases are found also in the Arabic tradition.

This Aristotelian-Theophrastean background is apparently the reason why we read in the compendium of Ḥunayn that the upper region not only 'cools' (*barrada*) the rising vapour but also that the vapour 'hits upon' (*istaqbala*) the cold up high, which 'repels' it (*radda*), thus combining Aristotle's account and Theophrastus's 'ceiling' case.[18] What we find in al-Kindī, in turn, is a

are blocked by the mountains and condense. His dependence on Theophrastus is evident with his analogy to hot bathhouses (*ḥamāmāt*; to be mentioned shortly below) where the risen vapours that gathered on the ceilings (*suqūf*) are transformed into water and fall again as drops (*Tafsīr Alumfīdūrūs*, 99.23–100.3).

16 In Theophrastus's account, the pressure is related to the influence of certain winds that accumulate the vapours; see al-Kindī, *Risāla fī l-ʿilla allatī lahā takūnu baʿḍ al-mawāḍiʿ lā takādu tumṭar*, in M.ʿA. al-H. Abū Rīda (ed.), *Rasāʾil al-Kindī l-falsafiyya* (Cairo: 1950–1953) 72.10–73.3; al-Ṭabarī in al-Ṣiddīqī M.Z. (ed.), *Firdausu'l-Ḥikmat: Paradise of Wisdom* (Berlin: 1928) 1.10, 24.9–10; see also n. 20.

17 Conversely, warmth and heat prevent cloud formation; see Aristotle, *Meteorology* 1.3, 340a24–341a12.

18 *Ǧawāmiʿ Ḥunayn ibn Isḥāq* 2, 33.52–55. Another possible interpretation of Ḥunayn's text would be to read the repelling of the rising vapour as an instance of *antiperistasis*, which would, then, potentially reduce the Theophrastean influence at least on the Ḥunayn

combination of Aristotle and Theophrastus's 'wall' case, for al-Kindī constantly emphasises the need for the 'confinement' (*inḥiṣār*) of vapours – best achieved in mountainous or valley regions – alongside cooling conditions such as the higher, cold regions of the atmosphere.[19] Other authors, such as the Iḫwān al-Ṣafāʾ, by contrast, use Theophrastus's mechanical explanation for cloud formation and introduce the Aristotelian thermic factor of coldness only to account for rain, while Job of Edessa (d. c. 220/835) and Ibn al-ʿAmīd (d. 360/970) seem to do the reverse.[20] Others again, such as Abū l-Ḥasan ʿAlī ibn Sahl Rabban al-Ṭabarī (d. c. 256/870), apparently give sole preference to Theophrastus's view and reserve the cold as an explanatory factor not even for rain but for colder occurrences, such as snow and hail.[21] Interestingly, it is not clear whether al-Ṭabarī's omission of Aristotle's thermic explanation of cloud formation is due to his actually preferring Theophrastus's mechanic account or whether it is due to his silence (by and large) on clouds as such.[22]

At any rate, with cloud formation, we have reached the apex of the water cycle; no wet or moist substance is able to rise higher than this point at which coldness, in addition to other causes, prevents its further ascent. At this high point – which is located perhaps somewhere between 16,000 cubits (according to the Iḫwān al-Ṣafāʾ) and 16 stadia (as discussed by al-Kindī) above the ground – the direction water takes turns downward, falling as various forms of precipitation.[23]

---

compendium. However, Ḥunayn will later, in the discussion of hail, refer to *antiperistasis* quite explicitly, using different terminology (*tanāfur*); see *Ǧawāmiʿ Ḥunayn ibn Isḥāq* 3, 35.74.

19 See, for instance, al-Kindī, *Risāla fī l-ʿilla al-fāʿila li-l-madd wa-l-ǧazr*, in ʿAbd al-Hādī Abū Rīda, *Rasāʾil al-Kindī l-falsafiyya* (Cairo: 1950–1953) 96.7–97.3, 98.20–21 (trans. Adamson – Pormann) 212–213, 215; *Risāla fī l-ʿilla allatī lahā takūnu baʿḍ al-mawāḍiʿ lā takādu tumṭar*, 72.10–73.5; see Lettinck P., *Aristotle's Meteorology and its Reception in the Arab World: With an Edition and Translation of Ibn Suwār's Treatise on Meteorological Phenomena and Ibn Bājja's Commentary on the Meteorology*, Aristoteles Semitico-Latinus 10 (Leiden – Boston – Cologne: 1999) 107–111.

20 *Rasāʾil Iḫwān al-Ṣafāʾ* XVIII.8, 215.6–216.5; Job of Edessa's *Book of Treasures: Encyclopaedia of Philosophical and Natural Sciences as Taught in Bagdad* A.D. *817*, ed. and trans. A. Mingana (Athens: 2010) V.1 198–199; Ibn al-ʿAmīd, *Letter I* §2, 22.38–24.45. In the present context, the Iḫwān al-Ṣafāʾ do in fact mention the 'cold of the *zamharīr*', but do so for mechanical reasons, as the cold 'hinders' (*māniʿ*) – but does not cool – the rising vapours.

21 al-Ṭabarī, *Firdaws al-ḥikma* I.10, 24.8–12; see also n. 16.

22 Apparently, al-Ṭabarī shares this feature with al-Kindī. Yet, al-Kindī still occasionally refers to the condensation and thickening of the vapours (using words derived from the roots *ġ-l-ẓ* and *k-ṯ-f*), whereas al-Ṭabarī, as we shall see shortly, does not.

23 *Rasāʾil Iḫwān al-Ṣafāʾ* XVIII.5, 200.9; al-Kindī, *Risāla fī l-ʿilla allatī lahā yabrudu aʿlā al-ǧaww*, 90.12, trans. Adamson – Pormann 209. The figure in al-Kindī's text – 16 stadia which, following his own conversion, amounts to 22,400 cubits – is ultimately denied

1.2 *Cloud Formation*

We may well expect the cloud to be the link between rising and falling water and, hence, a central factor in all forms of precipitation, i.e., of rain, snow, and hail as well as dew and frost.[24] As Ps.-Olympiodorus notes: 'The transformation of water into the nature of air is by the intermediary of vapour, while the transformation of air into the nature of water is by the intermediary of the cloud'.[25] Historically, however, the situation is a bit more complicated, and the complication already begins with the question of what a cloud actually is. For some authors under consideration here, a cloud does not seem to be a 'thing' or an 'object' at all but should be understood more along the lines of vapour that is exceptionally dense. A low density, then, would be the normal state of vapour, whereas 'mist' (*ḍabāb*) lies somewhere in-between. Ps.-Olympiodorus, again, puts it nicely when he writes that 'mist is in the genus of cloud' (or: 'a kind of cloud'), subsequently echoing Aristotle's remark that a cloud is a sign of rain, whereas mist is a 'barren cloud' (and actually a sign of good weather).[26]

Another interesting remark is offered by al-Kindī, who in his *Treatise on the Cause of the Generation of Mist* (*Risāla fī ʿillat kawn al-ḍabāb*) writes that 'mist is nothing different from a cloud (*ġamām*) that sunk to the Earth's surface', however one that was, then, 'dispersed by the warmth of the air at ground level'. A little later, al-Kindī describes what happens when one, standing on a mountaintop, climbs down and through a cloud. This actually feels like the mist down on the ground, he writes, yet thicker and replete with moisture, so that one's breathing may be severely impaired.[27]

There seems to be, then, a continuous spectrum of vapour from thin to thick – with exceptionally thick vapour also being called 'cloud'. If we think of vapour, mist, and clouds in this way, i.e., in terms of degrees in density, this may also explain why we find so many different terms that describe this process of condensation in our texts. It may also explain why *explicit* theories of

by al-Kindī, asserting that the exact altitude cannot be determined with precision, as it depends on various circumstances (96.13–14, 98.8–11, 100.4–8); see also Sersen W.J., *Arab Meteorology from Pre-Islamic Times to the Thirteenth Century* A.D. (Ph.D dissertation, University of London: 1976) 173–180.

24 Also for Aristotle, clouds are of no less than crucial importance, as the discussion in *Meteorology* I.3 shows; see also n. 17.

25 Ps.-Olympiodorus, *Tafsīr Alumfīdūrūs* 100.5–6.

26 Aristotle, *Meteorology* I.9, 346b32–35, trans. Webster; *Tafsīr Alumfīdūrūs*, 100.6–8; see Olympiodorus, *In Aristotelis* Meteora *commentaria* 84.25–26.

27 al-Kindī, *Risāla fī ʿillat kawn al-ḍabāb* 76.13–14, 77.11–13. On mist, see also Ibn al-Biṭrīq, *Kitāb Arisṭūṭālīs fī l-Āṯār al-ʿulwiyya* I.7, 37.275–276; al-Ṭabarī, *Firdaws al-ḥikma* I.10, 23.21–22, 24.14–15; Job of Edessa, *Book of Treasures* V.3, 204.

clouds (*saḥāb*, sometimes *ġaym* and *ġamām*) are not necessarily included in scientific accounts of vapour and its atmospheric products. The fact that most of the terminology utilised when it comes to cloud formation are verbs or verbal nouns emphasises, arguably, the processual character of cloud *formation* instead of the final product, i.e., the cloud itself. The most prominent terms for cloud formation are words from the root *k-ṯ-f* such as *kaṯufa* and *takāṯafa* ('to become dense' or 'to condense') and *ġaluẓa* ('to become thick'), which can both be found in various texts.[28] Ibn al-Biṭrīq and al-Kindī emphasise that a confinement (*inḥiṣār*) of the vapours catalyses the condensation process.[29] Another key term is *inqabaḍa*, found in Ḥunayn and Ibn al-ʿAmīd, a word that appears to describe not so much the structure of the vapour as dense and thick but rather a process of contraction and shrinking (the result of which, of course, is a dense structure).[30] Compared to *takāṯafa* and *ġaluẓa*, the (passive) verb *inqabaḍa* may describe a condensation due to an inner contraction, in addition to an outer force, and thus could be motivated, once more, by combining an Aristotelian thermic explanation with the Theophrastean mechanical account: while Theophrastus explicitly described how heavy wind pushes the vapours against mountain walls and how in bathhouses the vapours keep rising, thus creating outside pressure, Aristotle's thermic explanation lets us think of the stiff fingers that we experience during cold winters – and so, consequently, once a dense accumulation of vapours additionally also cools down, it *contracts* from the inside (and not only through outside pressure).[31]

Perhaps most interesting is the account in Ibn al-ʿAmīd's first letter to the Būyid *amīr* ʿAḍud al-Dawla (d. 372/983). He describes, first, how the double exhalations of smoke and vapour rise together. Reaching the colder regions of the atmosphere, the hot and dry smoke is less affected and continues to rise, while the hot and wet vapour slows and cools down, thus becoming separate from the (still) hot smoke.[32] So, vapour, we may well assume, rather suddenly loses most of its heat, which initiates a multi-stage process: 'it mats and becomes dense (*talabbada wa-kaṯufa*), then it contracts and is squeezed

28 See for example Ibn al-Biṭrīq, *Kitāb Arisṭūṭālīs fī l-Āṯār al-ʿulwiyya* I.7, 35.272: *ġaluẓa wa-takāṯafa; Ǧawāmiʿ Ḥunayn ibn Isḥāq* 2, 33.55: *takāṯafa*; al-Kindī, *Risāla fī l-ʿilla allatī lahā takūnu baʿḍ al-mawāḍiʿ lā takādu tumṭar* 72.9: *ġallaẓa wa-kaṯṯafa*; Ibn al-ʿAmīd, *Letter I* §2, 24.44: *kaṯufa*; *Rasāʾil Iḫwān al-Ṣafāʾ* XVIII.8, 215.10: *ġaluẓa* and – after a possible emendation – *kaṯufa* (read instead of *kaṯura*).

29 Ibn al-Biṭrīq, *Kitāb Arisṭūṭālīs fī l-Āṯār al-ʿulwiyya* I.7, 35.272; al-Kindī, *Risāla fī l-ʿilla allatī lahā takūnu baʿḍ al-mawāḍiʿ lā takādu tumṭar* 72.12.

30 *Ǧawāmiʿ Ḥunayn ibn Isḥāq* 2, 33.55; Ibn al-ʿAmīd, *Letter I* §2, 24.44.

31 This is also clearly the thrust of Job of Edessa's explanation in *Book of Treasures* V.1, 198.

32 Ibn al-ʿAmīd, *Letter I* §2, 22.38–24.44; §6, 28.97–30.104.

(*inqabaḍa wa-nʿaṣara*), then it liquefies and falls' (*sāla wa-qaṭara*) as rain. Note here the double separation of the steps: the particle *ṯumma* ('then') creates a three-step division, in which each step consists of two terms that should not be taken as simply synonymous, as the clearly different *sāla* and *qaṭara* indicate. Thus, it should perhaps be read as a consecutive six-stage process, so that 'to mat' is not merely identical with 'to thicken' and 'to be contracted' not with 'to be squeezed' but describe different phases in the process of cloud formation.[33]

Also noteworthy, although for the opposite reason, is al-Ṭabarī, as he does not mention any of these terms at all but simply writes that the vapours 'gather' (*iǧtamaʿa*) in the air. There may be two reasons for him doing so. The first is that he may think of a Theophrastean explanation of cloud formation, as perhaps his immediate reference to winds which 'violently blow at them' indicates.[34] The other is that he may not regard clouds as proper objects of scientific inquiry as he – having noted that the vapours 'gathered' – quickly turns to the explanation of rain and snow without making any immediate use of clouds. In fact, even these violent winds are employed by al-Ṭabarī not so much for cloud formation but as the driving forces behind the transformation from vapour to rain and – under the influence of cold – to snow.[35] When al-Ṭabarī does mention clouds elsewhere, he compares them to mist and explains that mist and clouds are vapour that condensed (*takāṯafa*) or, a little later, that mist is actually a cloud which became dense but is now in the process of being dispersed.[36] This lack of explicit references to clouds and the direct move from rising vapours to falling water is shared also by Ḥunayn. Explaining the thermic and mechanic processes of condensation, he writes that the vapour

33 This also appears to be Daiber's reading in his German translation. Incidentally, the verbal noun *talabbud* is an attested translation (by Isḥāq ibn Ḥunayn) of *pilêsis* at Aristotle, *Physics* IV.9, 217a15.

34 al-Ṭabarī, *Firdaws al-ḥikma* I.10, 24.9. Incidentally, the verb *iǧtamaʿa* occurs prominently in Theophrastus Arabus, *al-Qawl fī l-Āṯār al-ʿulwiyya* (e.g., in the seventh chapter on 'the causes of clouds'); see al-Kindī, *al-Ibāna ʿan al-ʿilla al-fāʿila al-qarība li-l-kawn wa-l-fasād* 231.20, 232.5; see also Ibn al-Biṭrīq, *Kitāb Arisṭūṭālīs fī l-Āṯār al-ʿulwiyya* I.8, 47.364.

35 al-Ṭabarī, *Firdaws al-ḥikma* I.10, 24.10–11, 24.23–24. Interestingly, the emphasis on the role of violent winds for the production of rain may, again, be of Theophrastean origin, who likewise explained that winds bring about (or intensify) rain; see Theophrastus Arabus, *al-Qawl fī l-Āṯār al-ʿulwiyya* 8, 237.31–33; see also n.16. However, it should be noted that in the Arabic version of Theophrastus, the winds are explicitly said to squeeze 'the cloud' (*al-ġamāma*), and that this is the cause of heavy rain, thus marking a difference to al-Ṭabarī in that respect.

36 al-Ṭabarī, *Firdaws al-ḥikma* I.10, 23.21–22, 24.14–19; see also the reference to the 'vapours confined in thick air' in the explanation of thunder and lightning (*Firdaws al-ḥikma* I.10, 24.19–23).

immediately 'becomes water (*ṣāra mā'an*) and falls down' (*inḥadara*). What follows upon this is a very systematic account of dew, rain, frost, and snow – and at the core of this account is not the cloud but the water which now is 'falling down' (*munḥadir*).[37]

While this may be the view of the majority, there were nevertheless thinkers who did indeed regard clouds not only as proper meteorological objects but also as complex enough to warrant extensive theoretical elaboration. Job of Edessa, as a prime example, offers a more complicated and robust account of clouds, featuring a theory on their formation and their internal structure. Emphasising that there are *two* kinds of vapour – one thick and the other thin – Job writes that clouds are formed from a cold-induced condensation of the thick vapour. Subsequently, the condensed vapour encloses and 'imprisons' the thin vapour – and it is this imprisonment of the thin vapour, sometimes intensified through further compression due to strong cold, which turns the enclosed thin vapour into rain. In consequence, it is (i) this complexity of the cloud with a thicker outer layer and a thinner core; (ii) the dynamics *inside* the cloud; and finally (iii) the explicit reference to a sponge that creates the impression that for Job, the cloud is more than merely one extreme on a continuous spectrum of vapour density.[38] For him, apparently, the cloud is more of a proper object. This impression is further corroborated by the fact that in the subsequent chapter, Job discusses the different colours of clouds – from white to black, and even red – thus examining specific properties and the particular behaviour of the cloud as an object in its own right.[39]

2 Falling Water

Regardless of whether a given meteorological theory includes a separate treatment of clouds as proper, even complex objects, the water cycle follows its natural course, and at this stage, water is falling. Theories on precipitation – commonly discussing rain, snow, dew, and frost as well as hail – faced the task of formulating a unified and coherent theory in the face of the sheer variety of its different forms.

---

37 *Ǧawāmiʿ Ḥunayn ibn Isḥāq* 2, 33.55–35.63; see Aristotle's explanation in *Meteorology* I.11, 347b16–24.

38 Job of Edessa, *Book of Treasures* V.1, 198–199. Whether Job's relating a cloud to a sponge was inspired by Aristotle's description of mountains as sponges is an open question.

39 Job of Edessa, *Book of Treasures* V.2.

One first approach to tackling this task and, indeed, a formulation that resounds through the tradition starting from Aristotle regarding rain, snow, dew, and frost is that they are often related to one another in the form of an analogy: rain is to dew as snow is to frost.[40] By framing them in this way, (i) rain and snow can be grouped together and set against their terrestrial counterparts of dew and frost on one axis, while (ii) rain and dew could be grouped together and set against their frozen counterparts of snow and frost on another axis. While the former axis emphasises the quantity of vapour and the place of origin, the second stresses the role of coldness in the process of formation.

Explaining the characteristic features which differentiate rain and snow against dew and frost, Ḥunayn first refers to the *quantity* and concentration of rising or accumulating vapour. It is not surprising that both rain and snow are said to require a greater amount of vapour than dew and frost.[41] The idea seems to be simply that the amount of rising vapour must have been greater in the cases of rain and snow to account for the obvious fact that a considerably greater amount of water falls as rain or snow from the clouds above than that which settles as dew and freezes as frost close to the surface. This difference, being merely one of quantity, was expressed in even more direct language by Ibn al-Biṭrīq, who formulates the relation reciprocally, saying that 'dew is slight rain and rain is much dew'.[42] For Ibn al-Biṭrīq, then, rain and dew – and likewise snow and frost – can be defined simply in terms of their quantity compared to their counterpart.

However, the notion of quantity in this regard could be ambiguous. On the one hand, there is an acute, momentary, and perhaps regional large amount of gathering vapour that may give rise to a likewise locally confined resulting phenomenon, much like we would expect dew and frost to occur more frequently, consistently, and intensely in the vicinity of a lake, for example. On the other hand, we may consider the entire atmosphere and the quantity of vapour 'stored' therein over longer periods of time and, in fact, during entire seasons, so that we experience throughout the year periods of more rain and snow (but not so much periods of more dew and frost). Ps.-Olympiodorus provides an interesting remark in this direction which also introduces a different way to differentiate between the various forms of precipitation in terms of everyday locally occurring quantities and in terms of seasonal quantities throughout the year. To that end, Ps.-Olympiodorus states that 'while rain and snow occur from vapour that has risen from year to year (*min sana ilā sana*), dew and

40 Aristotle, *Meteorology* I.11, 347b16–17 (*nota bene* the term *analogon* in 347b14).

41 *Ǧawāmiʿ Ḥunayn ibn Isḥāq* 2, 35.64–65.

42 Ibn al-Biṭrīq, *Kitāb Arisṭūṭālīs fī l-Āṯār al-ʿulwiyya* I.7, 39.306.

frost are from vapour that gathers together each day' (*fī kull yawm*).[43] Going on to align the accumulation of vapour in the atmosphere and the resultant rain and snow with the annual motion of the Sun, and the nightly occurrence of dew and frost with its daily motion, Ps.-Olympiodorus seems to indicate a differentiation between a seasonal concentration of humidity in the atmosphere in addition to daily rising and precipitating vapour with no long term effect. By day, the Sun approaches and gives rise to vapour, and in its absence at night, this daily vapour precipitates as dew or, depending on the temperature, as frost. Similarly, leading up to summer, the Sun approaches zenith and causes a constant rise of considerable amounts of vapour that reaches higher strata and enriches the atmosphere over the course of the season. With the departure of the Sun away from the zenith as winter sets in, this accumulated vapour descends again as rain or snow.[44] Thus, both Ḥunayn and Ps.-Olympiodorus stress the sheer quantity of rising vapour as a factor in categorising the various forms of precipitation.

In addition, Ḥunayn mentions a second feature differentiating rain and snow from dew and frost: the place of origin. As he explains, dew and frost are vapour turning into water 'in the air and not in clouds' (*fī l-hawāʾ lā fī l-saḥāb*), whereas rain and snow originate 'from clouds' (*min saḥāb*).[45] It seems that his intended meaning by 'in the air' indicates the air close to the surface of the Earth rather than far above the surface where clouds form. Again, Ps.-Olympiodorus apparently agrees, as he states:

> Dew and frost are analogous to rain and snow. That is because the [same] thing that is rain in the higher region is found as dew in the region close to Earth, while that which is snow in the higher region is found as frost in the region close [to Earth].[46]

Among the interesting points in this passage is Ps.-Olympiodorus's insistence that rain and dew as well as snow and frost are actually 'the same thing' and mainly differentiated in terms of their region of formation. What that 'thing' is, however, remains open: is it vapour? And if it is vapour, what differentiates rain and dew from snow and frost? Apparently, we are required to incorporate the second of the above-mentioned axes to fully understand the picture – and

---

43 *Tafsīr Alumfīdūrūs* 101.10–2; see Olympiodorus, *In Aristotelis* Meteora *commentaria* 79.16–20; see also Aristotle, *Meteorology* I.9, 346b35–347a8.

44 *Tafsīr Alumfīdūrūs* 101.1–8; see Olympiodorus, *In Aristotelis* Meteora *commentaria* 79.20–29.

45 *Ǧawāmiʿ Ḥunayn ibn Isḥāq* 2, 35.65–66.

46 *Tafsīr Alumfīdūrūs* 100.20–101.1.

that means we must focus a bit more on the role our thinkers attribute to coldness in the formation of precipitation.

2.1 *The Role of Coldness*

For Ḥunayn, as well as for others, the second axis required to distinguish all four forms of precipitation from each other, aside from their quantity and region of formation, is coldness (*bard, burūda*). Rain, snow, dew, and frost were all seen as requiring coldness to occur, only that snow and frost required a more intense cold, rain and dew a less intense cold. Ps.-Olympiodorus, for instance, makes the observation that, 'the cause for the vapour's freezing and for its coming-to-be snow is an intense degree of coldness, whereas its transformation into the nature of water and of its coming-to-be rain is a lesser degree of coldness'.[47] He goes on to explain that the cold is likewise necessary for the formation of dew and frost, only that the cold required for frost is greater; dew, therefore, requires air that is cold, but not so cold that frost forms. The following matrix can serve to summarise what we have gathered thus far:

	requires less coldness	**requires more coldness**
forms in the air and less in quantity	dew	frost
forms in clouds and greater in quantity	rain	snow

In his letter to ʿAḍud al-Dawla, Ibn al-ʿAmīd explains the formation of clouds as but one step in a complex condensation process from rising to falling water, whereupon he introduces the cold which may exert a strong influence over the cloud, ultimately resulting in both snow and hail. Apparently hail, so far rarely mentioned in our survey, is an additional, fifth form of precipitation that needs to be taken into consideration. Indeed, since snow and hail are apparently both the frozen outcome of 'wet materials' (*al-mawādd al-raṭba*) up high – thus sharing the causal factors of both axes mentioned above: quantity

47 *Tafsīr Alumfīdūrūs* 100.14–15. Moreover, in the passage just discussed above, he emphasises the role of seasonal as well as daily changes of the Sun's course, so that the vapours in the atmosphere may cool down, to various degrees, both during winter and at night (*Tafsīr Alumfīdūrūs* 101.3–6).

and locus of formation as well as reliance on intense coldness – we may well ask what, now, accounts for *their* difference.

For Ibn al-ʿAmīd, the answer lies in the complex mixture of watery and airy parts in the accrued vapour, i.e., in the cloud.[48] Consequently, if the accumulation of vapour is such that the watery and airy parts are not yet sufficiently separated from one another, such that 'what is freezing is imbued with vapour', the result is snow. If, however, the wetness (*ruṭūba*) already 'escaped' (*taḫallaṣa*) the airy substance and is, we may assume, already on its way downward – as rain – toward the Earth's surface, the result is hail rather than snow.[49] Thus, it is not only the place of origin and the amount of gathered vapour or the influence of severe coldness that differentiates between snow and hail; it is also the specific state of the mixture and the ratio of watery and airy parts in the moment of the cold's influence that has an effect on the outcome. The 'fluffy' nature of snow requires an admixture of air, whereas the solidity of hail presupposes pure water drops – apparently already on their way to Earth as rain – to freeze. On this point, Ibn al-ʿAmīd's account is highly reminiscent of Theophrastus's explanation, in particular of the latter's insistence that 'we can see with our own eyes that in snow much air is confined', for it is precisely the admixed air which explains the snow's whitish complexion.[50]

Job of Edessa agrees with the view that snow requires some admixture of air but adds the interesting point that – for this very reason – snow also requires wind. After all, it is the cold and dry wind, he explains, which penetrates the moisture – presumably of the cloud – and 'expands' it, by which he apparently means that air is pushed deep into the moisture, before it freezes under the influence of coldness.[51] Moreover, since the specific concentrations of accumulated vapour – which we have seen emphasised by Ḥunayn and Ibn

48 Here, Ibn al-ʿAmīd explicitly uses the term *ġamām* ('clouds'), and earlier he already used *saḥāb* (*Letter I* §2, 24.51 and 24.45).

49 Ibn al-ʿAmīd, *Letter I* §2, 24.51–56. Interestingly, al-Kindī similarly describes not the difference between snow and hail but the difference between two kinds of snow: *ṯalǧ* and *zamharīr*.; see al-Kindī, *Risāla fī ʿillat al-ṯalǧ wa-l-barad wa-l-ṣawāʿiq wa-l-raʿd wa-l-zamharīr* 84.12–85.3); see also Lettinck, *Aristotle's Meteorology* 110, referring to Sersen's paraphrase in *Arab Meteorology* 226; on another usage of *zamharīr*, see n. 20.

50 Theophrastus Arabus, *al-Qawl fī l-Āṯār al-ʿulwiyya* 9, 238.2–11. Similarly, Aristotle described snow as a frozen cloud in *Meteorology* I.11, 347b23.

51 Job of Edessa, *Book of Treasures* V.3, 203, trans. Mingana. Job's repeated emphasis on dryness is surprising. One interpretation would be to assume that Job accepts Aristotle's account of wind, and thus accepts that wind is caused by the dry exhalation, so that his explanation of snow requires dryness by virtue of its reliance on wind. However, Job discusses and explicitly rejects Aristotle's account of wind elsewhere in *Book of Treasures* IV.14. Additionally, Job states that hail also requires dryness but also the *absence* of wind.

al-ʿAmīd or, additionally, also certain dry and cold winds as in Job – are only realised at certain locations it emerges that not only the stratigraphic but also the geographic (or rather: topographic) place of origin may have an influence on the phenomenon in question. In light of the large variety of regional conditions and, additionally, the complex dynamics of the atmosphere, two things can be explained: the great number of different forms of precipitation *and* the typical meteorological conditions of certain areas.[52] Apparently, the relevance of the phenomena's 'place of origin' is not exhausted by a mere opposition 'inside or outside the cloud'.

However, there may also be a different line of interpretation of Ibn al-ʿAmīd's account, as one could consider the formation of precipitation not so much as dependent on the *place* of origin but on the exact *moment* of the cold's influence upon the 'wet matters' (*al-mawādd al-raṭba*). As previously mentioned, for Ibn al-ʿAmīd, hail results from a cloud when the wetness already 'escaped' the airy substance, whereas it is snow when the wetness is still mixed with the air. Thus, we said, hail is produced outside and snow inside the cloud. Alternatively, though, we may consider the formation of snow and hail as parts of a rather linear process of transformations with *temporally* subsequent stages, for the wetness which has *already* fallen out of the cloud and which, then, through the cold, is turned into hail would have been turned into snow had the cold been able to act upon it *earlier* while it was still in the cloud. Since there is a congruence between the spatial location of the wetness (inside vs. outside), its composition of watery and airy parts, and its temporal stage (*earlier* while inside vs. *later* while already falling outside) within the entire course of vapour transformation – from rising to slowing and stopping, then to accumulating and condensing, then to falling and, perhaps, to freezing – what we find here described in Ibn al-ʿAmīd may, in fact, be more like a linear (or: circular) process of subsequent stages.

This is especially obvious in the case of al-Ṭabarī, who explicitly describes such a process:

> When the vapours that gather in the air are wet and the winds violently blow at them, they transform and become rain; and when they encounter from the air some heavy cold, the rain becomes snow; and when the cold is together with the very dry, this snow (*hāḏā l-ṯalǧ*) becomes hail.[53]

---

52 For example, al-Kindī explains in his eponymous treatise *Risāla fī l-ʿilla allatī lahā takūnu baʿḍ al-mawāḍiʿ lā takādu tumṭar* why it hardly ever rains in some places and Ibn al-ʿAmīd responds in his fifth letter to the *amīr*'s request to learn about the 'reason of the scarcity of rain in the climes of Egypt *vis-à-vis* the abundance of rain in the lands of India'.

53 al-Ṭabarī, *Firdaws al-ḥikma* I.10, 24.9–12.

Notice the shift of subjects undergoing these processes: first it is 'the vapours' that become rain, then it is explicitly 'the rain' that becomes snow and again explicitly 'this snow' which is turned into hail. Even though immediately afterwards, al-Ṭabarī devotes a few more words on the peculiar case of hail, thereby emphasising the special place of hail's origin inside the cloud, his explanation overall describes a linear process with subsequent stages. Taken as a whole, this process is an elaborate series in which each form of precipitation is required for the next, together forming an entire segment in the circular water cycle.

2.2 *The Curious Case of Hail*

Before we can turn to the conditions of flowing water, a few more words on the curious case of hail are in order. Hail in and of itself is a perplexing phenomenon: why is it that relatively large chunks of what seems to be pure ice are falling from the sky – and why, in heaven's name, does this happen in seasons other than winter? If anything, the various known forms of ice commonly arise in winter, whereas hail is depicted by ancient and medieval authors largely as a phenomenon occurring mostly in spring and autumn, and even in the summer.

However, there are also exceptions. Above we have already seen that Ibn al-ʿAmīd explains hail as frozen rain: as the cold exerts its influence upon the atmosphere and as the accrued vapours in the cloud form a complex mixture of watery and airy parts in various densities, it is either the vapour inside the cloud which freezes (in the case of snow) or the already condensed drops of water that have fallen out of the vapourous cloud (in the case of hail). What is interesting here in Ibn al-ʿAmīd is that it is the intense cold that does all the work – and the only season that brings such an intense cold is winter. Thus, although Ibn al-ʿAmīd is fully aware that others described hail as a phenomenon that occurs precisely *not* (or rarely) in winter (and although he knows and describes the phenomenon of *antiperistasis*, which most others use in one way or another in their accounts of hail as will be explained shortly), he later explicitly emphasises that it is the 'rains of winter' (*amṭār al-šitāʾ*) that hit upon the cold while falling and which, then, freeze either up in the air (leading to small and round hailstones) or closer to the Earth (resulting in bigger and coarser hailstones).[54] So, for Ibn al-ʿAmīd, hail is quite surprisingly a winter phenomenon. We find what appears to be the same view of hail as frozen rain in Ps. Olympiodorus and in the epistles of the Iḫwān al-Ṣafāʾ.[55] Such explanations could well be the result of Aristotle's own wavering between arguing that

---

54 Ibn al-ʿAmīd, *Letter I* §3, 24.58–26.61 (whether or not hail is a winter phenomenon); §5, 28.86–95 (knowledge of *antiperistasis*); §7, 30.106–111 (final account with reference to the 'rains of winter').

55 *Tafsīr Alumfīdūrūs*, 101.10–17, 102.4–6; *Rasāʾil Iḫwān al-Ṣafāʾ* XVIII.8, 217.5–218.3.

water turns into hail inside the cloud or does so only later while already falling as rain, but the emphasis on the season of winter lacks a correspondence in the Greek Aristotle, even though winter comes a bit more to the fore in Ibn al-Biṭrīq's Arabic version of Aristotle's *Meteorology*.[56]

It seems that we could add to this group also Job of Edessa and al-Ṭabarī, as they also stress the importance of coldness for the production of hail and, thus, point toward the colder rather than the warmer seasons. While Job explicitly asserts that hail is a winter phenomenon, occurring 'either at the end or at the beginning of winter', al-Ṭabarī mentions spring and autumn. Since he is also mentioning winter (but not summer), we may perhaps assume that al-Ṭabarī thinks of *early* spring and *late* autumn, which would, then, coincide with Job's statement. Moreover, Job and al-Ṭabarī both explain hail through the – perplexing – influence of dryness. In doing so, Job relies on our experience down here, on Earth, where we witness the formation of ice when there is a 'severe and dry cold'.[57] Since, according to Job, also snow requires dryness it appears that it may be a basic feature of his natural causality that frozen phenomena come about (at least more readily) through dryness alongside the cold, for both together are responsible for the contraction of the moisture in the upper strata.[58]

Additionally, since dryness and wetness are contraries, it seems that Job and al-Ṭabarī implicitly refer to the antagonism of contraries usually subsumed under the notion of *antiperistasis*. Further support for this reading lies in al-Ṭabarī's additional reference to the heat which drives the cold wetness

---

56 However, see below, 49. Having mentioned (apparently critically) a simple explanation of hail as a mere freezing of cloudy vapour after rising up to colder strata, Aristotle explains hail as the result of a cloud descending into warmer regions: as the outside warmth reacts with the colder cloud, the cold core of the cloud condenses to water and freezes to hail before it could fall as rain; see *Meteorology* I.12, 348a14–20, 348b12–18. Subsequently, he adds that hail could also freeze a little later somewhere in the air while already falling; see *Meteorology* I.12, 348b18–22; Ibn al-Biṭrīq, *Kitāb Arisṭūṭālīs fī l-Āṯār al-ʿulwiyya* I.7, 41.315–45.345.

57 Job of Edessa, *Book of Treasures* V.3, 203–204, trans. Mingana; al-Ṭabarī, *Firdaws al-ḥikma* I.10, 24.12–14. Nonetheless, al-Ṭabarī explicitly denies that hail occurs in (the middle of) winter, for it requires for its formation the 'heat of the air' (*ḥarārat al-hawāʾ*), as will be mentioned shortly.

58 Two other interpretations of the notion of dryness involved are similarly possible: since the Arabic root for 'dryness' can also have the connotation of 'stiffness' or 'rigidness', the meaning here could be a result of 'dry' coldness in the sense that whatever is exposed to coldness becomes stiff and rigid ('dry'), as our fingers in cold winter tend to be stiff. Alternatively, the factor of dryness in Job's and al-Ṭabarī's accounts could also point toward the smoky exhalation instead of (or in addition to?) *antiperistasis*; see Lettinck, *Aristotle's Meteorology* 109; Sersen, *Arab Meteorology* 225.

deeper into the cloud, thus spelling out the antagonism between the contraries of heat and cold, which is after all the paradigmatic case of *antiperistasis*. This Greek term – which is also well known in a somewhat different sense from Aristotle's account of ballistics from the *Physics* – describes peculiar dynamic effects of intensification in a conflict of contraries. For example, cold air is driven away by warm air. If, however, the cold air is surrounded by warm air, where should the cold air flee? The only possible way is to flee to its own centre and, thus, to condense or contract and, thereby, to *intensify* by accumulating a large amount of coldness at an increasingly confined spot. At the heart of *antiperistasis* is always the idea of a large difference between the two contraries. If a mild cold meets a mild warmth, they merely mix and balance each other out. Yet, if extreme cold hits upon extreme heat, a rather strong effect can be witnessed – and this is, according to some, what we perceive in hail. Thus, the solution for the puzzle why something very cold (such as the ice of a hailstone) is formed especially when the air is rather warm (e.g., during summer) is exactly this: hail is the result of a great *difference* in temperature which triggers the phenomenon of *antiperistasis*. It is especially when it is warm that the remaining coldness that is already caught up in a cloud is put under further pressure, leading to rapid condensation and resulting in ice. In that spirit, Aristotle states according to Ibn al-Biṭrīq's version:

> There is hail only at hot times, rather than cold times, on account of the heat counteracting (*yuḍaddu*) the cold, so that the cold contracts (*yanqabiḍu*) toward the inside of the cloud through the large amount (*kaṯra*) of the air's heat at that moment, and so what is in it in terms of water congeals (*yağmudu*).[59]

So it is the 'large amount' of heat during warmer times of the season that drive together the already retreating cold deep into the core of the cloud, so much so that the cold vapours readily freeze to large clods which cannot be held by the fluffy nature of the cloud and, thus, fall toward the Earth. Ḥunayn's compendium agrees with this explanation, but it is interesting to see Ḥunayn's specific emphasis: he stresses that hail comes about during the 'hot season and not in winter' (*fī l-faṣl al-ḥārr lā fī l-šitā'*); he explicitly highlights that this occurs 'because of mutual conflict between heat and coldness' (*li-tanāfur al-ḥarāra wa-l-burūda*); and he underscores that the water vapours are compressed (*inqabaḍa*) 'inside the cloud through the large amount (*kaṯra*) of the vapour's

59 Ibn al-Biṭrīq, *Kitāb Arisṭūṭālīs fī l-Āṯār al-ʿulwiyya* I.7, 43.324–326.

heat, and so congeal (*yaǧmudu*) in the depth of the cloud'. All this, he writes 'is due to the contrariety (*muḍādda*) of heat with cold' – an account which closely corresponds, even in wording, with Ibn al-Biṭrīq's prose.[60]

With *antiperistasis* being the central factor in the accounts of hail, we are more and more encouraged to reread and revisit Job's and al-Ṭabarī's aforementioned explanations. Although they see dryness at work in the production of hail, they might rely on *antiperistasis* – not, however, between heat and cold but between dry and wet. This being the case, al-Ṭabarī additionally argues that it is the 'heat of the air' (*ḥarārat al-hawāʾ*) which drives the wetness deep into the cloud, even though he insists that there it turns into hail by being dried (*taybisu*). Perhaps what he has in mind is a double-*antiperistasis*, i.e., one that in the final analysis relies on all four basic qualities.[61]

Overall, the various forms of precipitation are described as the result of material necessities within the dynamism of the natural world – and especially the strong case of hail could appear as a questionable ingredient in an otherwise providential world. Just some factors to be considered in the dynamic systems conditioning precipitation are the influence of seasonal change effectuated by the motion of the Sun especially, the amassing of humidity in the atmosphere, and the influence of different topographical conditions. In this dynamic natural world, there are rising vapours, cold regions of air which catalyse condensation and contraction, and sooner or later wet matters – frozen or not – which come falling down (*inḥadara*).

3 Flowing Water

One obvious feature of forms of falling water is its irregular downfall. Even though there may be whole seasons that are determined by stronger affinity to rain or to snow, it simply does not always rain in the same places at the same times. There is, then, an unevenness in the frequency of precipitation spatially and temporally (even if nightly dew and frost may seem to break the mentioned irregularity to some extent). Already Aristotle began his *Meteorology* with a remark that meteorological phenomena are natural but 'more irregular'

60 *Ǧawāmiʿ Ḥunayn ibn Isḥāq* 3, 35.71–37.76.

61 al-Ṭabarī, *Firdaws al-ḥikma* 1.10, 24.12–14. Comparing this passage with the explanation of precipitation quoted earlier (above, 46), it is unclear whether or to what extent al-Ṭabarī's account is coherent and can avoid contradiction.

(*ataktoteran*) than celestial occurrences.[62] Ḥunayn, for instance, elaborates on this irregularity of the objects studied in meteorology, observing in his first chapter that some winters are colder than others, and certain days of the winter are colder than others in a way that is unsystematic (*ġayr niẓām*) and disorderly (*lā tartība wāḥida*).[63]

There are, however, some related features of the natural world that are more regular or that even appear to be constant. Among these are two natural facts that occupy the remainder of this contribution: the flow of water in rivers – especially the constant flow of perennial rivers – and the salinity of the sea, both of which were objects of inquiry traditionally addressed in the science of meteorology. Up to this point, we have been discussing the processes to which vapour is subject in the atmosphere, clearly in closer alignment to the domain of objects studied in modern meteorology and so perhaps for us more intuitively 'meteorological'. In the remainder of this survey, we will learn just why theories on rivers and seas traditionally fell to the hands of meteorological theorists. Starting with the constant flow of water in perennial rivers, the observable – and puzzling – fact is that these rivers continue to flow even when it is not raining, a natural given which indicates that certain underlying conditions must be in place that are needed to explain such a consistent flow which, apparently, does not (or not only) rely on irregular rainfall.

One option to account for the apparently constant flow of water in rivers is to assume a constant, welled-up supply of fresh water at the head of these rivers, i.e., a subterranean reservoir which is the supply for this flowing water. This ancient theory is mentioned by Aristotle in his *Meteorology* only to be subjected to refutation.[64] Among the main reasons for rejecting the reservoir theory is that Aristotle favours an inherently dynamic understanding of the natural world in which all phenomena are the result of the interplay of certain causal factors and the transformation of certain natural substances into one another. It is this dynamic understanding of the natural world that also contains the Aristotelian answer to the puzzle about the constant flow of water in rivers and springs. Yet, in order to appreciate Aristotle's ingenuous, *meteorological* explanation and its reception in Arabic treatises, we ought first to become better acquainted with the theory he refutes.

62 Aristotle, *Meteorology* I.1, 338b20. It is important to emphasise that the meteorological events are irregular but *natural*. In other words, 'irregular' precisely does *not* mean 'random' or 'chaotic' but rather 'less even' and, hence, 'more complex'.

63 *Ǧawāmiʿ Ḥunayn ibn Isḥāq* I, 31.28.

64 See Aristotle, *Meteorology* I.13, 349a20–26, 349b3–350a17, 350b22–351a18.

3.1 *Reservoirs Explained*

The reservoir theory has it that rains (especially during winter) feed subterranean cavities in the Earth which in turn are the supply of water that flows in rivers only to rise and fall again. The greater the cavity, the more water it can store up to, then, flow through the year until the next winter rain, in which case the river is perennial. This theory has the advantage that flowing water is directly tied into a cycle precisely of *water*: water flows, rises, falls, is stored in cavities, and breaks to the surface to flow and rise again. Consequently, the overall amount of water in the natural world is constant – so much so that Aristotle emphasises that this theory conceives the circuit of water in the world as if it was *one* big river.[65] On this understanding, the reservoir is no more than an interim stopping point, a mere welling-up of falling water that is made to flow.

Despite his criticism, and ultimate refutation of the reservoir theory, based on the conviction that there is a more complex set of meteorological mechanisms at work at each stage of the water cycle (including the production of fresh water) than merely the circulation of pre-existing water masses, Aristotle does admit with certainty that 'many reservoirs of this kind do exist in many parts of the earth'.[66] This concession could explain why the reservoir theory was apparently met with favour by some in the Arabic tradition, including those writers whose works we would expect to be closely modelled on the Aristotelian text (and, thus, to preserve his critical stance toward the reservoir theory). On the other hand, the reservoir theorists to whom Aristotle referred did not offer any explanation of how these subterranean cavities are supposed to have formed in the first place. Supplying such an explanation could have been yet another incentive for his Arabic followers to engage more directly with the theory. This, in fact, is exactly what happened in the case of Ḥunayn, whose compendium on Aristotle's *Meteorology* clearly demonstrates fluency with Aristotle's theories, more often than not faithfully summarising key tenets of Aristotelian meteorology – and, with deference to Aristotle, Ḥunayn follows the order of topics in Aristotle's *Meteorology* almost exactly. This makes it all the more surprising that Ḥunayn appears to have adopted and even furthered the reservoir theory that Aristotle refuted.

What is conspicuously absent in Ḥunayn's compendium is precisely the evidence adduced by Aristotle both against the reservoir theorists as well as in favour of his own alternative, entirely omitting Aristotle's solution to the

---

65 See Aristotle, *Meteorology* I.13, 349a25–26, 349b5–6; see also Aristotle, *Mete*orology I.9, 347a2–3.

66 Aristotle, *Meteorology* I.13, 349b19–20. For more details on Aristotle's refutation, see below, 60f.

puzzle of flowing water.[67] Rather, what we find is that Ḥunayn presents a theory which is easily reconcilable with the reservoir theory that Aristotle explicitly rejected. Beginning his section 'On Rivers, Springs, and Valleys', based at least thematically on chapters I.13–14 of Aristotle's *Meteorology*, Ḥunayn quite plainly writes: 'the cause of these is rain'.[68] He goes on to specify that 'much rain' (*maṭar kaṯīr*) is required to have fallen upon the Earth for there to be springs and rivers. While it may seem that Ḥunayn simply states that precipitation is a necessary and sufficient cause for flowing water, he details a set of processes that answer to the puzzle of perennially flowing water. Down here on Earth, the fallen waters 'gather' (*iǧtamaʿa*) into a downward course, carving out channels in the earth. Underneath, the gathered waters erode away soft earth until they come to encounter firmer, more impenetrable earth. Provided there is enough rain, these waters continue to erode more and more soft earth, thereby hollowing out ever greater reservoirs that can store ever greater quantities of water. Ultimately, this process of erosion leads to an essential occurrence, one that constitutes for many the link between falling water and flowing water:

> Whenever the waters become abundant, and eat away at the soft earth around them (*akalat mā ḥawlahā min al-arḍīn al-līna*) until these waters pierce through a [certain] spot and go out from there, then that place is what is called a 'spring'.[69]

The spring is characterised here as the result of flowing waters *and* their erosive power. Although water that has fallen back down to Earth, as it begins to gather together in a downward course, does indeed flow, it is important to realise that precipitation and the gathering of water alone is insufficient to produce anything like a perennially flowing river. What is required in addition to the mere phenomenon of precipitation and the flow of water is the formation of underground cavities that store up the water to such an extent that even at the end of the dry season there is still enough water contained in the reservoirs below so that springs and rivers continue to flow until the seasonal rains return to fill them anew. Ḥunayn's account precisely offers us a more detailed exposition of how these cavities originate. So, we see that the erosive processes

67 See Lettinck, *Aristotle's Meteorology* 134; see also Ḥunayn's later chapter on the sea, which comes close to describing the water cycle as one big river circulating; see *Ǧawāmiʿ Ḥunayn ibn Isḥāq* 5, 41.124–129.

68 *Ǧawāmiʿ Ḥunayn ibn Isḥāq* 4, 39.93, 39.101–106.

69 *Ǧawāmiʿ Ḥunayn ibn Isḥāq* 4, 39.98–100.

of freshly fallen flowing water lead to a storing-up of water in abundance, producing what is perhaps a novel account of how these reservoirs are first formed and, we might add, how reservoirs are continually *being* formed.

Yet, with his explanation, Ḥunayn is *twice* removed from Aristotle's discussion: he is in alignment neither with Aristotle's refutation nor even with his depiction of the reservoir theory, since Aristotle's report has the reservoir theorists assume the pre-existence of reservoirs filled with actual water. The advantage of Ḥunayn's *developed* reservoir theory is that it supplies an explanation of how such reservoirs come to exist in the first place, their formation being a direct outcome of the erosive power of flowing water – a key point in Ḥunayn's compendium conspicuously absent in Aristotle. Additionally, Ḥunayn describes the origination of the spring, being precisely that outlet where the abundant supply of fresh water has pierced through the earth, making the flow of water consistent and perennial. Thus, Ḥunayn seems to have self-consciously endeavoured to present what is by all appearances an advancement to reservoir theory.[70]

Ps.-Olympiodorus similarly emphasises the erosive power of water, a natural fact that Aristotle simply had not incorporated into his *Meteorology*. Also, Ps.-Olympiodorus sets out to make the erosive power of water his prime explanation for the welling-up of water in the earth, providing us again with a more detailed and systematic process whereby water erodes away at earth until it finally finds an outlet for its welled-up supply. Stating that he will prove that there is, indeed, a great amount of water within the earth, he argues as follows:

> That is because rivers continue to plunge through the [earth] until its course is hindered because of something that stands in its way, preventing it from finding a path which it would [otherwise] follow to the sea. So, it is forced to return (*yarǧiʿu qasran*) and hollows out another path,

70 Consequently, we are faced with the question regarding his overall motivation for adopting and refining the reservoir theory against Aristotle. Two possible explanations come to mind. First, seeing as Aristotle had admitted that many such reservoirs do indeed exist, Ḥunayn may have sought to devise a scientific account of how such cavities in the earth come to be. Having done so, making the erosive power of water central to his account of reservoir-formation, he secondly may have felt that his new reservoir theory was an adequate explanation for flowing water, developed enough to retain some level of autonomy and no longer in need of Aristotle's alternative – hence his deliberate exclusion of evidence levelled against the reservoir theory and his omission of Aristotle's entire positive theory of flowing water. Perhaps also anti-eternalist inclinations played a role for the Christian Ḥunayn. Yet, we should remind ourselves that it is not entirely clear whether Ḥunayn was the original *author* of the compendium or merely its *translator*. In the introduction to his edition, Daiber argues for the second option.

> suddenly plunging into the depth of the earth and flowing beneath it. As for the emergence [of these waters], however, it is when they become abundant within the earth such that they burst forth. In their moving by force (*tataḥarraku qasran*) downward to make a path, [these waters] rise again and suddenly emerge.[71]

Ps.-Olympiodorus seems to consider a process with three distinct phases. First, (i) the water flows obliquely downward across the surface of the Earth to the sea, and after it has encountered some hindrance, 'returns'. In need of an outlet, (ii) the water takes its erosive activity vertically downward. As it flows downward, it again encounters a hindrance which forces it back up on itself, and then (iii) bursts forth, emerging suddenly. For Ps.-Olympiodorus, these opposite directions in the motion of water especially force the water into a perpendicular direction, laterally and vertically. The plunge down into the earth allows for subterranean channels of water which would likewise carry with them erosive powers – he twice refers to 'force' (*qasr*) – until, when there is no other outlet, they burst forth, flowing on the surface and with a sure supply for consistently flowing rivers. What sets Ḥunayn and Ps.-Olympiodorus apart from other writers more in line with Aristotle's own theory is their shared emphasis on the need for flowing water, even if inconsistently fed by falling water, in the formation of springs. These springs, in turn, are but the outlets of a massive supply of fresh water stored up in those cavities which had been formed through erosion.

Despite the prominence of erosion in both of their treatises, the main difference between Ps.-Olympiodorus and Ḥunayn is that the former shares Aristotle's stance against the reservoir theorists (and echoes Olympiodorus's four critical arguments), whereas the silent and (arguably) deliberate omission of Aristotle's refutation as well as of Aristotle's positive account of flowing water makes Ḥunayn come across as a proponent of the reservoir theory.[72]

Ps.-Olympiodorus's text also contains a further element that the reservoir theorists had not factored into their account: according to Aristotle's observation, precipitant water effectively *saturates* the earth. Aristotle even provides us with a rather intuitive proof for that fact, writing that '[w]hen men construct irrigation works, they collect the water in pipes and trenches, as if the earth in the higher ground were sweating the water out'.[73] This idea that

---

71 *Tafsīr Alumfīdūrūs* 102.14–18.

72 *Tafsīr Alumfīdūrūs* 102.19–103.8; Olympiodorus, *In Aristotelis* Meteora *commentaria* 99.12–100.6, 103.24–30.

73 Aristotle, *Meteorology* 1.13, 349b35–350a2.

the earth 'sweats out' water, what is subsequently referred to as 'oozing', is an observation that is critical for his positive theory of the role of mountains in the production of flowing water, as we shall see.[74] Ps.-Olympiodorus reiterates the proof for the earth's being saturated and reproduces Aristotle's example of irrigation trenches:

> Those who dig [trenches] (*yaḥtafirūn*) seeking water undoubtedly do not find waters gathered [there already]. Rather, they first encounter moist and dewy places predisposed to the production of waters (*mustaʿidda li-tawlīd al-miyāh*). Ultimately, that trench will become full by way of gradual oozing (*taḥallub*) and secretion (*rašḥ*).[75]

The water that saturates the earth does not exist as a readily available supply of fresh drinkable water but rather, as in irrigation, water gradually oozes out of the earth into the trenches and collects into greater masses. This observation of Aristotle's – that waters are secreted from the earth in irrigation – proved to be an incentive also to the great scientist and philosopher al-Kindī, who provides a unique and sophisticated classification of different terrestrial waters brought about by 'secretion' (*rašḥ*) in his *Treatise On the Efficient Cause of Ebbing and Surging* (*Risāla fī l-ʿilla al-fāʿila li-l-madd wa-l-ǧazr*).[76]

Now, those more in line with Aristotle's own theory, i.e., those who do not depend on there being flowing water for springs to form and who rely on an entirely different mechanism to explain flowing water, start off from the observation that 'the greatest rivers flow from the greatest mountains', 'waters flow from great mountains', or 'springs are only ever found in in the vicinity of mountains', as Aristotle, Ibn al-Biṭrīq, and Miskawayh asserted, respectively.[77]

3.2 *The Role of Mountains*

That the answer about the origin of flowing water lies, for many thinkers, in raised land is surely not surprising. After all, for water to flow, it must flow *downward*, which requires elevation. Indeed, even Ḥunayn briefly mentions mountains in the explanation of rivers and spring formation. Yet, we need to distinguish how reservoir theorists discuss the role of mountains from how non-reservoir theorists do. Although both start from the observation that springs and rivers derive from mountains, the former consider mountains

---

74 See Aristotle, *Meteorology* I.13, 350a7–9.

75 *Tafsīr Alumfīdūrūs* 103.3–5.

76 See al-Kindī, *Risāla fī l-ʿilla al-fāʿila li-l-madd wa-l-ǧazr* 111.13–112.9.

77 Aristotle, *Meteorology* I.13, 350a14–15; Ibn al-Biṭrīq, *Kitāb Arisṭūṭālīs fī l-Āṯār al-ʿulwiyya* I.8, 49.384; Miskawayh in al-Tawḥīdī – Miskawayh, *The Philosopher Responds*, vol.2 272.17.

merely as conduits for the *gathering* of waters that flow downward, collecting and steering them in their oblique courses both across and beneath the earth.[78] This downward flowing water in turn, through the erosive power that hollows out soft earth, forms the reservoirs in which fresh water is welled-up. Thus, cavities additionally rely on solid and rocky earth, i.e., earth that defies the erosive power and defines the walls of the cavities. Otherwise, the erosive powers would continue to eat away the earth and a stable cavity capable of holding water would never form.

Non-reservoir theorists working under the influence of Aristotle entertain a more complex picture of mountains, considering them as essential not for the mere redirecting of falling waters, but for the *production* of flowing terrestrial water. Already Aristotle explained why a mere gathering of precipitant water alone is insufficient to explain springs and rivers. Given the massive size of rivers – and, in effect, the sheer abundance of constantly flowing water on Earth – underground reservoirs of water would be far too small to feed all streams in the world. If flowing water simply were to gush forth from reservoirs, then these would have to be 'larger than the Earth or, at any rate, not much smaller'.[79] Additionally, accounts that rely on the new production of water may also help to explain why spring waters are fresh and drinkable, as opposed to stale and spoiled.[80]

Perhaps the most striking feature of the more Aristotelian oriented theories of spring formation is that they consider the mechanisms of water production as the terrestrial, even subterranean, counterpart to atmospheric condensation. Just as up in the air, vapour is made to gather together as a cloud and, further, to condense and transform into water on account of the coldness of the surrounding air, subsequently falling back to Earth as rain, so it happens that moisture confined inside the earth or subterranean formations would transform into water if it was subjugated by a strong influence of coldness. This at any rate is the first of three explanations found in Aristotle's *Meteorology*: 'it is unreasonable for anyone to refuse to admit that air becomes water in the earth for the same reason as it does above it', exclaims Aristotle.[81]

---

78 See *Ǧawāmiʿ Ḥunayn ibn Isḥāq* 41.111–114; Ibn al-ʿAmīd, *Letter IV* §6, 58.376–60.386; see also Miskawayh in al-Tawḥīdī – Miskawayh, *The Philosopher Responds*, vol. 2, 272.16–17.

79 Aristotle, *Meteorology* I.13, 349b17–19, trans. Webster, modified; see also Aristotle's criticism that reservoir theorists deny that 'water [...] is generated' and his insistence that what is required is not only a generation of water but a *continuous* generation (*Meteorology* I.13, 349b5–6, 349b25–27).

80 For other accounts of the sweetness of water, see Job of Edessa, *Book of Treasures* IV.12; Ibn al-ʿAmīd, *Letter IV* §§1–5, 48.286–58.372; see also *Ǧawāmiʿ Ḥunayn ibn Isḥāq* 4, 41.118–119.

81 Aristotle, *Meteorology* I.13, 349b21–23, trans. Webster.

Several points are interesting about this passage. First, this passage in particular, which contains Aristotle's most explicit analogy between atmospheric and terrestrial water production through cold-induced condensation, has *not* been retained in either Ibn al-Biṭrīq's translation or Ḥunayn's compendium. While it may be seen easily why the latter – our probable reservoir theorist Ḥunayn – may have deliberately passed it over in silence, there is no clear reason why the former did so, too, but it surely is characteristic for his style of 'translation'.[82]

The second interesting point is that even though these two important sources for Aristotle's views on meteorology lack that precise analogy, Ibn al-ʿAmīd's fourth letter to the Būyid *amīr* ʿAḍud al-Dawla very clearly and precisely reproduces it:

> If the air is as we have said it, it is not possible that the rule governing it (*ḥukmuhū*) be different outside the earth and inside it. Inside the earth are cavities in which air condenses, and it transforms recurrently into water, for it would be ignoble to accept that [the air] outside the earth be governed by a [certain] rule which entails a cause but, then, let it inside the earth not be governed by that same [rule], even though they both share in this cause.[83]

Unfortunately, Ibn al-ʿAmīd's explanation of the air's natural behaviour in this epistle is not entirely without its problems. Yet, we have already seen above that in his other letters, he emphasises coldness as a key factor in explaining precipitation, and so there is no reason to doubt that what he is aiming at – without explicitly naming it – is air's capacity to transform into water under the influence, precisely, *of coldness*.[84] If that is right, then Ibn al-ʿAmīd's analogy is a very precise reproduction of Aristotle's expressed view, and with his elaboration on these cavities in the earth, substantiates the validity of precisely the parallel Aristotle was drawing between atmospheric and terrestrial condensation.

---

82 See the crippled remnant of Aristotle's passage in Ibn al-Biṭrīq, *Kitāb Arisṭūṭālīs fī l-Āṯār al-ʿulwiyya* I.8, 47.371–372.

83 Ibn al-ʿAmīd, *Letter IV* §5, 58.365–369.

84 See above, 44f. The problems about elemental transformation and the way he claims to 'have said' the air to be in this fourth letter pertain to the interpretation of §3. Daiber is much more straightforward in his commentary than the text appears to allow, even though Daiber seems to be right overall that Ibn al-ʿAmīd makes use of 'die aristotelische Elementarlehre' and explains 'die Verwandlung von Luft zu Wasser unter der Einwirkung der Kälte' (*Naturwissenschaft bei den Arabern im 10. Jahrhundert n. Chr.*, 139).

The atmospheric-terrestrial parallel concerning the conditions for falling and flowing water in Ibn al-ʿAmīd's passage arguably extended beyond what was stated in Aristotle's formulation. Ibn al-ʿAmīd does account for these processes by relying on the presence of the cavities in the earth, already admitted by Aristotle, as we have seen. It even appears within this passage that Ibn al-ʿAmīd regards the conditions in subterranean cavities to be like a 'micro-water cycle', since he argues that the natural processes that occur therein are, like their atmospheric counterpart in the grand water cycle, recurrent. Still, one may wonder – and this is the third interesting point – where *precisely* this subterranean water production is supposed to take place. Aristotle merely hinted that air becomes water 'in the earth' without further specification. Ibn al-ʿAmīd clarifies for his readers that these processes occur in 'cavities [...] inside the earth', even if he does not go further into detail concerning their location. Now, if these cavities were deep down within the earth, producing water, which – as we have seen above – tends to seek depth and, thus, to flow downward, then this would create the problem of having to account for how water could first come to flow upward, out of these subterranean cavities, to then flow downward again, above and outside the earth, across its surface. Thus, and in order to avoid this undesirable situation, it seems that for Ibn al-ʿAmīd, the formation of those waters that feed the rivers takes place in cold cavities up and inside mountains, where fresh water is produced through a cold-induced condensation process that mirrors the formation of rain through clouds.

This is an important result, because it contextualises a passage in Aristotle which this time was retained in Ibn al-Biṭrīq's version. Shortly after the analogy between atmospheric and subterranean water production, Aristotle emphasises the importance precisely of mountains, likening them to 'dense sponges' hanging over the earth. Like a sponge, the mountains 'make the water ooze out and trickle together', and they 'receive a great deal of water falling as rain'.[85] This passage is reproduced by Ibn al-Biṭrīq as follows:

> In lowlands, the waters are less, whereas the higher areas of mountains stand as a dense sponge (*bi-manzilat al-asfanǧ al-kaṯīf*) which absorb waters and gather them [together] little by little.[86]

---

85 Aristotle, *Meteorology* I.13, 350a7–9; see Wilson M., *Structure and Method in Aristotle's Meteorologica: A More Disorderly Nature* (Cambridge: 2013) especially 161–164.

86 Ibn al-Biṭrīq, *Kitāb Arisṭūṭālīs fī l-Āṯār al-ʿulwiyya* I.8, 49.382–383.

It is striking that Ibn al-Biṭrīq does not here mention coldness and merely describes what seems to be a simple process of absorption and gathering.[87] In other words, we may be surprised that at the heart of the Aristotelian rejection of the reservoir theorists, the most important piece is missing here in Ibn al-Biṭrīq's equivalent passage: the cold-induced water production facilitated by mountainous regions. The impression that Aristotle's passage in the Arabic version of the *Meteorology* may fall short of what it wants to achieve is reinforced in Ibn al-Biṭrīq's version which entirely lacks even that brief reference to coldness and merely states in leading up to the sponge analogy:

> I say that the waters confined within the earth are not dispersed as they are dispersed above the earth. Instead, parts of [the confined waters] gather together with others just as a droplets of rain gather together with others, and so there comes from that much water. Thus, the waters confined within the earth are gradually squeezed (*tanʿaṣiru*) and gather (*yağtamiʿu*) together, there coming from this the emergence of the water that flows in rivers.[88]

That this is not just a reservoir theory in disguise becomes clear only once we realise that what is required for the droplets inside the mountain to contract or 'squeeze' and to gather, much like with vapour above the earth that is transformed into falling water, is probably nothing other than coldness. In consequence, Ibn al-Biṭrīq's version is more convincing and coherent if we read it as presupposing the role of coldness as given – perhaps so much so that he abstained from explicitly referring to it.

This interpretation is reinforced by a later passage in which the relation of mountains to coldness and, thus, to cold-induced water production is finally made explicit. There, in the subsequent chapter I.14, Aristotle reengages with the reservoir theorists, calling on their account for how rivers come to run dry. For these reservoir theorists, according to Aristotle, the consistency of flowing water corresponds directly to the size of the reservoirs that feed a river. The smaller the reservoir is, the sooner will the river that it feeds dry up, and the greater the reservoir is in size, the more continuously will its waters flow in rivers. Aristotle, apparently content that he has refuted this theory already and

87 Only later, when Aristotle reengages with the reservoir theorists in *Meteorology* I.14, Ibn al-Biṭrīq mentions the coldness of mountains as a factor in the production of flowing water (*Kitāb Arisṭūṭālīs fī l-Āṯār al-ʿulwiyya* I.8, 57.445).

88 Ibn al-Biṭrīq, *Kitāb Arisṭūṭālīs fī l-Āṯār al-ʿulwiyya* I.8, 47.373–49.376; *tanʿaṣiru* in Schoonheim's edition is to be preferred over *tanquṣu* in Badawī's edition.

favouring a more complex set of causal factors, reformulates his own position once more, stating that it is not the size of the reservoirs that corresponds to the continuity of flowing water, but the size (*megethos*), density (*pyknotês*), *and coldness* (*psychrotês*) of the mountains which receive (*dechontai*), preserve (*stegousin*), *and produce* (*poiousin*) most water.[89] Ibn al-Biṭrīq's version fairly faithfully renders Aristotle's argument, thus emphasising the more complex set of conditions that must be in place to guarantee a river's perennial flow:

> As for us, we state that the cause of rivers is the magnitude (*ʿiẓam*), density (*kaṯāfa*), and coldness (*bard*) of elevated places in which there is water. That is so because places that exist in that way are receptive (*qābila*) to a great amount of waters and they are those preserving (*ḥāfiẓa*) them, so that nothing dissolves [from the waters] and so the waters become abundant there.[90]

While this passage finally emphasises the importance of coldness for mountains to play their part, the only missing detail – and, in fact, missing keyword – is that of *producing* water. We do not see any explanation for why Ibn al-Biṭrīq failed to reproduce it. Additionally, one may ask what we are supposed to make of that notion that mountains are somehow especially 'receptive' to precipitant water – especially when compared to other regions on Earth.

One possible answer to this latter question might be found in the philosophical exchange between the two Baġdādī intellectuals ʿAlī Abū Ḥayyān al-Tawḥīdī (d. 414/1023) and the aforementioned Miskawayh. Being asked by his correspondent about the 'wisdom' (*ḥikma*) pertaining to the existence of mountains, Miskawayh stresses that mountains – and likewise cavities – facilitate the production of water not only because of their coldness but because they seal the gathering vapours off from the winds which would otherwise disperse them before they could transform into water. Whenever vapour rises in mountainous regions, it comes immediately under the influence of coldness and is caused to 'condense' or 'freeze' (*yuǧammidu*) and 'to congeal' (*yuʿqidu*), and 'then they are squeezed (*yaʿṣiru*) and return – being transformed – to water'.[91] The greatest bulk of water production, according to Miskawayh, is accomplished through atmospheric condensation among mountains, reiterating the

89 Aristotle, *Meteorology* 1.14, 352b2–16.

90 Ibn al-Biṭrīq, *Kitāb Arisṭūṭālīs fī l-Āṯār al-ʿulwiyya* 57.444–446; see Aristotle, *Meteorology* 1.14, 352b4–16.

91 Miskawayh in al-Tawḥīdī – Miskawayh, *The Philosopher Responds*, vol. 2 272.6, trans. Vasalou.

Aristotelian conviction that the perennial flow of rivers relies on fresh waters and that these waters are produced – some way or another – through coldness as well as integrating the Theophrastean trait that vapours need to be confined on a grand scale, best achieved through mountain walls.

In this same response to his correspondent, Miskawayh likewise mentions the analogy of mountains as sponges which are soaked in water, so that we gain a theory from Miskawayh in which mountains are indeed like the dense sponges Aristotle compared them to, and are therefore, at least according to the analogy, more receptive to falling water than lands for which the sponge analogy is less suitable. Miskawayh, taking Aristotle's hint, provides a scientific account for how water comes to rise and fall while remaining among mountains such that these rains presumably keep the mountains soaked in water, thereby guaranteeing a constant supply that solves the puzzle of flowing water.

On the whole, we have seen two distinct theories regarding the natural principles of flowing water. We also perceived that both of them have been developed further in the Arabic tradition: the reservoir theory gained further plausibility through the formulation of theories of erosion, which in turn are meant to show how such reservoirs are formed in the first place, while on the side of Aristotle's meteorological theory, al-Kindī further develops forms of secretion in his classification of terrestrial waters and Miskawayh provides more detailed conditions by which mountains are indeed fed with rain more so than other regions (in which the rising vapour is dispersed). We now have a better idea of where these flowing waters come from, but to where do they flow? Let us now follow this stream of flowing water downward to its end.

4 Deep Water

As we make our way down to the ultimate destination of the river along which we trek, we begin to wonder why it is that, seeing as the river water flowing beside us is fresh, the sea is not, then, filled to the brim with fresh water, too. Why, after all, is the sea salty and why must such nourishing freshwater rivers come to such a bitter end? This natural fact of the sea's salinity was a particularly perplexing puzzle for those who endeavoured to explain it for this reason, namely, that both the sources flowing into it and the masses that dissipate from it are a different kind of water entirely, namely, fresh water, which both ancient and medieval philosophers called 'sweet' (gr. *glykus*, ar. *ʿaḏb*). 'Sweet water' denotes water that is drinkable, while salt water is diametrically opposed to this, possessing a quality of bitterness (gr. *almyros*, ar. *murr*, *marāra*) to taste.

Indeed, the properties of seawater make it not comestible, with adverse effects on health and well-being on those who attempt to drink it. We are surely not surprised that Ibn al-ʿAmīd, being asked once more by the Būyid *amīr* ʿAḍud al-Dawla on the nature of water, in particular on the puzzling question of why it happens that we find sweet water even on mountainless islands that are surrounded by the salty sea, immediately speaks of an 'important topic of research' (*ǧalīl mubāḥaṯā*). What is more, Ibn al-ʿAmīd refers to a longer treatise he has earlier composed for the education of the *amīr*'s municipal architects in which he 'has already explained a lot about the principles of this question', so that he can be more brief and concise in his present epistolary response.[92] We may well assume that the treatise he referred to contained an explanation of the conditions of natural water insofar as is necessary to know for the construction of what is no less than essential for the survival of urban societies: aqueducts, conduits, and wells.

However, the saltiness of seawater was not only known for its bitter taste and negative effects on humans and animals; it was also known for positive features within the world. For instance, the salinity of the sea was traditionally thought to also impart heaviness to its waters. The heaviness of salty water was proven by observing that ships of the same cargo sooner sink in freshwater rivers than in the sea.[93] Ps.-Olympiodorus explicitly uses this observation as evidence (*dalīl*) that salty waters are heavier than fresh, sweet waters, stating that 'the ships employed for use in rivers are lightweight due to the sweetness of the rivers' water, whereas those employed for use in the sea [may be] heavy due to the salinity of its water'.[94] So, due to its salinity seawater is heavier than fresh water, and while fresh water itself rises only to fall and flow again, saline water seeks depth, ensuring that the sea remains in its place. In Ibn al-Biṭrīq's Arabic version of the *Meteorology*, we find the following lines, which correspond nearly verbatim to Aristotle:

> I say that sweet waters rise because of their lightness above the salty waters, whereas salty waters remain because of their heaviness below.[95]

---

92 Ibn al-ʿAmīd, *Letter* IV §1, 48.286–50.297.

93 Aristotle, *Meteorology* II.3, 359a7–11. The passage is not retained in Ibn al-Biṭrīq's version of Aristotle's *Meteorology*.

94 *Tafsīr Alumfīdūrūs* 111.22–112.3; see Olympiodorus, *In Aristotelis* Meteora *commentaria* 159.11–23, 163.2–5.

95 Ibn al-Biṭrīq, *Kitāb Arisṭūṭālīs fī l-Āṯār al-ʿulwiyya* 67.517–518; see Aristotle, *Meteorology* II.2, 355a32–34; see *Ǧawāmiʿ Ḥunayn ibn Isḥāq* 5, 41.122–124.

What is clear is that sweet, drinkable water is light in weight, and this lightness corresponds neatly to its becoming airborne to drift with the wind, whereas salt water, due to its salinity, is inherently heavy and this heaviness explains its remaining earth-bound *and* its stability in place: ponds and lakes may evaporate and even dry out, but saline seas do not. Apparently, the seas' regional stability to remain in place is tied with its persistence through time. Ps.-Olympiodorus, for instance, takes the salinity of the sea as a way to succinctly show why it perseveres and how it staves off corruption and passing away out of existence. After stating that the salty sea is 'not susceptible to corruption' (*ġayr qābil li-l-fasād*), Ps.-Olympiodorus adds a further, general note:

> That is because the salty bodies remain for a greater [period of time]. For that reason we find the Egyptians, if they wished that [certain] bodies not putrefy or become corrupted, apply salt to them.[96]

Explaining the saltiness of the waters in the sea, then, becomes all the more pressing a concern, since it could be shown to be useful as an explanation both for the sea's apparent steadfastness in place and time. It could even be surmised that the apparent persistence of the salty sea *vis-à-vis* the more fleeting and airborne condition of sweet water is the reason why, according to Ibn al-ʿAmīd, some ancient scholars have opined the bitter, salty taste of seawater to be 'the original taste' (*al-ṭaʿm al-awwal*) of water, and that, thus, seawater is the original state of water.[97]

Yet, even if salinity may help to explain persistence, what is it that explains the sea's salinity itself? Surely, the just mentioned ancient scholars in Ibn al-ʿAmīd's letter would perhaps simply avoid this question altogether and argue that since salt water is the natural, original state of water, it hardly requires an explanation.[98] However, Ibn al-ʿAmīd brings in a compelling line of reasoning why they erred. Arguing that what is simpler is also more original and invoking experiments which attest to heavy residues remaining when salty seawater evaporates, he states that sweet water – and not salt water – is the pure form of water, and that the bitter or salty taste is the result of a mixture or composition (*tarkīb*) of water with ashes and salty soil. Conversely, he describes evaporation as a dissolution (*taḥlīl*) from the more composite to the simple. Seawater,

96 *Tafsīr Alumfīdūrūs* 114.1–2.

97 Ibn al-ʿAmīd, *Letter* IV §2, 52.309–310; see Aristotle, *Meteorology* II.2, 355a34–355b6; Ibn al-Biṭrīq, *Kitāb Arisṭūṭālīs fī l-Āṯār al-ʿulwiyya* 67.518–69.521.

98 Surely, though, these scholars would, then, owe us an account of how the sweetness of fresh water comes about.

Ibn al-ʿAmīd argues, is neither pure nor original but 'spoiled'.[99] In a way, he counters those ancients who believed seawater to be the original form of water contending precisely that because seawater is heavier, it cannot be original.

While Ibn al-ʿAmīd's account of salinity asserts *that* something has happened to the sea, he does still not provide an explanation of *why* it has happened that the biggest water masses on Earth acquired that bitter, salty taste. Surely, he may be excused, as the Būyid *amīr* did, indeed, ask for something else, and so Ibn al-ʿAmīd quickly moves on to his theory of natural percolation to explain the occurrence of fresh water lakes on islands. Looking to others sources to find answers to the why-question, we shall, in what follows, acquaint ourselves with three distinct modes of explaining the salinity of the sea, one *natural*, which explains the sea's salinity as deriving from natural causes; one *theological*, which explains the sea's salinity as deriving from divine wisdom; and a third *mixed* account, which explains the sea's salinity with natural mechanisms, but viewed in theological terms. Eventually, we shall see that there is another factor involved in the sea's salinity, one already discernible in the above quotation: apparently salt water is not only heavy and stable but also has important preserving, even sanitising powers (despite being not comestible).

4.1 *Natural Explanations*

We begin with the natural mode of explaining the salinity of the sea and restrict ourselves to Aristotle's positive solution to the salinity puzzle as it was known among medieval philosophers in the Islamic world inspired by his *Meteorology*. The main features of the natural explanation (already indicated by Ibn al-ʿAmīd's above reference to 'ashes') are that it is the application of heat that ultimately explains the sea's salinity, especially its bitter taste, and that the thermal causes for salinity are best understood with a natural analogue likening the salty sea to residues of the digestive process, specifically urine and the excretion of sweat, which are likewise naturally salty or bitter. In both cases, it was thought that heat was necessary to account for this bitter salinity of the residual liquid, so it is worthwhile to take a close look at the natural processes thought to underlie the sea's salinity.

In Ibn al-Biṭrīq's Arabic version of the *Meteorology*, after paraphrasing Aristotle's invocation of the digestive analogy, we find the following lines:

> Something burnt likewise becomes bitter upon its being burnt, such as ashes and other similarly burnt things. Some people have said that the

99 Ibn al-ʿAmīd, *Letter IV* §4, 54.340–56.351.

> water of the sea only becomes salty because the earth is burnt and then was salinized, and that the like [happens] with the salt of water. Now, I state that it would be base for us to claim that the earth is burnt. Rather, I state that it [only] resembles being burnt, but not that it is burnt itself. We can also discern that there is some salinity in the water of rains, and that is because, therefore, the hot dry exhalation intermixes with it, for when the hot dry exhalation intermixes with the cold moist [exhalation], from which there are clouds and rain, it makes it salty.[100]

The invocation of the hot and dry exhalation – referred to as 'smoke' (*duḫān*) – to explain the sea's salinity is significant. To reiterate from what we have already discussed, the two exhalations – moist and hot vapour, which rises from a body of water, and dry and hot smoke, which rises from the earth itself – are the principles particular to the science of meteorology. The inclusion of the exhalations in explaining the salinity of the sea, then, has the consequence that the sea's saltiness undoubtedly has been considered as a *meteorological* phenomenon alongside earthquakes, the formation of stones, and volcanism. Therefore, a discussion of the sea's salinity properly belongs to the science of meteorology.

That said, Ps.-Olympiodorus's endorsement of Aristotle's natural explanation, which attributes the salinity of the sea to the admixing of smoke, thereby effectively singling out the hot and dry exhalation explicitly as its first and foremost cause, was a special, rather isolated case.[101] Other thinkers of the medieval Islamic period considered the invocation of the exhalations less significant, even inessential, in explaining the salinity of the sea. From the admixture of hot exhalation, they moved to application of heat itself. While this may seem to be only a minor variation it entailed a major shift in the perspective on the process as such: if the sea's salinity resulted from a change caused by heat itself, the phenomenon as such is chemical, rather than meteorological, in kind. Considered in this way, the salinity of the sea was, then, thought by many to result directly from the application of excessive heat by the celestial bodies, especially the Sun, on the water masses on Earth. Without the need for exhalations, the celestial heat radiating first and foremost from the Sun changes the quality of the waters of the sea imbuing them directly with salinity.

100 Ibn al-Biṭrīq, *Kitāb Arisṭūṭālīs fī l-Āṯār al-ʿulwiyya* II.3, 75.567–573; see Aristotle, *Meteorology* II.3, 358a3–27.

101 For Ps.-Olympiodorus's elaboration on how smoke interacts with the water of the sea, see *Tafsīr Alumfīdūrūs* 114.3–11; see also Aristotle, *Meteorology* II.1, 353b13–16; II.3, 357b21–26.

Both the thermic cause of saltiness and bitterness and the analogues of urine and sweat were known among philosophers of medieval Islam, an exemplar case of which is from Ḥunayn's compendium on Aristotle's *Meteorology*, which apparently deviates once more in an important way from its source:

> As for the salinity and bitterness of seawater, it is due to the frequent passing of the Sun over it. Moisture, when it is intermixed with heat, becomes salty, and when [said intermixture] is excessive, [the water] becomes bitter. An exemplar case of that being sweat and urine which are both salty due to the action of heat on them.[102]

Ḥunayn mentions the examples of sweat and urine, derived from Aristotle's *Meteorology*, accepting them as suitable analogues to the process of salinisation. We also find that the thermal cause is fully operative, the consequential difference being that this heat acts on the water itself. A divergence from the original Aristotelian explanation, the more chemical-oriented explanation in Ḥunayn finds another voice in Miskawayh, who retains this thermic process acting on the water itself, albeit with a different analogy. Explaining to his correspondent why the sea is salty, Miskawayh states:

> That is only on account of the closeness of the Sun to the surface of the water and its ability to cook it. It is of the nature of water that, if the heat of cooking is incessant upon it, then what of it is fine is dissolved into vapour while what remains receives an affection of saltiness. If the heat increased and was continual, that water would become of considerable saltiness and, thereupon, would ultimately end in bitterness. Alchemists (*aṣḥāb al-ṣanʿa*) contrive a water of theirs on fire and do so frequently repeating its being on the fire, and so, through this, the water becomes hot water that inclines toward bitterness.[103]

So, into the eleventh century, we find the natural explanation being advocated in Baġdād, one of the most vibrant cities of learning, in one of its most exclusive circles. A kind of natural burning or scalding process – even one that can be imitated by experimental alchemists – is portrayed as explaining the salinity of the sea and its bitter taste, the common analogy being that the nature of the sea is best understood as comparable to the liquid excreta of warm organisms. While philosophers and scientists had their natural explanations for the

102 *Ǧawāmiʿ Ḥunayn ibn Isḥāq* 5, 43.135–138.

103 Miskawayh in al-Tawḥīdī – Miskawayh, *The Philosopher Responds* vol. 2, 278.8–12.

sea's salinity and their natural analogues to illustrate their theories – one meteorological, taking recourse to the two exhalations, one chemical, relying on direct heat – an alternative interpretation would rather have us see the sea as a consequence of God's wisdom in designing and ordering the sublunary realm. What, one might very well ask, does God have to do with the salinity of the sea? In answering this question, we will also return to the positive attributes of salt, to which we already alluded.

4.2 *Theological and 'Mixed' Explanations*

The earliest extant Arabic treatise of which we are aware that endeavors to explain the salinity of the sea in theological terms is from the famous Sabian polymath Ṯābit ibn Qurra (d. 288/901).[104] In his *Discourse on the Reason for Which the Waters of the Seas Were Made Salty* (*Qawl fī l-sabab allaḏī ǧuʿilat lahū miyāh al-biḥār māliḥa*), Ṯābit argues that there is a divine purpose behind the salinity of the sea, arguing that the saltiness of the sea is no less than essential for the flourishing of life on the face of the Earth. In this treatise, he insists on the following point, pivotal to his argument:

> We find that sweet waters, except those concealed, whenever they are stationary for a long time, become putrid and corrupt and have an altered smell to the point that they corrupt the surrounding air, bringing about, through this cause, many grave illnesses, plague, and widespread death.[105]

Seeing as stationary sweet water has these effects, the image which Ṯābit draws with these few words is disastrous. Just imagine if that immense expanse of stationary water on the Earth was actually sweet water, stale and rotten, affecting the air, with winds blowing a foul and stinky stench to the mainland, flies and mosquitos cultivating that pit of mire we would like to call the sea. On this logic, the flourishing of life would be prevented, perhaps on a global scale, pestilence and illnesses spreading.

One point to bear in mind, and apparently also one that Ṯābit considered, is that to ask the question why seawater is salty, also means to ask, *au contraire*, why seawater is not sweet. Ṯābit ibn Qurra, focusing on the latter, answers the question, as we have seen, by pointing out that if seawater was sweet, then this would prevent life from flourishing – a view in apparent conflict with his

---

104 See Roshdi Rashed's collective volume, *Thābit ibn Qurra. Science and Philosophy in Ninth-Century Baghdad*, Scientia graeco-arabica 4 (Berlin – New York: 2009), comprising both primary and secondary literature.

105 Ṯābit ibn Qurra, *Qawl fī l-sabab allaḏī ǧuʿilat lahū miyāh al-biḥār māliḥa*, fol. 197r, ll. 3–5.

conviction that we are, in fact, living in a providential world. Hence, seawater must possess a nature that allows or is even conducive to the prospering of human life on the face of the Earth, in other words, something that prevents corruption, sicknesses, and widespread plague. The only point left, needed to solidify his argument, is to explicitly state that salinity in water has the positive attribute which prevents all this from happening. Ṯābit states this point as follows:

> As for water which is salty to the taste, it is among that which staves off corruption and putrefaction; this is something known by all people.[106]

On the theological interpretation, the salinity of the sea has a divine purpose, namely, to support the emergence and flourishing of life on Earth, a point that even common sense could discern. We have seen that on the natural explanation, regardless of whether this salinity was considered the result of meteorological or chemical processes, what was required was the application of heat. On the theological explanation, at least in the case of Ṯābit ibn Qurra's treatise, these natural explanations, it seems, are for the most part dispensed with. What Ṯābit emphasises instead is that the sea in all its salinity is the product of divine wisdom so as to allow for the thriving of God's creation of mankind.

The question that remains is whether the natural and the theological explanations are indeed incompatible or whether they could be combined. One possible way to combine them, would be to argue that it was God who brought about those very causes by virtue of which the sea was and is being salinised. God, thus acting through natural causation, indirectly brings about salinity in the sea. Such an interpretation of the sea's salinity that combines both possible explanations can be found embedded in the epistles of the *Iḫwān al-Ṣafāʾ* (the afore-mentioned 'Brethren of Purity'). There, in the nineteenth epistle *On the Explanation of the Minerals*, we find the following statement:

> As for the cause of salinity belonging to the waters of the entirety of the seas, it is by the providence of the Creator, great be His glorification, and by the divine wisdom, in whom there is complete goodness and universal benevolence. This [cause] is that the exhalations, rising from [the seas] into the atmosphere, whenever their parts intermix with the air and undulate in [various] directions, clear [the seas], salinize them, and

106 Ṯābit ibn Qurra, *Qawl fī l-sabab allaḏī ǧuʿilat lahū miyāh al-biḥār māliḥa*, fol. 198r, ll. 2–3.

> prevent them from putrefaction, alteration, and corruption. Were that not [the case], air-breathing animals would perish at once.[107]

In this passage, we can discern both modes of explanation being present together. For the Iḫwān al-Ṣafāʾ, what is required to explain the sea's salinity is an invocation of the exhalations, and by all appearances, they seem to regard the sea's salinity as a residual by-product of rising water. This is much like Aristotle envisaged it, and indeed more in line with him than even Ḥunayn in his compendium. Additionally, though, the salinity of the sea serves a divine purpose, too, much like it did in Ṯābit ibn Qurra's treatise, preserving life on Earth. Yet, in contrast to Ṯābit, the processes which render the sea salty are, for the Iḫwān al-Ṣafāʾ, natural outcomes of God's wisdom and benevolence toward living creatures.

So, there are in the textual record at least three distinct modes of explaining the sea's salinity. The first is to say that the causes behind the sea's salinity are strictly natural – either the exhalations or the radial heat of the Sun. That view is, then, commonly but not exclusively, supplemented with the natural analogue of liquid excreta, especially urine and sweat. An alternative theory, the strictly theological explanation, would have us see the salinity of the sea as resulting directly from God's benevolence in bringing about and sustaining life: God created a saline sea. The third explanation mixes the natural and theological and maintains that God brings about those natural processes which inevitably salinise the sea, doing this so that life may flourish. On both the theological and the 'mixed' theories, the salinity of the sea is to be seen in a broader view which has it that the natural inanimate world in which we live is arranged precisely so as to be conducive to life. In turn, for intelligent life to understand that habitable world is not only to grasp the natural processes which make it possible, but, what is more, to discern the wisdom of God therein.

5 Conclusion

Water is an element of great variety: it rises to great heights as vapour, falls again as precipitation in any number of forms, oozes from the earth, gushes forth from springs, and flows in river channels to the salty sea only to rise again. Just as there is variety in the element of water, so there is variety among

107 *Rasāʾil Iḫwān al-Ṣafāʾ* XIX.VI, 276.2–8.

theorists facing puzzles about this particular element. Both the puzzles and the observable variety of water in the world we inhabit gave rise to no less variegated theories propounded almost invariably in the context of the science called 'meteorology'.

For thinkers whose theories were more firmly based on Aristotle's *Meteorology*, the puzzles for each stage in the water cycle were often solved on the basis of the core principles of the science: the exhalations. The vapour that rises is precisely the material out of which clouds and precipitation are constituted. Falling water is seen as vapour now condensed and cold. Flowing water similarly relies on the exhalations. Sponge-like mountains provide specific conditions for water to slowly gather into minute quantities and into a flowing mass, as if mountains were the terrestrial counterpart to clouds. At the end of this flowing water stands a sea apparently steadfast in its place and over time, the salinity of which is explained for Aristotle and those influenced by his *Meteorology* with recourse to the smoky exhalation. Vapour and smoke, as core principles of the science, provided theorists with basic but powerful tools by which the meteorological system could be made theoretically intelligible and offered pillars on which to firmly establish theories about the natural world around us. The exhalations, then, in a sense unify these theorists who otherwise differed in their solutions to the puzzles surrounding water.

All of the thinkers surveyed in this contribution rely on and incorporate the exhalations for their explanatory power; yet, while this is a scientific commitment that unified these thinkers, the precise application and scope of these principles to explain discrete meteorological occurrences was a matter in which they differed, and this sometimes led to a limiting of the principles, as with Miskawayh's explanation of the salinity of the sea. In other cases, the application of the principles was extended and led to furthering key Aristotelian terms so much so that entirely unchartered (meteorological) territory was being newly explored, as with al-Kindī's elaboration on terrestrial waters that occur by secretion. The ideas and the principles were acquired from Aristotle, but their application and that application's scope were open to fruitful discussion.

When it comes to water, the task was to formulate theories that were simple enough to grasp such a basic fact of the natural world around us as there being water, but complex enough to reckon with the complexity of its forms and occurrences. More often than not, thinkers of the medieval Islamic tradition produced theories that achieve an admirable balance between the complexity and simplicity of water. The historian of philosophy is faced with a task similar to that of meteorological theorists addressing the element of water, insofar as what we find is a diverse and variegated set of theories that at times compete with each other in terms of cogency and argumentative strength. The next great systematic thinker of the period, Avicenna, had a wide array of materials to draw on for devising his own account of meteorology including the observable forms of water in the natural world around us. Which views he furthered and which arguments he adopted remains to be seen, but it is more than apparent that the centuries immediately preceding him contributed greatly to his vision and, thus, constitute a historical epoch in the history of meteorology.

Bibliography

Manuscripts

Istanbul, Topkapı Sarayı Museum, MS Ahmet III 3342.

Primary Sources

Aristotle, *Aristotle's Meteorology in the Arabico-Latin Tradition: A Critical Edition of the Texts*, ed. P.L. Schoonheim, Aristoteles Semitico-Latinus 12 (Leiden – Boston, MA – Cologne: 2000).

Aristotle, *Fī l-samāʾ wa-l-Āṯār al-ʿulwiyya*, ed. Badawī A. al-R. Dirāsāt Islāmiyya 28 (Cairo: 1961).

Aristotle, *Meteorologicorum libri quattuor*, ed. F.H. Fobes (Cambridge, MA.: 1919).

Aristotle, *Meteorology*, trans. E.W. Webster, in Barnes J. (ed.), *The Complete Works of Aristotle: The Revised Oxford Translation*, Bollingen Series 71.2 (Princeton, NJ: 6th ed. 1995), vol. 1 555–625.

Aristotle, *Otot Ha-Shamayim: Samuel ibn Tibbon's Hebrew Version of Aristotle's Meteorology*, ed. and trans. R. Fontaine, Aristoteles Semitico-Latinus 8 (Leiden – New York – Cologne: 1995).

Ḥunayn ibn Isḥāq, *Ein Kompendium der aristotelischen Meteorologie in der Fassung des Ḥunain Ibn Isḥāq (Ǧawāmiʿ Abī Zayd Ḥunayn ibn Isḥāq al-ʿIbādī li-kitāb Arisṭūṭālīs Fī l-āṯār al-ʿulwiyya)*, ed. and trans. H. Daiber, Aristoteles Semitico-Latinus: Prolegomena et Parerga 1 (Amsterdam – Oxford: 1975).

Ibn al-ʿAmīd, Abū l-Faḍl, *Naturwissenschaft bei den Arabern im 10. Jahrhundert n. Chr. Briefe des Abū l-Faḍl ibn al-ʿAmīd (gest. 360/970) an ʿAḍudaddaula*, ed. and trans. H. Daiber, Islamic Philosophy, Theology and Science: Texts and Studies 13 (Leiden – New York – Köln:1993).

Iḫwān al-Ṣafāʾ, *On the Natural Sciences: Epistles 15–21*, ed. and trans. C. Baffioni, Epistles of the Brethren of Purity (Oxford: 2013).

al-Kindī, Abū Yaʿqūb ibn Isḥāq. *al-Ibāna ʿan al-ʿilla al-fāʿila al-qarība li-l-kawn wa-l-fasād*, in M.ʿA. al-H. Abū Rīda (ed.), *Rasāʾil al-Kindī l-falsafiyya* (Cairo: 1950–1953) vol. 1 214–237.

al-Kindī, *Risāla fī l-ʿilla al-fāʿila li-l-madd wa-l-ǧazr*, in M.ʿA. al-H. Abū Rīda (ed. and annotated with an Introduction), *Rasāʾil al-Kindī l-falsafiyya* (Cairo: 1950–1953) vol. 2 110–133.

al-Kindī, *Risāla fī l-ʿilla allatī lahā takūnu baʿḍ al-mawāḍiʿ lā takādu tumṭar*, in M.ʿA. al-H. Abū Rīda (ed.), *Rasāʾil al-Kindī l-falsafiyya* (Cairo: 1950– 1953) vol. 2 70–75.

al-Kindī, *Risāla fī l-ʿilla allatī lahā yabrudu aʿlā al-ǧaww wa-yasḫunu mā qaruba min al-arḍ*, in M.ʿA. al-H. Abū Rīda (ed.), *Rasāʾil al-Kindī l-falsafiyya* (Cairo: 1950– 1953) vol. 2 90–100.

al-Kindī, *Risāla fī ʿillat kawn al-ḍabāb*, in M.ʿA. al-H. Abū Rīda (ed.), *Rasāʾil al-Kindī l-falsafiyya* (Cairo: 1950– 1953) vol. 2 76–78.

al-Kindī, *The Philosophical Works of al-Kindī*, trans. P. Adamson – P.E. Pormann, Studies in Islamic Philosophy (Oxford – New York: 2012).

Ibn al-Nadīm, *Kitāb al-Fihrist*, ed. G. Flügel (Leipzig: 1871–1872).

Olympiodorus, *In Aristotelis* Meteora *commentaria*, ed. W. Stüve, Commentaria in Aristotelem graeca XII.2 (Berlin: 1900).

Ps.-Olympiodorus, *Tafsīr Alumfidūrūs li-Kitāb Arisṭūṭālīs fī l-Āṯār al-ʿulwiyya*, in Badawī A. al-R. (ed.), *Šurūḥ ʿalā Arisṭū mafqūda fī l-yūnāniyya wa-rasāʾil uḫrā: Commentaires sur Aristote perdus en grec et autres épitres*, Recherches publiées sous la direction de l'Institut de Lettres Orientales de Beyrouth, Nouvelle série A: Langue arabe et pensée islamique 1 (Beirut: 1971) 83–190.

Philoponus John, in Vitello G. (ed.), *Aristotelis Physicorum libros tres priores commentaria*, Commentaria in Aristotelem graeca XVI (Berlin: 1887).

al-Ṭabarī, Abū l-Ḥasan ʿAlī ibn Sahl Rabban, *Firdausu'l-Ḥikmat: Paradise of Wisdom*, ed. M. Zubayr al-Ṣiddīqī (Berlin: 1928).

Ṯābit ibn Qurra, *Qawl fī l-sabab allaḏī ǧuʿilat lahū miyāh al-biḥār māliḥa*. Ms. Istanbul, Topkapı Sarayı Müzesi, Ahmet III 3342, 195v–201v.

al-Tawḥīdī, ʿAlī Abū Ḥayyān – Abū ʿAlī Aḥmad Miskawayh, *The Philosopher Responds: An Intellectual Correspondence from the Tenth Century*, ed. B. Orfali – M.A. Pomerantz, trans. S. Vasalou – J.E. Montgomery, Library of Arabic Literature (New York: 2019).

Theophrastus, *On Winds*, ed., trans. and commented, with an introduction, R. Mayhew, Philosophia Antiqua, A Series of Studies on Ancient Philosophy 147 (Leiden – Boston: 2018).

Theophrastus, *al-Qawl fī l-Āṯār al-ʿulwiyya*, ed. and trans. H. Daiber, in Daiber H., *The Meteorology of Theophrastus in Syriac and Arabic Translation*, in Fortenbaugh W.W. – Gutas D. (eds.), *Theophrastus: His Psychological, Doxographical and Scientific Writings*, Rutgers University Studies in Classical Humanities 5 (New Brunswick – London: 1992) 166–293.

Theophrastus, *Der syrische Auszug der Meteorologie des Theophrast*, ed. and trans. E. Wagner, with annotations by P. Steinmetz, Akademie der Wissenschaften und der Literatur: Abhandlungen der Geistes- und Sozialwissenschaftlichen Klasse 1 (Wiesbaden: 1964).

Secondary Sources

Lettinck P., *Aristotle's Meteorology and its Reception in the Arab World: With an Edition and Translation of Ibn Suwār's Treatise on Meteorological Phenomena and Ibn Bājja's Commentary on the Meteorology*, Aristoteles Semitico-Latinus 10 (Leiden – Boston, MA – Cologne: 1999).

Peters F.E., *Aristoteles Arabus: The Oriental Translations and Commentaries on the Aristotelian Corpus*, Monographs on Mediterranean Antiquity 2 (Leiden: 1968).

Rashed R. (ed.), *Thābit ibn Qurra. Science and Philosophy in Ninth-Century Baghdad*, Scientia Graeco-Arabica 4 (Berlin – New York: 2009).

Sersen W.J., *Arab Meteorology from Pre-Islamic Times to the Thirteenth Century A.D.* (Ph.D Dissertation, University of London: 1976).

Sezgin F., *Geschichte des arabischen Schrifttums*, 8 vols. (Leiden: 1967–1984).

Steinmetz P., *Die Physik des Theophrastos von Eresos*. Palingenesia: Schriftenreihe für klassische Altertumswissenschaft 1 (Bad Homburg v.d. Höhe: 1964).

Wilson M., *Structure and Method in Aristotle's Meteorologica: A More Disorderly Nature* (Cambridge: 2013).

CHAPTER 2

From the 'King of the Waters' to Curative Manuscripts: Water and Medicine in Medieval Irish Textual Culture

Deborah Hayden

In the opening section of his *Ecclesiastical History of the English People*, completed in 731 CE, the scholar and historian Bede famously refers to the practice of curing snake-bite by giving the patient a drink of water that contains the scrapings of leaves from Irish manuscripts:

> Denique uidimus, quibusdam a serpente percussis, rasa folia codicum qui de Hibernia fuerant, et ipsam rasuram aquae inmissam ac potui datam talibus protinus totam uim ueneni grassantis, totum inflati corporis absumsisse ac sedasse tumorem.

> For instance we have seen how, in the case of people suffering from snake-bite, the leaves of manuscripts from Ireland were scraped, and the scrapings put in water and given to the sufferer to drink. These scrapings at once absorbed the whole violence of the spreading poison and assuaged the swelling.[1]

This passing observation, which is set within the wider context of Bede's generally laudatory comments about the healthfulness and plentifulness of the island of Ireland in comparison to Britain,[2] encapsulates two interrelated themes that are central to the following discussion. On the one hand, it evokes the literal concept of water as an element essential for the restoration and preservation of bodily health – a substance that can be mixed with other ingredients and ingested, often as part of a medicinal remedy measured and prepared by a healer. On the other hand, the fact that Bede's anecdote can be read as an allegory for the power of the Word of God to combat the evil of Satan

1 Colgrave B. – Mynors R.A.B. (eds. and trans.), *Bede's Ecclesiastical History of the English People* (Oxford: 1969) I, i.20–21.

2 On this motif and its context, see e.g. Byrne A., *Otherworlds. Fantasy & History in Medieval Literature* (Oxford: 2016) 143–151.

 | DOI:10.1163/9789004696501_004

(the serpent) suggests that it might also be understood to express the deeply-rooted symbolic function of water as a means to attain not only physical, but also spiritual purity.[3] Bede's specification that scrapings from the leaves of Irish manuscripts should be used in the preparation of this curative drink may allude, moreover, to a particular association between Ireland and the production of sacred books that were thought to have acquired certain miraculous qualities by virtue of having come into contact with their saintly owners.[4]

A wide range of early Irish written sources survive that enrich our understanding of Bede's account insofar as they illustrate the profound interconnectedness between physical and spiritual modes of healing in early Christian culture, as well as the centrality of water in medieval approaches to curing both body and soul. While much evidence for these themes can be found in well-known (or at least better-studied) Irish-language literary genres of the early medieval period, such as hagiographical, penitential or narrative texts, this preliminary survey of the topic will also introduce some previously unpublished material from the large, but comparatively neglected, corpus of Irish vernacular medical writing, which likewise offers much valuable insight into the function and representation of water as a medicinal element in the Middle Ages.

1 The *Christus medicus* Metaphor and Saintly Healing in Medieval Irish Literature

An explicit connection between spiritual and physical healing can be traced to an early period in Christian culture. The metaphorical depiction of Christ

3 For the allegorical interpretation of Bede's comment, see i.e. Brown M.P., "Hagiography or History? Early Medieval Approaches to Establishing Origin and Provenance for Insular Copies of Scripture", in Moss R. – O'Mahony F. – Maxwell J. (eds.), *An Insular Odyssey: Manuscript Culture in Early Christian Ireland and Beyond* (Dublin: 2017) 24–41, 29.

4 The connection made by Bede between Ireland and the curing of snake-bite is also reflected in early English medical texts, where ritual incantations containing garbled Irish linguistic forms are found for ailments thought to have been inflicted by snakes or 'worms'. It has been argued that references to the 'demon-worm' in these contexts may have had an eschatological significance in relation to sicknesses reflecting a spiritual punishment of some kind: see Arthur C., *"Charms", Liturgies and Secret Rites in Early Medieval England* (Woodbridge: 2018) 73–80. On the influence of Irish scholars and liturgical books on other aspects of ecclesiastical culture in early Britain, see e.g. Campbell J., "The Debt of the Early English Church to Ireland", in Ní Chatháin P. – Richter M. (eds.), *Irland und die Christenheit: Bibelstudien und Mission* (Stuttgart: 1987) 332–346, and Hughes K., "Evidence for Contacts between the Churches of the Irish and English from the Synod of Whitby to the Viking Age", in Clemoes P. – Hughes K. (eds.), *England Before the Conquest: Studies in Primary Sources Presented to Dorothy Whitelock* (Cambridge: 1971) 49–67.

the Physician – the Divine Healer of mankind's spiritual diseases (*Christus medicus*) – is a frequent *topos* in the works of influential figures such as Jerome, Ambrose and Augustine, although it appears to have been introduced by Tertullian, who described Christ's work on earth as that of a preacher (*praedicator*) and healer (*medicator*).[5] Thus Augustine wrote in his *De doctrina Christiana* (1.14.13) that 'the wisdom of God, in healing mankind, has employed Himself to cure it, being Himself Physician and Medicine both in one'.[6] In several of his sermons, Augustine sought to establish a 'contrast between the humility and omnipotence of the Divine Physician on the one side, and the pride and frailty of man on the other', and emphasised Christ's devotion to his spiritual patients by comparing the saviour's ordeal on the cross to the reassurance offered by a physician who agrees to taste unpleasant medicine first:

> The Physician Himself, though no such medicine is necessary for Him, drinks what He does not need, in order to cheer up the patient. Addressing, as it were, the reluctant patient, and encouraging him in his state of trepidation, He drinks first. 'The cup', He says, 'of which I am about to drink (Matt. 20:22; Mark 10:38). In Me there is no disease which could be cured by this cup, and yet I am going to drink of it so that you, who are in need of drinking of it, may not refuse it'. You see now, brethren, whether, after receiving such a great medicine, the human race ought still to be sick. Now God is humble, while man is still proud.[7]

Comparable medical metaphors abound in penitential texts that were written in Ireland as early as the sixth century, where penance is seen as the cure for sins and other spiritual ailments. This concept is most elaborately set out in the 'B-text' of the Latin *Penitential of Columbanus*, most probably written at some point after 575:

> Diuersitas culparum diuersitatem facit paenitentiarum. Nam et corporum medici diuersis medicamenta generibus conponunt. Aliter enim uulnera, aliter morbos, aliter tumores, aliter liuores, aliter putredines, aliter caligines, aliter confractiones, aliter conbustiones curant. Ita igitur etiam spirituales medici diuersis curationum generibus animarum uulnera

5 Arbesmann R., "The Concept of 'Christus Medicus'", *Traditio* 10 (1954) 1–28, 6.

6 Augustine, *De doctrina Christiana*, ed. H.J.S.Vogels, *Aurelii Augustini episcopi Hipponensis De doctrina christiana libros quattuor, Florilegium patristicum* 24 (Bonn: 1930) 1.14.13 (See Vogels, 8; and trans. Arbesmann, "The Concept of 'Christus Medicus'" 13).

7 Ibidem 11 and 15, citing the sermon edited by J.J. Migne in *Patrologia Latina* 38.781. On the reception of the *Christus medicus* theme in late-medieval England, see also the discussion by Rawcliffe C., *Medicine and Society in Later Medieval England* (Stroud: 1995) 18–25.

> morbos [culpas] dolores aegritudines infirmitates sanare debent. Sed quia haec paucorum sunt, ad purum scilicet cuncta cognoscere, curare, ad integrum salutis statum debilia reuocare, uel pauca iuxta seniorum traditiones et iuxta nostram ex parte intellegentiam (ex parte namque prophetamus et ex parte cognoscimus) aliqua proponamus.
>
> Diversity of offences causes diversity of penances. For doctors of the body also compound their medicines in diverse kinds; thus they heal wounds in one manner, sicknesses in another, boils in another, bruises in another, festering sores in another, eye diseases in another, fractures in another, burns in another. So also should spiritual doctors treat with diverse kinds of cures the wounds of souls, their sicknesses, [offences], pains, ailments, and infirmities. But since this gift belongs to few, namely to know to a nicety all these things, to treat them, to restore what is weak to a complete state of health, let us set out even a few prescriptions according to the traditions of our elders, and according to our own partial understanding, for we prophesy in part and we know in part.[8]

The voluminous corpus of hagiographical literature that survives from medieval Ireland demonstrates that saints, as earthly representatives of the Divine Healer, were naturally seen to perform a similar role to that of Christ the Physician. While Wendy Davies has noted that comparatively few of the miracles depicted in Irish hagiographical texts relate specifically to healing,[9] those which are found in such narratives often depict water as a key element in the curative process.[10] For example, in what may have been intended to echo of Jesus's transformation of water into wine for the Wedding at Cana (John 2:1–11), the author of the ninth- or mid-tenth-century *Vita Tripartita* (a bilingual *Life* of Patrick, the preeminent saint of Ireland) describes how Patrick blessed a vessel full of water, which then turned into honey and was used to heal ailments: 'Ro ucc Patraic iar sin lestar lais docum ind uisci, ⁊ ro lín ⁊ ro sén in n-uisce coro sóad i mmil, ⁊ ro ícc cech ngalar ⁊ cech n-ainces forsa tardad (.i. roboí do chretaib léo)'. (Then Patrick brought a vessel with him to the water, and he filled it, and blessed the water so that it was converted into honey, and it

8 Bieler L., *The Irish Penitentials* (Dublin: 1975) 98–99; for further examples of such medical metaphors in the Irish penitential texts, see ibidem 46–47.

9 Davies W., "The Place of Healing in Early Irish Society", in Ó Corráin D. – Breatnach L.- McCone K. (eds.), *Sages, Saints and Storytellers: Celtic Studies in Honour of Professor James Carney* (Maynooth: 1989) 43–55, 44–46.

10 On the representation of water in early medieval hagiographical literature of the Insular world, see also Hugh Magennis's chapter in this volume.

remedied every disease and every ailment to which it was applied, that is, they held it as a blessed object).[11] As Niamh Wycherley has noted, the text explains that this water was regarded as a *cretair*, a term which seems to have referred to a consecrated object or relic believed to possess special powers.[12]

Natural and man-made sources of water, such as wells, springs and rivers, feature prominently in early Irish literature as symbols of ritual cleaning and rebirth, and in the extant texts this symbolism is closely associated with the use of such sources for baptismal purposes.[13] It is therefore not surprising to find that Irish saints are portrayed as having the ability to instil healing properties into wells and other bodies of water. For example, a text written in the first half of the twelfth century claims that St. Monenna, having immersed herself in a cold spring while chanting the psalter, was able to order the water from this source to follow her up a mountain. She then created a 'permanently hot' bath from it that had the power to heal invalids, and which seems to have subsequently come to be associated with St Kevin:

> Et accessit sancta Monenna ad frigidum fontem suum, quo solitis noctibus sedens usque ad mamillas in aqua totum decantare solebat psalterium, et imposita benedictione cum baculo suo, iussit aquam in nomine Dei paths omnipotentis sequi illam. Et secessit inde ad culmen proximi montis, quo preparauerat balneum, post se trahens suum bacul um. Et aqua secuta est eam usque ad summum. Et uirtute sancte uirginis aqua detracta de imis, in balneo recepta, gratia Dei fit semper calida. De inde usque ad hunc diem multi habentes infirmitatem in nomine sancte uirginis per balneum recipiunt sanitatem. Et dixit sancta uirgo ad Cheuin: 'Ecce balneum tibi offero in honore tuo et non deficiet aqua in eo'. Et uocatum est balneum Cheuin quia sancta Monenna dedit illud tiranno ut eriperet eum a diabolo.

> And St Monenna went to her own cold fountain in which she used to chant the whole psalter on accustomed nights, sitting up to her breasts in water, and casting a blessing with her staff she ordered the water in the name of God the Father Almighty to follow her. And she withdrew from there to the top of the nearby mountain on which she had made ready

11 Mulchrone K. (ed. and trans.), *Bethu Phátraic* (Dublin: 1939) 8 (ll. 170–172).

12 Wycherley N., *The Cult of Relics in Early Medieval Ireland* (Turnhout: 2015) 125.

13 Bitel L.M., *Isle of the Saints: Monastic Settlement and Christian Community in Early Ireland* (Cork: 1990) 73. On the connection between holy wells and baptism, see also Ray C., *The Origins of Ireland's Holy Wells* (Oxford: 2014) 89–92.

> a bath drawing her staff behind her. And the water followed her right to the top. And the water being drawn by the power of the holy virgin from the depths and received in the bath by the grace of God became permanently hot. Thereafter right up to the present day many with an infirmity recover their health through the bath in the name of the holy virgin. And the holy virgin said to Kevin: 'Behold I offer you the bath in your honour, and the water in it will not fail'. And the bath is called Kevin's because St Monenna gave it to the chieftain in order to snatch him from the devil.[14]

In some instances, the saint's intervention at an aqueous source is framed as an effort to convert and condemn unchristian practices. Thus Patrick's seventh-century biographer, Tírechán, recounts how the saint visited a certain well in Findmag that was referred to by the local people both as *Slán* ('good health') and as the *rex aquarum* ('king of the waters') because it was thought to have certain supernatural powers. The narrative explains that 'some wise man' had made a shrine for himself under the water in order to preserve his bones from burning:

> Et uenit ad fontem Findmaige qui dicitur Slan, quia indicatum illi quod honorabant magi fontem et immolauerunt dona ad illum in modum dii. Fons uero quadratus fuit et petra quadrata erat in ore fontis (et ueniebat aqua super petram .i. est per glutinationes) quasi uestigium regale, et dixerunt increduli quod quidam profeta mortuus fecit bibliothicam sibi in aqua sub petra, ut dealbaret ossa sua semper, quia timuit ignis exust<ion>em; quia adorabant fontem in modum dii. Et indicata est Patricio causa adorationis, et ipse zelum Dei habuit de Deo uiuo, et dixit: 'non uerum quod dicitis, quod rex aquarum fons erat' (quia dederunt illi nomen "aquarum rex"').

> And he came to the well of Findmag, which is called Slán, because he had been told that the druids honoured the well and offered gifts to it as to a god. The well was of square shape and the mouth of the well was covered with a square stone (and water flew over the stone, that is through ducts closed with cement) like a regal trail (?), and the infidels said that some

14 Ulster Society for Medieval Latin Studies, "The Life of Saint Monenna by Conchubranus: Part I", *Seanchas Ardmhacha: Journal of the Armagh Diocesan Historical Society* 9.2 (1979) 250–273, 268–269; on the background to this text, see Esposito M., "The Sources of Conchubranus's Life of St. Monenna", *English Historical Review* 35.137 (1920) 71–78. The episode cited here is also noted in Bitel, *Isle of the Saints* 187.

> wise man had made for himself a shrine in the water under the stone to bleach his bones perpetually because he feared the burning by fire; and they worshipped the well as a god. And Patrick was told the reason for its worship, and he had the zeal of God for the living God, and said: 'It is not true what you say that it was the king of the waters' (for this is the name they gave the well: 'king of the waters').[15]

Having thus intimated that only the Christian God could be king of the elements, Patrick then proceeded to bless the stone over this well and lift it away before the eyes of 'the magicians and pagans of that region' (*magi et gentiles regionis illius*), revealing only water beneath. The site subsequently became a shrine to the Christian saint himself.[16]

There is an echo of this hagiographical episode in the Irish saga-narrative *Cath Maige Tuired* ('The [Second] Battle of Mag Tuired'), argued to have first been composed in the ninth century and later reworked in the eleventh or twelfth.[17] The tale recounts the war between the supernatural races of the Fomoire and the Túatha Dé that were believed to have inhabited Ireland before the arrival of the first Gaelic settlers, and the victory of the Túatha Dé over the Fomoire is explained as having resulted in part from the actions of the healer-figure Dían Cécht and his children, Octriuil and Míach, who were able to restore mortally-wounded men to life by casting them into a well called *Sláine* ('health') and chanting incantations over it. Thus here again, an aquatic source is explicitly linked to the concept of healing. In this case, however, the imagery of the scene evokes the opposing element of fire in a paradoxical fashion, since it is claimed that the warriors were thrown into the well in order to

---

15 Bieler L., *The Patrician Texts in the Book of Armagh* (Dublin: 2004) 152–153.

16 According to Catherine Swift ("Tírechán's Motives in Compiling the *Collectanea*: an Alternative Interpretation", *Ériu* 45 (1994) 53–82, 58, n. 29), the church site associated with this well in the thirteenth century was named 'Slánpatrick', and has been identified by Kenneth Nicholls with a church in the townland of Ballynew, parish of Aglish, Co. Mayo: see also the latter's "Tobar Finnmhuighe – Slán Pádraig", *Dinnseanchas* 2 (1966–1967) 97–98.

17 Only a single copy of the text survives in the sixteenth-century London, British Library MS Harleian 5280; see Gray E. (ed. and trans.), *Cath Maige Tuired. The Second Battle of Mag Tuired* (London: 1982) 54–55. On the dating of the work, see also Murphy G., *Saga and Myth in Ancient Ireland* (Dublin: 1955) 19, and Ó Cathasaigh T., "*Cath Maige Tuired* as Exemplary Myth", in de Brún P. – Ó Coileáin S. – Ó Riain P. (eds.), *Folia Gadelica: Essays Presented by Former Students to R. A. Breatnach* (Cork: 1983) 1–19, 1. For a more recent discussion of the so-called 'Mythological Cycle' of tales, see Carey, *The Mythological Cycle of Medieval Irish Literature*.

'kindle' them (*doberiud bruth*) so that they would be 'more fiery' (*ániu*) the following day:

> Is edh dano doberiud bruth isna hógaib nogontais ann, comtar ániu íarnauhárach: fo bíth roboí Díen Cécht & a dí mac & a ingen .i. Ochttríuil & Airmedh & Míach oc dícetul foran tibrait .i. Sláine a hainm. Focertdidis a n-athgoíte indte immairlestis; botar bí notégdis esde. Batir slán a n-athgoíte tre nert an dícetail na cethri lege robátar imon tibrait.
>
> Now this is what used to kindle the warriors who were wounded there so that they were more fiery the next day: Dían Cécht, his two sons Octriuil and Míach, and his daughter Airmed were chanting spells over the well named Sláine. They would cast their mortally-wounded men into it as they were struck down; and they were alive when they came out. Their mortally-wounded were healed through the power of the incantation made by the four physicians who were around the well.[18]

The healing power of the well is underscored by the narrator's subsequent comment that *ainm n-aild dano din tibroid-sin Loch Luibe, ar dobered Díen Cécht ind cech losa rouhótar a n-Éri* ('another name for that well is *Loch Luibe* ["Lake of the Herbs"], because Dían Cécht put into it every herb that grew in Ireland').[19]

It is apparent from many Irish sources that purificatory healing rituals of this kind were perceived as a natural extension of everyday bathing practices. For example, Patrick is said to have baptised the two daughters of the pagan king Lóegaire when 'they came to the well [called Clébach], as women are wont to do, in the morning to wash, and they found the holy assembly of bishops with Patrick beside the well' (*et ecce duae filiae regis Loiguiri Ethne alba et Fedelm rufa ad fontem* [*qui dicitur Clebach*] *more mulierum ad lauandum mane uenierunt et senodum sanctum episcoporum cum Patricio iuxta fontem inuenierunt*).[20] That bathing the sick was a regular practice in medieval Irish society is likewise indicated by commentary on the seventh-century medico-legal tract *Bretha Crólige* ('Judgements concerning blood-lying'), which specifies that bathing facilities, a bed and the services of a doctor should be among

18 Gray, *Cath Maige Tuired* 54–55.

19 Ibidem 56–57.

20 Bieler, *Patrician Texts* 142–143 (§26). As Elizabeth Boyle has pointed out to me, this episode evokes the scene depicted in John 4:6–53, where Jesus first conversed with a woman of Samaria who had come to draw water at a well, and then went on to heal the son of a ruler in Cana.

the things provided for a person on sick-maintenance: 'Geall .vii. cumal fri tairechtain a legha ⁊ a focruicthi ⁊ a leptha, ⁊ a letheneclann do dia nelaidther impo'. (A pledge worth seven cumals [is due] for providing him with leech and bathing and bed, and half his honour-price [is due] to him if there be evasion of these [duties].)[21]

The healing function of baths is also reflected in Irish saga-narratives such as the famous epic *Táin Bó Cúailnge* ('The Cattle Raid of Cooley'), in one version of which the duelling warriors Cú Chulainn and Ferdia are said to have had curative baths containing clay (*foithraicedh criadh*) prepared for them during a truce in their contest.[22] A more unusual bath is prepared for the wounded warrior Cethern mac Fintain, who, after having rejected the fatal prognosis of fifty different doctors, was finally examined by the king Conchobor's physician Fingín. The latter gave him two options: one to accept a treatment that would take a whole year but would save his life; and the other to accept a cure that would make him fit to return to battle immediately, but would only allow him to survive for another three days. In the fashion of a true warrior, Cethern chose the latter course, and so Fingín ordered him to immerse himself in a 'marrow bath' (*smirammair*) prepared from the pulped flesh, bones and hides of cattle.[23]

The *Christus medicus* metaphor is again evidenced in analogous episodes from hagiographical literature, where saints are portrayed as healing the ill or injured by bathing them in water that they themselves have already used. For example, St Moling is said to have restored a demented youth to sanity by washing him in his own bath, while St Comgall cured a leper by immersing him in a tub in which he himself had bathed.[24] The latter account no doubt reflects a wider *topos* of washing the feet or ulcerated bodies of lepers that is well attested in sermons, homilies and hagiography from other contemporary literary traditions. Carole Rawcliffe has noted that, although this procedure 'performed an important therapeutic function, the overwhelming purpose of

---

21 Binchy, "Bretha Crólige" 44–45; the passage is also noted by Lucas A.T., "Washing and Bathing in Ancient Ireland", *Journal of the Royal Society of Antiquaries of Ireland* 95.1/2 (1965) 65–114, 67. For the healing function of baths in the Middle Ages, see Chapter 3 of this volume.

22 Best, "Comhrag Fir Diadh & Chon Chulainn. Táin Bó Cúailnge", *Zeitschrift für Celtische Philologie* 10 (1915) 274–308, 303; cited in Lucas, "Washing and Bathing" 67–68.

23 O'Rahilly C. (ed. and trans.), *Táin Bó Cúalnge from the Book of Leinster* (Dublin: 1967) 101–105 and 235–240; cited in Lucas, "Washing and Bathing" 67.

24 Both of these episodes are noted by Lucas, "Washing and Bathing" 68. For the tale concerning St Moling and the demented youth, see de Smedt C. – de Backer J., *Acta Sanctorum Hiberniae ex Codice Salmanticensi* (Edinburgh – London: 1888) 822–823; for that concerning St Comgall and the leper, see Plummer C. (ed. and trans.), *Vitae Sanctorum Hiberniae*, 2 vols (Oxford: 1910) II, 19.

the exercise was spiritual rather than hygienic'; she illustrates this by reference to the legend of the Magdalen, who had bathed Christ's feet with her tears both during his life (John 13:4–15) and again when he hung *quasi leprosus* on the cross, prompting a further outpouring of grief.[25] Rawcliffe also observed that the curing of leprosy by immersion in water is prefigured in the Old Testament by the cleansing of Naaman the leper in the Jordan seven times, once for each of the seven deadly sins (2 Kings 5).[26]

A comparable association between washing and carnal sin is found in an anecdote from the Old Irish ecclesiastical document known as the 'Monastery of Tallaght', in which the Clonmacnoise anchorite Laisrén, described as 'quite naked and free from sin, with nothing on his conscience', but also 'infirm with disease' (*imnocht imdilmain cen ní for a cubus* […] *hi luibri galair*) is taken home by a student (*mac cleirchib*) who puts a cloak under him. That night Laisrén sees a 'carnal vision' (*aislingi coildnidi*) for the first time in his life; as he subsequently performs a penitential vigil, an angel visits him and explains that this sinful vision had resulted from the fact that the cloak he slept upon had not been washed since it was used by a wedded couple – 'for any cloak that is taken from a lustful couple, a demon accompanies it so long as it is not washed' (*ar nach brat berair do áos étraich conamteit demmun airet nat negar* 7 *cetera*).[27] The purificatory function of water as an element understood to be capable of washing away carnal sin was clearly an enduring concept: in her 2019 account of the Magdalene laundries, or institutions for so-called 'fallen women' and babies run by the Roman Catholic orders in Ireland until the late twentieth century, Caelainn Hogan interviewed a woman adopted from Bessborough House in 1988, who

> […] was told that when she was brought home to her adoptive parents' house, her granny had insisted on washing her. Catherine[28] suspected this was a 'moral thing', rather than a result of her infant body being actually dirty after her stay in the institution. She was baptized twice, first

25 Rawcliffe C., *Leprosy in Medieval England* (Stroud: 1995) 142–144. On the origins and enduring concept of 'Christ-as-leper', see ibidem 60–64.

26 Ibidem 125.

27 Gwynn E.G. – Purton W.J., "The Monastery of Tallaght", *Proceedings of the Royal Irish Academy* 29 (1911/1912) 115–179, 155. On this episode and its parallels in a medical context, see Hayden D., "A Sixteenth-century Irish Collection of Remedies for Ailments of the Male Reproductive Organs", *Celtica* 33 (2021) 248–276, 260–261.

28 A pseudonym.

at Bessborough, and then a second time in her parents' local parish – another form of cleansing.[29]

2 Water and Healing in Medieval Irish Medical Texts

Evidence comparable to that outlined above from Irish penitential, hagiographical, and narrative literature can also be found in the less well-studied corpus of Irish-language medical texts. The latter comprise a substantial number of vernacular translations of Latin works that circulated across western Europe during the late-medieval period, as well as many collections of herbal cures, instructions for making textual amulets, and verbal charms, some of which may ultimately derive from the early medieval period. For example, a remedy book compiled in the early sixteenth century by the North Connacht physician Conla Mac an Leagha, who was probably drawing on a range of sources of varying dates, includes several cures that specify the use of *uisce choisricthe* ('consecrated' or 'holy' water) in the preparation of a drink that is to be ingested by the patient:[30] 'Eōl*us* do togai*rm* na fola m*ista* ⁊ a*r* galur fūail .i. l*us* na c*r*oici do b*r*uit ar medg ⁊ u*í*s*ce* coisregta do cratad ai*r* ⁊ ibed īa*rum* ⁊ īc*caid*'. ('A prescription for inducing the menses and for urinary disease, i.e. boil fennel in whey and sprinkle holy water on it and let him drink it then, and it heals').[31] Two other references to *uisce choisricthe* are found elsewhere in this same collection, one in a section concerned with ailments of the stomach (*gaile*), and the other in a chapter of cures for urinary disease; in both cases, the holy water is to be 'shaken' or 'sprinkled' (*crothaid*) either on the other ingredients of the recipe or on the patient themselves.[32] In yet another remedy

29 Hogan C., *Republic of Shame: Stories from Ireland's Institutions of 'Fallen Women'* (Dublin: 2019) 191.

30 On the scribal background, contents and dating of this remedy book, see i.e. Hayden D., "Three Versified Medical Recipes Invoking Dían Cécht", in Ahlqvist A. – O'Neill P. (eds.), *Fír fesso: A Festschrift for Neil McLeod* (Sydney: 2018) 107–123; and "Attribution and Authority in a Medieval Irish Medical Compendium", *Studia Hibernica* 45 (2019) 19–51; and Nic Dhonnchadha, "An Irish Medical Treatise on Vellum and Paper from the 16th Century", in Ó Macháin P. (ed.), *Paper and the Paper Manuscript: A Context for the Transmission of Gaelic Literature* (Cork: 2019) 111–125.

31 RIA MS 24 B 3, p. 75.15–16. In this and the following citations from unpublished manuscript sources, expansions are indicated by italics, missing letters are supplied in square brackets, and length-marks, where not found in the manuscript, are marked using a macron over vowels. Word-division and punctuation are editorial.

32 See RIA MS 23 N 29, fols. 3r12–13 (which also specifies the use of *uisce ferthuna*, 'rainwater') and fol. 8r38 (where the verb *crothaid* is implied), respectively.

for impotence, the instructions are more ritualistic in nature: after reciting an incantation, the water is to be placed in a yew-wood vessel (*soitheach iobhair*) and then sprinkled over the patient morning and night for nine days.[33]

A fourth cure from Conla Mac an Leagha's collection illustrates the use of consecrated water for healing both physical and spiritual afflictions in an even more obvious way. This remedy is claimed to work not only against an 'ailment of the head' (*cendgalar*, perhaps referring to some kind of mental illness in this context), but also against harm of a supernatural variety, including 'injury caused by evil eye' (*aidmilled*)[34] and other attacks by demons: 'Ar cendgalur ⁊ ar aidmill*ed* ⁊ ar g*ach* n-urc*ad* donīt demna ar dai*n*eb .i. ūir *ceithri* roil*igi*, lint*ur* drocta dīb ⁊ u*isce* *ceithri* prīmtipr*at* a*r*na coiserg*ad* faiur ⁊ fer dā tin*n*saigi*n* ⁊ folc*ad* as'. ('Against headache and injury from the evil eye and every [harmful] cast that demons make upon people, i.e. [take] the clay of four graveyards, fill a tub with it and [put] consecrated water from four eminent springs on it, and pour it on the man and wash him with it.')[35] Here it is noted that the consecrated water (*uisce* [...] *arna coisergad*) to be used for the washing should be drawn from 'four eminent springs' or 'wells' (*ceithri prīmtiprat*) and mixed with the 'soil of four graveyards' (*ūir ceithri roiligi*) – a detail that recalls the reference to the aforementioned 'baths containing clay' (*fothraicedh crīadh*) that were prepared for the wounded warriors Cú Chulainn and Ferdia during the truce from their duel in *Táin Bó Cúailnge*. The specification that the clay should come from four graveyards suggests, however, that the earthly substance may have been understood to have been sanctified by association with the remains of saints who had been buried near church structures, which were themselves often located in close proximity to holy wells. Celeste Ray has observed that 'as soil from saints' graves was often considered curative and blessed (and is still packed in the mouths of those suffering toothaches in the twenty-first

33 For an edition and translation of this text, see Hayden, "A Sixteenth-century Irish Collection of Remedies" 267; the same publication (258) also notes an implicit parallel for the remedy in an episode from the Old Irish hagiographical *Life of Brigit*. The language of the oral incantation is considered in Hayden D., "The Context and Obscure Language of Medical Charms in a Sixteenth-century Irish Remedy Book: Four Case Studies", in Kobel C. (ed.), *Obscuritas in Medieval Irish and Welsh Literature* (Dublin: forthcoming).

34 The verbal noun *aidmilled* has the general meaning of 'complete destruction' or 'ruin', but is also attested in several instances with the more specific sense of supernatural harm or 'bewitchment', perhaps from a cast of the evil eye; this is what seems to be the meaning here. On comparable attestations of the verbs *millid* and *aidmilled* in other Irish literary sources, see Borsje J. – Kelly F., "The Evil Eye and Early Irish Literature and Law", *Celtica* 24 (2003) 17–20.

35 RIA MS 24 B 3, 43.27–9 (my translation). I am grateful to Siobhán Barrett for drawing my attention to this example.

century), waters flowing close to a saint's pure remains might also have been consumed without concern for disease, but rather in hopes of the opposite'. In illustration of this, Ray cites the legend pertaining to St Odhrán, who was thought to have been buried alive 'to purify the land of Iona from any devilish genius in preparation for the construction of Colum Cille's church there'.[36] A possible play on this theme, in which the terrestrial element to be mixed with water takes the form of saintly excrement, is found in an anecdote from the *Fragmentary Annals of Ireland* concerning the seventh-century king of the Northern Uí Néill, Áed Uaridnach. Áed is said to have stopped to wash in a river that flowed through the monastic town of St Muru in Donegal, but just as he was raising a handful of water to his face, he was cautioned by one of his men, who warned him that the clergy's privy was located over the water of the river. Their dialogue continued as follows:

> 'An ann," ar an rí, "téid an clereach féin ar imthelgudh?'
> 'As ann go deimhin,' ar [an] t-ócclach.
> 'Ní namá," ar an rí, "cuirfead fom aighidh, acht cuirfead um bél, ⁊ ibhadh (ag ól trí mbolgoma de), úair as sacarbaicc leam an t-uiscce i ttéid a imthelgun'.

> 'Is that where the cleric himself [St Muru] goes to defecate?' asked the king.
> 'It is indeed', said the young man.
> 'Not only will I put it on my face', said the king, 'but I'll also put it in my mouth, and I'll drink it' – drinking three gulps of it – 'for the water his faeces go into is a sacrament to me'.[37]

Áed's display of piety naturally results in promises of rewards from St Muru, and the anecdote has been read as 'an allegorical statement on the proper relations between king and cleric, or (to speak anachronistically) between church and state', since 'it is evidently only proper that Áed, future high king of Ireland, should wash in and drink from the wastewater of that representative of the church, St Muru'.[38] In light of these examples, it might be argued that

36 Ray, *Holy Wells* 103–104, citing MacLeod Banks M., "A Hebridean Version of Colum Cille and St Oran", *Folklore* 42.1 (1931) 55–60, and Herbert M., *Iona, Kells, and Derry: The History and Hagiography of the Monastic Familia of Columba* (Dublin: 1988) 261.

37 Davies M.T., "Kings and Clerics in Some Leinster Sagas", *Ériu* 47 (1996) 45–66, 45, citing (with minor adjustments) Newlon R.J (ed. and trans.), *Fragmentary Annals of Ireland*, (Dublin: 1978) 6–7. My thanks to Elizabeth Boyle for bringing this example to my attention.

38 Davies, "Kings and Clerics in Some Leinster Sagas" 48.

the specification in the medical remedy cited above, according to which the elements of water and earth should each be sought from four different sources, simply echoes the tetradic structure of the Gospels – a symbolism that would have no doubt had a strong resonance in the context of a cure intended to combat the ill effects of demons (*demna*) in particular.

There is some uncertainty regarding the precise meaning of the second element in the term *prīmtipra* ('eminent spring' or 'well') used in Conla Mac an Leagha's aforementioned remedy for *cendgalar*. The translations given in the historical dictionary of the Irish language (eDIL), s.v. *tipra*, include 'well', 'spring', 'fountain' and 'source', perhaps suggesting that fresh water is implied.[39] In this regard, it is noteworthy that a number of other remedies in Conla's collection emphasise the 'purity' of the water to be drawn from a *tipra* by prefixing the word with the element *fír-* (lit. 'true' or 'veracious'); this prefix is also attested in other contexts where it evidently denotes a substance that it is in a state of 'freshness', as opposed to being stagnant or stale.[40] The concept of 'fresh' water is also often linked to that of water that is 'moving', and the further association of the latter with a state of good health is clearly evident in other vernacular medical traditions of the medieval period, where running water appears to have been thought to serve the function of carrying away disease. For example, it is specified in an Old English ritual cure for an erysipelatous swelling, found in two separate collections of medical remedies, that one should 'first take a stick of hazel or elder wood, inscribe your name on it, cut three incisions on (the erysipelatous spot), fill the name with your blood, throw it over your shoulder or between your thighs into running water. Cut the incisions and do all this in silence'.[41]

Similar imagery is found in a collection of Irish medical remedies in British Library MS Harley 546, dated to 1459, where a cure for epilepsy (*epilenncia*) recommends the following:

---

39 See *eDIL*, s.v. *tipra* (dil.ie/40932). An Electronic Dictionary of the Irish Language, based on the Contributions to a Dictionary of the Irish Language (Dublin: 1913–1976). Accessed May 14, 2024. www.dil.ie.

40 Cf. *eDIL*, s.v. 1 *fír* A(c) (dil.ie/22171), which defines *fírthipra* as 'a spring', and also cites the examples of *fíruisce* 'fresh water' (opp. to brine), 'spring water', and *fírlemnacht* 'fresh milk'. I have identified eight examples of the collocation *uisce fírtibrat* (some with slight variations of spelling) in the medical collection considered here.

41 Storms G., *Anglo-Saxon Magic* (Gravenhage: 1948) 74–76. The remedy in question is included in the collection of cures known as 'Bald's Leechbook', found in British Library MS Regius 12 D XVII, fols. 1r–108v (probably compiled in the mid-tenth century), as well as in the compilation of remedies, charms and prayers known as the 'Lacnunga', preserved in London, British Library MS Harley 585, fols. 130r–193r (written around the year 1000).

> Coilech derg ocus a cenn do buain d'éinbéim de ocus a ae do bein as ocus ben díb an domblas ocus ben a cridi as ocus cuir in domblas ann ocus cengal snáithi fó bél in cridhi ocus bruith galún d'uisgi tráighes ar an croidi ocus tabair cúic fiuch*tha* nó a sé air ocus eabar re nái lá ocus re nái naidhchi ocus curtar a fuigill re sruth ocus re fairgi ocus adnaicther in cailech a comrac na ceithre sligeadh mar nach mbenfa nech ris ocus ícaid.
>
> Take a red cock, and at a stroke dock him of his head; take out his liver, thence extract the gall and insert into his heart [already taken out ad hoc]; the orifice of which draw close with a thread. Upon the heart [so prepared] let seethe a gallon of water drawn from a fallen tide, and let it have five boils or six. For nine days and nine nights be this potion drunk, and the remnant cast out to be carried off by the sea-tide. For the cock, have him buried at the meeting of four ways, where none shall meddle with him, and it cures.[42]

The suggestion here that one should sacrifice a cock in order to cure an afflicted patient no doubt draws upon a long medical tradition of using animal parts in treating various kinds of diseases, typically on the basis of a perceived correspondence between some feature or attribute of a given animal and the ailment to be cured.[43] In the case of epilepsy, for example, one might compare the remarks of the fourteenth-century English physician John of Gaddesden, who noted that the incinerated ashes of a cuckoo might be served in a drink to an epileptic patient, and that some people were inclined to carry a cuckoo's head around the neck, because 'the cuckoo has epilepsy every month, so that some say that it has the property of curing this disease by drawing the matter to itself'.[44] However, the additional observation in the Irish text that a

42 British Library MS Harley 546, fol. 18r2–6; O'Grady S. (ed. and trans.), *Catalogue of Irish Manuscripts in the British Library [Formerly British Museum], Volume 1* (London: 1926; repr. Dublin: 1992) 182–183. The fully digitised manuscript and a more recent catalogue of its contents can be viewed online at www.bl.uk/manuscripts/FullDisplay.aspx?ref=Harley_MS_546 (last accessed May 14, 2024).

43 For a general overview of the subject, see e.g. MacKinney L., "Animal Substances in materia medica: a Study in the Persistence of the Primitive", *Journal of the History of Medicine and Allied Sciences* 1.1 (1946) 149–170.

44 Demaitre L., *Medieval Medicine: the Art of Healing, from Head to Toe* (Santa Barbara, CA: 2013) 143–145. The treatment of epilepsy by sacrificing a cock or hen is also attested in later folklore; for example, in the nineteenth century Arthur Mitchell recorded instances of a black cock being buried alive as a remedy for the disease in Wester Ross and elsewhere in the north of Scotland, and suggested that such practices had their roots in Antiquity: see Miller A., "Superstitions in the North-west Highlands and Islands of Scotland, especially

medicinal potion for epilepsy should be made with *uiscgi tráighes* (lit. 'water which ebbs') that has been boiled five or six times may again indicate a common perception that it was best to use pure or fresh water in the preparation of medicinal remedies. It is possible that the occurrence of the phrase in this context alludes to the seashore wells that are well found across many parts of Ireland and often associated with the curing of diseases.[45] Patrick Logan cites the following description of one such well in Moyarta parish, Co. Clare, made in the first half of the nineteenth century by Eugene O'Curry in his letters for the *Ordnance Survey of Ireland*:

> About a furlong east of the [church of Teampull-an-Aird in Kilcradaun townland], at the bottom of a cleft in the very face of the headland, is the fresh water Holy Well of St. Cradaun, ranking among the most popular wells in Ireland for the cure of all diseases, but more especially diseases of the eyes and limbs. The well is sunk in the solid rock and is overflown by the salt sea at every full tide but the moment the tide recedes the water in the well is as pure and fresh as ever. There is a small cave or recess in the cliff behind it in which people are in the habit of spending whole nights in prayer in fulfilment of vows made in times of danger from sickness, drowning, etc.[46]

Here 'fresh' or 'pure' water is considered to be that which is left in the source after the salt water of the sea has subsided.[47] The instruction in our remedy for epilepsy to cast out the remnants of the potion so that they can be carried off

in Relation to Lunacy", *Proceedings of the Society of Antiquaries of Scotland* 4 (1860–1862) 251–288, 274–276.

45 Siobhán Barrett has drawn my attention to a separate remedy in RIA MS 24 B 3, 44.29, for a lye to cure *galar cinn* ('ailment of the head'), which similarly specifies that *uisce tipraide tráiges* ('water from an ebbing source') should be used if the ailment occurs in the spring (*a n-errach*). She has speculated that the phrase might refer to a turlough, a disappearing or seasonal lake found mostly in limestone areas in the west of Ireland; this type of source floods annually in autumn through springs and fissures in the limestone, but drains in the springtime. It is unclear, however, whether the water from such sources was historically understood to have any special healing powers.

46 Logan, *The Holy Wells of Ireland* 57–58; for the text, see O'Donovan J. – O'Curry E. – O'Flanagan M., *Letters Containing Information Relative to the Antiquities of the County of Clare: Collected during the Progress of the Ordnance Survey in 1839, with Letters and Extracts Relative to Ancient Territories in Thomond, 1841*, 3 vols (Bray, Co. Wicklow: 1928), accessed May 14, 2024, https://www.clarelibrary.ie/eolas/coclare/history/osl/moyarta2_st_cradanns_well.htm.

47 On the description of tides and the consistency of water in early medieval sources, see the Introduction to this volume 1-27.

by the sea-tide (*curtar a fuigill re sruth ocus re fairgi*) is, moreover, reminiscent of the aforementioned Old English ritual cure for an erisypelatous swelling, where the reader is advised to symbolically cast away the disease in running water.

Another comparable example of this theme is found in a separate early English cure for a 'holy drink' intended to prevent 'tricks of elves' and 'every temptation of the fiend'.[48] This recommends that 'an immaculate person silently [...] fetch half a jar of running water against the current' (*hat unmælne mon gefeccean swigende ongean streame healfne sester yrnendes wæteres*) and mix this with certain herbs in a paten on which have been inscribed the following words, drawn from the Gospels of John and Matthew and from the Psalms:

> In principio erat verbum usque non comprehenderunt et plura ('"In the beginning was the Word" up to "they did not comprehend" and again'; John 1:1–5)
>
> Et circumibat Jesus totam Galileam docens usque et secuti sunt eum turbae multae ('"And Jesus went about all Galilee, teaching" up to "and great multitudes followed him"'; Matthew 4:23–5)
>
> Deus in nomine tuo usque in finem ('"O God, by thy name" up to the end'; Psalm 53)
>
> Deus misereatur nobis usque in finem ('"May God have mercy on us" up to the end'; Psalm 66)
>
> Domine deus in adjutorium usque in finem ('"Help, O Lord God" up to the end'; Psalm 69)

These words would clearly have resonated in the context of a medicinal remedy to combat demons, since the passage from the Gospel of Matthew describes how Jesus healed 'all manner of sickness and every infirmity' among the people of Galilee and Syria, including those who 'were possessed by devils, and lunatics', while Psalm 53 is a prayer for help in distress. The reader of this ritual cure is subsequently instructed to 'carefully wash the writing from the paten' into the water (*and þweah þæt gewrit of ðan husldisce þær in swiðe clæne*), which has been mixed with certain herbs and hallowed wine; the resulting

48 The text of this cure is found in the aforementioned collection of remedies, charms and prayers known as the *Lacnunga*, on which see above, n. 41.

substance is then to be consumed by the patient in church after singing Masses and reciting various litanies.[49]

The idea that a potable medicine could be created by bringing water into contact with powerful words or letters such as these is also attested in the following Irish remedy for *cuthach*, a term which literally means 'madness' or 'frenzy', but was probably understood to indicate epilepsy:[50]

> Ar c*hu*[th]ach[51] .i. rūsc critaich ⁊ gilcaide, b*er*bt*ur* ⁊ do-berur a cūa ⁊ sc*r*ib*th*ur naoi mūti isin cūa, id ōn .b.c.d·:g.p.t·:f.k.q·: ⁊ ai*n*m i*n* dui*n*e *eter* ga*ch* dā mūit dīb ⁊ sec*ht*m*a*d ūat*haid* āis ēsca, *nó* i *secht*mad *nó* .21. do-nīter, ⁊ ibed in dui*n*e in lind-sin asin cūa gu cend naomaide.[52]

> For *cuthach* ('rage, madness') i.e. bark of a poplar tree and of a reed are boiled, and it [i.e. the liquid] is poured into a cup, and nine mute consonants are written in the cup, that is *b, c, d – g, p, t – f, k, q* and [write] the name of the person between every two of those letters and on the seventh day of the lunar cycle – it is done either in a seven or a 21-day period – and let the person drink that liquid from the cup for nine days.

The explicit liturgical instructions found in the aforementioned English remedy are absent from this Irish cure, as is any specification that the water to be used should come from a fresh or moving source. The Irish text is comparable, however, insofar as it recommends combining certain herbs with water and then drinking the concoction from a cup on which something has been written (in this case, a sequence of letters and the name of the afflicted individual).[53] The act of consuming powerful words or letters, either by eating

49 Storms, *Anglo-Saxon Magic* 232–235; the passage is found in British Library MS Harley 585, fols. 137r–138r, digitised at The British Library, accessed November 3, 2023, https://www.bl.uk/manuscripts/Viewer.aspx?ref=harley_ms_585_f001r. For discussion of a similar remedy (albeit without any reference to running water) from the early English medical compilation known as 'Bald's Leechbook', see Jolly K., *Popular Religion in Late Saxon England: Elf Charms in Context* (Chapel Hill, NC: 1996) 119–120.

50 It is included in a chapter on cures for this disease in Conla Mac an Leagha's remedy collection.

51 The intervocalic 't' appears to have been written in the manuscript with a mark of lenition above it, but the symbol is now so faded as to be almost illegible. The reading given here is supported, however, by the headings of the remedies which immediately precede and follow this passage.

52 RIA MS 24 B 3, 92.29–93.4.

53 On the significance of the letters cited in this particular example, see Hayden, "The Context and Obscure Language of Medical Charms".

or drinking them, also has many parallels in other early medical traditions, and was probably based on the rationale that the liquid or food in question would have absorbed the magical power of the symbols; similarly, the act of washing powerful words or letters with water is clearly analogous to the way in which water was poured over holy relics, shrines and tombs of the saints, collected in a bottle, and then drunk by the faithful.[54]

The healing power of sacred words is also reflected in an Irish context by the use of entire manuscripts, or parts thereof, as associative relics on the basis that they were thought to have once come into contact with saintly figures, and therefore acquired certain miraculous properties.[55] A number of surviving early Insular religious manuscripts, such as the Book of Mulling (Dublin, TCD MS 60, a pocket Gospel-book), the *Cathach* of Colum Cille (Dublin, RIA MS 12 R 33, a psalter), and the Stowe Missal (Dublin, RIA MS D ii 3, a sacramentary) were enshrined in precious metalwork containers because they were believed to have once been in the possession of a particular saint. The perceived apotropaic function of these objects is indicated, for example, by accounts of how the later medieval hereditary keepers of the *Cathach* carried its shrine into battle with them in order to ensure divine favour.[56] In a comparable way, vernacular Irish medical remedy collections contain many instructions for making textual amulets that could be worn on the body in order to protect the bearer against enemies or heal particular afflictions.[57]

Water again plays a central role in several accounts of such associative relics, as is illustrated by legends concerning manuscripts that appeared intact and undamaged after being retrieved from a river or other source, much as the saint with whom they were linked was said to have survived trials by drowning.[58] Textual amulets made from fragments of vellum could also indirectly serve as

54 Skemer D.C., *Binding Words: Textual Amulets in the Middle Ages* (University Park, PA: 2006) 137 n. 30.

55 Wycherley, *Cult of Relics* (Turnhout: 2015) 38, has noted that the use of such associative relics is increasingly evidenced in Irish sources from the eighth and ninth centuries onwards.

56 Brown, "Hagiography or History?" 24–27.

57 For a comprehensive study of textual amulets in medieval culture more widely, see Skemer, *Binding Words*. For discussion of some Irish examples, see Nic Dhonnchadha A., "Michael Casey's Medical Transcripts in Gilbert MS 147", *Éigse* 60 (2019) 112–113 (on the amulet known as the 'Seven Sleepers of Ephesus') and Rempt, "'And Straightaway the Fountain of Her Blood Dried Up' (Mark 5: 29): Constructing a Template for Late Medieval Irish Obstetric Charms".

58 Brown, "Hagiography or History?" 28–29; see also Ó Floinn R., "Insignia Columbae I", in Bourke C. (ed.), *Studies in the Cult of Saint Columba* (Dublin: 1997) 136–161, 150–151, where it is noted that Adomnán 'recounts two instances of how books recovered from rivers in

healing agents after coming into contact with water: thus the seventh-century cleric Adomnán, in his hagiographical account of the life of Saint Columba, reported that the saint was able to heal a broken hip by sprinkling water on it in which a box containing a blessing inscribed on vellum had been dipped.[59] Almost a thousand years later, in his 1627 translation of the *Annals of Clonmacnoise*, Conell Mageohegan gave the following account of Columba's activities:

> He wrote 300 bookes wth his one hand. They were all new Testaments, left a book to each of his Churches in the Kingdome, which Bookes have a strange property which is that if they or any of them had sunck to the bottom of the Deepest waters they would not lose one letter, signe, or character of them, w[ch] I have seen partly myselfe of that book of them which is at Dorow in the K[s] County, for I saw the Ignorant man that had the same in his Custody, when sickness came upon cattle, for their Remedy putt water on the booke & suffered it to rest there a while & saw alsoe cattle returne thereby to their former or pristin state & the book to receave no loss.[60]

The saintly relic to which Mageohegan refers here is most likely the seventh-century codex known as the Book of Durrow (Dublin, Trinity College MS A 4. 5), which contains an illuminated copy of the Gospels. Its association with Columba, the supposed founder of the Durrow monastery, evidently led to a perception that the codex had a kind of salvific power capable of healing bovine affliction.

3 Conclusion

Although Bede's anecdote concerning the preparation of an antivenin in the first book of his *Ecclesiastical History* might at first be viewed as a mere passing remark set in the context of a much lengthier account of the conversion of the English people to Christianity, it nonetheless serves to shine a spotlight on the rich textual culture of medieval Ireland, as well as on the close connection

Ireland were ruined, save only those parts written in Columba's own hand which were miraculously protected as if kept in a cover or case'.

59 Bitel, *Isle of the Saints* 186–187, citing Anderson, A.O. – Anderson M.O. (eds.), *Adomnán's Life of Columba* (London: 1961) 336–338.

60 Murphy, *The Annals of Clonmacnoise, being Annals of Ireland from the Earliest Period to A.D. 1408* 95–96.

between spiritual and physical modes of healing in the medieval world more broadly. This discussion has aimed to highlight some aspects of these themes by examining the representation of water as a curative element in medieval Irish textual sources. As a symbol of spiritual cleansing closely associated with baptismal practices, hydrological imagery played a central role in early Irish hagiographical, penitential and narrative sources that sought to demonstrate the power of Christianity and of the saints who were its earthly representatives. Such symbolic representations of water were never far removed, however, from its physical significance as an element used for cleansing the body of impurities. The nexus between these literal and symbolic portrayals is perhaps best illustrated by material drawn from the comparatively unexplored realm of Irish-language medical writing, where liturgical practices often melded seamlessly with practical advice concerning the administration of natural substances for healing purposes. It is clear from this survey that there is still much to be discovered with regard to how the full range of manuscript sources extant from medieval Ireland might inform our understanding of the intellectual history of that region and its relationship to other contemporary cultures and literary traditions.

Bibliography

Manuscripts

Dublin, Trinity College MS A 4. 5.

London, British Library MS Harley 546. British Library Digitised Manuscripts. Accessed November 3, 2023. www.bl.uk/manuscripts/FullDisplay.aspx?ref=Harley_MS_546.

London, British Library MS Harleian 5280.

London, British Library MS Harley 585. British Library Digitised Manuscripts. Accessed November 9, 2023. https://www.bl.uk/manuscripts/Viewer.aspx?ref=harley_ms_585_f001r.

London, British Library MS Regius 12 D XVII.

Royal Irish Academy. MS 23 N 29

Royal Irish Academy. MS 24 B 3

Primary Sources

Anderson A.O. – Anderson M.O. (eds.), *Adomnán's Life of Columba* (London: 1961).

Best R.I. (ed.), "Comhrag Fir Diadh & Chon Chulainn. Táin Bó Cúailnge", *Zeitschrift für celtische Philologie* 10 (1915) 274–308.

Bieler L. (ed. and trans.), *The Irish Penitentials* (Dublin: 1975).

Bieler L. (ed.), *The Patrician Texts in the Book of Armagh* (Dublin: 2004).

Binchy D.A. (ed. and trans.), "Bretha Crólige", *Ériu* 12 (1938) 1–77.
Byrne A. *Otherworlds. Fantasy & History in Medieval Literature* (Oxford: 2016).
Colgrave B. – Mynors R.A.B. (eds. and trans.), *Bede's Ecclesiastical History of the English People* (Oxford: 1969).
de Smedt C. – de Backer J., *Acta Sanctorum Hiberniae ex Codice Salmanticensi* (Edinburgh – London: 1888).
Gray E. (ed. and trans.), *Cath Maige Tuired. The Second Battle of Mag Tuired*, Irish Texts Society 52 (London: 1982).
Mulchrone K. (ed. and trans.), *Bethu Phátraic* (Dublin: 1939).
Murphy D. (ed.), *The Annals of Clonmacnoise, being Annals of Ireland from the Earliest Period to A.D. 1408, translated into English A.D. 1627 by Conell Mageohegan* (Dublin: 1896).
Newlon Radner J. (ed. and trans.), *Fragmentary Annals of Ireland* (Dublin: 1978).
O'Rahilly C. (ed. and trans.), *Táin Bó Cúalnge from the Book of Leinster* (Dublin: 1967).
Plummer C. (ed. and trans.), *Vitae Sanctorum Hiberniae* 2 vols. (Oxford: 1910) II.
Vogels H.J. (ed.), *S. Aurelii Augustini episcopi Hipponensis De doctrina christiana libros quattuor, Florilegium patristicum* 24 (Bonn: 1930).

Secondary Sources

Arbesmann R., "The Concept of 'Christus Medicus' in St. Augustine", *Traditio* 10 (1954) 1–28.
Arthur C., *"Charms," Liturgies and Secret Rites in Early Medieval England* (Woodbridge: 2018).
Bitel L.M., *Isle of the Saints: Monastic Settlement and Christian Community in Early Ireland* (Cork: 1990).
Borsje J. – Kelly F., "The Evil Eye and Early Irish Literature and Law", *Celtica* 24 (2000) 1–39.
Brown M.P., "Hagiography or History? Early Medieval Approaches to Establishing Origin and Provenance for Insular Copies of Scripture", in Moss R. – O'Mahony F. – Maxwell J., (eds.), *An Insular Odyssey: Manuscript Culture in Early Christian Ireland and Beyond* (Dublin: 2017) 24–41.
Campbell J., "The Debt of the Early English Church to Ireland", in Ní Chatháin P. – Richter M. (eds.) *Irland und die Christenheit: Bibelstudien und Mission* (Stuttgart: 1987) 332–346.
Carey J., *The Mythological Cycle of Medieval Irish Literature* (Cork: 2018).
Davies M.T., "Kings and Clerics in Some Leinster Sagas", *Ériu* 47 (1996) 45–66.
Davies W., "The Place of Healing in Early Irish Society", in Ó Corráin D. – Breatnach L. – McCone K. (eds.,), *Sages, Saints and Storytellers: Celtic Studies in Honour of Professor James Carney* (Maynooth: 1989) 43–55.
Demaitre L., *Medieval Medicine: the Art of Healing, from Head to Toe* (Santa Barbara, CA: 2013).

eDIL 2019: An Electronic Dictionary of the Irish Language, based on the Contributions to a Dictionary of the Irish Language (Dublin: 1913–1976). Accessed November 9, 2023. www.dil.ie.

Esposito M., "The Sources of Conchubranus's Life of St. Monenna", *English Historical Review* 35.137 (1920) 71–78.

Gwynn J. – Purton W.J., "The Monastery of Tallaght", *Proceedings of the Royal Irish Academy* 29 (1911/1912) 115–179.

Hayden D., "Three Versified Medical Recipes Invoking Dían Cécht", in Ahlqvist A. – O'Neill P. (eds.), *Fír fesso: A Festschrift for Neil McLeod,* Sydney Series in Celtic Studies 17 (Sydney: 2018) 107–123.

Hayden D., "Attribution and Authority in a Medieval Irish Medical Compendium", *Studia Hibernica* 45 (2019) 19–51.

Hayden D., "A Sixteenth-century Irish Collection of Remedies for Ailments of the Male Reproductive Organs", *Celtica* 33 (2021) 248–276.

Hayden D., "The Context and Obscure Language of Medical Charms in a Sixteenth-century Irish Remedy Book: Four Case Studies", in Kobel C. (ed.), *Obscuritas in Medieval Irish and Welsh Literature* (Dublin: forthcoming).

Herbert M., *Iona, Kells, and Derry: The History and Hagiography of the Monastic Familia of Columba* (Dublin: 1988).

Hogan C., *Republic of Shame: Stories from Ireland's Institutions of "Fallen Women"* (Dublin: 2019).

Hughes K., "Evidence for Contacts between the Churches of the Irish and English from the Synod of Whitby to the Viking Age", in Clemoes P. – Hughes K. (eds.), *England Before the Conquest: Studies in Primary Sources Presented to Dorothy Whitelock* (Cambridge: 1971) 49–67.

Jolly K., *Popular Religion in Late Saxon England: Elf Charms in Context* (Chapel Hill, NC: 1996).

Logan P., *The Holy Wells of Ireland* (Gerrards Cross, Bucks: 1980).

Lucas A.T., "Washing and Bathing in Ancient Ireland", *Journal of the Royal Society of Antiquaries of Ireland* 95.1/2 (1965) 65–114.

MacKinney L.C., "Animal Substances in materia medica: a Study in the Persistence of the Primitive", *Journal of the History of Medicine and Allied Sciences* 1.1 (1946) 149–170.

MacLeod Bank M., "A Hebridean Version of Colum Cille and St Oran", *Folklore* 42.1 (1931) 55–60.

Miller A., "Superstitions in the North-west Highlands and Islands of Scotland, especially in Relation to Lunacy", *Proceedings of the Society of Antiquaries of Scotland* 4 (1860–62) 251–288.

Murphy G., *Saga and Myth in Ancient Ireland* (Dublin: 1955).

Nic Dhonnchadha A., "Michael Casey's Medical Transcripts in Gilbert MS 147", *Éigse* 60 (2019) 112–113.

Nic Dhonnchadha A., "An Irish Medical Treatise on Vellum and Paper from the 16th Century", in Ó Macháin P. (ed.), *Paper and the Paper Manuscript: A Context for the Transmission of Gaelic Literature* (Cork: 2019) 111–125.

Nicholls K., "Tobar Finnmhuighe – Slán Pádraig", *Dinnseanchas* 2 (1966–1967) 97–98.

Ó Cathasaigh T., "*Cath Maige Tuired* as Exemplary Myth", in de Brún P. – Ó Coileáin S. – Ó Riain P. (eds.), *Folia Gadelica: Essays Presented by Former Students to R. A. Breatnach* (Cork: 1983) 1–19.

O'Donovan J. – O'Curry E. – O'Flanagan M., *Letters Containing Information Relative to the Antiquities of the County of Clare: Collected during the Progress of the Ordnance Survey in 1839, with Letters and Extracts Relative to Ancient Territories in Thomond, 1841*, Typescript edition, 3 vols. (Bray, Co. Wicklow: 1928); text available online at Clare County Library. Accessed November 3, 2023. https://www.clarelibrary.ie/eolas/coclare/history/osl/moyarta2_st_cradanns_well.htm.

Ó Floinn R., "Insignia Columbae I", in Bourke C. (ed.), *Studies in the Cult of Saint Columba* (Dublin: 1997) 136–161.

O'Grady S. (ed. and trans.), *Catalogue of Irish Manuscripts in the British Library [Formerly British Museum, Volume 1* (London: 1926; repr. Dublin: 1992).

Rawcliffe C., *Medicine and Society in Later Medieval England* (Stroud: 1995).

Rawcliffe C., *Leprosy in Medieval England* (Woodbridge: 2006).

Ray C., *The Origins of Ireland's Holy Wells* (Oxford: 2014).

Rempt M., "'And Straightaway the Fountain of Her Blood Dried Up' (Mark 5:29): Constructing a Template for Late Medieval Irish Obstetric Charms" (Unpublished R. M. A. dissertation, Utrecht University: 2019).

Skemer D.C., *Binding Words: Textual Amulets in the Middle Ages* (University Park, PA: 2006).

Storms G., *Anglo-Saxon Magic* (Gravenhage: 1948).

Swift C., "Tírechán's Motives in Compiling the *Collectanea*: an Alternative Interpretation", *Ériu* 45 (1994) 53–82.

Ulster Society for Medieval Latin Studies, "The Life of Saint Monenna by Conchubranus: Part I", *Seanchas Ardmhacha: Journal of the Armagh Diocesan Historical Society* 9.2 (1979) 250–273.

Wycherley N., *The Cult of Relics in Early Medieval Ireland*, Studies in the Early Middle Ages 43 (Turnhout: 2015).

CHAPTER 3

Bathing in Medieval Europe

Elisa Ramazzina

The purpose of this chapter is to provide an overview on the evolution of balneology in medieval Europe through the analysis of a selection of medical texts, including English, Italian and German treatises, and poems.[1] It will show that two phases can be distinguished in medieval hydrotherapy. In the early Middle Ages, corresponding to the first phase, water plays a fundamental role in patient care, despite references to baths being scattered and bathing being always part of more complex remedies, which included drinking potions and the application of salves. During this phase, the instructions for the preparation of medicinal waters and their application are very brief and general. The chapter will focus mainly on the second phase, corresponding to the late Middle Ages, when bathing started playing an increasingly vital role in healing practices, so that entire sections of medical texts were devoted to remedies involving baths, and specific treatises on their curative virtues and on the preparation of waters were produced. The therapeutic properties of baths[2] and their role in helping the patients regain their psycho-physical wellbeing will thus be examined in relation to the elemental qualities of water and to other remedies, including potions, diets, and exercise, thus demonstrating that the increasing relevance of balneotherapy in the late Middle Ages is directly connected to the elemental and humoral theories that permeated the medieval worldview.

The importance of hydrotherapy has been known since Antiquity, when Hippocrates and Galen, who deeply influenced medieval medicine,[3] discussed its benefits, describing the chemical and organoleptic qualities of thermal water and the therapeutic effects of hot and cold baths on the human body. Hippocrates claimed that all human diseases were due to an imbalance of the

1 This publication is part of the results of the *Cleansing Water: Water and Baptism in Old English Poetry* project that has received funding from the European Union's Horizon 2020 Research and Innovation Programme under the Marie Skłodowska-Curie grant agreement No. 744615.

2 For the therapeutic properties of baths in the medieval Irish tradition see Hayden D., "From the 'King of the Waters' to Curative Manuscripts: Water and Medicine in Medieval Irish Textual Culture" 75–98.

3 For the history and sources of early medieval medicine see Wallis F. (ed.), *Medieval Medicine: A Reader* (Toronto: 2010) 3–46.

 | DOI:10.1163/9789004696501_005

bodily humours and that to cure the body physicians had to restore that balance. His prescriptions, foundational to medieval *regimina sanitatis*, included changes in habits and environment, bathing, perspiration, and exercise.[4] Unsurprisingly, bathing was considered an important part of medieval medical practice and was related to the elemental and humoral theories that dominated the worldview of the time.[5] Since diseases were thought to be caused by an excess of bodily fluids, the element 'water', through bathing, was considered as fundamental to restore humoral balance.

Isidore of Seville was one of the first medieval authors to appreciate different kinds of water according to their chemical composition and their different qualities, recognising their therapeutic virtues and relating them to baths. In his *Etymologies* he explains how aluminous and sulphurous waters are naturally hot, associating them with another element – fire. According to Isidore, cold water, traveling from its source, draws the heat from the fire contained in the minerals it carries, which are in turn fuelled by the movement itself:

> Aquarum naturae diversitas multa est: aliae enim salis, aliae nitri, aliae aluminis, aliae sulphuris, aliae bituminis, aliae curam morborum adhibentes. [...] Iam vero in multis locis aquae manant perpetim ferventes, tanta vi ut balnea calefaciant. Quaedam enim terrae sunt quae multum sulphuris et aluminis habent. Itaque cum per venas calentes aqua frigida venit, vicino sulphuris calore contacta excandescit, nec talis ab origine effluit, sed permutatur dum venit. Sulphur enim alumenque secum ferunt aquae; utramque materiam igne plenam minimisque motibus incalescentem. (XIII:XIII:1;11).[6]

> There is a great diversity in the qualities of water, for some water is salty, some alkaline, some with alum, some sulfuric, some tarry, and some

4 Gianfaldoni S. – Tchernev G. – Wollina U. – Roccia M.G. – Fioranelli M. – Gianfaldoni R. – Lotti T., "History of the Baths and Thermal Medicine", *Open Access Macedonian Journal of Medical Sciences* 5.4 (2017) 566–568, 567; see also Galen, *On Hippocrates. On the Nature of Man*, eds. and trans. W.J. Lewis – J.A. Beach, Ancient Medicine /Medicina Antiqua website, especially part one. Accessed November 3, 2023. https://www.ucl.ac.uk/~ucgajpd/medicina%20antiqua/tr_GNatHom.html.

5 For medieval elemental and humoral theories see Cesario – Magennis – Ramazzina, "Medieval Ideas on the Elements", in Cesario M. – Magennis H. – Ramazzina E. (eds.), *The Four Elements in the Medieval World: Interdisciplinary Perspectives. 'Earth'* (Leiden – Boston, MA: forthcoming).

6 Isidore, *Etymologiarum sive Originum*, ed. Lindsay, Digitised Edition @University of Chicago, 2010; (XIV.I.1). Accessed May 14, 2024, https://penelope.uchicago.edu/Thayer/L/Roman/Texts/Isidore/13*.html#12.

> containing a cure for illnesses. […] Indeed, in many places waters now flow with perpetual boiling of such vigor that they make baths hot. There are some lands that have a great deal of sulfur and alum. And thus, when the cold water comes through hot passages, it is affected by the adjacent heat of the sulfur and is itself heated; it does not flow from its source at such a temperature, but is changed as it travels. Indeed, waters may carry sulfur and alum with them; both materials are full of fire and grow hot with the slightest movement.[7]

In Book 15, Isidore connects bathing to the re-establishment of psychological wellbeing explaining the etymology of 'bath' as 'the lifting of sorrow' as it helps the spirit in getting rid of anxiety. In his discussion of the structure of bathhouses, Isidore relates bathing to exercise, food, and drinks in a process restoring psycho-physical wellbeing that also includes massaging the body with ointments, that will be found in late medieval *regimina sanitatis*:

> Thermas appellatas quod caleant; Graeci enim θερμὸν calorem vocant. Balneis vero nomen inditum a levatione maeroris; nam Graeci βαλανεῖον dixerunt, quod anxietatem animi tollat. Haec et gymnasia dicuntur, quia ibi athletae uncto corpore et perfricato manibus exercitantur; nam γυμνάσιον Graece, Latine exercitium dicitur. […] Propina Graecus sermo est, quae apud nos corrupte popina dicitur: est autem locus iuxta balnea publica, ubi post lavacrum a fame et siti reficiuntur. Unde et propina et propinare dicitur. Πείνα enim Graece famem significat, eo quod hic locus famem tollat. (XV:11:39–42).[8]

> Hot baths (*thermae*) are so called because they are warm, for the Greeks call heat ζερμον. Baths (*balneum*) are assigned their name from the idea of the lifting of sorrow, because the Greeks called it βαλανειον (cf. βαλλειν, 'cast away'; ανια, 'grief'), since it takes away one's anxiety of spirit. Gymnasiums (*gymnasium*) are so called because there athletes are trained, with their bodies anointed and massaged, for γυμνασιον in Greek means 'training' in Latin. […] Propina is a Greek word, which now among us has been corrupted to popina. It is a place next to a public bath where after

7 All translations from the *Etymologies* are taken from Barney S.A. – Lewis W.J. – Beach J.A. – Berghof O. (trans.), *The Etymologies of Isidore of Seville* (Cambridge: 2006) 276–277.

8 Isidore, *Etymologiarum sive Originum*, Digitised Edition @University of Chicago, 2010; (XIV:1:1); accessed May 14, 2024, https://penelope.uchicago.edu/Thayer/L/Roman/Texts/Isidore/15*.html#2 .

bathing one may be refreshed of one's hunger and thirst. From this are named both propina and propinare ('to drink a toast'), for πεινα in Greek means 'hunger' – because this place relieves hunger.[9]

1 Early Medieval Balneology

The first phase of balneotherapy, in which references to baths are scattered and always accompanied by other remedies, reflects the nature of the early medieval medical corpus, with a predominance of pharmacological writings, i.e., collections of remedies arranged either alphabetically or according to ailments affecting various bodily parts, following the Hippocratic order *de capite ad pedem*.[10] The eighth-century German *Lorscher Arzneibuch*,[11] for example, contains drug recipes and dietary advice in Latin, and bathing is often suggested as part of other treatments to enhance the healing power of salves and potions. The structure of the remedies is quite repetitive and schematic, with a list of ingredients usually followed by the instructions for their preparation. Book II, for instance, prescribes a remedy against body spots and leprosy that is very specific regarding the quantities of the ingredients to be added to water and when they must be added. Dried mint and sulphur should be reduced to a very fine powder and rubbed on the patient's skin when he starts sweating so that the pores can better absorb healing substances: 'Mentae siccae ˜ 1, sulphuris uiui ⁀ ½ in puluerem mollissimum redigis et in balneo cum sudare coeperis, super cutem aspargis et ex eo puluere perfricabis […]'.[12] (Crush 1 ounce of dried mint and ½ ounce sulfur to very fine powder, sprinkle it in the bath at the beginning of sweating on the skin and rub it in vigorously […]').[13] Similarly, drinking a potion made of a mixture of herbs and white wine twice a day while bathing will help urinary stones dissolve. Once again, the quantities of the ingredients for the preparation are specified in great detail, as is when the potion and the bath should be administered, i.e., morning and evening (IV:92).

Similarly, in early medieval England baths were considered as constituents of leechdoms, which also included a series of healing practices to cure a variety

9 Barney – Lewis – Beach – Berghof, *The Etymologies* 307–308.
10 See also Beccaria A., *I codici di medicina del periodo presalernitano* (Rome: 1956) 29–31.
11 Bamberg, Staatsbibliothek medicinalis MS 1.
12 Stoll U. (ed. and trans.), *Das Lorscher Arzneibuch: Ein medizinisches Kompendium des 8. Jahrhunderts* (*Codex Bambergensis Medicinalis 1*) (Stuttgart: 1992) II:176, 202.
13 Translations from the *Lorscher Arzneibuch* are mine.

of diseases.[14] Bathing water was often prepared with different ingredients (mostly herbs) that varied depending on the symptoms of the patient. In the first book of 'Bald's Leechbook',[15] one of the earliest medical texts produced in England, bathing is prescribed quite often and for a variety of conditions in relation to other remedies, that are organised according to the usual head-to-foot order. Here, balneotherapic instructions are usually brief, simple and indicate whether individual bodily parts or the whole body should be immersed and how to prepare the medicinal bath: the type of water, sometimes its temperature, or the liquids that should be employed as an alternative to water, along with other ingredients. For instance, to cure sore knees the patient has to drink and bathe in the same concoction of woodwax, hedgerife and ale; if the pain persists, he should bathe in a decoction made with different herbs and water: 'Ƿiþ cneoƿ ƿærce· ƿudu ƿeaxe· ⁊ hegerife gecnuƿa þa togædere ⁊ do ón ealu læf licgean neahtenne sele him þæt þonne drincan beþe mid ⁊ lege ón· Ƿiþ þon gíf cneoƿ sar sie· gením ƿealƿyrt ⁊ clufþung· reade nettan aƿyl ón ƿætere beþe mid' ('For knee pain; pound together woodwax and hedgerife, and put into ale; let it lie for a night, give him then that to drink, bathe with it, and lay it on. In case that a knee be sore, take wall wort and cloffing, and red nettle, boil in water, bathe therewith').[16] The remedy for elephantiasis contains a series of detailed instructions for the preparation of two different medicinal baths: one prescribes the immersion or fumigation of only one part of the body, while the other seems to suggest the immersion of the whole body. Firstly, the patient should sit on a somewhat ancestor of the modern bidet (*bydene*), a keel-shaped bowl in which a mixture of boiled water and herbs has been poured. Subsequently, similarly to the remedy for sore knees, he should drink a potion while bathing in the same liquid. Specific ingredients, which include a variety of herbs (such as helenium, ivy, mugwort, henbane, mallow, betony, red nettle, lupin, and sage, among others), should be boiled in foreign ale when making the decoction. To cure the ailment, the patient must consume the decoction and pour it into a basin, avoiding any touch with the steam from the bath as this could change its qualities and jeopardise their recovery (XXXII.4).

Baths were also associated with specific diets according to the patient's symptoms as the *Lacnunga*,[17] a collection of remedies in Old English and

14 For the evolution of balneotherapy in medieval England see Ramazzina E., "Therapeutic Baths in Medieval English Medicine", in Hayden D. – Baccianti S. (eds.), *Medicine in the Medieval North Atlantic World: Vernacular Texts and Traditions* (Turnhout: 2024).

15 Contained in London, British Library, Royal MS 12 D XVII (tenth century).

16 Cockayne T.O. (ed. and trans.), *Leechdoms Wortcunning, and Starcraft of Early England, vol. 2* (London: 1864–1886; reprinted Cambridge: 2012) XXIV, 66–67.

17 Contained in London, British Library Harley MS 585 (late tenth-early eleventh century).

Latin, shows. For example, against heart ache the patient must be treated with a stone bath and should eat southern radish with salt: 'Ƿið heortece [...] þonne stan bæð ⁊ on þem ete suþerne rædic mid sealte þy mæg seo ƿund ƿessen gehealed [...]' (For heart ache [...] then work him a stone bath, and in it let him eat southern radish with salt; by that may the wound be healed. [...]). (Remedy 115).[18] Against a swollen stomach, *Leechbook II* (IIII)[19] suggests a combination of remedies in which bathing combines with dietary advice. The patient should eat easily digested foods, such as game and specific vegetables, and drink white wine; this must be associated with draining, astringent, and soothing ointments, which have to be *þenda* ('moist'), referring to one of their elemental qualities according to humoral theory. Baths feature twice in this remedy: firstly, combined with the application of a salve on the patient's body; secondly, a decoction must be prepared with apples, wine, pepper and honey, which has to be both ingested by the patient and used for their bath.

Meanwhile, in ninth-century Italy, in Salerno, an important medical school destined to revolutionise medieval European medicine, and consequently also balneotherapy, had been founded.[20] Due to the complex political history of the region – a Roman colony, occupied by Goths, Byzantines, and, in the period of the medical school, firstly by the Langobards and subsequently by the Normans – the school benefitted from a multi-cultural environment, with local communities living with Jewish, Greek, and Arabic ones.[21] Furthermore, the presence of a busy international port that brought a large number of traders, pilgrims, travellers, and scholars to the city, promoted transfer of knowledge and facilitated scholarly debate.[22] The school, which attracted both local and foreign students and physicians, had a solid reputation by the tenth century and flourished between the eleventh and thirteenth centuries, peaking in its manuscript production in the twelfth. It turned out to be the most significant

18 Cockayne, *Leechdoms, vol. 3* 115 74–75.

19 Cockayne, *Leechdoms, vol. 2* IIII 182–183.

20 The Salernitan school was a medical institution founded in the southern Italian city of Salerno. Its foundation date has been the object of scholarly debate for a long time. For a summary of this scholarly debate, see Ramazzina, "Therapeutic Baths". Nevertheless, starting from the ninth century, in Salerno gathered what Green has defined 'an informal community of masters and pupils who [...] developed more or less formal methods of instruction and investigation' (Green M.H. (ed. and trans.), *The Trotula. A Medieval Compendium of Women's Medicine* (Philadelphia: 2001) 9), which then flourished between the eleventh and thirteenth centuries.

21 Drell J.H., *Kinship and Conquest: Family Strategies in the Principality of Salerno during the Norman Period, 1077–1194* (Ithaca, NY: 2002) 3.

22 Skinner P., *Health and Medicine in Early Medieval Southern Italy* (Leiden – Boston, MA: 1997) 131.

hub for the introduction and dissemination of Greek and Arabic medicine into western Europe. There, Greek texts, including the works of Hippocrates, Galen, and Dioscorides's plant-lore from the library of Montecassino were translated into Latin.[23] Arabic medical-lore, available through contact with Sicily and North Africa, was integrated into medical practice, thus placing the school at a crossroads between the materiality of the local southern Italian culture, and the scientific knowledge, and the empiricism inherited from the Greek and the Islamic worlds. As noted by Green, around the turn of the twelfth century, due to shared culture and language, the texts produced at Salerno and Montecassino travelled quickly to Normandy and Anglo-Norman England, which played a fundamental role in the dissemination of the 'Salernitan corpus', with up to twenty-five percent of the manuscripts being produced there.[24] The influence of the Salernitan school on European medicine is universally recognised, not least in being 'the forerunner of modern university medical schools',[25] with the teaching of medicine based on the medical curriculum called *articella* developed by the masters of Salerno.[26] Moreover, as illustrated below, the school gave a fundamental impulse to the systematisation of balneotherapic practices as related to diet, exercise, and psycho-physical wellbeing, which progressed in the following centuries with the development and the expansion of university medical schools.

2 An Intermediate Phase in Medieval Balneology

One of the products of the Salernitan school, the so-called *Trotula*, marks an intermediate phase in medieval balneotherapy. This twelfth-century compendium on women's medicine, which circulated widely throughout medieval Europe,[27] deals mainly with gynaecological problems, such as menses, (in)

23 See also Smith V., *Clean. A History of Personal Hygiene and Purity* (Oxford: 2008) 152.

24 Green M.H., "Salerno on the Thames: The Genesis of Anglo-Norman Medical Literature", in Wogan-Browne J. (ed.), *Language and Culture in Medieval Britain: The French of England c.1100-c.1500* (Woodbridge: 2009) 220–232, 221.

25 de Divitiis E. – Cappabianca P. – de Divitiis O., "The 'Schola Medica Salernitana': The Forerunner of the Modern University Medical Schools", *Neurosurgery* 55.4 (Oct 2004) 722–744, 739.

26 For an overview on the main authors and texts of the school, see Jacquart D. – Paravicini Bagliani A., *La scuola medica salernitana: gli autori e i testi* (Florence: 2007).

27 Green, *The Trotula* XIV. The *Trotula* was for a long time erroneously attributed to a female physician named Trota, practising in Salerno; however, as Green observes, it was composed by at least three distinct authors with different perspectives on female diseases (Ibidem XII).

fertility, childbirth, as well as cosmetics, and consisted of three books – *Liber de sinthomatibus mulierum* (Book on the Conditions of Women), *De curis mulierum* (On Treatments for Women), *De ornatu mulierum* (On Women's Cosmetic) – that, after circulating independently for a period, were bound together towards the end of the twelfth century.[28] In *Trotula*, baths are still considered as part of wider treatments to heal specific gynaecological conditions and for cosmetic purposes but it is clear how they are given more and more importance: as Green observes, 'The frequency with which medicinal and cosmetic baths are prescribed in the *Trotula* suggests that both public and private baths in Salerno would have been put to extensive use'.[29] Baths are indeed prescribed for a variety of conditions, including haemorrhoids; male and female barrenness and strangury; vaginal prolapse; ailments of the womb, such as pain, wind, ulcerations, and prolapse; disorders related to childbirth, such as excessive bleeding, difficulties in the delivery, and the preservation of the foetus. Balneotherapy is also used as part of the regimen for the care of new-borns and for cosmetic purposes, as to make women's skin softer. The frequency of the prescription of baths in *Trotula* attests to the relevance of balneotherapy at least for the cure of gynaecological ailments within the Salernitan school.[30]

Elemental and humoral theories, at the basis of the treatments, are mentioned more explicitly than in the early Middle Ages, and the primary qualities of humours and ingredients are part of diagnoses and therapies. For example, to facilitate conception, phlegmatic, fat women should be treated with baths prepared with hot herbs and rainwater. The women must lie therein until they sweat to favour the absorption of active ingredients through dilated skin pores, and, after bathing, they must rest, well covered in bed. Particular attention should be paid to refreshments as they should be fed with warm foods and moderately drink wine as they are meant to counterbalance the excess of phlegm to restore health:

> [...] Si ergo fuerit fleumatica et grossa, faciamus ei balneum de aqua maris et salsa ad moderationem cum aqua pluuiali inponimus, scilicet iuniperum, nepitam, pulegium, laureolam, absinthium, arthimesiam, ysopum, et huiusmodi herbas calidas. In hoc balneo faciat moram quousque satis sudet, post recipiatur in lecto caute, et bene sit cooperta, et si alium

28 Ibidem.

29 Ibidem 6.

30 Despite its disputed authorship, as Green notes, '[...] both the genesis and the later uses of the *Trotula* are shown to be intimately tied to the milieus that shaped the production of medical knowledge and the delivery of healthcare in twelfth-century Salerno and in medieval Europe more broadly' (ibidem 1).

> cibum pecierit, detur ei in principio rosata nouella. Detur etiam cibus bonus et sanus et calidus, et uinum optimum modicen sumptum (*De curis mulierum*: 142).

> If she is phlegmatic and fat, we should make her a bath of seawater, moderately salty, with rainwater. We put in [various herbs], that is, juniper, catmint, pennyroyal, spurge laurel, wormwood, mugwort, hyssop, and hot herbs of this kind. In this bath she should stay until she sweats sufficiently; afterward let her be received in bed carefully and let her be well covered. And if she desires some food, let her at first be given rosata novella. Also, let her be given good and wholesome and warm food, and wine of the best quality taken moderately.

The fumigations should also be repeated three or four times a day to be effective. The same remedy can also be used to help men with a cold nature conceive; in this case, however, instead of fumigations, they should be given warm strengthening medicines to re-establish their humoral balance.[31]

In the *Liber de sinthomatibus mulierum*, some treatments for uterine prolapse are listed. Firstly, possible causes are carefully reported: beyond the strain of childbirth, the two main reasons are the weakening of the uterus ligaments and an excess of cold humours, both of which are due to contact of the vagina with something cold – air entering the orifices, a bath in cold water, or sitting on a cold stone (Ch. 52).[32] This demonstrates that bathing in cold water can be extremely harmful for women. A series of cures is subsequently prescribed: firstly, aromatic herbs are applied to the patient's nose; then fumigations with fetid substances are applied; the belly is anointed and covered with cooked grain, and the womb is restored manually. Afterwards, the woman must immerse herself in a decoction of bark, fruit, lentils, and nuts. Finally, a steam bath is prepared on live coals for the fumigation of the vagina. The mention of Dioscorides' medical authority testifies to the effectiveness of the remedy: since uterine prolapse is caused by an excess of cold humours, hot baths and fumigations are undoubtedly effective cures to restore bodily balance and health:

> Postea mulier intret aquam in qua cocta sint balaustie, rose, cortices mali granati, galle, sumac, mirtilli, glandes et folia et cortex quercus, et nuces cypressi, et lenticula.

31 Ibidem 120–123.

32 Ibidem 86–87.

> Deinde fiat eis stupha que multum ualet, quia Dioscorides precepit eis stupham facere de buxo posito in olla super carbones uiuos, et mulier sedeat desuper cooperta et fumum recipiat interius (*Liber de sinthomatibus mulierum*: 56–57).[33]

> Afterward, let the woman enter water in which there have been cooked pomegranate, roses, rind of pomegranate, oak apples, sumac, myrtleberries, the fruit and leaves and bark of oak, and juniper nuts, and lentils.
>
> Then let there be made for them a steambath, which works very well. For Dioscorides prescribes that there be made for them a steambath of box-wood placed in a pot upon live coals, and let the woman, covered on top, sit on it, and let her receive the smoke inside [her vagina].

The list of remedies concludes with dietary advice, which includes the consumption of cold and styptic foods, the avoidance of pungent ones, and the recipe of a potion.

As it is evident from the examples analysed, the treatment of baths in *Trotula*, marks a transition into a new phase of balneology. If, on one hand, baths are still components of complex treatments also entailing the use of potions, the application of ointments, salves, compresses, etc., while on the other, the use of balneotherapy is much more frequent than in the early Middle Ages, as its restorative function of humoral balance is gradually becoming more explicit. Moreover, the association between baths and diet already present in the first phase, in *Trotula* is based on the four humours and is becoming part of a sort of 'balneotherapic routine'.

3 Late Medieval Balneology

In the Late Middle Ages, corresponding to the second phase of medieval balneology, bathing started playing a vital role in medical practice so that entire sections of medical treatises were devoted to remedies and specific treatises on the therapeutic virtues of water were produced. Bathing practices were also related to elemental and humoral theories more explicitly; advice on the appropriate season to bathe according to temperaments, and specific diets to enhance the effects of balneotherapy depending on different ailments were included in treatises.

33 Ibidem 87–89.

The revival of thermalism, which started in Italy in the twelfth century and was devoted to the study of mineral components and therapeutic virtues of waters, influenced greatly the development of this specialised literature. As noted by Zuccolin,[34] between the twelfth and the sixteenth centuries, thanks to Italian medicine, Europe witnessed a progressive 'medicalisation of thermalism', in that, thermal baths from simple *recreatio corporis* (physical recreation) became indispensable therapeutic tools for *preservatio and curatio sanitatis* (health's preservation and treatment) and the curative properties of different thermal springs started to undergo careful mineral analysis. In turn, thermalism was closely connected with elemental and humoral theories and, as will be shown below, balneotherapic literature insists on how baths are fundamental remedies in restoring humoral balance or can be harmful for some patients, depending on their temperament and therefore on their predominant humour. As explained by Nicoud,[35] water, whether consumed as a drink or in the form of baths, is one of the components of dietetics as conceived by Arab Galenism. As illustrated in the chapters dedicated to baths in health *regimina*, together with drinks and food, waters are part of the six non-naturals,[36] the harmony of which is essential for the preservation of health and for its restoration. Indeed, water is used to cook foodstuffs; drinking it quenches thirst and, in the form of saliva, water plays a fundamental role in digestion, as it allows food to be reduced into a uniform bolus that can be thus swallowed. Subsequently, it allows digested foods to be sent to different parts of the body. Used for bathing, water substitutes for exercise or enhances its effects, toning the body, favouring the expulsion of the residues of the third digestion,[37] which

---

34 Zuccolin G., "Ruolo ed evoluzione della balneoterapia nel pensiero scientifico-medico in Italia dal XII al XVI secolo", in Ciardi M. – Cataldi R. (eds.), *Il calore della terra: contributo alla storia della geotermia in Italia* (Pisa: 2005) 98–115, 99.

35 Nicoud M., "Les vertus médicales des eaux en Italie à la fin du Moyen Âge", in Guérin-Beauvois M. – Martin J.M. (eds.), *Bains curatifs et bains hygiéniques en Italie de l'Antiquité au Moyen Âge*, Collection de l'École française de Rome (Rome: 2007) 321–344, 332–334.

36 According to Galen, health depends not only on humoral balance, but also on external factors, that is, on the so-called six non-naturals, that is, ambient air; food and drink; sleep and wakefulness; exercise and rest; excretion and retention (for instance of feces, urine, and semen); sadness and joy. Health is the result of the harmony between the humours and the six non-naturals, and any excess would compromise it. See also Pormann P.E., "Medical Conceptions of Health from Antiquity to the Renaissance", in Adamson P. (ed.), *Health: A History* (Oxford: 2019) 43–74, 51.

37 Three steps were thought to be involved in the synthesis of humours during digestion in the late Middle Ages. In the initial stage, food was broken down into chyle by the stomach. After that, the chyle was sent to the liver for 'second digestion', meaning reducing it into its constituent elements and producing the four humours. Phlegm was not thought to have a particular home in the body, despite the fact that it tended to collect in the brain.

occurs sweating. It converts superfluous substances into fluids, and makes humours thinner, favoring their expulsion through the pores of the skin, which are relaxed by its humidity and opened through bathing. For these reasons, physicians must be very vigilant about the characteristics of the waters, which are determined by their origin: stagnant, rain-, sea-, river-, well-waters have different qualities, hence their different applications. Additionally, depending on the ailment, patients should bathe during a precise season[38] or at a certain time of day. The patient should also follow a specific diet avoiding foods with certain elemental qualities according to the superfluous humour, and ingesting others with opposite characteristics.

A case in point is the *Regimen sanitatis Salernitanum*,[39] the first set of rules concerning health, containing advice on diet, exercise, and hygiene, produced in the second half of the twelfth century.[40] It is constituted by a series of epigrams in Latin of varying length, describing humoral imbalances, and matching them with their contrasting therapies, thus initiating a tradition of versified medicine that will continue in the following centuries. Dietary advice includes a variety of herbs, foods, and drinks, with their elemental qualities enhanced by cooking and spicing: light foods as opposed to heavy ones, hot foods to contrast diseases caused by cold humours, and cold foods to counter ailments

---

Blood was thought to be kept in the liver, yellow bile was sent to the gall bladder, and black bile to the spleen. Humours were thought to be the body's food, supplying nutrients to muscles, nerves, sinews, and organs. Vascular blood, a dark, viscous blend of the four humours, was pushed through the veins by the liver. Humours were released during the third digestion to nourish the organs. A range of waste materials were thought to be the end products of the digestion processes: excrement from the first digestion, urine from the second digestion, and a variety of substances from the third digestion, such as skin flakes, sweat, ear wax, and mucus from the nose, etc. See also Luft D., "Uroscopy and Urinary Ailments in Medieval Welsh Medical Texts", *Transactions of the Physicians of Myddfai Society* (2018) 1–14, 3.

38 Depending on the season, each humour becomes predominant: blood in spring, yellow bile in summer, black bile in autumn and phlegm in winter. See also Mamiani M., *Storia della scienza moderna* (Bari: 2002) 63.

39 Henceforth *Regimen*.

40 The *Regimen*, the authorship of which is uncertain, was probably translated from or inspired by Arab sources and spread throughout Europe over the following centuries, so much so that it is contained in more than two hundred extant manuscripts. During its textual transmission, it circulated in various versions: some fragmentary and shorter ones, others containing additions, reaching more than a thousand lines and 278 chapters. In most of its versions, the text is generally divided in five thematic parts concerning the natural components of the body, the non-naturals, things against nature, recipes, and cures against diseases. See de Frutos-Gonzáles V. (ed.), *Flos medicine: Regimen sanitatis Salernitanum* (Valladolid: 2010) 144–146.

caused by hot humours. To relieve humoral excess, diets are combined with other remedies, most frequently with bloodletting, baths, and exercise. The *Regimen* also prescribed a very thorough daily hygienic routine which became so influential, that, as Smith argues, 'it was via these regimens that the morning grooming session was institutionalised in medieval life'.[41] According to the theory of the six non-naturals, to relieve the brain, one should not sleep excessively and wake at the appropriate time, then clear the evacuations (including dejections, nose, throat, and mouth). Subsequently, one should cleanse the night residue from their face:

> Lumina mane manus surgens gelida lavet aqua
> Hac illac modicum pergat, modicum sua membra
> Extendat, crines pectat, dentes fricet : ista
> Confortant cerebrum, confortant cetera membra.
> Lote cale, sta paste vel i frigesce minute.[42]

> In the morning, when you wake up, wash the eyes and the hands in cool water
> Then run a little here and there, a little your limbs
> Stretch out, comb your hair, brush your teeth: these things
> Invigorate the mind and the spirit, invigorate the other bodily parts.
> Heat the bath, and after the meal rest or enjoy the cool for a little while.

Facial and dental hygiene are combined with exercise, as some stretching is suggested to gradually re-mobilise the body after the night's rest. As mentioned in the poem, these activities, due to their restorative power on both body and mind, play a key role in man's psycho-physical wellbeing.[43] Another important component in this routine is bathing with warm water, which must precede breakfast and must be followed by some rest.

Baths are also mentioned in Chapter 93, which lists a series of contraindications for bloodletting: in fact, this practice must be avoided when the person has a cold nature, after bathing, eating and after sexual intercourse. Chapter 103 is devoted to the four seasons: for each season its elemental qualities are mentioned, and a few beneficial activities are suggested to get rid of 'humores

41 Smith, *Clean* 154.

42 Ll. 10–14. Magenta P. (ed. and trans.), *La scuola salernitana ossia precetti per conservar la salute* (Pavia: 1835) 2. The translations are mine.

43 The invigorating function of hand washing is reiterated at Chapter XXIII, where in addition to cleansing, it improves eyesight (ll. 64–66, ibidem 12).

nocuos' ('harmful humours', l. 353) while warning against unhealthy ones.[44] Among these, bathing is advised in spring, being a hot and wet season, to purge the body from medicaments, together with sweating and bloodletting.[45] In summer, which is hot and dry, instead, baths are not recommended as the prevalent humour is red bile; bloodletting should be performed rarely, and the patient should drink moderately and get much rest.[46]

Another important text, inspired by the *Regimen* and consolidating the tradition of versified medicine, is the *De balneis Puteolanis,*[47] a poem in Latin describing the baths of Pozzuoli and Baiae near Naples, written by Peter of Eboli at the end of the twelfth century at the Hohenstaufen's Swabian court. The text is composed of thirty-seven epigrams composed of elegiac couplets, one of which is a prologue, and the final one is a dedication to the emperor. The remaining 35 are devoted to the baths: in each, the bath's name is explained, its therapeutic properties are described, and examples of successful treatments are reported.[48] Although some scholars[49] have defined the *De balneis* as a non-medical text, and whether its author was a physician as well as a man of letters has not been determined yet,[50] the use of a specific medical jargon that includes anatomical terminology and refers to physiological processes and to therapeutic indications,[51] together with the fact that Peter operated at the Swabian court, suggest a possible influence from the Salernitan medical school. Indeed, Peter of Eboli with *De balneis,* innovates medical literature: as De Angelis[52] argues, not only can he be considered the inventor of the 'poetic balneotherapic genre', but he incorporates for the first time Aristotelian medicine into traditional and popular therapeutic treatments, such as

44 Ibidem 60.

45 Ll. 355-359; ibidem 60.

46 Ll. 360–364; ibidem 60.

47 Henceforth *De balneis.*

48 The text had a rich manuscript tradition: there are 28 extant manuscripts of which 13 are richly illustrated, various vulgarisations and numerous printed editions; some versions contain fewer epigrams. See De Angelis T. (ed. and trans.), *Pietro da Eboli: De Euboicis aquis* (Florence: 2018) 32. Scholars have long debated about the dedication, trying to determine whether the dedicatee of the poem is Emperor Frederick II or his father, Henry VI. See Danzi M. "Le terme in Europa tra letteratura e medicina", *Quaderns d'Italià* 22 (2017) 43–56, 49. For a complete introduction on Peter of Eboli, on the *De balneis*, its sources and its manuscript tradition see De Angelis, *Pietro da Eboli* 1–109.

49 See for example Danzi, "Le terme" 49.

50 As De Angelis notes, the title of 'magister' Peter used did not necessarily correspond with a university medical education (De Angelis, *Pietro da Eboli* 29).

51 See also Garcia-Ballester L. – French R. – Arrizabalaga J. – Cunningham A., *Practical Medicine from Salerno to the Black Death* (Cambridge: 2004) 30–31.

52 De Angelis, *Pietro da Eboli* 29–30.

balneotherapy, for informational purposes. Furthermore, as Garcia-Ballester[53] observes, Peter's lexicon is innovative in that he uses a philosophical-natural and medical terminology that arose in the circles of medieval Galenism, one of which was in Salerno.

In the prologue to the poem, the author celebrates thermal water, 'aqua feruens' ('boiling water', l. 5) as a gift from God helping men and declares his intent: the text will praise the names of the different sources and their therapeutic properties.

Inter opes rerum Deus est laudandus in illis
In quibus humane deficit artis opus.
Res satis est dictu mirabilis horrida visu
A tormentorum prouenit ede salus:
Nam que defunctos aqua feruens punit in ymis,
Nec eadem nobis missa ministrat opem
Cetera cum rebus curantur regna syrupis
Balnea, que curant, terra laboris habet.
Vos igitur, quibus est nullius gutta metalli,
Querite, que gratis auxiliantur, aquas
Quarum virtutes et nomina Maxime Cesar,
Presens pro vestra laude libellus habet (I:1–12).[54]

Among the multitude of things God is to be praised for those
in which human action is deficient.
The thing is extraordinary to say, frightening to see:
the healing came from the seat of torment:
in fact, the boiling water, which punishes the dead in the otherworld,
is the same which, sent to us, helps us.
The other kingdoms are cured with medicines and syrups;
the Land of Work has baths that cure.
You, then, who have not even a drop of metal,
look for the waters that help for free.
The therapeutic properties and the names of these waters, Supreme Caesar
are present, to praise you, in this booklet.

53 Garcia-Ballester, *Practical Medicine* 31.

54 De Angelis, *Pietro da Eboli* 114. The translations from the *De balneis* are mine.

The boiling water refers to thermal springs but could also be a reference to the river Styx, the waters of which, according to classical mythology, have a double function: a curative one, in that they make the hero Achilles partially invulnerable, and harmful, as they punish the dead in the otherworld. The author describes water as a wonder ('mirabilis', l. 3) operated by God, who is generous and deserves to be praised for having given such a cure to the inhabitants of the Swabian kingdom. The abundance of thermal baths is described as a peculiarity of the area and accessible to all social classes, including the poor who cannot afford to pay for medicaments (l. 9).

According to Peter, the waters of the Phlegrean Fields are beneficial for several diseases. Most of them can heal various ailments according to humoral theory,[55] as for instance in the description of Sudatorium, where the bath 'Evacuat chymos leue corpus reddit; in ipso / […] Hec aqua languentes restaurat et ylia sanat. / Vlcera desiccat sub cute si qua latent' (II:7–10) ('it expels humours, makes the body light; / [….] This water restores the weak and heals the intestines. / It dries up ulcers, if there are any latent ones under the skin'). However, certain waters are harmful for some conditions as for example the bath at Culma, which is described as Culma nocet sanis, morbida membra iuuat' (XIX:10) ('Culma harms the healthy, benefits the sick limbs').[56] The water of Pratum, then, should be avoided if one is sweaty and warm: 'In sudore madens fugiat pro tempore frigus; / Nec potum sumat, dum sua membra calent' (XIII:11–12) ('those who are drenched in sweat should temporarily avoid the cold; / and not drink this water as long as their limbs are warm').

The narrator, presumably the poet himself, intervenes in various epigrams as an eyewitness to testify to the success of the balneotherapic treatments he describes. For example, the bath at Petra heals the urinary system, and the author reports having seen many people drinking the warm water of the spring and then expel kidney stones: 'Quamplures uidi calidam potare petrosos, / In quibus urina post lapidosa fuit' (VIII:9–10) ('I have seen many kidney stone sufferers drink it warm / in which urine subsequently becomes stony').

Additionally, Peter often underlines how balneotherapic treatments, to be successful, must be repeated very frequently; for instance, when describing the bath at Bracula, he claims that 'Non opus est intrare semel; quia balnea quantum / Quis magis ingreditur, tam magis ipsa iuuant' (XXVIII:11–12) ('It is not enough to enter once: in fact the more / you enter the baths, the more useful they are').[57]

55 Some baths are beneficial to the heart, some others to the kidneys or the digestive system; other cure skin or ophthalmological diseases; other are helpful in case of osteoarticular ailments, some other for gynaecological and andrological diseases, etc.

56 Ibidem 116; 152.

57 Ibidem 128; 138; 170.

Some of the waters are not only used for balneotherapy but also for hydropinic treatments, as in the case of Arcus (XV), whose waters are sweet like those of Foris Criptae (VII) and are recommended for similar therapies. While in the latter the poet explicitly mentions water 'dulcissima potu'[58] ('very sweet to drink', VII:3), in epigram XV the poet only describes the water's taste, which is similar, without explicitly referring to its use as a drink:

> Dulce satis lauacrum quod nomen sumit ab arcu,
> Virtutem magne comoditatis habet.
> Hec aqua consumptos restaurat corporis artus.
> Dat stomacho vires; arida membra rigat.
> Si quis in extremis patitur, festinet ad undam.
> Omnia, ne dubidet, inferiora iuuat:
> Non tamen hiis prodest tumidi quos sarcina ventris
> Aggrauat atque dolent splene tumente iecur.
> Rem loquor expertam proprio quam lumine vidi; [...]
> Vidi consumptum tantum cum pelle relictum
> Tempore non longo restituisse cutim (XV:1–12).[59]
>
> The very sweet bath, which takes its name from the arch,
> has the power of great benefit.
> This water invigorates the worn limbs of the body;
> gives strength to the stomach; irrigates the arid limbs.
> If someone has a pain in the lower limbs, get to the waves quickly:
> do not doubt that it benefits all the lower parts of the body.
> However, it does not help those who suffer from stomach heaviness
> It aggravates, or rather harms the swollen spleen and the liver.
> I am telling [you] a thing I experienced, which I have seen with my own eyes [...];
> I saw someone so sick, reduced to skin and bones,
> Not long after, the skin was restored.

The poet insists on how these waters are particularly curative for diseases affecting the lower limbs, as it is remarked at ll. 3, 5 and 6. The corresponding illustrations enrich the text with meaning. For instance, the illustration at fol. 8r of Milan, Biblioteca Ambrosiana MS I 6 inf.[60] (Figure 3.1) is divided into

58 Ibidem 126.
59 Ibidem 142.
60 The parchment codex dates to 1471 and was illustrated by Cola Rapicano, a Calabrian miniaturist, active at the Aragonese court at the end of the fifteenth century. See Toscano

three parts. In the upper one, two male figures are portrayed against a mountainous background behind which stands a city, represented by a tower. Both wear a crossbody cylindrical container, which could be interpreted as a flask for collecting thermal water. The man on the right wears a pointed hat, while the one on the left wears an ocher tunic and with his hand gestures invites the viewer to follow him. In the lower left part, a domestic scene is depicted: a sick man, dressed in a purple robe, is lying in bed. Another man, presumably a doctor, wearing a pointed hat and a lilac gown, holds a jug of thermal water in his hand and gives a glass to a third figure, perhaps an attendant or an assistant. Some details of the clothing – the hat and the ocher garment – seem to suggest that the standing figures are the same as the ones in the upper register. The lower right part depicts the actual bathhouse, an arched structure in which three of the four men portrayed immerse only the lower part of the body, to visually represent the therapeutic advice mentioned in the text.

The visual depiction of the specific bodily parts treated by the bath is a very common iconographic motif in MS I 6 inf. and in other illustrated manuscripts of *De balneis* and probably served to immediately convey medical information on the waters, even before reading the epigram. Possibly, as in a modern catalogue, the viewer could leaf through it and, depending on their afflictions, choose a bath according to the illustrations, subsequently reading the epigram to obtain medical information.

An interesting example is the illumination at fol. 17r representing St. George's bath (XXIV).[61] In its upper part a rural scene is represented, while in the lower one two subterranean constructions are depicted, as the bath is located 'underground' (l. 4), so that to access it one must walk through an unlit road. To the right, an underground gallery is portrayed, with a man facing left, fumbling towards the bath, and another one walking out of the bath with his head covered by a white cloth. The cloth probably conveys two meanings: on the one hand, the man is drying his head, i.e., the afflicted part, as the forehead is one of the bodily parts treated by these waters (l. 9); on the other, it emphasises the darkness of the place, so much so that the patient holds a lamp ('flamma') in his hand to illuminate the way (l. 3). On the left, the actual bath is depicted: inside the tub, the man to the back-left seems to be touching his pubic area, possibly to indicate that this water breaks kidney stones, is diuretic and heals the

G., "Rapicano Cola", in Bollati M. (ed.), *Dizionario biografico dei miniatori Italiani* (Milan: 2004) 893–896, 895.

61 The poem narrates that a vein hidden under the ground brings Georgica, a water that can only be accessed by using a flame as a guide. The more the man's mind is afraid of entering the bath, the more pleased he is when he leaves it. This water is used to break kidney stones, liquefy urine, and relieve joint pain. A witness reported that when a man tried to urinate after bathing in it, his penis was forced to expel two stones (XXIV:1–12).

FIGURE 3.1 Milan, Biblioteca Ambrosiana, MS I 6 inf. fol. 8r, *De Balneis puteolanis*, XV: Arcus.
REPRODUCED WITH PERMISSION OF ©VENERANDA BIBLIOTECA AMBROSIANA

FIGURE 3.2 Milan, Biblioteca Ambrosiana, MS I 6 inf. fol. 17r, *De Balneis puteolanis*, XXIV: St. George.

pelvis (ll. 7,9). To indicate that water cures the forehead, the figure on the right touches his forehead and cheek, the one in front on the left washes his head, and the one in the centre points to his face. The man in the lower right portion, on the other hand, is touching his chest while tilting his head to the side, as if to convey the fear (ll. 4–5) he experienced during the descent to the bath.

As will be shown below, *De balneis* proved to be an extremely influential text. It provided a reference model for subsequent balneotherapic treatises in that the physician, present at the baths, witnesses the effectiveness of the thermal water's healing power. The doctor himself seems to be the author of the treatise, thus creating a relation of immediacy and truthfulness between direct experience and patient. Following the model of *De balneis*, in the following centuries, thermal water will be described as mirabilia that heals as if by miracle, through God's will.[62]

The translations from Arabic and Greek of the masters of the Salerno inspired several adaptations of Arabic scientific, medical, and philosophical texts also in the surrounding court environments, especially in the Angevin one.[63] It is in this context that the *Tacuinum sanitatis*,[64] a fourteenth-century picture book depicting scenes of daily life and containing medical information,[65] was first translated into Latin.[66] The text, then, copied in the Lombard area, had a great diffusion throughout Europe between c.1350 and 1450, and treats the six non-naturals in a very concise way, listing environmental and dietary factors that influence human health, describing their properties.[67] It was probably intended for an aristocratic public interested in maintaining health through a short text decorated with sumptuous illustrations.[68] Indeed, each section[69] consists of an illustration accompanied by its explanation: the miniature occupies most of the folio, while the text is short and simple. As Cogliati Arano elucidates,

62 When describing Balneolum, Peter associates the great power of those waters with the presence of God so that even the sick can perceive it (VI:3–4; Ibidem 124).

63 See also Ventura I., "Cultura medica a Napoli nel XIV secolo", in Alfano G. – Perriccioli Saggese A. – D'Urso T. (eds.), *Boccaccio angioino* (Bern: 2012) 251–288, 255.

64 Henceforth *Tacuinum*.

65 Witthoft B., "The Tacuinum Sanitatis. A Lombard Panorama", *Gesta* 17.1 (1978) 49–60, 50.

66 The original text by Ibn Butlan el Bagdadi dates to the eleventh century and was translated into Latin in the thirteenth century. The translation's author has been long debated; however, it seems to have been realised at the Swabian court of King Manfredi in Sicily or at the Angevin court in Naples, which, thanks to its connections with northern Italy and France, might have contributed to the dissemination of the text. (See Ventura, "Cultura medica a Napoli" 271, 286–287).

67 Witthoft, "The Tacuinum Sanitatis" 50.

68 Ibidem 50–51.

69 Several non-illustrated manuscripts are also extant.

> [...] representations of herbs, flowers, and fruits predominate in the *Tacuina* illustrations. However, there are also folios devoted to natural elements, foodstuffs, winds, and seasons. Moreover, the changeable conditions of human emotions, from anger to joy, as linked to the medico-chemical theory of humors upon which herbal compounds are considered to have such a profound influence, are certainly not neglected. [...] The influence of the stars, planets, and, therefore, the climate upon the places where a man's life is lived, and also the contingent usefulness of each practice, drink, or particular kind of food, is precisely depicted on every folio.[70]

In some versions of the *Tacuinum,* the theory of the six non-naturals is so relevant that a prefatory chapter elucidating its importance for health's preservation has been added. The text's purpose, aiming at being brief and summarising medical advice for a healthy life, is also stated. A case in point is Paris, BNF Latin MS 9333, a fifteenth-century codex produced in the Rhineland, where each chapter in Latin is followed by the Middle German translation:

> [...] Der Sechs Stŭcksen. Die zŭ erhalltŭng gŭeter gsŭndthait. dem Mennschen von nöten. wie man dieselbigen corrigieren. unnd was dero wurchsŭng. ist der lŭfft daß erst. [...] Das annder ist. wie man speisz und tranncks beraitten soll. Das dritt. wie man sich geben soll. Das vierdt. was weisz der geslaaff und das wacsen dem Menschen nŭß oder gŭet seie. Das fünnfft. mann manpŭrgieren soll oder nitt. Das sechst. wie man sich in freüd und traŭrikait. sorn. forerst unnd engsten halten soll. Dann wer diese Stŭcks. mit maasz waisz zŭgebraŭchen. der lebt wol. wer nit wŭrd krannck s [...].[71]

> [...] the six things that are necessary for every man in the daily preservation of good health, about how to correct the same and their effect. The first is the air [...]. The second is: how to advise foods and drinks. The third: how one should behave. The fourth: what is known about sleep and wakefulness and what harms or what is good for man. The fifth: when one should be purged or not. The sixth: how one should behave in joy, anger, distress, and fear. So, whoever knows how to use these elements in moderation lives well and does not get sick [...].

70 Cogliati Arano L. (ed.), *The Medieval Health Handbook Tacuinum Sanitatis*, trans. O. Ratti – A. Westbrook (London: 1976) 9.

71 F. Dv. The transcriptions and translations from MS 9333 are mine.

As the extract shows, moderation and the right balance between the retention and expulsion of humours, as suggested by the term 'manpŭrgieren' ('purge'), are key factors for wellness.

In the *Tacuinum*, among the recommendations to maintain health, bathing features quite prominently as it is used to neutralise the negative properties of several activities. For example, if one oils their body while bathing it counterbalances the effects of terrestrial hunting, as this activity dries the body out.[72] Baths are also a remedy against the southerly wind, which dulls the senses;[73] in autumn, which is especially dangerous for temperate constitutions, pouring moistening substances into the water nullifies its dangers.[74]

A few chapters are devoted to bathing and to different kinds of water used for balneotherapy, attesting to how it was considered helpful in the preservation of human health. For example, a section is devoted to warm water, which can be both useful and harmful to the digestive system; the use of rose water is suggested to counterbalance its negative effects.[75] In the chapter devoted to baths, bathing 'alternately and in sweet water' is advised to 'all men' as the best way to benefit from it. It is contraindicated in case of 'acute diseases', but, according to humoral theory, this can be overcome with 'extremely cold substances'.[76]

Some kinds of water can have a negative effect on the body and need to be mixed with other ingredients to neutralise their harmfulness. For instance, 'aqua salsa' ('salt water'), while 'free[ing] the body', dries it and makes it itchy and, to avoid this, good clay must be added to the bath.[77] The corresponding chapter in MS 9333 presents an interesting variant if compared to other versions of the *Tacuinum*: this water dries the body but is also harmful for the eyes. Instead of clay, it is advisable to mix good wine into the water, possibly for a greater availability of this ingredient. Other effects of this water are

72 Ibidem Black and white plates 71.

73 Ibidem Black and white plates 80. Paris, BNF Nouvelle acquisition latine MS 1673, fol. 101v. The suggestion to introduce a moistening ingredient is mentioned in the following sentence since it is implied that the southerly wind is drying.

74 Ibidem Black and white plates 86. Rome, Biblioteca Casanatense, MS 4182, fol. CII.

75 New York, Public Library, Spencer Collection MS 65, a codex produced in Lombardy at the end of the fifteenth century, contains an Italian version of the *Tacuinum*. The corresponding illustration at fol. 196 depicts a noblewoman washing her feet while being attended by a maid.

76 Ibidem Black and white plates 76.

77 Ibidem Black and white plates 78, Bibliothèque de l'Université de Liège, MS 1041, fol. 76v.

mentioned: it causes thirst and constipation but fights the cold and the elderly who live in cold areas benefit from it.[78]

Among the extant manuscripts of the *Tacuinum*, two feature additions that reveal a specific interest in balneology and in thermalism. Paris, BNF NAL MS 1673,[79] contains a chapter on thermal water. In proximity to the chapter on *Aqua fontum* ('spring water', fol. 96), *Aqua aluminosa* ('aluminous water', fol. 98) is treated: after having listed its proprieties, the text prescribes drinking it for gastrointestinal ailments while warning against its draining powers: '[...] melior ex ea. que potari potest. Iuuamentum. emoroidibus 7 fluxiu necessitatis 7 sanguis. nocimentum. disicat corpora. Remotio nocimentum. cum rebus humidus. 7 uino dulci' ('[...] Optimum: that it is drinkable. Usefulness: haemorrhoids and serious diarrhoea. Dangers: it drains the bodies. Removal of dangers: with moist things and sweet wine').[80]

New York, Public Library, Spencer Collection MS 65[81] features another interesting addition: the manuscript's compiler has in fact added a series of treatises to the *Tacuinum*. After the sections on rain and hot water (fol. 95r) and the ones on spring water and on pleasantly warm water (fol. 97v), starting from fol.96r, six texts on thermal baths have been added, each of which is preceded by an illustration, as in the case of the *Tacuinum*. Of these texts, only the first three are extensive ones, with each of them describing an individual spring located in a specific geographical area in Italy.[82] In these texts, a series of

78 'De saltzen wasser. ist warm unnd trŭcken im anndern grad. das nit bitter und fliessend ist das besst. Es laviert unnd trŭcknet den leib. Es macht das beissen unnd se badt den Oŭgen. Darŭmb soll mans mit gŭetem wein vermischen unnd darnach daraŭff baden. Es macht dŭrst und verstopfft. fugt mer den kallten unnd feiechten allten in kallten landen' (fol. 86; 'Of salt water. It is warm and dry in the second degree. The one that is not bitter and is running is the best. It washes and dries the body. It causes itchiness and washes the eyes, so that it should be mixed with good wine and afterwards bathe in it. It causes thirst and obstructs. It cures the cold and invigorates the elderly in cold lands').

79 Produced in the first half of the fifteenth century around Pavia or Milan.

80 The transcription and the translation are mine.

81 This paper manuscript was most likely copied in the years 1460–1470 at Ferrara, or at Giovanni Cadamosto's workshop in Lodi, Italy. See Paris H.- Daunay M.C. – Janick J, "The Cucurbitaceae and Solanaceae Illustrated in Medieval Manuscripts Known as the *Tacuinum Sanitatis*", *Annals of Botany* 103 (2009) 1187–1205. It contains a botanical treatise (fols. 3r–63v), featuring information on the medicinal and occasionally alchemical uses of the plant species described; a version of the *Tacuinum* (fols. 64r–95v), which concludes with the chapters on water; six treatises on baths (fols. 96r–102r); and a treatise on falconry (fols. 102v–107v).

82 Fol. 96r marks the beginning of a tract on a bath at *Aquario*, identifiable with Aquaria near Sirmione, or Quara in Toano, near Reggio Emilia. See Bouisseuil D. – Nicoud M., "L'invention d'une source, Les *Bagni della Porretta*, les médecins et les autorités publiques",

healings proving the bath's effectiveness is listed; subsequently, the therapeutic qualities of its waters are described; finally, advice on how to face thermal treatments is given. As regards the remaining texts,[83] the copyist of this manuscript or of one of its antigraphs seems to have hastily dismissed the treatment of the baths with a few lines. The tract on falconry and the medical *consilium* following the bath section, are tidily written and richly illustrated and do not convey the idea of a hurried codicological compilation. It can thus be inferred that the copyist, probably bored by the repetitive style of these balneotherapic treatises, by the redundancy of their structure and of the technical and medical terminology, has chosen to devote only a few lines to these three texts, providing only essential information and citing *Bagni della Porretta* for any reference.[84]

The analysis of some chapters of the *Tacuinum* demonstrates how balneology was an important therapeutic tool for maintaining health and the balance between the humours and the six non-naturals. It can be noted how, in the *Tacuinum*'s textual transmission, the interest in balneotherapy has grown: MS 1673 features an interest in water mineral composition and, consequently, in its chemical analysis and in its therapeutic use, so much so that aluminous water is recommended for hydropinic treatments. MS 65, with the addition of an entire section on baths, attests to the affirmation and importance of spa treatments within medical practice. It also identifies the tract on *Bagni della Porretta* as one of the most popular in the field, as it is considered as a reference text for the entire balneological section.

Among late medieval texts showing an interest in balneology is the Middle English poetic rewriting of Palladius's *De re rustica*, originally a prose agricultural calendar in twelve books, one for each month of the year,[85] in which

in Bouisseuil D. – Nicoud M. (eds.), *Séjourner au bain: le thermalisme entre médecine et société* (*XIV*ᵉ–*XVI*ᵉ *siècle*) (Lyon: 2010) 63–98, note 99); fols. 97r-98r are devoted to the description of the virtues of *Bagni della Porretta*, near Bologna, which seems to be the most important text on balneology in the manuscript as it is used as a reference in the shorter texts. Fols. 99r–99v contain a treatise on *Bagno de Ayqui*, discussing the waters of Aqui Terme, near Alessandria in Piedmont.

83 Fol. 100r contains a text on the baths at 'Val trescho', possibly identifiable with Trescore Balneario, near Bergamo; fol. 101ris devoted to the waters at 'Petrioli' (Petriolo, in Tuscany), while fol. 102r mentions the spring at 'San filipo' (San Filippo, Tuscany).

84 As Bouisseuil and Nicoud argue in "L'invention d'une source", *Bagni della Porretta* was a very popular text probably due to its practical and useful nature: several manuscripts and incunabula, different editions and vernacular translations were produced. It also greatly influenced later literature, being widely used by physicians in the fifteenth century.

85 Palladius's treatise, originally dating back to the late 4th-early 5th century, is preserved in over one hundred manuscripts, as well as glosses and summaries, and has been translated

a practical interest in baths connected to health can be noted. As can be observed in MS Hunter 104, for example, two individual chapters are dedicated to the theme: one to the construction of baths (see Figure 3.3) and the other to their maintenance, thus attesting to the relevance of this practice in the late medieval domestic and rural environment. After having introduced the four main elements constituting husbandry in Ch. 2, 'ayer · watir londe and gouernaunce'[86] (l.2, fol. 1r, 'air, water, land and management'), the poet argues that knowing the source of the water used for agriculture and domestic activities is fundamental, as it must be pure and not harmful, and suggests which waters to use and how to identify healthy springs.

Ch. 39 contains a detailed list of instructions to build a bath. Despite the practical topic of the text, the relation between bathing and health is clarified at the onset of the section, where the poet attributes thoughtfulness to the householder as, thanks to him, bathing, seen both as a therapeutic practice and as a leisure activity, brings psycho-physical wellbeing into the household: 'Hit is not strange if watir wul suffice /An husbonde on his bath to be bithought · / ffor therof may plesaunce · and helthe arise' ('It is not strange, rightly it is adequate for the needs / For a house master to be considerate of his bath / For thereof pleasure and health may arise').[87] Therefore, in the bath's construction nothing can be left to chance, from the geographical position to the building materials,[88] to the measurements of each individual component. Elemental theory plays a key role in this as the temperature and lighting of the rooms are determined according to cardinal directions and to the soil's primary qualities.[89]

The procedure for building a bath is complex as it is made of different chambers: there are summer and winter rooms and the furnace must be slightly inclined so that flames go upwards warming them up (ll. 1086–1088). The chambers should be built in the same way as the cistern (ll.1106–1107); the winter house, instead, requires more effort as, being lower in the ground, it takes

from Latin into various vernacular languages, including Middle English. The Middle English poetic rewriting dates to around 1420 and is preserved in three manuscripts, two of which are incomplete: Glasgow, University Library, MS. Hunter 104 and Oxford, Bodleian Library, MS Additional A. 369; Oxford, Bodleian Library, MS. Duke Humfrey d.2 is the only complete codex. See Delgado C.M.J., "An Edition of MS. Glasgow, University Library, Hunter 104" (Master's dissertation, University of Glasgow: 2000) 1–9.

86 Ibidem 60. The translations from *De Re Rustica* are mine.

87 Ll. 1079–1081; Delgado, "An Edition" 102.

88 '[…] And vppon hem thou sprede / A marbul floor · or tile hit yet for need / A myliair of leed the bottom brasse' (ll. 1091–1093, Ibidem 103. '[…] and upon you lay / a marble foor or tile according to need / a water tank of lead, the bottom of brass').

89 Ll. 1082–1084; ibidem.

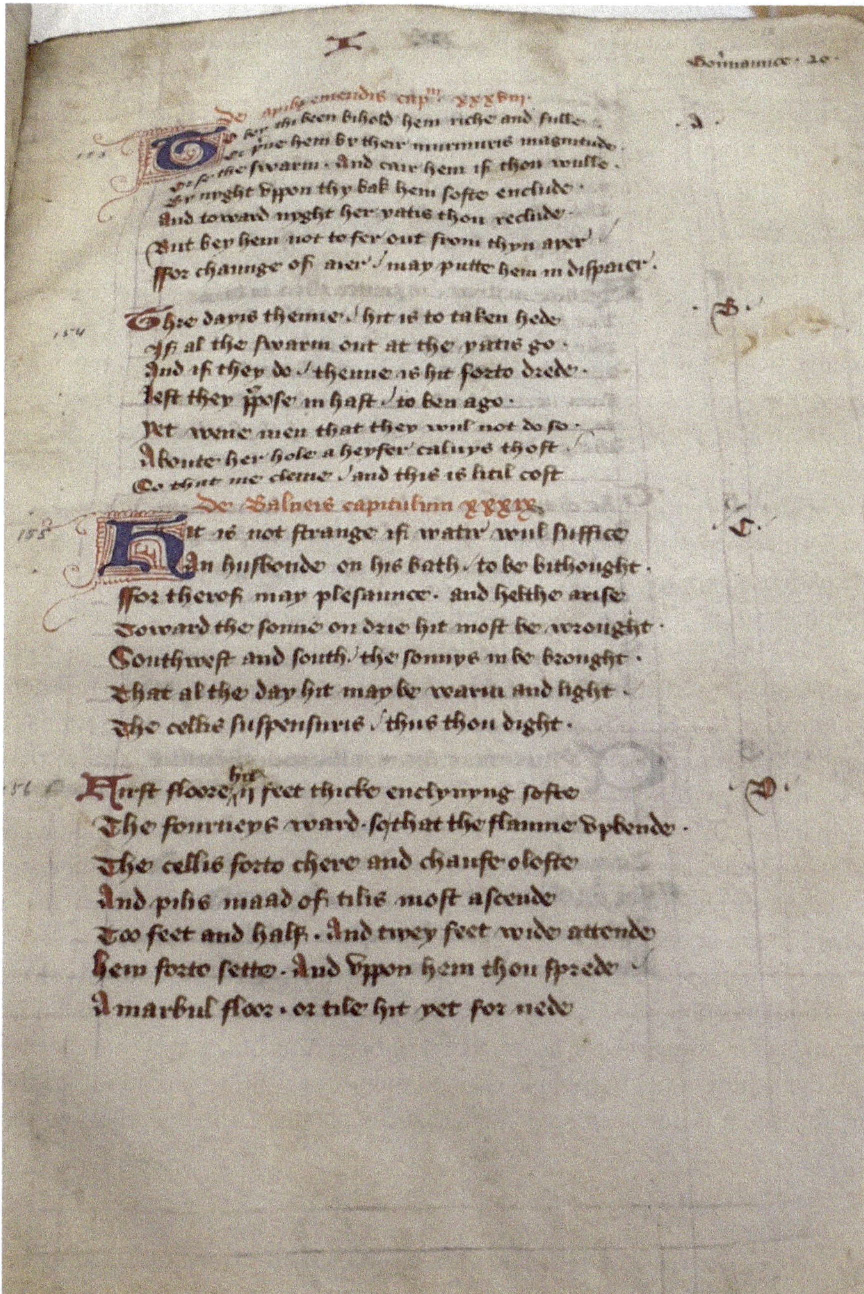

FIGURE 3.3 Glasgow, University Library, MS Hunter 104, fol. 18r.

all the warmth from it (ll. 1110–1112). The bath also needs a plumbing system: since the seat is placed on the outside of the furnace, so that it is constantly heated, both warming and cooling pipes are needed: 'Anend the setis sette hit so withoute / The fourneys · and the fir ther vndir passe · / A condit colde into hit brynge aboute · / Mak pipis · watir warm inward tospoute ·' ('Next to the seat set it on the outside of / the furnace and the fire beneath it passes / A pipe brings about cold into it / Make the pipes splash warm water inward').[90]

It is interesting to note how the bath's design in terms of lighting, space heating, and use are 'eco-sustainable', so much so that the wastewater is recycled for irrigation (ll. 1104–1105).[91] Ch. 41 in the Humfrey MS, dealing with milling, even recommends conserving wastewater from baths and reusing it as a driving force for water mills,[92] thus highlighting the connection between daily activities, bathing, nutrition (represented by the production of flour) and health.

Ch. 40 deals with bath maintenance and mortars for repairing damage and cracks that form with wear: cooling or heating cements are used depending on the bath temperature.[93] The Humfrey MS version provides different recipes for the preparation of mortars, whose lexicon is reminiscent of medical remedies, as if the householder, repairing the bath and allowing its continued use, were compared to a physician. For example, one of the recipes for heating mortars is compared to a remedy that 'al the woundis leche' ('heals all the wounds', l. 1125), whereas another one for a cooling cement is described as a salve that must be smeared over the 'wounded' bath: 'But malthis colde in other craft let founde: / Ox blood with picche and syndir al toframe, / And make hit lyk a salue, and ouerflame Vche hole and chene'. ('But cold mortar in another way is made: / ox blood with wood tar and ash mix all together / And make it like a salve and smear over / each hole and crack').[94]

As the passage shows, the worn bath is compared to an afflicted, ulcerated body needing the cures of a physician to be restored. Indeed, the bath's 'health' is fundamental, for, as stated at the beginning of Ch. 39, the wellbeing and the health of the household depend on it. As a reflection of man, i.e., the microcosm, it is not surprising, then, that the planning, the building, and the

90 Ll. 1094–1097; ibidem.

91 Ibidem 104.

92 Ll. 1142–1145; Liddel M. (ed.), *The Middle English Translation of Palladius De Re Rustica* (Berlin: 1896) 64–65.

93 The poem claims that one of the most important concepts in the maintenance of a bath is that both hot and cold mortars are subject to cracking, clefts or damage, so gum ammoniac is added to make it more resistant (ll. 1114–1120).

94 Ll. 1137–1141; ibidem 64.

maintenance of baths are based on elemental theory, with reference to cardinal directions, seasons, and primary qualities.

4 The Apogee of Balneology

The late medieval development of balneology culminates in the fifteenth century, when not only treatises devoted to baths are written, but also entire collections dealing with this topic are produced. An interesting example is Pavia, Biblioteca Universitaria, Aldini MS 488, a paper manuscript entitled *Collectio de balneis*,[95] which testifies to how the interest in balneotherapy has grown as as to provide demand for a dedicated anthology. The collection, arranged according to a geographical rationale, starts with Savonarola's treatise, describing various Italian thermal sites and listing their nature and their properties, explaining their therapeutic uses according to the various afflicted bodily parts arranged compatibly with the head-to-toe order, and then moves on to texts devoted to individual baths. It demonstrates how much balneology is now a very complex science from different points of view. Firstly, the texts of MS 488 are much more chemical in focus than the ones analysed so far: the mineral composition of the waters is now fundamental to determine their correct therapeutic use. Chemical analysis is performed through distillation, a process that physicians draw from alchemy. These texts, therefore, also demonstrate how alchemy has influenced thermalism by providing it

95 As Nicoud observes, the manuscript probably dates to 1425 and features mostly the same texts as Paris, BNF, NAL MS 211 (c.1454–1469), also entirely dedicated to thermalism. These *libri de balneis* somehow anticipate the collection of texts on water published in Venice in 1553 by Tommaso Giunti, entitled *De Balneis*. (See Nicoud, "La scrittura dei *Consilia* nella medicina medievale", in Centro Italiano di Studi sul Basso Medioevo (ed.), *La medicina nel basso medioevo: tradizioni e conflitti* (Todi: 2019) 189–212.200). Casagrande, Mazzoli and Vecchio list the contents of the manuscript as follows: fol. 1: Savonarola Michele, *De balneo et termis naturalibus omnibus Italiae sicque totius orbis proprietatibusque earum*; fol. 45: Ugolino Da Montecatini, *De balneis mineralibus et artificialibus*; fol. 61: *Epigrammata de balneis puteolanis*; fol. 66: *Consilium pro balneis de Corsena in comitatu luchano* pro domino Lanzaloto De Crotis ducali consiliario; fol. 67: Tura di Castello, *Regula et tractatus balnei de poreta*; fol. 68v: *Tractatus pro balneis de aquis* per Petrum de Tussignano; fol. 70v: Antonii Guaynerii papiensis, *De balneis aquis ciuitatis antiquissime que in marchionatu montisferrati sita sunt tractatus*; fol. 74v: *De balneis* secundum Petrum de Ebano; fol. 75v: *Tractatus de balneis* secundum Gentillem; fol. 76v: *De balneis de Burmio* secundum magistrum Petrum de Tussignano; fol. 77v: *Regula balnei loci de Aquaria in territorio regii*; fol. 78v: *De balneo aque porrete*. See Casagrande C. – Mazzoli M.A. – Vecchio S., "Ms Aldini 488 (128) no 79. Miscellanea De Balneis", in Casagrande C. – Mazzoli M.A. – Vecchio S. (eds.), *Pavia biblioteca universitaria: Fondo Aldini* (Verona: 1993) 251–254.

with a fundamental analytical tool to determine both risks and benefits for the patient.[96] For instance, Guaineri's *De balneis aquis ciuitatis* focuses on the mineral composition of the waters at Acqui Terme. To prove the truthfulness of the information provided, the author cites a series of examples of illustrious healings he himself witnessed, according to the tradition inaugurated by Pietro of Eboli. The author narrates that, accompanying some famous characters to the baths, he witnesses the water's healing virtues and notes how the sick are amazed by it, as if it were more divine than natural. Like a modern scientist, Guaineri trusts empirical observation and investigates the water's therapeutic virtues by observing its distillation, as the term 'sublimatione' clearly suggests:

> [...] ut uirtuem earum potius inuestigarem pluries ex aqua fontium balneorum sublimatione facta feces conspexi: quarum ut quantitatis comparationem faciam: tres partes salis continent nitri unam aluminis parum sulfuris uero quo ad apparentiam: ipsum enim eum sale an nitro cum quo symboleitatem maiorem habet est commixtum.[97]

> [...] I should rather investigate their virtue, I have seen several times the residues from the sublimation of the water of the fountains of the baths, of which I will make a comparison of the quantity: three parts of salt contain sodium, one part of alum contains a little sulphur, but as to appearance, for it is salt or sodium with which it has a greater quantity it is mixed.

Having determined the water's mineral composition, Guaineri, using a technical jargon, elaborates on the prevalence of salt or saltpeter in the water to prove that the combination of minerals determines its curative properties: 'Sic ex sale, nitro, alumine, & quae in aquis istis reperiuntur: quaedam sit commixtio, per quam complexio una resultat: virtute cuius balnea illa sic virtuosa fiunt' ('Thus from salt, saltpetre, and alum, and what are they found in these waters: what is the mixture, by which any combination results, by virtue of which those baths become so virtuous').[98]

The author provides a series of therapeutic prescriptions according to chemical notions, including when one should bathe: the water is always curative,

---

96 Bodiš, Božiková and Mackových argue that 'history of chemical analysis of water began obviously with alchemy, which had prepared some background for scientific analytical chemistry'. See Bodiš D. – Božiková J. – Mackových D., "Mineral Waters of the Slovak Spas – Chemical Analysis, History and Present", *Slovak Geology Magazine* 16.2 (2016) 41–56, 41.

97 Ch. 1; fol. 70v. The transcriptions and the translations from MS 488, unless otherwise stated, are mine.

98 Ibidem.

however, its minerals are subject to the astrological influence of the sun and the stars, which enhance their healing virtues. Therefore, there are periods in which these waters are more effective: this is why the water, usually very clear and transparent, becomes greener in the month of May:

> Clara est balneorum aqua: & adeo peruia, ut denarius in fundo proiectus, uideri queat. Sed de mense Maij eius color ad aliqualem uiriditatem mutatur: quod est quia tunc de minera plus cum aquis illis commiscetur: propter quod grossioris ac maioris fiunt ponderis. Nam quaelibet mineralis uirtus a sole ab alijsque stellis uirtutem recipit: quae illo mense quam alijs in nostrorum balneorum minera multo fortior reperitur [...].[99]

> The baths' water is clear, but from the month of May its colour changes to a certain greenness: that is because then more of the minerals is mixed with those waters: for which reason they become coarser and of greater weight. For every mineral receives power from the sun and other stars: which in that month than in any other is found in the minerals of our baths much stronger [...].

As elucidated by Guaineri, distillation leads to the awareness that different minerals mixed in variable quantities give different types of water. The physician is thus essential to evaluate which waters are suitable for which afflictions and which are incompatible with certain temperaments to ensure that balneotherapic treatments are effective. For example, Crotti in *Consilium pro balneis de Corsena*, after having described the mineral composition of the Tuscan baths, the quality of their clear waters, their hot temperature, having tasted and smelled them and finding them pleasant, having explained their functioning and listed its beneficial effects on health, discusses how to deal with balneotherapic treatments.

> Et volentes vti dictis balneis primo optime purgent corpora ipsorum ab humore precaute de consilio periti medici per aliquos dies ante accessum ad dicta balnea. Et vtatur debito regimine cibi et potus et aliarum sex rerum non naturalium toto tempore, quo balneis vtitur, maxime a coitu sibi caueat tanquam a re sibi multum nociua et inimica.[100]

---

99 Ibidem fol. 71r.

100 Sudhoff K. (ed.), "Ein Gutachten zweier Ärzte von Lucca über die Heilquellen von Corsena", *Archiv für Geschichte der Medizin* 16.1/2 (1924) 74–76, 75.

> And those who wish to take the said baths, first, cleanse their bodies of the humours carefully, on the advice of an expert physician, for some days before approaching the said baths. And they vow a due regime of food and drink and six other non-natural things during the whole time that they vow to bathe, especially to beware of intercourse, as from something very harmful and inimical to them.

As the passage shows, the patient must free his body from excessive humours by relying on an expert doctor. The author underlines the importance of the physician in preserving the patients' health, indicating how to approach the baths and how to prepare in terms of diet and lifestyle, thus illustrating that balneology has become an extremely complex science. During the treatments, the physician must guide the patient to the moderation of food, drink, and the other six non-naturals, including abstention from sexual intercourse, as it is extremely harmful. Tura di Castello shares this idea, and in *Bagni della Porretta* even argues that the doctor's presence is necessary at the baths, a very modern concept, which we find in modern spas and thermal stations. Without medical advice, an improper use of baths could be ineffective or even harmful to the patient. Thus, Tura writes a *regula*, a set of rules the patient must follow for the duration of the thermal treatment, including dietary advice and prescriptions regarding exercise, rest, and sexual intercourse, i.e., the six non-naturals. Observing these rules, the water of Porretta, used as drink and in baths for three consecutive years, guarantees a long healthy life.[101] It is therefore no coincidence that the assembler of MS 488 arranged the *Corsena* and the *Porretta* treatises consecutively. Indeed, these rules become twelve canons in *De balneis de Burmio* by Peter of Tussignano, Gian Galeazzo Visconti's personal physician at the court of Pavia. Describing the baths of Bormio in Valtellina, he briefly mentions the waters' analysis through the alembic, finding them predominantly nitrous and partly sulphureous and emphasizing their astringent and drying properties, especially in cases of illnesses caused by cold humours. He lists their therapeutic virtues for many diseases and sets the rules for their use: treatments must alternate immersion in the bath and distillation on the head or on the sick bodily part, so that a higher dosage of minerals can target the area. He also provides advice on food and lifestyle, reaffirming the need to follow twelve rules, set out precisely in the final part of the treatise, to prevent the baths from being ineffective: 'Siquis diceret quod aliqui accedunt ad balneum dictum, nihil tamen iuvamenti percipiunt: respondendum,

101 See also Bouiseuil – Nicoud, "L'Invention d'une source".

quod non est culpa balnei, sed qui accedunt ad ipsum non servant servanda et sunt XII canones quos servare oportet' ('If someone were to say that some people enter the said bath, yet they do not perceive any benefit: the answer is that it is not the bath's fault, but those who enter it do not respect the rules, and there are twelve canons which must be respected').[102] Among these, the third one indicates the cases in which the waters of Bormio are harmful, as for instance in patients prone to inflammation, emaciated and recovering from fever.[103] The eighth canon instead consists in a series of instructions for after the baths: patients must enter the bed and wait there until they sweat, then get dry and dressed and take some fresh air until the bath's heat lasts. Since they entered the bath on an empty stomach, they should avoid drinking until they have eaten.

MS 488 also presents a more complex relationship between balneology and elemental and humoral theories than discussed so far. For example, Michele Savonarola in *De Balneo et termis naturalibus* compares two of the four elements as those mostly affecting human health: on the one hand, the air brings disease, and on the other, water, just as powerful, cures it.

> [...] caliditas in subiecto densiori existet sic prodicere poterit si convenienter applicata fuerit potet sippositum in caliditate in corpum humanum ex aere & in illa quae ex igne in eo accensis in corpus ipsum producitur. ex quo infertur quo caliditas balnearum non est minus potens humoris corporis commouere calafacere et alterare & ad putredinem praeparare quam caliditas aeris. Infertur secundo quod ad varias egritudines sic corpus nostrum poterit caliditas balnei commouere ut caliditas aeris.[104]

> [...] heat exists in a denser substance so it can produce it if it is suitably applied. From which it is inferred that the heat of the baths is no less powerful in moving, warming and altering the body's humours, and preparing them for putrefaction, than the air's heat. It is inferred, secondly, that for various illnesses our body will be able to react to the temperature of the bath as to the temperature of the air.

In the extract, Savonarola provides a physical explanation for the baths' effectiveness: heat exists in dense substances; water is denser than air; applying

---

102 Pedrana C. (ed.), "Il *Liber de Balneis Burmi* di Pietro da Tossignano", *Bollettino Storico Alta Valtellina* 16 (2013) 123–143, 135.

103 See also ibidem 128–132.

104 MS 488, fol. 36v.

heat to bath water will in turn produce more heat, which will fight the excess humours. To Savonarola, then, the bath's temperature is a fundamental therapeutic factor. Consequently, the relationship with the seasons is also essential as the climate alters the water's temperature. Furthermore, even the ingredients for the preparation of bathing 'artificial waters' must be carefully chosen according to their primary qualities after having been matched with the corresponding humoral quality – hot or cold – causing the disease. The choice of the ingredients is thus so complex that Savonarola tried to simplify it through a table, as seen in Figure 3.4. As can be observed, for example, for diseases of the chest caused by hot humours nettle, oregano, laurel, among other herbs are recommended; instead, for ailments caused by cold humours violet, poppy, purslane, etc. are used.

Furthermore, as the author explains, through the effects of temperature, the noxious humours are dissolved in bathing water; from here, however, through temperature's action, they can transfer to the air and infect other people. For these reasons, he considers seasonal changes with their sudden temperature variations as the most common causes of diseases. He thus devotes chapter v to prognostics, listing a series of factors to understand which types of baths are good and which are bad according to the seasons, the humours and their properties.

5 Readership and Conclusion

In the early Middle Ages baths are always part of more complex remedies that include the ingestion of a potion, the application of salves, compresses, plasters, and diets, with scattered reference to humoral theory. Examples of this are found in the German *Lorscher Arzneibuch* and in the Old English *Leechbooks* and *Lacnunga*. The medical school in Salerno and the surrounding Italian courts, with several translations of Greek and Arabic scientific, medical, and alchemical texts, revolutionised balneology.

The late Middle Ages witness the production of different textual genres either entirely devoted to balneotherapy or featuring one or more chapters dealing with bathing waters: medical poetry, thermal and husbandry treatises share a similar interest for the use of water for health preservation and the healing of diseases, with an explicit connection to humoral theory and the four seasons. Starting from the twelfth century, Italian medicine contributes to the medicalisation of balneology, so that many of the most influential texts for European thermalism are produced or translated there, such as the *Regimen sanitatis Salernitanum*, *De balneis Puteolanis*, and the *Tacuinum sanitatis*.

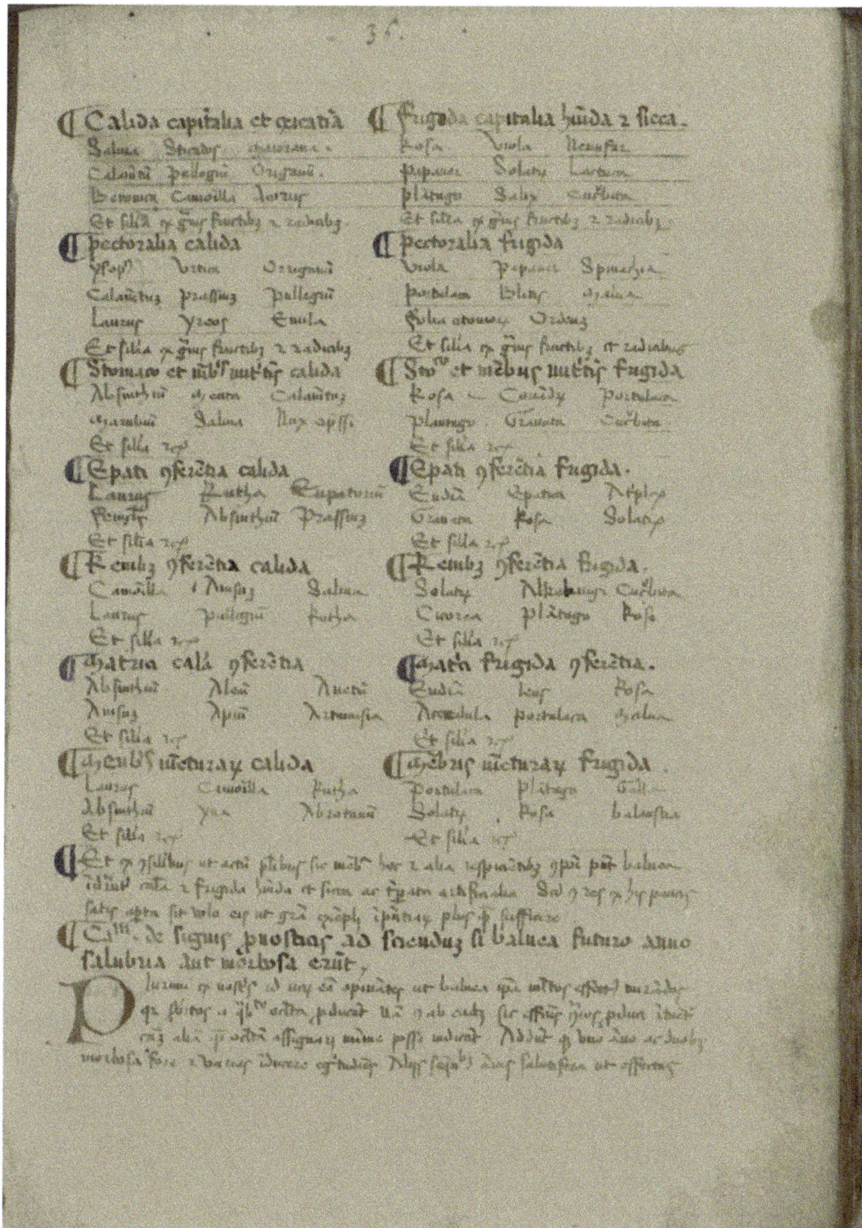

FIGURE 3.4 Pavia, Biblioteca Universitaria, Aldini MS 488, fol. 36r. Michele Savonarola, *De Balneis et Termis Naturalibus Omnibus Italiae.*

MS 488, being an anthology entirely devoted to the topic, marks the culmination of balneology, affirming it as an extremely complex science. It also asserts the waters' chemical and mineralogical analysis and the influence of alchemy using distillation, the alembic, and alchemical methods. It also features a more complex relationship between balneotherapy and elemental and humoral theories in the use of different ingredients to prepare bathing waters according to their elemental qualities, the diseases' humoral properties and the seasons.

An interesting and particularly complex issue, hitherto largely overlooked and worthy of further discussion in the future, concerns the readership of the late medieval treatises. This topic cannot be dissociated from the long and complex manuscript tradition of the treatises and is related to their dedicatee and their courtly milieu. A case in point is the *Regimen*: as Croke[105] noted, several copies contain a dedication to an unspecified *rex anglorum* ('king of the English'), whom he identifies with Robert of Normandy, son of William the Conqueror. These manuscripts derive from an antigraph written by a physician of the Salerno school at the end of the eleventh century aiming at the preservation of Robert's health during his presence in Italy.[106] According to Croke,[107] the poem addressed the king, who had no medical skills, while aiming at being a popular text on the preservation of health rather than the cure of diseases. The scholar also identifies a group of manuscripts in which the variants of the *Regimen*, attributed to a French doctor, are dedicated to Charlemagne and allude to his victory at Roncesvalles,[108] without, however, providing an extensive explanation for this dedication. It is possible that it was intended to celebrate the first great emperor of the Holy Roman Empire in homage to the Swabian court, as discussed above, already connected with the production of *De balneis* and of the *Tacuinum*. Nevertheless, aside from the dedication, the *Regimen* seems to be intended for an audience of educated nobles who understand Latin, enjoy poetry, and want to remain healthy. Similarly, the *Tacuinum* addressed an audience of aristocrats interested in maintaining their wellbeing while indulging in lavish illustrations. A point of encounter between the two

105 Croke A. (ed.), *Regimen Sanitatis Salernitanum, a Poem on the Preservation of Health in Rhyming Latin Verse. Addressed by the School of Salerno to Robert of Normandy, Son of William the Conqueror*, trans. J. Harington (Oxford: 1830) 29–30.

106 Croke provides a detailed account of Robert of Normandy's travels to Italy, his relationship with the family of Robert Guiscard and the episode of a poisoned arrow wound degenerated into a fistula which led him to be treated by the Salernitan doctors. His condition would have led to the composition of the *Regimen*. Ibidem 19–23.

107 Ibidem 29–30.

108 Ibidem 43.

treatises is represented by *De balneis*, which, as indicated, consists of epigrammatic verses, and, in many of his versions, comes with lavish illustrations and sometimes contains detailed medical information. As De Angelis indicates,[109] Peter of Eboli, in composing this didactic text can perfectly blend the medical and poetic arts, configuring himself as a court poet with good medical knowledge. His poem has a twofold goal: being both scientific and instructive, it employs a simple style of versification that allows readers of average cultural background to understand the treatise.[110] The use of technical and medical terminology, present constantly and consistently but not in an insistent and thorough manner, confirms the average culture Peter is aiming at.[111] The analysis conducted in this chapter seems to partially confirm De Angelis's hypothesis: on the one hand, *De Balneis* would target both the (fairly) educated aristocracy and doctors, who can thus advise patients on balneotherapic treatments in the Phlegraean Fields while enjoying poetry and illustrations, on the other, as suggested above, patients, looking at the illustrations and subsequently reading the epigrams, can choose, as from a modern catalogue, which thermal source is most suitable for their ailments.

Dissimilarly, the Middle English rewriting of *De re rustica* is aimed at an audience of small landowners and yeomen interested in administering, cultivating, and maintaining their estate as well as preserving the wellbeing of their household members according to elemental theory, while indulging in the reading of rhyming poetry. Arguing for the influence of this text, Reinert[112] contends that for several centuries it was foundational to the entire English husbandry genre. More generally, Jones states that 'husbandry establishes the link between high scholastic theory and the practicalities faced daily by the medieval famer' arguing that that 'consideration of the elements was not restricted to the intellegensia but was accepted and understood by the common man at least by the thirteenth century'.[113]

As discussed, the texts in MS 488 also have a practical purpose interconnected with elemental and humoral theories. In fact, Nicoud[114] asserts that balneotherapic treatises constitute a separate textual genre, in that they provide

---

109 De Angelis, *Peter of Eboli* 47.

110 Ibidem 49.

111 Ibidem 41.

112 Reinert L., "A Middle English Text on Planting and Grafting in Cambridge, Trinity College, O.5.26", *American Notes & Queries* 19.1 (2006) 9–16, 12.

113 Jones R., "Elemental Theory in Everyday Practice: Food Disposal in the Later Medieval English Countryside", in Klápšte J. – Sommer P. (eds.), *Food in the Medieval Rural Environment: Processing, Storage, Distribution of Food* (Turnhout: 2011) 145–154, 149.

114 Nicoud, "Les médecins italiens" 31.

medical information while aiming at practical applications, as they often contain *consilia*, recipes, canons and other procedures. As the scholar argues, it is not unusual for a compiler to bring together individual balneological treatises to form an anthology about the waters, as in the case of MS 211[115] and MS 488. The readership of these balneotherapic anthologies, however, has yet to be determined. In the case of MS 488, the use of Latin and the specialised technical language of many of the treatises, which draw on the jargons of anatomy, medicine, alchemy, but also mineralogy, and what could be defined as 'protochemistry' suggest that the recipient could be a physician specialised in balneotherapy. Since the thermal springs treated in the codex are scattered throughout Italy and several texts insist on the importance of the doctor's presence at the baths, one could hypothesise that MS 488 could be the book of a court physician who traveled, accompanying and assisting his lord in the balneotherapic treatments. Another hypothesis suggests that this collection and similar anthologies may serve as teaching manuals for the training of doctors specialised in balneotherapy.

Bibliography

Manuscripts

Bibliothèque de l'Université de Liège, MS 1041

Glasgow, University Library, MS. Hunter 104

Milan, Biblioteca Ambrosiana, MS I 6 inf. Pietro da Eboli "De Balneis Quae in Baiae Ora Patent, Henrico 6. Dicatus". Accessed May 14, 2024. http://213.21.172.25/0b02da8280051bc2

New York, Public Library, Spencer Collection MS 65. "Taccuinum Sanitatis" New York Public Library Digital Collections. Accessed May 14, 2024. https://digitalcollections.nypl.org/items/b248e800-1326-0131-509d-58d385a7bbd0

Paris, BNF Latin MS 9333. BNF Gallica. Accessed May 14, 2024. https://gallica.bnf.fr/ark:/12148/btv1b105072169/f1.item

Paris, BNF Nouvelle Acquisition Latine MS 1673. BNF Gallica. Accessed May 14, 2024. https://gallica.bnf.fr/ark:/12148/btv1b105380445

Pavia, Biblioteca Universitaria, Aldini MS 488

115 Nicoud (ibidem) describes MS 211 as the most comprehensive collection on thermalism probably prior to seeing the previous MS 488, which presents the same sequence of texts as MS 211 but also contains the popular *Bagni della Porretta*.

Rome, Biblioteca Casanatense, Ms 4182. Collezioni Digitali della Biblioteca Casanatense. Accessed May 14, 2024. https://casanatense.contentdm.oclc.org/digital/search/searchterm/Ms.4182

Primary Sources

Barney S.A. – Lewis W.J. – Beach J.A. – Berghof O. – Hall M. (trans.), *The Etymologies of Isidore of Seville* (Cambridge: 2006).

Cockayne T.O. (ed. and trans.), *Leechdoms, Wortcunning, and Starcraft of Early England, vol. 2.* (London 1864–1866; reprinted Cambridge: 2012).

Cockayne T.O. (ed. and trans.), *Leechdoms, Wortcunning, and Starcraft of Early England, vol. 3* (Cambridge: 2012).

Cogliati Arano L. (ed.), *The Medieval Health Handbook Tacuinum Sanitatis*, trans. O. Ratti – A. Westbrook (London: 1976).

Croke A. (ed.), *Regimen Sanitatis Salernitanum, a Poem on the Preservation of Health in Rhyming Latin Verse. Addressed by the School of Salerno to Robert of Normandy, Son of William the Conqueror*, trans. J. Harington (Oxford: 1830).

De Angelis T. (ed. and trans.), *Pietro da Eboli: De Euboicis aquis* (Florence: 2018).

de Frutos-Gonzáles V. (ed.), *Flos medicine: Regimen Sanitatis Salernitanum* (Valladolid: 2010).

Delgado C.M.J., "An Edition of MS. Glasgow, University Library, Hunter 104" (Master's dissertation, University of Glasgow: 2000).

Galen. *On Hippocrates, On the Nature of Man*, ed. and trans. W.J. Lewis – J.A. Beach Medicina Antiqua. Accessed May 14, 2024. https://www.ucl.ac.uk/~ucgajpd/medicina%20antiqua/tr_GNatHom.html.

Green M.H. (ed. and trans.), *The Trotula: A Medieval Compendium of Women's Medicine* (Philadelphia: 2001).

Isidore, *Etymologiarum sive Originum*, ed. W.M. Lindsay (Oxford: 1911). Digitised Edition @ University of Chicago, 2010. Accessed May 14, 2024. https://penelope.uchicago.edu/Thayer/E/Roman/Texts/Isidore/home.html.

Liddel M. (ed.), *The Middle-English Translation of Palladius De Re Rustica* (Berlin: 1896).

Magenta P. (ed. and trans.), *La scuola salernitana ossia precetti per conservar la salute* (Pavia: 1835).

Pedrana C. (ed.), "Il *Liber de Balneis Burmi* di Pietro da Tossignano", *Bollettino Storico Alta Valtellina* 16 (2013) 123–143.

Stoll U. (ed. and trans.), *Das Lorscher Arzneibuch: Ein medizinisches Kompendium des 8. Jahrhunderts* (*Codex Bambergensis Medicinalis 1*) (Stuttgart: 1992).

Sudhoff K. (ed.), "Ein Gutachten zweier Ärzte von Lucca über die Heilquellen von Corsena", *Archiv für Geschichte der Medizin* 16.1/2 (1924) 74–76.

Secondary Sources

Beccaria A., *I codici di medicina del periodo presalernitano* (Rome: 1956).

Bodiš D. – Božiková J. – Mackových D., "Mineral Waters of the Slovak Spas – Chemical Analysis, History and Present", *Slovak Geology Magazine* 16.2 (2016) 41–56.

Boisseuil D. – Nicoud M., "L'Invention d'une source: Les *Bagni della Porretta*, les médecins et les autorités publiques", in Boisseuil D. – Nicoud M. (eds.), *Séjourner au bain: le thermalisme entre médecine et société* (*XIV*e*-XVI*e *siècle*) (Lyon: 2010) 63–98. Open Edition Books. Accessed November 4, 2023. http://books.openedition.org/pul/5417

Casagrande C. – Mazzoli M.A. – Vecchio S., "Ms Aldini 488 (128) no 79. Miscellanea De Balneis", in Casagrande C. – Mazzoli M.A. – Vecchio S. (eds.), *Pavia biblioteca universitaria: fondo Aldini* (Verona: 1993) 251–254.

Cesario M. – Magennis H. – Ramazzina E., "Medieval Ideas on the Elements", in Cesario M. – Magennis H. – Ramazzina E. (eds.), *The Four Elements in the Medieval World: Interdisciplinary Perspectives. Earth* (Leiden – Boston, MA: forthcoming).

Danzi M., "Le terme in Europa tra letteratura e medicina", *Quaderns d'Italià* 22 (2017) 43–56.

de Divitiis E. – Cappabianca P. – de Divitiis O., "The 'Schola Medica Salernitana': The Forerunner of the Modern University Medical Schools", *Neurosurgery* 55.4 (Oct 2004) 722–744.

Drell J.H., *Kinship and Conquest: Family Strategies in the Principality of Salerno during the Norman Period, 1077–1194* (Ithaca, NY: 2002).

Garcia-Ballester L. – French R. – Arrizabalaga J. – Cunningham A., *Practical Medicine from Salerno to the Black Death* (Cambridge: 2004).

Gianfaldoni S. – Tchernev G. – Wollina U. – Roccia M.G. – Fioranelli M. – Gianfaldoni R.- Lotti T., "History of the Baths and Thermal Medicine", *Open Access Macedonian Journal of Medical Sciences* 5.4 (2017) 566–568.

Green M.H., "Salerno on the Thames: The Genesis of Anglo-Norman Medical Literature", in Wogan-Browne J. (ed.), *Language and Culture in Medieval Britain: The French of England c.1100–c.1500* (Woodbridge: 2009) 220–232.

Jacquart D. – Paravicini Bagliani A., *La scuola medica salernitana: gli autori e i testi* (Florence: 2007).

Jones R., "Elemental Theory in Everyday Practice: Food Disposal in the Later Medieval English Countryside", in Klápšte J. – Sommer P. (eds.), *Food in the Medieval Rural Environment: Processing, Storage, Distribution of Food* (Turnhout: 2011) 145–154.

Luft D., "Uroscopy and Urinary Ailments in Medieval Welsh Medical Texts", *Transactions of the Physicians of Myddfay Society* (2018) 1–14.

Mamiani M., *Storia della scienza moderna* (Bari: 2002).

Nicoud M., "Les médecins Italiens et le bain thermal à la fin du moyen âge", *Médiévales* 43 (2002) 13–40.

Nicoud M., "Les vertus médicales des eaux en Italie à la fin du moyen âge", in Guérin-Beauvois M. – Martin J.M. (eds.), *Bains curatifs et bains hygiéniques en Italie de l'Antiquité au moyen âge* (Rome: 2007) 321–344.

Nicoud M., "La scrittura dei *Consilia* nella medicina medievale", in Centro Italiano di Studi sul Basso Medioevo (ed.), *La medicina nel basso medioevo: tradizioni e conflitti* (Todi: 2019) 189–212.

Paris H. – Daunay M.C. – Janick J, "The Cucurbitaceae and Solanaceae Illustrated in Medieval Manuscripts Known as the *Tacuinum Sanitatis*", *Annals of Botany* 103 (2009), 1187–1205.

Pormann P.E., "Medical Conceptions of Health from Antiquity to the Renaissance", in Adamson P. (ed.), *Health: A History* (Oxford: 2019) 43–74.

Ramazzina E., "Therapeutic Baths in Medieval English Medicine", in Hayden D. – Baccianti S. (eds.), *Medicine in the Medieval North Atlantic World: Vernacular Texts and Traditions* (Turnhout: 2024).

Reinert L., "A Middle English Text on Planting and Grafting in Cambridge, Trinity College, O.5.26", *American Notes and Queries* 19.1 (2006) 9–16.

Skinner P., *Health and Medicine in Early Medieval Southern Italy* (Leiden – Boston, MA: 1997).

Smith V., *Clean. A History of Personal Hygiene and Purity* (Oxford: 2008).

Toscano G., "Rapicano Cola", in Bollati M. (ed.), *Dizionario biografico dei miniatori italiani* (Milan: 2004) 893–896.

Ventura I., "Cultura medica a Napoli nel XIV secolo", in Giancaro A. – Perriccioli Saggese A. – D'Urso T. (eds.), *Boccaccio Angioino* (Bern: 2012) 251–288.

Wallis F. (ed.), *Medieval Medicine: A Reader* (Toronto: 2010).

Witthoft B., "The Tacuinum Sanitatis: A Lombard Panorama", *Gesta* 17.1 (1978) 49–60.

Zuccolin G., "Ruolo ed evoluzione della balneoterapia nel pensiero scientifico-medico in Italia dal XII al XVI Secolo", in Ciardi M. – Cataldi R. (eds.), *Il calore della terra: contributo alla storia della geotermia in Italia* (Pisa: 2005) 98–115.

PART 2

Images of Seas and Rivers

CHAPTER 4

Opicinus de Canistris's Demonic Seas

Karl Whittington

The sea is ubiquitous in the art and literature of medieval Europe, and increasingly in the modern scholarship on the period. Amidst the current turn towards the study of cultural exchange and interaction within and beyond medieval Europe and the Mediterranean world, the sea has emerged as a unifying metaphor, demonstrating the fluid exchange and movement of ideas, people and things. As multiple humanistic disciplines work to connect previously disparate geographies of study, the sea often becomes the medium and stage for that connection; the emerging dialogues, book series, and job postings on the study of the 'Atlantic World' or 'Mediterranean World' reveal the ways that these (and other) seas come to act in the humanities as connecting zones. The Mediterranean, in particular, has long been cast as the site of a shared cultural zone by scholars such as Fernand Braudel, Peregrine Horden, Nicholas Purcell and many others.[1]

But the sea is a multivalent symbol, and not always a positive one for medieval thinkers. Often staged by scholars today in historical terms, either as a site of connection or a space of conflict, it is imperative as well to explore the sea as a site of imagination and projection for medieval thinkers and artists. The drawings and writings of the fourteenth-century Italian priest Opicinus de Canistris offer a revealing case study of the contradictory ways in which the sea was conceptualised both as a stable space for navigation and also as a dangerous, even demonic force. The study of the sea as a symbol in medieval art, philosophy, literature and theology is a vast topic, and one particularly well-studied by literary historians; what I hope this modest examination of Opicinus's work will reveal are the ways in which radically different ideas about the sea and its depths coexisted and, in Opicinus's drawings, were literally superimposed.

1 Braudel F., *La Méditerranée et le Monde Méditerranéen a l'époque de Philippe II* 3 vols. (Paris: 1949), and Horden P. – Purcell N., *The Corrupting Sea: A Study in Mediterranean History* (Hoboken: 2000) are two of the most well-known large-scale studies of the subject, but the bibliography is vast. An excellent reference source, with extensive bibliography and thematic essays, is Horden P. – Kinoshita S. (eds.), *A Companion to Mediterranean History* (Oxford: 2014). In art history, Brill has a new book series on "Mediterranean Art Histories", to name just one of many examples.

 | DOI:10.1163/9789004696501_006

Opicinus's strange and idiosyncratic work is slowly becoming more familiar to modern audiences through a series of recent books, including my own 2014 study of his drawings and maps.[2] Opicinus was born in Lomello, a small town west of Pavia in Lombardy, on December 24, 1296. His large family, the Canisters, were prosperous Pavians, actively involved in the political and religious life of the city. He was directed towards a religious career from an early age and received extensive education. He was ordained in Parma in 1320, at the age of 24, and in 1323 he took over the Church of S. Maria Capella in Pavia. In July 1328, Opicinus left Pavia and his parish, presumably as a result of further political turmoil in the city. After a period of transience, he settled in Avignon in 1329, where he describes meeting Pope John XXII, whom he greatly admired; in December 1330, he was appointed as a scribe in the papal penitentiary, where he would stay, as far as we can tell, for the rest of his life. It was four years later in Avignon, in 1334, that Opicinus fell ill, and underwent what has been variously characterised by scholars as a serious physical illness, a psychotic break followed by schizophrenic episodes, a series of visionary experiences, or a religious conversion (it was probably a mix of several of these, though in my own work I have not found the lens of mental illness particularly helpful in analysing his work). In the years following his illness, he continued to work at the papal court, in addition to making drawings in his spare time. Nothing is known about the circumstances of his death, sometime between 1352 and 1355.

Opicinus's drawings survive in two manuscripts, both kept today in the Vatican Library in Rome. One is a kind of day-book or diary, Vat. Lat. 6435, usually called the 'Vaticanus' by modern scholars. It is a medium-size book on paper, containing 90 folios. The first 48 contain little visual material besides a few marginalia, while the second half of the book includes some text-only pages, some full-page drawings, and some smaller drawings with extensive text on or around them. In all, the manuscript contains 36 substantial drawings and maps. The other surviving drawings are found in Pal. Lat. 1993, called the

---

2 For my own work on Opicinus, see Whittington K., *Body-Worlds: Opicinus de Canistris and the Medieval Cartographic Imagination* (Toronto: 2014), and Whittington, "Experimenting with Opicinus de Canistris", *Gesta* 51.2 (2012) 147–173. Other significant publications on Opicinus include Salomon R., *Opicinus de Canistris: Weltbild und Bekenntnisse eines Avignonesischen Klerikers des 14. Jahrhunderts*, 2 vols. (London: 1936), Kris E., *Psychoanalytic Explorations in Art* (New York: 1952); Morse V., *A Complex Terrain: Church, Society and the Individual in the Works of Opicinus de Canistris* (Ph.D. dissertation. Berkeley University: 1996); Roux G. – Laharie M., *Art et folie au moyen âge: aventures et énigmes d'Opicinus de Canistris* (Paris: 1997); Roux, *Opicinus de Canistris: prêtre, pape et Christ ressuscité* (Paris: 2005); Laharie M., *Le journal singulier d'Opicinus De Canistris* (Vatican City: 2008); and Piron S., *Dialectique du monstre: enquête sur Opicino de Canistris* (Brussels: 2015).

'Palatinus'. This 'manuscript' is a collection of 27 huge unbound parchment sheets, averaging about two feet by three, although some are even larger. On either side of each sheet is a single large drawing or diagram. There is every reason to expect that Opicinus made many more drawings that those contained in these two collections, but there is no trace of them in the historical record; indeed, there is no real record of anyone having read or viewed his work in subsequent decades or centuries until they were rediscovered in the Vatican Library in the early twentieth century.

The most distinctive feature of Opicinus's drawings is that they combine maps with human bodies; as Opicinus himself recorded in his diary, he had a vision during his illness during which 'my interior eyes were opened to discern the images of the earth and the sea'.[3] These 'images' were visions of continents and oceans transformed into human figures – the forms that I call 'body-worlds'. In these visions, Opicinus believed that he had been given insight into the workings of both the structure of the universe but also the complexion of the human soul; by experimenting with the core image of his revelation, he hoped that he would develop a new kind of representational system that would explain the meaning of divine mysteries, which in turn would help to renew the faith of others.[4] At the core of Opicinus's images was a single map, likely copied from an example he owned. Opicinus's maps were based on the most modern and technically accomplished cartography of his day – mariners' sea-charts, which we call portolan charts. To explain what the body-worlds 'mean', one must explore how and why Opicinus harnessed these practical maritime maps to a completely different purpose from that for which they were created. As I explored in my earlier work on Opicinus, he used this technical, practical, scientific cartography to probe deeper into the nature of God and the created world.

A quick introduction to one of Opicinus's drawings from the Vaticanus reveals the basics of their pictorial system. In a detail of folio 79v we see a map of the Mediterranean world – Europe, North Africa, Anatolia and part of the Near East are left the white colour of the paper, and the seas around them are tinted with a reddish-brown wash [Figure 4.1]. Inside of, above, or coexistent

3 This passage is from Vatican Library Vat. Lat. 6435, fol. 53r. The full passage reads, '[...] aliquantulum apertis oculus meis interioribus ad discernendas ymagines terre et maris'. Opicinus describes this same revelation in several other passages, on fols. 75v and 85v. Unless otherwise noted, all translations are my own, and all transcriptions of the Latin text are from Laharie's 2008 edition of this manuscript's text. See Laharie, *Le journal singulier*.

4 On Palatinus fol. 4v, Opicinus highlighted this point through his audacious assertion that his work could be considered the 'last gospel', *'Ecce evangelium novissimim sempiternum'*. See Morse, *A Complex Terrain* 106.

FIGURE 4.1 Opicinus de Canistris, The Mediterranean world. Vatican, Vatican Library, Vat. Lat. MS 6435, fol. 79v, detail. 14th century.

with the continents, we begin to see the human bodies. In this example, Europe is embodied as a man – his head occupies the Iberian Peninsula, his chest and stomach lie in France (where we see a beast in the ocean that tries to bite off his shoulder – more on that shortly), his arm arches up through the lowlands and Germany, and his legs occupy the Italian Peninsula and the Dalmatian Coast. Across the Mediterranean, we see the figure of Africa, in this case depicted as a woman. Her face is to the west, shown in profile as she seems to whisper into the ear of the European figure across the Straits of Gibraltar. Her pointing hand lies in modern-day Tunisia, and her legs and feet occupy Egypt.

The texts and labels written on and around these body-worlds assign identities to the figures; each drawing that includes these bodies and maps changes their identities in order to explore a different theme and series of relationships. Here, the relationship between the two figures is one of violence: a caption near the European figure begins with the words, in Latin, 'behold the defeated temple of the Lord, whose rivers have changed into corruptible blood'.[5] In Europe, we see this figure of defeat – his eyes are closed, his mouth frozen, his arm hangs limply at his side, and Europe's rivers are depicted as red, bloody wounds. The perpetrators of this violence can be found on all sides – in the huge beast that gnaws at his shoulder, in the tiny dragon that bites his heel, and most of all in the figure of Africa. Her caption identifies her as the 'armed woman, who represents the descendants of the corrupted world, sowing division within the Lord's temple'.[6] She sits in calm triumph with her sword at her side. Thus, the two main figures are personifications, and they stage an encounter that is, on the surface, straightforward. Corrupt human beings (Africa) have sowed division in the church (Europe), wounding it and bringing it to its knees.

1 Stable Seas

The kinds of allegories that Opicinus stages within these unusual drawings – their sources, potential meanings, and strange formats – have been explored elsewhere by myself and others. But as we begin to think in this chapter about the role of the sea in Opicinus's work, the first thing to note is the incredible empirical accuracy of the maps from which Opicinus copied his cartographic

5 Vatican Library Vat. Lat. 6435, fol. 79v: '*Ecce prostratio Templi dominici, cuius flumina conuertuntur in sanguinem corruptibilem* […]'.

6 Vatican Library Vat. Lat. 6435, fol. 79v: '*Ecce…quelibet corruptibilis progenies mundi, quasi mulier armata que arripiens sagittam Herculis…pro cuius ambitione totum templum Domini point in scismata uel scissuras*'.

forms. The briefest glace at the drawings reveals that their forms are copied from portolan sea-charts; no other cartographic tradition existed at the beginning of the fourteenth century whose forms were so accurate in shape, scale and proportion.[7] Medieval chart-makers had begun producing portolan charts in maritime centres like Venice, Majorca, Genoa and Constantinople sometime in the twelfth or thirteenth century.[8] An example made by Angelino Dulcert in Majorca in 1339 demonstrates the tradition in Opicinus's day (Figure 4.2).[9] The chart's coverage ranges from Gibraltar to Jerusalem, Scotland to Carthage, and its orientation is seaward, with place-names arranged along the coastlines while the interiors are less detailed. Networks of grids, called rhumb lines, cover the charts, indexing the land below and providing

7 The literature on early portolan charts is extensive. The two foundational reference texts for modern scholars are Campbell T., "Portolan Charts from the Late Thirteenth Century to 1500", in Harley J.B. – Woodward D. (eds.), *The History of Cartography* (Chicago: 1987) 371–463, and Pujades I Bataller R., *Les cartes portolanes: la representació medieval d'una mar solcada* (Barcelona: 2007), the text of which is translated into English in the last portion of the book, as *Portolan Charts: The Medieval Representation of a Ploughed Sea.* Both authors review the terminology, origins, features, production, use, toponymy and dating of the charts during this period, and Pujades includes the most up to date catalogue of extant examples and an extensive bibliography. The most important early texts in the field are Nordenskiöld A.E., *Periplus: An Essay on the Early History of Charts and Sailing Directions*, trans. F. Bather (Stockholm: 1897); Kretschmer K., *Die Italienische Portolane des Mittelalters* (Berlin: 1909); Stevenson E.L., *Portolan Charts: Their Origin and Characteristics with a Descriptive List of Those Belonging to the Hispanic Society of New York* (New York: 1911), and Taylor E.G.R., *The Haven-Finding Art* (New York: 1971). Important recent contributions include Freiesleben H.C., "The Still-Undiscovered Origin of Portolan Charts", *Navigation* 36 (1983) 194–199; Du Jourdin M.M. – de la Roncière M., *Sea Charts of the Early Explorers* (New York: 1984); Lanman J., *On the Origin of Portolan Charts* (Chicago: 1987); Casey E., *Representing Place: Landscape Painting and Maps* (Minneapolis, MN: 2002); Monmonier M., *Rhumb Lines and Map Wars* (Chicago: 2004); Edson E., *The World Map, 1300–1492: The Persistence of Tradition and Transformation* (Baltimore, MD: 2007), Falchetta P., "The Use of Portolan Charts in European Navigation during the Middle Ages", in Baumgärtner I. – Kugler H. (eds.), *Europa im Weltbild des Mittelalters: Kartographische Konzepte* (Berlin: 2008) and Pujades, *La carta de Gabriel de Vallseca de 1439*. A key recent reference text on the charts' continuation into the Renaissance is Astengo C., "The Renaissance Chart Tradition in the Mediterranean", in Woodward D. (ed.), *The History of Cartography: Cartography in the European Renaissance* (Chicago: 2007) 174–262.

8 Dalché pushed back the possible date for such an operation considerably. He discovered a text, which he dates to 1160–1200, called *Liber de Existencia Riveriarum et Forma Maris Nostri Mediterranei*, which he argues was made in Pisa, and was accompanied by a chart, now lost. See Dalché P., *Carte marine et portulan au XIIe siecle* (Rome: 1995). Edson, *The World Map* 43–47, contains an excellent summary in English of Dalché's findings.

9 The Dulcert Chart is now GE B 696 in the Bibliothèque Nationale in Paris. See Pelletier M., "Le Portolan d'Angelino Dulcert, 1339", *Cartographica Helvetica* IX (1994) 23–31; Campbell, "Portolan Charts" 378; and Du Jourdin and De La Ronciere, *Sea Charts* 201.

FIGURE 4.2 Angelino Dulcert, *Dulcert Chart*, 1339. Paris, Bibliothèque Nationale, Ge. B 696 (photo: BnF). REPRODUCED WITH PERMISSION OF THE BNF

the means to measure both distance and direction, with the aid of the six scale bars that lie across its surface. These new maps were based on empirical science: their measurements were taken using the new technology of the nautical compass, and the scientific principles behind their construction were shared by scientists, geometers, and urban engineers whose work paved the way for the perspective paintings and geometrically-planned urban spaces of the fourteenth and fifteenth centuries.[10]

10 The portolan charts' empiricism was long debated in the scholarship. But many of these questions may be settled in the wake of Pujades' landmark 2007 book, which explains in compelling and convincing detail the ways in which the portolans were based on scientific measurement, particularly with compasses. On these connections between art, geometry and engineering in the fourteenth century, see especially Trachtenberg M., *Dominion of the Eye* (Cambridge: 1997), Edgerton S., "Florentine Interest in Ptolemaic Cartography as Background for Renaissance Painting, Architecture, and the Discovery of America", *The Journal of the Society of Architectural Historians* 33 (1974) 275–292, and Edgerton, *The Renaissance Re-discovery of Linear Perspective* (New York: 1975).

Thus, the portolan charts that Opicinus used as the raw material for his drawings convey a particular understanding of the sea as an entity that was stable and measurable. Indeed, while the charts were produced on land, the viewpoint of their data was a mariner's-eye view. All of the detail in the portolans ultimately comes from the viewing and measuring of the sea and coastlines; the incredible overall accuracy and proportionate scale of the landforms (which is the portolans' most notable feature for modern viewers) is a convenient by-product of their accurate mapping of the shapes of the seas (more accurate in the Mediterranean and Black Seas, and progressively less accurate in the North Atlantic and North Sea). This measurability and empiricism are not only contained within the portolans' landforms, but are also further emphasised by the grids placed across their surfaces. These networks of rhumb lines are usually arranged around two sixteen-point circles that cover the Mediterranean region; the sixteen points correspond to the winds: eight primary winds and eight half winds.[11] Through the grids created within the intersections of these lines, the space of the chart is advertised to the viewer as quantifiable. Samuel Edgerton and others have argued that portolan charts were primarily used to calculate angular directions, but Lanman's and Pujades's analyses of their origins, and the recognition of the grids on the chart, suggests that they precisely measured space as well.[12]

Opicinus's own writings provide evidence of how the charts were used, and the confidence with which navigation of the sea was undertaken. On fol. 59r of the Vaticanus manuscript, he writes:

> In the map of the navigable ocean drawn up by the Genoese and the Majorcans, one can calculate exactly the number of feet with the aid of the miles indicated on the corners of the map. Each point corresponds to 10 miles (that is to say, the distance between two points). After this, five

11 In the Italian system, the eight primary winds were known by traditional names: *Tramontane* (N), *Ostro* (S), *Levante* (E), *Ponente* (W), *Greco* (NE), *Sirocco* (SE), *Maestro* (NW) and *Libeccio* (SW). Other systems were also in use; Taylor briefly discusses the interesting example of Matthew Paris's wind-rose drawing which attempted to combine a 12-point Latin system with a 16-point Norman-French system. See Taylor, *The Haven-Finding Art* 98–99.

12 Edgerton stated that portolan charts contained 'good approximations of angle and direction, but not of distance'. See Edgerton, "Florentine Interest in Ptolemaic Cartography as Background for Renaissance Painting, Architecture, and the Discovery of America" 282, and Edgerton, *Linear Perspective*. In the first chapter of *Body-Worlds*, I demonstrated the ways in which the scale bars on these charts could be used in tandem with the rhumb-line grids to quickly estimate distances on the map.

> of these spaces, there is an elongated rectangle, and the space between each of these is 50 miles. And these rectangles, multiplied by seven, permit one to calculate with precision a distance of 7 times 50, or 350 miles, divided into feet. And if one can precisely make an estimation of a distance in the sea, it is no more difficult to measure a distance on land.[13]

The measurability and accuracy of these maps of the sea is taken for granted; Opicinus's concluding point shows that he is also aware of how accurate their landforms were as well. Their networks of rhumb-line grids allowed that space to be measured or estimated while their empirical 'look' also suggested that rational space to the viewer more abstractly as well, as I have demonstrated elsewhere.[14]

This vision of the sea and its waters as measurable, traversable, and under the control of the mariner's vision and tools was the foundation for the creation of the portolan charts; indeed, the incredible (and still mysterious) process by which the first examples were compiled/created took this stable conception of the sea as its starting point and then demonstrated it through acts of making and representation. And the creation of the portolan charts, of course, coincides with the periods in which the kinds of travel and cultural exchange were increasingly taking place which themselves are often used as evidence to characterise seas as places of dynamic exchange in the Middle Ages. In the twelfth, thirteenth, and fourteenth centuries, the powerful Italian maritime republics of Pisa, Venice, and Genoa, along with their Islamic and Norman counterparts, formalised trade routes across the Mediterranean and into the Atlantic, partly through the increased use of pelagic (open sea) routes that took mariners far from the safety of the shore, emboldened by the new nautical compasses and charts.[15] All of this is implicit within Opicinus's representation system, based as it was on the empiricism of the portolan charts, of which we know he was well aware.

---

13 Vaticanus fol. 59r: '*In mappa maris navigabilis secundum Ianuenses et Mairicenses, habetur certa taxatio passuum per miliaria que in lateribus huius carte. Per quodlibet punctum denotat X miliaria (scilicet in spacio inter punctum et punctum). Deinde per V spacia ipsa facit unam quadraturam oblongam, que per quinquies X computat L. Que quadrature per* VII *faciunt certam taxationem in longum habentem septies L, id est* CCCL *miliaria passuum. Habita certa descriptione maris, sine dubio intelligitur certa descriptio terre*'.

14 See Whittington, *Body-Worlds* Chapter 1.

15 While most of the portolans that survive from this period are luxurious display copies, references to their use (by authors such as Opicinus and many others) suggest that more utilitarian copies must have been used on board ships at the time.

2 Demonic Waters

This conception of the sea as a stable and measurable space, however, directly abuts another vision of the sea that surfaces within a number of Opicinus's drawings, which itself has a long history in the region: the sea as an unpredictable and even evil space, both because of its real dangers and its cultural symbolism as a space of demons, dragons, and sin.[16] While Europe and Africa are embodied as a variety of different figures throughout Opicinus's drawings, the Mediterranean sea is only ever given one identity, as a devil or monster; indeed, Opicinus foreclosed any other possibility when writing on folio 83v of the Vaticanus that 'the Mediterranean can take no other form except that of the diabolical sea'.[17] The devil rises out of the sea in about one-quarter of Opicinus's drawings of the body-worlds; in the rest the sea is left blank, though the devil's presence there is sometimes implied textually. The incredible visual appearance and presence of the Mediterranean devil (as well as several other beasts that emerge out of the sea) are one of the most captivating and mysterious elements of Opicinus's drawings.

A typical example (if such a thing exists in Opicinus's complicated oeuvre) is the drawing on Vaticanus folio 69v [Figure 4.3]. Here, in a complex cartographic structure that involves two different portolan charts partly superimposed, we see the figure of the devil emerging in red in the negative-space sea of the lower map. His head is in profile in the eastern Mediterranean, with a single eye and grimacing mouth pointed towards Anatolia; the unruly beard forms the Aegean Sea; his bent left elbow forms the sharp bend northward at the border of modern-day Libya and Tunisia; his right fist plunges in a violent gesture forming the Adriatic Sea (between the two legs of the European figure); his crossed legs occupy the Western Mediterranean between the Iberian Peninsula and north-west Africa; and his oversized penis abuts the coast of France and Spain from Marseille to Barcelona, ejaculating or urinating onto the 'chin' of the Europe-figure in Spain. With this figure, the empty space of the sea between the landforms is activated as site of demonic power and instability. The sea-devil in this example is actually labelled as 'Lucifer' in small text off the coast of the Holy Land.

---

16 This idea of a dualistic sea, both peaceful and terrifying, is a trope also discussed in the introduction of Sobecki's book, *The Sea and Medieval English Literature*. See also, Karstens S., "Imagine Oceans: The Many Meanings of the High Seas in the Time of Early European Expansion" 199–202, 214–218.

17 Vaticanus folio 83v: '*Mediterraneum mare non potest formari in aliud nisi in diabolicum mare*'.

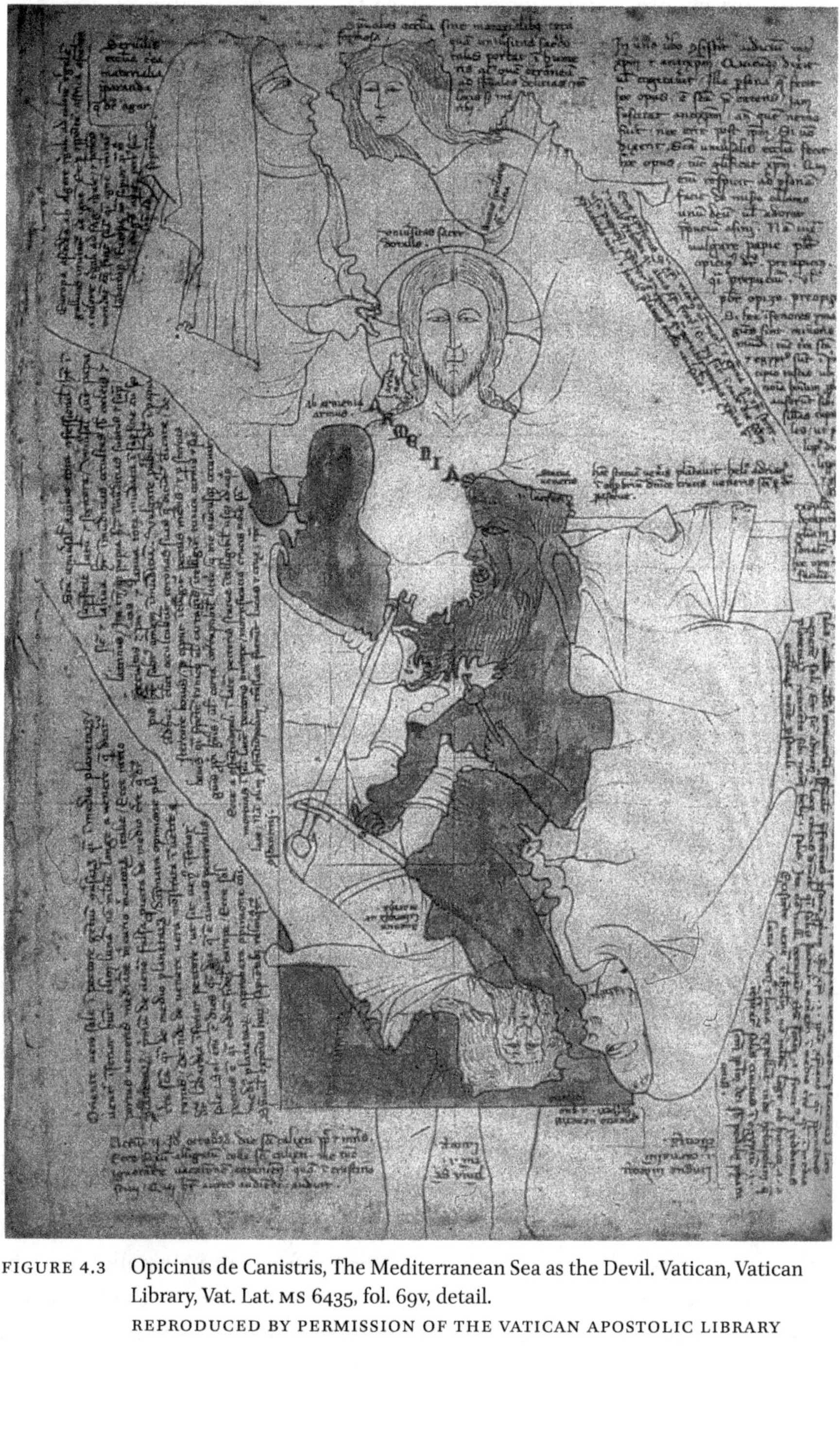

FIGURE 4.3 Opicinus de Canistris, The Mediterranean Sea as the Devil. Vatican, Vatican Library, Vat. Lat. MS 6435, fol. 69v, detail.

In this short chapter I don't have the space to fully unpack this, or any individual drawing; the extensive textual portions and complex superimpositions of each page make even a brief analysis nearly impossible. The basics of this drawing, however, create a contrast between what Opicinus labels as a 'spiritual church' and 'servile church' at the top of the page, both embodied as women and accompanied by a standing figure of Christ (not corresponding to any geography) labelled as the 'sacerdotal community', and a second map below consisting of the laity personified as the 'penitent nations' of Europe and the 'carnal clerics' of Africa. It is the sea between these more negative male continents at the bottom of the page (one of which is armed with a sword) that is embodied as Lucifer; the penitent and carnal figures are separated by this sea of sin. In another drawing with a similar format and theme on folio 61r, a map of 'spiritual Africa' and 'spiritual Europe' (with a pure, empty sea between them) is contrasted with a map of the 'natural' world: 'natural Africa' and 'natural Europe', with another devil between them [Figure 4.4]. This example also adds an allegory of the external versus internal senses which lead one to sin or to the understanding of God, respectively. Written across the face of the sea-devil are labels for the external senses that lead to sin: *auditus* near the ear, *gustus* near the mouth, etc. Again, we see the idea that it is the sea of a natural or carnal world that contains the devil, while the sea of a purer spiritual world is left pure.

In neither of these drawings is the devil the only monstrous force that emerges from the sea. On folio 69v [Fig. 4.3] we also see a beastly face in profile that bites at the 'shoulder' of Europe, with Brittany caught between its jaws. This is the Tarasque, a mythical demon usually associated with the Rhone River (it appears there on folio 61r, figure 4.4), who is most well-known as the beast tamed by Saint Martha, a story relayed in the *Golden Legend* (and from a region just adjacent to Avignon, where Opicinus lived). Though its head appears here looking more like that of a lion or bear, the Tarasque was usually understood to be a giant sea serpent, the offspring of the mythical Onachus and Leviathan (of the Biblical account in Job and elsewhere). On folio 69v the beast has left the river and now occupies the North Atlantic as it bites the shoulder of Europe; such a placement likely just came to Opicinus because of a similarity he happened to see between the landform of Brittany and a space caught between jaws, though in his texts and captions he tries hard to find meaning in these kinds of coincidences. In addition to the Tarasque, folio 69v also contains a small serpent along the coast of western North Africa, and a small dragon in north-eastern Greece, with its tail running up the north coast of the Black Sea. These beasts, which reappear on folio 61r and several other folios, are all within

FIGURE 4.4 Opicinus de Canistris, 'Natural' Africa and 'natural' Europe with the Mediterranean as the Devil. Vatican, Vatican Library, Vat. Lat. MS 6435, fol. 61r, detail.

or adjacent to the sea, and seem to constitute part of its demonic presence, though Opicinus never explicitly unites them in this way in his text.

In his extensive writings, Opicinus comments occasionally on these figures of devils in the sea (and he has much to say about sin, deception, and the devil in general): in one passage, for example, the Leviathan represents falsity, and Tarasque 'sets traps' for men (folio 52v). But the specifically sexual and aggressive aspects of the Mediterranean sea-devil are emphasised in unique ways in numerous drawings and captions, particularly on folios where the devil-sea is superimposed over another map, either of the entire region or sometimes just over a map of Opicinus's home town, Pavia; the locations of certain parts of his body are interpreted as indicating the corruption of those places which he overlaps. And the obscene, aggressive sexuality of the figure is always emphasised when he is present. His penis, near Spain, always directly abuts the face of the Europe figure (as either a man or a woman) and often seems to suggest the threat of corruption to that body. This is further signalled by the occasional presence of a second penis; we see it on folio 84r being pecked by a bird while clasped in the hand that punches up the Adriatic; Opicinus even references the later addition of the 'obscene member in the hand of the devil' in a caption near the top of the page [Figure 4.5]. Though this aspect of its content is not our focus, the conceit of this drawing (and two others that follow it) is that Opicinus has superimposed over the body-worlds a rectangular city-map of Pavia, and then seeks correspondences between the two. Commenting on the presence of this second penis near his hometown of Pavia, Opicinus writes in another caption of being unsure of whether he was 'inside or beside of the testicles of Leviathan', making it clear that Opicinus, like the patristic authors explored below, has conflated the biblical Leviathan with Satan/Lucifer.[18]

Twenty folios earlier, on 67v–68r, Opicinus seems to comment on this drawing from 84r, and in particular on the meaning of this sexual imagery in the sea. He writes that these erect penises symbolise the swollen pride of the devil (or of the sinner) and describes how as a result the 'abominable member' is 'hung' on the breast of Europe (referring to the Penis under the face of Europe in Spain). Elsewhere he writes that those who boast of their accomplishments become swollen like the sea; sexual relations, he indicates, can metaphorically be good when they take place between continents, but bad and even monstrous when performed with the sea.[19] This is part of a long and complicated passage, but in it we see some of the ways in which the sexuality of the sea-figure is

18 Vaticanus folio 84r: '*Additum de ignobli membro in manu diaboli anno renovationis, XV kalendas februarii*' and '*sine in mundo naturali fuerim intra vel iuxta testiculos Leviathan*'.

19 On these passages, see Morse, *A Complex Terrain* 218–219.

FIGURE 4.5 Opicinus de Canistris, The Mediterranean as the Devil. Vatican, Vatican Library, Vat. Lat. MS 6435, fol. 84r, detail.
REPRODUCED BY PERMISSION OF THE VATICAN APOSTOLIC LIBRARY

central to his sinful identity. On folios 33v–34r, Opicinus further describes the sexual aggression of the sea in relation to the body of Europe, suggesting that its insemination of Europe is one of the ways in which evil enters it: 'Now the diabolical sea [...] has impregnated this belly of Europe; it shall be seen on the day of the birth what kind of offspring it will be'.[20] It then follows that the context for this strange passages is an imaginative suggestion by Opicinus that the sea has impregnated Europe, leading to division within Christendom: the twin babies that are born are *another* version of the Africa/Europe body-worlds, this time diabolical (this relates to the imagery on the drawing on folio 87r). In similar imagery on folio 47r, he writes of the devil's ejaculate that, 'this diabolical seed, similar to the great sea, its father, immediately corrupts and divides the good seed [...] the diabolical seed that plants schism in the Christian people is obstinate affection for the corruptible flesh and blood of one's own house'.[21] Thus, the whole thing is an elaborate metaphor, but one which rhymes with the sexually aggressive forms and gestures assumed by the sea-devil in numerous drawings.

This idea of these sea-beasts as not only sexually sinful but also violent is apparent if we return to folio 79v, the drawing we started with in which Europe is depicted as wounded (figure 4.1). Opicinus wrote on the folio, 'Behold the defeated Temple of the Lord, whose rivers have changed into corruptible blood. Behold the numerous tears of flesh, the crushing of limbs and bones'. Here it is the Tarasque's attack on Brittany that is pictured, rather than the devil's Mediterranean body. But the pollution of blood from the rivers/wounds has reached the seas surrounding all of the figures, and the presence of these rivers/wounds in the Mediterranean and Black seas suggest that these coasts, too, were under attack from the water, though their attackers are not pictured. The sea is often pictured as a ruddy, bloody red colour, though this is the only drawing in which the rivers are explicitly depicted as the wounds that pollute it.[22]

While the representation of the sea as a series of aggressive monsters and devils and a space of sexual sin and corruption seems idiosyncratic (and, indeed, the ways in which they are visualised here *are* unusual), these ideas are based on a long-running understanding in Christian theology (and pre-Christian Judaic and pagan philosophy) of the sea as a demonic space,

20 Vaticanus fols. 33v–34r: '*Nunc diabolicum mare, ex consensus voluntatis eorum, impregnavit hunc ventrem Europe, cuius fetus in die partus videbitur qualis erit*'. See Morse, *A Complex Terrain* 224.

21 See Morse, *A Complex Terrain* 225.

22 In another passage, Opicinus describes the Europe-body as expelling a demon from within itself, through the birth 'canal' of Venice. See Morse, *A Complex Terrain* 221–230.

particularly among patristic authors. From his writings we know that Opicinus was incredibly well-read, appreciating the works of patristic authors like Augustine and many others, and I think it is likely that the presence of the devil within the sea was inspired by precisely this theological tradition. Muriel Laharie has pointed out places in the Vaticanus manuscript where Opicinus's language clearly evokes biblical descriptions of the Leviathan in Job 40, to name just one example.[23] The patristic understanding of the sea as demonic is outlined in meticulous detail by Hugo Rahner in his 1964 book on different symbolic or metaphorical conceptions of the church as a whole.[24] As part of a series of chapters on the symbolism of the Church as a ship, Rahner traces the origin and theological expression of the ocean as both 'the bitter sea' (a space of turbulence, danger, and death) and 'the evil sea' (a space of monstrosity and evil). The ship of the Church, built with the wood of the cross, is then understood as needing to navigate these turbulent waters. The metaphor is then elaborated to suggest that the waters are the sea of the world, of regular human life, which must be navigated through the aid of the Church to reach heaven.

The first conception, of 'the bitter sea' as a space of storms and chaos, has a long history in the philosophical, literary and theological texts of the ancient world, which is no surprise when thinking about a Mediterranean culture for whom sea storms were a relentless and unpredictable part of everyday life.[25] Tibullus offered the memorable phrase, 'Now the sea slays, and there are a thousand ways of sudden death', summarising a whole way of viewing the danger of the ocean's waters.[26] Sophocles wrote of the relentlessness of sea tides; Aeschylus wrote of 'floods of evil' in language echoed later by Ovid; Menander wrote that there were 'three evils in the world: the sea, fire, and women'.[27] In his *Consolations*, Boethius summarised this whole antique conception with his words, 'He who has ground proud fate beneath his heel / Calm in his own well-ordered life / And has looked in the face good and ill fortune / Still able to keep

23 Laharie finds these references on Vaticanus folios 33r and 38r. See Laharie, *Le journal singulier* 321 and 371. For an interesting summary of different conceptions and precursors of the Leviathan, see Wilson, "A Return to the Problems of Behemothand Leviathan" 1–14.

24 Rahner, *Symbole der Kirche* 272–303.

25 Sobecki also discusses some of these origin stories for a demonic conception of the sea in his introduction.

26 Tibullus, Elegy I, 3. Cited in Catullus, *Tibullus. Pervigilium Veneris* 209.

27 Aeschylus, "Sieben gegen Theben" 758, Ovid Trist. II, 99–102, Menander, Monostichoi 231. See Rahner, *Symbole der Kirche* 277.

erect his unconquered head / He shall not be troubled by the rage or threats of the sea / Driving the turning tide up from the deep'.[28]

Biblical imagery, particularly from the Old Testament, also makes use of the 'bitter sea' metaphor for a variety of images; as Rahner details, the image of the sea is used to express 'the power of the creator over nature, the roar of the people and the armies of war, the raging of the wicked, the depths of bitter pain, the death of the soul, and the nothingness of human life'.[29] But the primary image taken up by Christian and especially patristic authors is that of the rough sea as a metaphor for life itself (the metaphor of life as a sea voyage being traced by many classical authors back to Homer's *Odyssey*). Numerous pre-Christian authors speak of death as the final 'port' on the sea-journey of life – the 'port of Hades' or 'port of death', as Sophocles wrote in *Antigone*. Sea water's unique qualities, particularly its saltiness and bitterness, were used to play out this metaphor of life's journey as bitter, a link memorably described in Isidore's *Etymologies*, when he links 'amarum', the word for 'bitter', and 'mare', for 'sea'.[30]

But the trajectory that Rahner outlines so convincingly in the vast sea of patristic writings that followed casts the bitterness and danger of the ocean in more explicitly demonic terms, following numerous references in the Bible to either specific sea monsters or to more abstract language of evil (Rahner's argument is essentially that the view of the sea that emerges in the writings of the Church Fathers is a combination of classical/literary notions of the bitter sea and Judeo-Biblical notions of the evil or demonic sea).[31] Rahner demonstrates that the images of multiple demonic figures in the Bible become linked or overlaid in the image of the sea as monstrous: the serpent from Eden, the whale from Jonah, the Leviathan from Job, Isaiah and the Psalms, the dragon in Revelations, the Whore of Babylon, and even Satan more generally.[32] The beast of Revelations 13, for example, 'rise[s] up out of the sea, having seven heads and ten horns', and the Whore of Revelation 17 'sitteth upon the waters which are the peoples, multitudes, nations, and tongues' (here furthering the idea of the sea of this world and this life). In Isaiah 27, 'the Lord will punish with his sword […] Leviathan, the gliding serpent, Leviathan, the coiling serpent;

28 Boethius, *Consolations* 1, 4. Cited in Boethius, *Theological Tractates* 145.

29 See Rahner 278, for the citations of all of these biblical passages.

30 Isidore, Etymologies XIIIU.14.1.

31 These biblical images of sea demons certainly have roots, Rahner argues, in the classical world, related to sea gods like Typhon, a sea-god often described as a snake. See Rahner 285.

32 On the wide variety of biblical animals and beasts the become associated with Satan, see Russell J.B., *Lucifer: The Devil in the Middle Ages* (Ithaca, NY: 1984) 67; on the merging of these in patristic authors like Gregory, see Russell, *Lucifer* 97.

he will slay the monster of the sea'.[33] These specific figures merge with other more literary biblical passages that describe a 'great monster lying amongst your streams' or 'monster in the seas, thrashing about in your streams, churning the water with your feet, and muddying the streams' (Ezekiel: 29 and 32) or 'a beast that comes up from the abyss that will attack them, overpower them, and kill them' (Revelation: 11). Critical to many of these passages is imagery of 'the deep' or 'the abyss' – the idea of the sea as a place of darkness where evil lurks hidden in the depths. Sobecki demonstrates that the notion of the 'abyss', going back to Babylonian and Judaic creation myths, is conceived as a dark body of water that has existed since the beginning of time, and is a primordial element tamed by the waters of the surface, whose job it is to suppress it.[34]

Authors such as Augustine and Gregory take these biblical references and expand them into a whole theology of the sea as a space of sin, by turns demonic, general, and individual.[35] Parts of their writings correspond more generally to the notions of the sea that we have seen; in his eleventh homily, Gregory describes the ocean as 'deep eternal death', and Augustine wrote of the sea's 'deep impenetrable waters'. In his commentary on Psalm 103, Augustine links Job's discussion of the behemoth and Leviathan to the devil himself: 'when the devil fell from heaven he received this region [the sea]'.[36] Basil of Caesarea and John Chrysostom elaborate the image further, describing the struggles of the early Church against heresy as a great sea battle, between the ship of the Church and the 'great pirate', the devil.[37] Both Tertullian and Origen also link the devil with the dragon, the Leviathan, and the paradise serpent; most memorably, Augustine describes how 'this serpent then, our ancient enemy, glowing with rage, cunning in his wiles, is in the mighty sea'.[38] Once the devil himself is linked with these ancient sea-monsters, these authors are able to begin to conceptualise the sea more broadly as a space of sin, filled, as Augustine also wrote, 'with unbelievers still, dwelling in barren and bitter waters'.[39] In an extended passage in his commentary on Psalm 36, Augustine

33 On the Judaic origins of this imagery, see Pagels E., *The Origin of Satan* (New York: 1995) Chapter 2.

34 Sobecki, *The Sea* 34.

35 The notion of the sea as a space of sin is also found in the Old Testament: the book of Micah (7.19) describes how sins are 'cast into the depths of the sea'. Origen also describes the Sea as the seat of the Devil, and Hilary of Poitiers writes of 'the bottom of the sea as the seat of hell'. See Sobecki, *The Sea* 38.

36 Augustine, Commentary on Psalm 104. Tweed J.E. (trans.), *Nicene and Post-Nicene Fathers*.

37 Basil, *De spiritu Sancto* 30, 76 and Chrysostom, *Homilia in illud Vidi Dominum* 3.

38 Augustine, Commentary on Psalm 104. Tweed (trans.), *Nicene and Post-Nicene Fathers*.

39 Augustine, Commentary on Psalm 104. Ibidem.

writes of the abyss, 'Your judgments are like the great abyss. The abyss he calls the depth of sin, whither everyone comes by despising God [...] Because then they were proud and ungrateful, they were held worthy to be delivered up to the lusts of their own hearts, and became a great abyss [...]. In the Church of God you find an abyss, [but] you find also mountains [...] the mountains are few, the abyss broad [...] fly from the abyss, tend towards the mountains'.[40] As Rahner explains, it is out of these metaphors of the sea as a space of sin that must be navigated in life that Augustine and others create the metaphor of the Church as a ship; we have only scratched the surface of such metaphors, hundreds of which can be identified across a wide range of patristic authors.

3 Surfacing

It is this understanding of the sea, both as a monster or devil but also as a space of sin, that Opicinus explores so expressively in numerous drawings; as he summarises on folio 19v of the Vaticanus, the sea symbolises 'the human potential for wickedness'. This is both explored textually in his writings on the sources of sin and mechanisms of its spread, and in his evocative images of the sea as a series of devils or monsters. While imagery of sea monsters and marvels became common in later maps and sea charts from the fifteenth century onward, depicting monsters or devils in the sea was rare in medieval cartography; the Beatus Apocalypse world maps are the primary example but bear no visual relation to Opicinus.[41] Thus, his inspiration seems more likely textual in this case than visual. Sometimes the beast is the single Mediterranean figure labeled as Lucifer; other times it is a series of smaller dragons and beasts in the water or along the coast; and still other times the imagery of the sea seems simply to rhyme with Opicinus's extensive discussions of personal sin and the impossibility of avoiding it. As I have argued elsewhere, the strict dualism set up by so many of Opicinus's drawings, where the figure of Africa is sharply juxtaposed against Europe, is undermined by the very fact that the figures are drawn in/on maps: the sea between these figures is one that demands to be navigated, despite its dangers, and indeed many times the figures reach across or through this space of the sea to interact with, attack, or seduce each other.[42]

40 Augustine, Commentary on Psalm 36. Ibidem.

41 On the tradition of sea-monsters on maps, see van Duzer C., *Sea Monsters on Medieval and Renaissance Maps* (London: 2013).

42 See the discussion of Vaticanus fol. 61r in Whittington, *Body-Worlds* Chapter 2.

The concept of the water's surface can help bridge the two divergent conceptions of the sea in Opicinus's drawings: the stable world of the portolan chart and the demonic world of the sea. One vision looks at the sea's waters horizontally: the view from the ship towards the land, or the figure on the shore looking at the horizon.[43] It is the gaze of one who travels across these waters, who sees them as a medium for movement and transition. The other vision is one down *into* the ocean's waters, probing their depths and visualising the way those depths can come to the surface to wreak havoc. The insistent language of the abyss in biblical and patristic texts made a deep impression on Opicinus, I think, when he thought about the ways in which evil, demons, and sin come into the world; as he tried to imagine a new world-system linking microcosm and macrocosm (which is what his drawings attempted to do), it made sense to him to place evil, both the monstrous evil of the world and the evil of sin within the human soul, within the sea. We have to remember how truly unfathomable and unmeasurable the ocean's depths were in the premodern world; while sailors of course measured the shallow depths of coastal waters as part of navigation, using the technique of lead-and-line sounding, beyond a modest range (rarely over 20 fathoms, or around 36 meters) there was simply no way of knowing the extent of its depths. Into this uncertainty was projected an imagined darkness and savagery, stoked by sightings of creatures both massive and strange, both within the water and washed up on shore.

Opicinus conceived of the portolan charts as showing a view from a divine viewpoint; what appears to us as a measurable view from above, as if from a satellite, was conceived instead as a privileged view from God's perspective down onto the earth. In addition to the practical reason that he was trained in the making of portolan charts as a young man, this accuracy seems to have been a key reason that he based his cartographic drawings on portolans rather than medieval world maps, or *mappaemundi*, the forms of which were less based in scientific observation. We know from his writings that he perceived a real empiricism or truth-value in the shapes of the portolan chart – shapes that were invisible from the earth's surface and only revealed in the charts' new projection and viewpoint. Perhaps from this height, Opicinus suggests, it is possible to view down into the depths of the ocean to see the forms of these devils and monsters that emerge both in the depths and through the detailed shapes of the coastlines. Dante, Opicinus's near-contemporary, who he likely read, explicitly reads the depths of hell as the place of physical impact

43 On the image out to sea from land, see Baader H. – Wolf G., "A Sea to Shore Perspective: Littoral and Liminal Spaces of the Medieval and Early-Modern Mediterranean", *Mitteilungen des Kunsthistorischen Institutes in Florenz* 56 (2014) 2–15.

created by Satan when he fell from heaven. Canto 34 describes, 'This was the side on which he fell from Heaven; for fear of him, the land that once loomed here made of the sea a veil and rose into our hemisphere; and that land which appears upon this side – perhaps to flee from him – left here this hollow space and hurried upward'.[44] Inside the hole, Satan is frozen in ice, akin to the coldest depths of the sea (indeed, Canto 34 uses the term 'abyss' to describe its space). The earth that 'fled' from the impression Satan created was reconstituted as the lands that grew even higher out of the sea (part of which became the mountain of Purgatory). Though it is not stated explicitly, perhaps Opicinus was even inspired by Dante to interpret the Mediterranean as the space of depth imprinted on the earth in the form of the devil's body, with the Satan himself pictured as 'frozen' inside – bound and immobile by the landforms that create the boundaries of his body. Dante's is an allegorical image, but so is Opicinus's, despite being framed within a scientific map, making his ensuing visual metaphor all the more strange. We will never know the exact sources of Opicinus's thoughts on the sea, which he never traces directly in his writings. But what is clear is that his drawings merged both practical and symbolic interpretations of its waters, creating fantastical imagery whose forms still shock us but whose theologies are deeply enmeshed in a series of established ways of both measuring and imagining the sea.

Bibliography

Manuscripts

Rome, Vatican Library, Pal. Lat. MS 1993

Rome, Vatican Library, Vat. Lat. MS 6435

Primary Sources

Boethius, *Theological Tractates. The Consolation of Philosophy*, trans. H.F. Stewart – E.K. Rand – S.J. Tester, Loeb Classical Library 74 (Cambridge, MA: 1973).

Isidore of Seville, *The Etymologies of Isidore of Seville*, trans. S. Barney – W.J. Lewis – J.A. Beach – O. Berghof (Cambridge: 2006).

Kamal Y. (ed.). *Monumenta cartographica Africae et Aegypti*, 5 vols. (Cairo: 1926–1951).

Tibullus, *Catullus. Tibullus. Pervigilium Veneris*, trans. F.W. Cornish – J.P. Postgate – J.W. Mackail – G.P. Goold, Loeb Classical Library 6 (Cambridge, MA: 1913).

---

44 For a recent discussion of this famous passage, see Lachey T.J., "Lettura del canto XXXIV dell'Inferno", *Rivista di Studi Danteschi* 19.1 (2019) 40–58.

Tweed J.E. (trans.), *Nicene and Post-Nicene Fathers, First Series*, vol. 8, ed. P. Schaff (Buffalo, NY: 1888).

Secondary Sources

Astengo C., "The Renaissance Chart Tradition in the Mediterranean", in Woodward D. (ed.), *The History of Cartography: Cartography in the European Renaissance* (Chicago: 2007) 174–262.

Baader H. – Wolf G., "A Sea to Shore Perspective: Littoral and Liminal Spaces of the Medieval and Early-Modern Mediterranean", *Mitteilungen des Kunsthistorischen Institutes in Florenz* 56 (2014) 2–15.

Braudel F., *La Méditerranée et le Monde Méditerranéen a l'époque de Philippe II*, 3 vols. (Paris: 1949).

Campbell T., "Portolan Charts from the Late Thirteenth Century to 1500", in Harley J.B. – Woodward D. (eds.), *The History of Cartography* (Chicago: 1987) 371–463.

Casey E., *Representing Place: Landscape Painting and Maps* (Minneapolis, MN: 2002).

Cortesao A., *History of Portuguese Cartography*, 2 vols. (Coimbra: 1969–1971).

Du Jourdin M.M. – de la Roncière M., *Sea Charts of the Early Explorers* (New York: 1984).

Edgerton S., "Florentine Interest in Ptolemaic Cartography as Background for Renaissance Painting, Architecture, and the Discovery of America", *The Journal of the Society of Architectural Historians* 33 (1974) 275–292.

Edgerton S., *The Renaissance Re-Discovery of Linear Perspective* (New York: 1975).

Edson E., *The World Map, 1300–1492: The Persistence of Tradition and Transformation* (Baltimore, MD: 2007).

Falchetta P., "The Use of Portolan Charts in European Navigation during the Middle Ages", in Baumgärtner I. – Kugler H. (eds.), *Europa im Weltbild des Mittelalters: Kartographische Konzepte* (Berlin: 2008) 269–276.

Freiesleben H.C., "The Still-Undiscovered Origin of Portolan Charts", *Navigation* 36 (1983) 194–99.

Gautier-Dalché P., *Carte marine et portulan au XIIe Siecle* (Rome: 1995).

Horden P. – Kinoshita S. (eds.), *A Companion to Mediterranean History* (Oxford: 2014).

Horden P. – Purcell N., *The Corrupting Sea: A Study in Mediterranean History* (Hoboken: 2000).

Kretschmer K., *Die Italienische Portolane des Mittelalters* (Berlin: 1909).

Kris E., *Psychoanalytic Explorations in Art* (New York: 1952).

Laharie M., *Le journal singulier d'Opicinus De Canistris* (Vatican City: 2008).

Lanman J., *On the Origin of Portolan Charts* (Chicago: 1987).

Lachey T.J., "Lettura del canto XXXIV dell'Inferno", *Rivista di Studi Danteschi* 19.1 (2019) 40–58.

Monmonier M., *Rhumb Lines and Map Wars* (Chicago: 2004).

Morse V., *A Complex Terrain: Church, Society and the Individual in the Works of Opicinus de Canistris* (Ph.D. dissertation. Berkeley University: 1996).
Nordenskiold A.E., *Periplus: An Essay on the Early History of Charts and Sailing Directions*, trans. F. Bather (Stockholm: 1897).
Pagels E., *The Origin of Satan* (New York: 1995).
Pelletier M., "Le Portolan d'Angelino Dulcert, 1339", *Cartographica Helvetica* IX (1994) 23–31.
Piron S., *Dialectique du monstre: enquête sur Opicino de Canistris* (Brussels: 2015).
Pujades i Bataller R., *La carta de Gabriel de Vallseca de 1439* (Barcelona: 2009).
Pujades i Bataller R., *Les cartes portolanes: la representació medieval d'una mar solcada* (Barcelona: 2007).
Rahner H., *Symbole der Kirche: Die Ekklesiologie der Väter* (Salzburg: 1964).
Roux G. – Laharie M., *Art et folie au moyen age: aventures et énigmes d'Opicinus de Canistris* (Paris: 1997).
Roux G. – Laharie M., *Opicinus de Canistris: prêtre, pape et Christ ressuscité* (Paris: 2005).
Russell J.B., *Lucifer: The Devil in the Middle Ages* (Ithaca, NY: 1984).
Salomon R., *Opicinus de Canistris: Weltbild und Bekenntnisse Eines Avignonesichen Klerikers Des 14. Jahrhunderts* (London: 1936).
Sobecki S., *The Sea and Medieval English Literature* (Cambridge: 2008).
Stevenson E.L., *Portolan Charts: Their Origin and Characteristics with a Descriptive List of Those Belonging to the Hispanic Society of New York* (New York: 1911).
Taylor E.G.R., *The Haven-Finding Art* (New York: 1971).
Trachtenberg M., *Dominion of the Eye* (Cambridge: 1997).
van Duzer C., *Sea Monsters on Medieval and Renaissance Maps* (London: 2013).
Whittington K., *Body-Worlds: Opicinus de Canistris and the Medieval Cartographic Imagination* (Toronto: 2014).
Whittington K., "Experimenting with Opicinus de Canistris", *Gesta* 51.2 (2012) 147–173.
Wilson J.V.K., "A Return to the Problems of Behemothand Leviathan", *Vetus Testamentum* 25 (1975) 1–14.

CHAPTER 5

From Sign to Embodiment: the River Jordan and the Representation of Water in Western Art 400–c. 1500

François Quiviger

This chapter takes one of the most frequent aquatic subjects of Western art, the Baptism of Christ in the Jordan river as a point of access to approach the broad theme of water in the arts from ca. 400 up to the sixteenth century.[1] The Baptism of Christ is a recurring imagery in baptistries and baptismal fonts as well as in cycles of the life of Christ decorating spaces of public and private devotions from mosaics, fresco cycles and altarpieces to prayer books, prints, plaquettes and medals. This means that images of the Baptism of Christ from late Antiquity to early Modernity are examples of changes, fluctuations and continuities in the representation and perception of water. Baptismal waters are of course symbolic of cleansing, regeneration and rebirth, but the primary purpose of artists was not to represent doctrine but rather to produce recognisable figurations of an element that has neither stable shape nor colour.[2] Thus, observable transformations and shifts in the representation of the waters of the Baptism of Christ are not only episodes of Christian iconography but also screenshots of the cultural history of water perhaps symptomatic of changing attitudes towards the world.

With this broad hypothesis in mind the three parts of this chapter retrace and highlight aspects of water and rivers in art through the history of their representation in the Baptism of Christ up to the beginning of the sixteenth century. The first part examines the main conventions observed in the representation of water and compares medieval and Renaissance images of water as a standalone subject and as a personification in allegories of the four elements. Moving from personifications of water to personifications of rivers the second part focuses on the representation of the Jordan river which fades from Western art towards the late thirteenth century to re-emerge briefly in the Renaissance. Part three chronicles the solutions developed by medieval artists to represent

1 Matthew 3:13–17, Mark 1: 9–11, Luke 3:21–22, John 1:29–34.

2 On the symbolism of baptismal water, see Hidiroglou P., *L'eau divine et sa symbolique* (Paris: 1994) part 2: *du Jourdain au Baptistère*; for water in the Bible, see Francini M., "The Semantic Field of 'Water' in the Gothic Bible: The Letter and the Spirit" in this volume 253–283.

© KONINKLIJKE BRILL BV, LEIDEN, 2024 | DOI:10.1163/9789004696501_007 the river without resorting to personification. As we shall see these solutions display an increasing awareness of water materiality and dynamics to the point that by the sixteenth century aquatic attributes such as fluidity, movement and flow eventually rose as aesthetic categories. The conclusion briefly examines the concept of material imagination proposed by Gaston Bachelard in *L'eau et les rêves* in the context of the history of the representation of water.

1 Shapes of Water

Water is reflective, dynamic, colourless and shapeless, so representing it is not obvious. Even if there is no line in water the most common and recognisable visual signs of water in the arts are undulating lines, a somehow universal entry level signifier, the uninterrupted history of which extends from prehistoric artefacts to contemporary warning road signs. Many stories, histories, tales and Biblical narratives happen near a river or a shore, but aside from the Baptism of Christ, three subjects in particular challenged artists to focus on imagining water: the *Separation of Water and Firmament* (Genesis 1:7), water as an element in scientific illustrations and water as a personification, usually accompanied by personifications of the other three elements.

The *Separation of Water from Firmament* is possibly one of the most elemental themes of Christian art as it sets water and air side by side. The twelfth-century mosaic of the Palatine chapel in Palermo lays out the main solutions adopted by medieval artists well into the fourteenth century. In contrast to the monochrome grid-like regularity of the section assigned to the sky, the mosaists have signified water as irregular curls made out of at least five shades of blue, from dark to near white.

Blue, curly, wavy yet not always regular, these are the main conventions shared in the medieval West and the Byzantine East to indicate water. Stacks of ordered curls also abound in images of the *Flood*, of *Moses striking the Rock*, the *Rivers of Paradise* and the *Legend of Saint Christopher*. We also encounter this mode in traditional cosmological diagrams locating earth and water at the centre surrounded by concentric circles representing the other two elemental spheres of air and fire and the spheres of the seven planets. Such iconography, well known through paintings such as Giovanni di Paolo's *Creation of the World and Expulsion from Eden*, continued well into the age of prints.[3]

3 See Bertola F., *Imago mundi, la rappresentazione del cosmo attraverso i secoli* (Padua: 1997) 115–121, 158–159; Giovanni di Paolo, c. 1403–c. 1482. c.1484, *Expulsion from Paradise* (Metropolitan Museum of Art: New York).

In spite of the expansion of medieval mnemonic and its elaborate use of personifications, allegorical figures of water or of the four elements are few and far between until the sixteenth century.[4] In a twelfth-century *Aratea* water is a nymph riding a winged horse [figure 5.1]. The contemporary encyclopaedic compendium, the *Hortus deliciarum*, personifies water as a cross-legged bare-chest bearded Neptune holding a trident with one hand and a fish with the other [figure 5.2]. This male classical god here posing as an Indian ascetic – the trident is also an attribute of the god Shiva – served to represent the sea but had however little following as a personification of water.

While illustrations of the elemental spheres of the world were still produced in the age of print, signalling water as curly and blue when coloured, serial themes such as the *hours of the day*, the *four seasons*, the *five senses* or the *four elements* were rising to new prominence thanks to the expansion of the print market. The allegory of water from the prolific Antwerp workshop of Maerten de Vos (1532–1603) offers an excellent example [figure 5.3]. The male figure of the *Hortus deliciarum* has become a naked female personification, almost a living fountain with water gushing out of her nipples, a motif expressive of nourishment, abundance and fertility. In the same spirit her head is literally a fertile ground on which grow many plants.[5] The engraver has merged two ancient personifications of water in one: the river god, a male figure with a shaggy head of reeds and bulrush resting on an urn of gushing water, and the fountain nymph, a nude female figure holding a vase of flowing water.[6] The single fish of the medieval allegory has multiplied to express the variety and abundance of aquatic fauna flowing out of a cornucopia. The border of the image further expands this evocation in a format harking back to the illuminated margins of late medieval manuscripts such as the herbarium which decorates the margins of the *Grandes Heures* of Anne de Bretagne.[7]

The tendency of depicting, and therefore apprehending, the elements through the minerals, plants and animals they produce, rather than through their barren appearance rises throughout the sixteenth century. Two influential series of paintings more or less contemporary, by the Italian Giuseppe Arcimboldo (1526–1593) and the Flemish Joachim Beucklaer (c. 1535–1575), exemplify this bodily engagement with the elements in general and water in particular.

---

4 On Medieval mnemonic, see Carruthers M.J., *The Book of Memory: A Study of Memory in Medieval Culture* (Cambridge – New York: 1990).

5 The Renaissance association of rivers with abundance and fertility is discussed in Lazzaro C., "River Gods: Personifying Nature in Sixteenth-Century Italy", *Renaissance Studies* 25.1 (2011) 70–94.

6 On ancient river gods see, Gais R.M., "Some Problems of River-God Iconography", *American Journal of Archaeology American Journal of Archaeology* 82.3 (1978) 355–370.

7 *Grandes Heures d'Anne de Bretagne*, c. 1505–1508 (Paris: Bibliothèque Nationale; lat. 9474).

FIGURE 5.1 Herrad of Hohenbourg, *Hortus deliciarum*, ed. R. Green – T.J. Brown – K. Levy, Studies of the Warburg Institute 36 (London: 1979) pl. 3.
REPRODUCED BY PERMISSION OF THE WARBURG INSTITUTE

FIGURE 5.2 *Aratea*, c. 1150. Vienna, Österreichische Nationalbibliothek. Cod. 12600, fol. 30r.
REPRODUCED BY PERMISSION OF THE WARBURG INSTITUTE

FIGURE 5.3 Maerten de Vos, *The Element of Water*, c. 1560. Print. London, private collection

In Giuseppe Arcimboldo's well known series, the four elements are human heads composed of animals bred by each element: fish (water), birds (air), mammals (earth), only fire is represented by fire. This phenomenon of elemental embodiment takes a rather bourgeois turn in a series signed by Flemish painter Joachim Beucklaer where each of the four elements features as a market stall of sea-food (water), vegetables (earth), poultry (air) and meat (fire), emblematic as a whole of urban prosperity and merry dining.[8] The butcher stall assigned to fire alludes to the element's transformative power, turning raw meat into tasty food, and further highlights Beucklaer's evocation of the nourishing, if not mouth-watering, virtue of the elements.

Seen from the perspective of Arcimboldo and Beucklaer's versions, Marteen De Vos's print is an iconographic intermediary. It anticipates these later allegories prioritising the products of the elements over the elements themselves and connecting the idea of earth, water, air and fire to the viewer's trained taste buds. De Vos's allegory also combines core elements of the early iconography of the Baptism of Christ: personification, real water and fish. Indeed,

8 On Beucklaer, see Sullivan M.A., "Aertsen's Kitchen and Market Scenes", *The Art Bulletin* 81.2 (June 1999) 236–266.

sixteenth-century artists sat at the apex of more than a thousand years of uninterrupted depiction of water which can be chronicled through the history of the Baptism of Christ in art up to these early modern allegories.

This history unfolds in several periods partially corresponding to changes in the performance of baptismal liturgy.[9] The first period extends up to the twelfth century. While baptism was administered mostly to adults through full immersion, artists represented Christ naked and immersed sometimes up to his shoulders [figs. 5.4, 5.8]. During these early centuries, especially in England, missionaries adapted rivers and their ancient places of rituals to Christianity.[10] From the twelfth century onwards baptism of new-born children had become the norm and was performed by means of symbolic aspersion and ointment in baptismal fonts rather than through literal immersion in pools. In parallel to these changes artists began to cover Christ's genitalia with a loin cloth and lowered the water level of the river. By the fourteenth century the water level had dropped down to his knees [figure 5.11] and to his ankles in the fifteenth century. Piero della Francesca's version with Christ standing on the dry and sandy riverbank anticipates one of the main formulas adopted from the sixteenth century onwards.[11] By then Christ tended to be shown standing or kneeling on the dry riverbank, or if immersed, rarely higher than his ankles. Thus, as the performance of the sacrament became increasingly symbolic, the river level went down in images, artists left out the river personification and began developing means of conveying the fleeting appearances of tangible water.

2 River Gods

Together with high water level another steady feature of the first centuries of imaging of the Baptism of Christ is the presence of the Jordan river god.

---

9 Réau L., *Iconographie de l'art Chrétien*, T. 2, Pt. 2 in *Iconographie de la Bible. Nouveau Testament* (Paris: new ed. 1957) 296–301. Bloch A.R., "Baptism, Movement, and Imagery at the Baptistery of San Giovanni in Florence", in Freni G. – Zchomelidse N. (eds.), *Meaning in Motion: The Semantics of Movement in Medieval Art* (Princeton, NJ: 2011) 131–160. Sonne de Torrens H.M. – Torrens M.A. (eds.), *The Visual Culture of Baptism in the Middle Ages. Essays on Mediaeval Fonts, Settings and Beliefs* (Farnham, Surrey: 2013).

10 On the use of local rivers in Baptismal rituals see, Twomey C., 'Rivers and Rituals: Baptism in the Early English Landscape', in Twomey C. – Anlezark D. (eds.), *Meanings of Water in Early Medieval England* (Turnhout: 2021) 59–84.

11 Piero della Francesca, c. 1415/20–1492. 1450s. *The Baptism of Christ* (The National Gallery, London: Bought, 1861). For an historical overview, see plates I–XXII of Strzygowski J., *Ikonographie Der Taufe Christi* (Munich: 1885).

Well before Christianity the ancient Greeks and Romans represented rivers as reclining bearded male divinities crowned with reeds and leaves, sometimes surrounded by real or mythical animals emblematic of the lands they irrigate.[12] The Nile for instance reclines in the company of a crocodile and a sphinx while the Tiber rests by the she-wolf nurturing Remus and Romulus. In the case of the Jordan river, the obvious attribute is the Baptism of Christ.

The early fifth-century mosaists who decorated the Ravenna Orthodox Baptistery integrated a river god figure alongside their Baptism of Christ and even added an inscription to identify him as the Jordan [figure 5.4]. About a century later, on the ceiling mosaic of the nearby Arian Baptistry [figure 5.5], the river god reclines on blue watery curls holding a reed, and wearing crab claws on his head. He is larger than Christ himself whose nudity is a reminder that these images were seen by naked adult neophytes, themselves experiencing immersion in the sacramental water of the baptismal pool.[13]

Scholars have hypothesised that the presence of the personification of the Jordan echoes the ways in which Biblical and patristic authors address the river as a conscious being.[14] Nevertheless as time went by, and in spite of the permanent textual presence of Biblical and Patristic sources, the Jordan river god literally shrunk and disappeared from Western versions of the Baptism of Christ, only to return briefly in the sixteenth century. The Jordan is already smaller in the Ravenna Orthodox Baptistry. It stands shorter at the left of Christ on the door of the twelfth-century Sicilian Cathedral of Monreale, in the twelfth century Baptistry of the Palermo Palatine chapel, as well as in Herrad of Hohenbourg's contemporary *Hortus deliciarum*.[15] In these examples the river god has become a short stubby struggling individual, drowning under an overwhelming mass of ribbon-like waves of water.

The Jordan god shrinks further in Byzantine manuscripts, mosaics and frescoes. He is a dwarf in the Hosios Loukas church. In the Cypriot church of Asinou, decorated between the twelfth and fifteenth century, the Jordan has become a comical restless winged demon disrespectfully addressing Christ.[16] At the Church of the Archangel Michael (Panagia Theotokos, Galata, Cyprus) someone has even blanked out the face of the river god in a way reminiscent

12 Gais, "Some Problems of River-God Iconography" 367.

13 Wharton J., "Ritual and Reconstructed Meaning: The Neonian Baptistery in Ravenna", *The Art Bulletin* 69.3 (1987) 358–375, 362.

14 Ibidem 370.

15 Herrad of Hohenbourg, *Hortus deliciarum* 161 pl. 56.

16 Baptism of Christ. Fresco Painting twelfth to fourteenth century (Panagia Phorbiotissa, Asinou church, Panagia Phorviotissa, Interior).

FIGURE 5.4 *The Baptism of Christ*, mid-5th century. Dome mosaic. Ravenna, Orthodox (Neonian) Baptistery. REPRODUCED BY PERMISSION OF THE ARCIDIOCESI DI RAVENNA-CERVIA

of iconoclastic interventions and statements on the undesirability of pagan images.[17]

The Jordan river god survived in this altered state in Byzantium until the Fall of Constantinople but disappeared from Western images of the Baptism from the late thirteenth century onwards. Other classical river personifications survived throughout the medieval period, the Eridanus constellation for example, named after the river of the same name and traditionally depicted as another

17 Baptism of Christ. Fresco Painting (Church of Panagia Theotokos, Church of Archangel Michael, Interior).

FIGURE 5.5 *The Baptism of Christ*, c. 500 (detail). Cupola. Ravenna, Arian Baptistery. REPRODUCED BY PERMISSION OF THE MINISTERO DELLA CULTURA – DIREZIONE REGIONALE MUSEI DELL'EMILIA-ROMAGNA

reclining river god.[18] Nevertheless astrological imagery was not influential enough to keep alive or relevant the inclusion of a river god in the Baptism of Christ. To blur things further, the four rivers of Paradise were sometimes depicted as female personifications rather than river gods following the Patristic idea that the four rivers of Paradise replaced and surpassed the Pagan Castalian font of inspiration and were therefore personified like their predecessor as fountains nymphs holding urns. This contamination perhaps explains why images of the Eridanus constellation occasionally morphed from the features

18 Lippincott K., "The Problem with Being a Minor Deity: The Story of Eridanus", in Duits R. – Quiviger F. (eds.), *Images of the Pagan Gods: Papers of a Conference in Memory of Jean Seznec* (London: 2009) 43–96.

of a river god to those of a fountain nymph reclining on an undulating bed of stars.[19]

The Jordan river god, whose masculinity was never in doubt, briefly returned to sixteenth century Italy through the Renaissance revival of ancient art and iconography. The most spectacular version is undoubtedly that of Jacopino del Conte (1510–1598), a Florentine artist close to Michelangelo, whose Baptism of Christ decorates the Oratorio di San Giovanni Decollato in Rome [figure 5.6].[20]

FIGURE 5.6 Jacopino del Conte, *The Baptism of Christ*, 1541. Fresco. Rome, Oratorio di San Giovanni Decollato. https://en.m.wikipedia.org/wiki/File:Jacopo_del_Conte_001.jpg. Last accessed 14 December 2023

19 See Grabar A., "Les quatre fleuves du Paradis et la source de Castalie", *Bulletin de La Société Nationale Des Antiquaires de France* (1970) 45–53, 52. On Eridanus as female deity, see C. Julius Hyginus, *C. Iulii Hygini, Augusti liberti, Fabularum libe* (Lugd. Bat. – Amstel: Apud Gaasbekios, 1535) 88.

20 On Jacopino see, Cheney I.S., "Notes on Jacopino Del Conte", *The Art Bulletin* 52.1 (1970) 32–40.

Here the gigantic river god, the largest figure of the composition, rests on the border of the image, his massive elbow protruding out of the frame with some of the river's foamy flow gushing out of his massive urn to cover his genitals. Such vigorous embodiment of the generative power of baptismal water echoes the associations between rivers and fertility typical of the Renaissance understanding of classical river gods, which we have seen personified in Maerten De Vos's allegory of water examined at the beginning of this chapter.

The Renaissance is a time of rebirth of the Antique and consequently of the Ancient river gods. While the most famous sculptures of the Nile, the Arno and the Tiber had all been excavated, celebrated and honoured with installation of fountains in the gardens of popes and cardinals,[21] the renaissance of the Jordan river god in the iconography of the Baptism of Christ was relatively brief. By the end of the sixteenth century the proliferation of religious images promoted by the Counter Reformation seems to have definitively expelled ancient river gods from pictorial evocations of the banks of the Jordan. In the seventeenth century even Nicolas Poussin (1594–1665), who earned fame for his erudite reconstitutions of the Classical world, chose not to use river gods in his versions of the Baptism of Christ.

Thus, to recapitulate, during the first period of its history the iconography of the Baptism of Christ included a personification of the Jordan river which in the West progressively came into disuse by the thirteenth century. This disappearance might be a symptom of the decline of the relevance of classical forms during the Western Middle Ages, but for the purpose of this essay its most important consequence was to leave medieval artists with only visual means to signify water. It is to the second part of this history that we shall now turn.

3 Representing Water

Baptismal waters symbolise purification and regeneration and the fish frequently depicted swimming around Christ allude to the Christian community renewed by virtue of the sacrament.[22] Such ideas are the foundation of doctrinal thinking on the use, symbolism and meaning of baptism. Yet artists had to represent water, fish, and rivers rather than theological concepts. When requested to depict the *Baptism of Christ* part of their brief was to make the

21 Lazzaro "Personifying Nature" 75.

22 Daniélou J., "Le symbolisme de l'eau vive", *Revue des Sciences Religieuses* 32.4 (1958) 335–346, 341–342.

surface of the image assigned to the river Jordan recognisable as water. This situation, which left considerable room for experimentation, eventually led to formulas, solutions and approaches intended to convey aquatic attributes such as fluidity, movement and transparency.

In medieval mosaics, frescoes and miniatures of the Baptism, figurations of water expand and oscillate between several overlapping traditions. Some illustrators merely applied repeated patterns of small waves of alternating colours, some used undulating lozenges – turning into patterns the geometric shapes produced by ripples.[23] Many artists enclosed entire paragraphs of tight wavy lines in a shape reminiscent of a cocoon, a bee hive or a bell, others opted for a shape reminiscent of a pyramid or a mound.[24] These two formulas are already present in the wavy mass of blue flowing from the urn of the Jordan personification over Christ's lower body in the tenth-century Benedictional of St. Æthelwold [figure 5.7].

3.1 *Cocoons, Mounds and Ribbons*

A very common solution adopted in the eleventh-century Baptistry in Pisa is the tracing of wavy lines running from one bank of the river to the other and covering Christ's nudity [figure 5.8]. This method tends to produce a shape evocative of a hive or a cocoon out of which emerges the upper torso of Christ. Many medieval artists adopted this format, eventually turning the river from a geographical entity into a transparent container. Unsurprisingly representations of the Jordan were taking a shape evocative of the baptismal font they decorated.[25] This visual analogy further extended in the Spanish tradition of setting the Baptism of Christ in a container rather than a stream.[26]

Mosaists liked to treat water as floating ribbons in between which they sandwiched the figure of Christ. In Venice, in the tenth-century Baptistry of San Marco, water is an intense flow of white and blue ribbons wrapped around the body of Christ.

---

23 See, for waves, Psalter ms. add. 17687 F. *Baptism of Christ*, c.1240. Lozenges: early XIII. Psalter, fol. 9v (Bamberg: Lib., Staatsbibliothek; Bibl. 48 A II 47). Overview of main types in Strzygowski, *Ikonographie Der Taufe Christi* plates II–XXI.

24 Strzygowski, *Ikonographie Der Taufe Christi* plates XII–XVI.

25 See Nordström F., *Mediaeval Baptismal Fonts: an Iconographical Study*, Acta Universitatis Umensis 6 (Stockholm: 1984) 100.

26 Patton P., "Et Partu Fontis Exceptum. The Typology of Birth and Baptism in an Unusual Spanish Image of Jesus Baptized in a Font", *Gesta* l. 33.2 (1994) 79–92.

FIGURE 5.7 *The Baptism of Christ*, c. 971–984. Benedictional of St. Æthelwold, London, British Library, Add. MS 49598, fol. 25r.
REPRODUCED BY PERMISSION OF THE BRITISH LIBRARY BOARD

Back from Venice where he learned his art from Greek mosaists the Tuscan Andrea Tafi (1213–1294) applied and developed their methods in his version of the Baptism for the side walls of the Florentine Baptistery [figure 5.9].[27] Like

27 See Tafi's biography in Vasari G. – Bettarini R. – Barocchi P. (eds.), *Le vite de' più eccellenti pittori scultori e architettori nelle redazioni del 1550 e 1568* (Florence: 1966).

FIGURE 5.8 *The Baptism of Christ*, c. 1204. Marble. Pisa, Baptistery, east portal, architrave, relief: https://en.wikipedia.org/wiki/File:Pisa_Baptistery,_architraves_over_east_portal,_c1200.jpg Last accessed 10 January 2024

his predecessors he inserted the body of Christ between bundles of layered ribbons leaving his bust, right arm and blessing hand outside the first layer of ribbons while his left arm floats between the first and second layer. Departing from earlier traditions he arranged ribbons in such a way that, instead of ascending and descending, they seem to radiate and ripple from the centre of Christ's lower body, thus producing a hypnotic perspective expressing the assumed generative power of Baptismal waters. Andrea was expanding on another widespread visual evocation of the Jordan as a mound, if not a pyramid of waters with Christ's upper body at its apex.

FIGURE 5.9 Andrea Tafi, *The Baptism of Christ*, 13th century. Mosaic. Florence, Cathedral Baptistery: https://commons.wikimedia.org/wiki/File:Florence_baptistery_ceiling_mosaic_7247px.jpg. Last accessed 14 December 2023

In the same years the sculptors of the elaborate Hildesheim Cathedral baptismal font (c. 1250) adopted this very same compositional pattern but ended up generating a tangible pyramid of thick tubing [figure 5.10]. Instead of the multi-layered transparencies of the Florentine baptistry, in Hildesheim water completely hides the lower part of the body of Christ like a thick blanket of curvy cables.

3.2 *Thirteenth-Fourteenth Century: Transparencies*

By the fourteenth century, as Italian artists were beginning to compete in terms of naturalism, two particularly elusive aspects of the Jordan waters began mobilising their attention and skill: transparency and flow.

The illustrator of the Benedictional of St. Æthelwold used white lines to suggest the underwater presence of Christ's lower body [figure 5.7]. This way they

FIGURE 5.10 Baptismal font, c. 1250. Hildesheim, Cathedral: https://commons.wikimedia.org/wiki/File:Hildesheim_Dom_Taufe_02.jpg . Last accessed 10 January 2024

treated water very much like a drapery, revealing and amplifying the shapes it covers. Envisaging water in these terms led artists to expand the twelfth-century stone cocoon of their predecessors into a transparent and reflexive shroud wrapped around the body of Christ.

At the Arezzo Church, Pieve di Santa Maria, (c. 1220), the sculptors of the right portal of the Cathedral have augmented the motif of the bee hive with undulating lines with the relief of Christ's lower body.[28] Andrea Pisano (1290–1348), considered the most innovative sculptor of his generation, expanded this effect on the South bronze door of the Florence Baptistery, completed in

28 The Baptism of Christ c.1221. Tympanum (relief; sandstone) (Arezzo: Church, Pieve di S. Maria; right portal).

1336 [figure 5.11].[29] There he accentuated the presence of the immersed part of Christ through networks of gently undulating metallic furrows covering his lower body and achieved the sculptural tour-de-force of expressing transparency in bronze.

Transparency was easier to obtain for painters, mostly through additional layers applied with diluted pigments. The altarpiece of the Baptism of Christ by Niccolò di Pietro Gerini (d. c. 1415), for the Florentine church of Santa Croce exposes three layers of transparency: transparency of water treated as horizontal waves with fine white highlights; transparency of Christ's loin cloth, and double transparency of superposed water and loin cloth just at the level of Christ's knees [figure 5.12]. Gerini also spent much time and attention rendering the curls of the Baptist's fur but did not really pick up the kinship between the forms of water flow and that of hair curls. Some of his predecessors and successors did.

FIGURE 5.11 Andrea Pisano, *The Baptism of Christ*, 1336. Bronze. Florence, Baptistery, south door. https://fr.m.wikipedia.org/wiki/Fichier:Battistero_di_firenze,_porta_sud_di_andrea_pisano_10_battesimo_di_ges%C3%B9.JPG Last accessed 10 January 2024

29 On Andrea Pisano, see Pope-Henessy J., *Italian Gothic Sculpture* (London: 1972) 23–25.

FIGURE 5.12 Niccolò di Pietro, *The Baptism of Christ*, 1387, egg and tempera on wood, 160 × 76 cm.

REPRODUCED BY PERMISSION OF THE NATIONAL GALLERY, LONDON

3.3 *Fourteenth-Fifteenth Century: Hair Flow, Water Curls and Aquatic Auras*

Fourteenth-century images of the Baptism of Christ confirm that some artists began to envisage and depict hair and water flow in terms of formal kinship. Both Christ and the Baptist wore long hair and the Gospels specified that the Baptist covered himself with wild beasts' skins. Furthermore, the baptismal water he poured over the head of Christ could potentially inspire formal associations between water flow and long locks of hair.

Taddeo Gaddi (c. 1290–1366) picked up these analogies literally in his version of the Baptism of Christ for the sacristy of the Florentine Church of Santa Croce [figure 5.13]. Christ, unusually shown from the back, is kneeling through a cascade of transparent lines concealing his nudity. Taddeo has rotated the flowing lines, from the standard horizontal format to a diagonal orientation, and brought to the water a sense of downward cascading reminiscent of hair. A few years later Giovanni del Biondo expanded Taddeo's solution in the predella of the Baptism of Christ of the Rinuccini Politpych of 1379.[30] Giovanni's conspicuous lack of subtlety makes bluntly apparent his intention of treating the river flow as long thick translucid locks wrapping around Christ's legs as they gush towards the spectator's space.

Prints, paintings and even terracotta works confirm that from the fourteenth century onwards many artists interpreted the Jordan waters as translucid locks coiling around Christ's body.[31] This state of things undoubtedly prepared the path for the obsessive water research and experiments of Leonardo da Vinci (1452–1519). Throughout his life Leonardo observed and drew the shapes produced by flowing water from which he extracted 'natural patterns' which he applied to representing things as diverse as hair curls and floods [figure 5.14].[32] Next to a drawing that recorded this type of experiment he writes:

> Note the motion of the surface of the water which conforms to that of hair, which has two motions, of which one responds to the weight of the strands of hair and the other to the direction of the curls; thus the

30 Giovanni del Biondo, *The Baptism of Christ*, 1379. Rinuccini Polyptych.

31 See for example Giovanni della Robbia (Italian, 1469-(?)1529) 1510–1520. *The Baptism of Christ*, S. Cassiano, Pisa, Italy, or in print Martin Schongauer (1435/50–1491 Breisach). Baptism of Christ c. 1470–1474. Print, The Metropolitan Museum of Art.

32 Quiviger F., *Leonardo Da Vinci: Self, Art, and Nature* (London: 2019) 173–117.

FIGURE 5.13 Taddeo Gaddi, *Baptism*, 1330–1335. Painting. From Florence, Santa Croce Sacristy Panels, Life of Christ. Galleria dell'Accademia.
REPRODUCED BY PERMISSION OF THE GALLERIA DELL'ACCADEMIA DI FIRENZE

water makes turning eddies which in part respond to the impetus of the principle current, while the other responds to the incidental motion of deflection.[33]

33 Quoted by Kemp M., *Leonardo da Vinci: the Marvellous Works of Nature and Man* (Cambridge: 1981) 258.

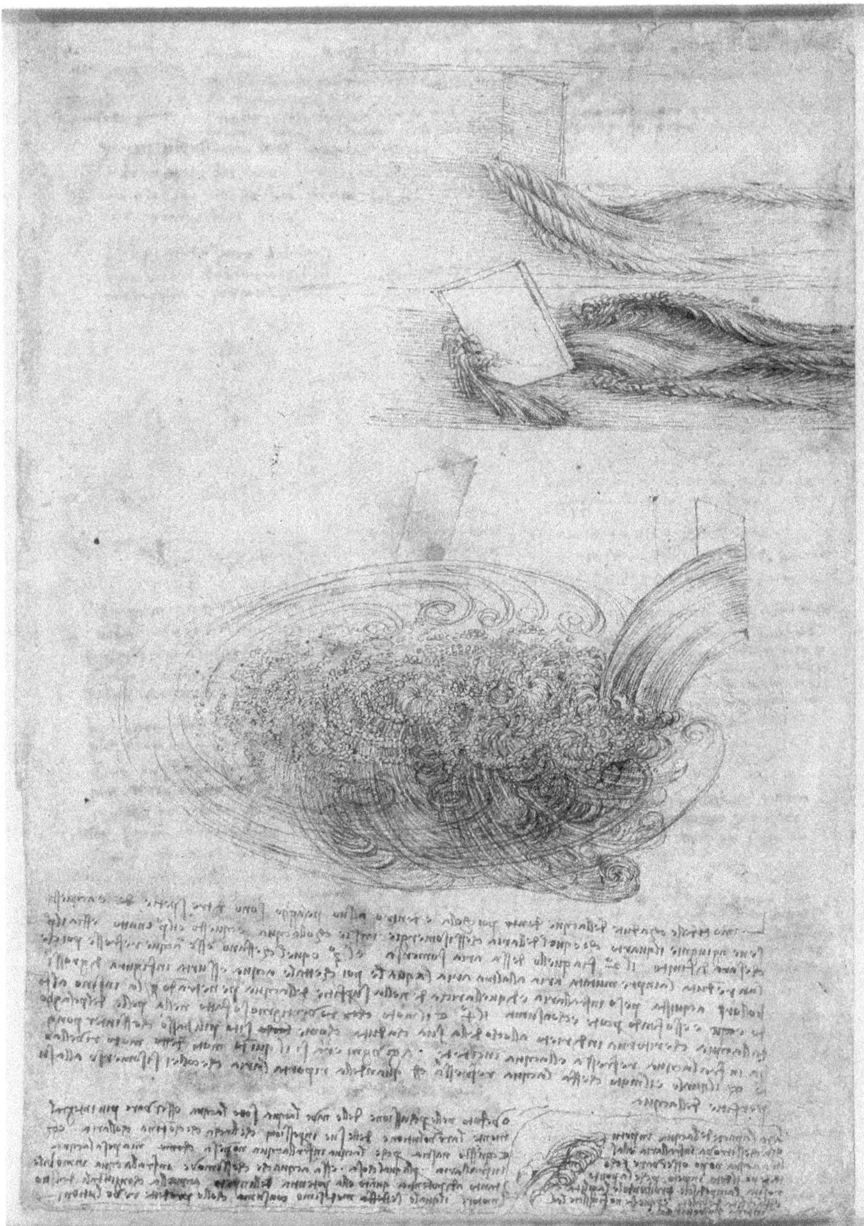

FIGURE 5.14 Leonardo da Vinci, *Movement of Water*, 1507–9. Windsor, Royal Library, 12660v.

By the 1480s Leonardo was already planning an extensive treatise on water now known as the Leicester codex.[34] His interest, which translated into drawings, paintings, experiments, speculations as well as various works of hydraulic engineering, is already apparent in his early years as an apprentice. One of his first recorded paintings happens to be a Baptism of Christ, commissioned around 1472–1475 in which his master, Andrea del Verrocchio, entrusted him the figure of Christ, one angel and the river. Leonardo took this opportunity to depict the reflective properties of water and the various shapes it takes when responding to movement. Here the concentric ripples radiating from Christ's ankles are particularly interesting, for they attest both to Leonardo's observation of nature as well as to his familiarity with Flemish painting, which he knew from his early acquaintance with the Medici collection.[35]

Flemish artists depicting the legend of St Christopher and the Baptism of Christ had already introduced such ripples. This tradition reaches its peak in Gerard David's version of 1502–05. In this most intense rendering the concentric ripples radiating from each of Christ legs on the transparent surface of the river eventually merge into a single rippling circle [figure 5.15].[36] These concentric waves suggestive of movement were also associated with the diffusion of sound for it was a common belief, well until the seventeenth century, that sound spreads in the air as ripples comparable to those caused by a stone thrown in water.[37] In this case then, artists adapted a natural and observable phenomenon to suggest a religious belief – the aura of saintly figures.

Emphasis on Christ's aquatic aura appears in works painted by Leonardo's contemporaries and colleagues such as Pietro Perugino (1445–1523) and Domenico Ghirlandaio (1448–1494). The latter version, at Santa Maria Novella, in the unforgiving medium of fresco, is particularly interesting since the painter has accentuated the round ripples radiating from Christ's immersed ankles with thick white lines.[38] By then this motif had spread to prints. In the

34 Ibidem 304.

35 See above 32.

36 Ninagawa J. – Slachmuylders R., "A Succession of Wishes: The 'Triptych of the Baptism of Christ' by Gerard David", *Zeitschrift Für Kunstgeschichte* 76.1 (2013) 13–50.

37 Burnett C., "Perceiving Sound in the Middle Ages", in Smith M.M. (ed.), *Hearing History: A Reader* (Athens, GA: 2004) 75–76; on this analogy in Leonardo see Kemp, *Leonardo da Vinci* 14.

38 Pietro di Cristoforo Vannucci, called Perugino (1445/46–1523), The Baptism of Christ 1500/1505, The Art Institute of Chicago, Mr. and Mrs. Martin A. Ryerson Collection 1023, Domenico Ghirlandaio. c. 1468–1470. The Baptism of Christ and the Madonna and Child Enthroned with Saints Sebastian and Julian. Fresco, Florence, Church of Santa Maria Novella.

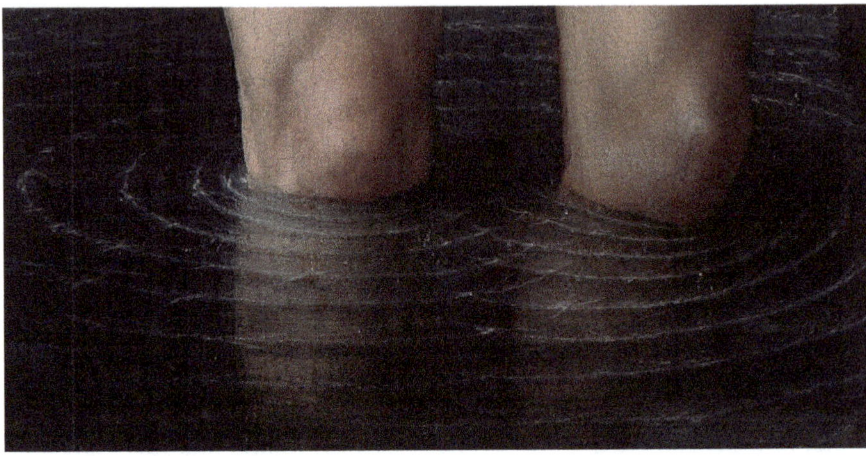

FIGURE 5.15 Gerard David, *The Baptism of Christ*; c. 1502–1508. Painted altarpiece. Bruges, Groeningemuseum: https://commons.wikimedia.org/wiki/File:Gerard_David_-_Triptych_of_Jan_Des_Trompes_-_WGA6031.jpg. Last accessed 14 December 2023

Nuremberg Chronicle, for example, the waves around Christ body are amplified to the point of radiating towards the edge of the image.[39]

The iconography of Saint Christopher is another indicator of the awareness of the shapes and movements of water since the *Golden Legend* speaks of a giant crossing a river with high currents. Medieval artists usually depicted the saint crossing a river with his legs and pole planted in a wavy ornamental pattern rather than generating ripple patterns.[40] From the early fifteenth century onwards, in versions coming from the ambiance of the Van Eyck workshop, the ripples caused by saint Christopher's stick and his massive legs received increasing painterly attention and eventually became a standard feature of the saint's early modern iconography.[41]

39 Hartmann Schedel, *Liber Chronicarum* (Nuremberg: Anton Koberger, 1593) 12.

40 See e.g. Book of Hours c. 1430–1440. Use of Paris., fol. 148r. Manuscript. Bodleian Library, University of Oxford, http://www.bodley.ox.ac.uk/. Last accessed May 14, 2024.

41 Boucicaut Master, fifteenth century: St. Christopher c.1410–1415. Book of Hours: MS. 2, fol. 28v. Musée Jacquemart-André https://library.artstor.org/asset/ARTSTOR_103_418220038 14389; Last accessed November 3, 2023. See e.g. Follower of Jan van Eyck. *c.* 1440–50. Saint Christophe, Philadelphia Museum of Art, Philadelphia, Pennsylvania, USA, John G. Johnson Collection, 1917, Cat. 342.

3.4 *Water and Style*

Thus, Leonardo's early ripples and naturalist approach to water are merely an influential relay in the history of the representation of water. His concrete engagement with the elements in general and water in particular has been seen as untypical of the approach characteristic of medieval science. The same applies to his art which in his time was hailed as the inauguration of modernity.[42] There are of course many reasons for this, but for the purpose of this essay I would like to focus on those pertaining to aquatic matters. For his first and most important biographer, Giorgio Vasari (1511–1574), Leonardo's art stands in sharp contrast with the 'dry' manner of his predecessors. In Vasari's *Lives* the recurring adjectives *secco e tagliente*, dry and cutting, insistently characterise the art of the past and distinguish it from the fluidity of the *maniera moderna*. This unflattering appellation of dryness features not only in individual lives but also functions as an umbrella concept for fifteenth-century art up to Leonardo.[43] Vasari's understanding of pre-Renaissance art did not contrast the dry and cutting style of the past to a wet avant-garde, but rather to a fluid avant-garde. What he understood as the attributes of the modern style inaugurated by Leonardo, were also aquatic qualities: movement, flow, fluidity and liveliness in sharp contrast to what he perceived as medieval clunkiness. Vasari and Leonardo's ideal was one of grace and facility, derived from the courtly ideal of *sprezzatura*.[44] The fact that dryness is its opposite hints at the emergence of aquatic qualities in Renaissance appreciation of the world.

From the rigidity of early medieval evocations to the Renaissance aesthetic of fluidity and liveliness, Leonardo's experiments and Vasari self-consciousness of a modern manner emancipated from a dry past, artists were giving increasing attention and visibility to the materiality of water. In this long history the

42 Park K., "Observation in the Margins, 500–1500", in Daston L. – Lunbeck E. (eds.), *Histories of Scientific Observation* (Chicago – London: 2011) 15–44, 35.

43 See examples in Vasari, *Le vite*, in particular the lives of Paolo Uccello, Alesso Baldovinetti, Lorenzo Costa, Andrea Mantegna (207, 277, 302, 341), as well as the introduction to the third part: '[...] che per lo soverchio studio avevano lasciata in questa arte Pietro della Francesca, Lazaro Vasari, Alesso Baldovinetti, Andrea dal Castagno, Pesello, Ercole Ferrarese, Giovan Bellini, Cosimo Rosselli, l'Abate di San Clemente, Domenico del Ghirlandaio, Sandro Botticello, Andrea Mantegna, Filippo e Luca Signorello; i quali, per sforzarsi, cercavano fare l'impossibile dell'arte con le fatiche, e massime negli scórti e nelle vedute spiacevoli, che, sì come erano a loro dure a condurle, così erano aspre a vederle'.

44 Castiglione B., *The Book of the Courtier* (London – New York: 1974, 1st ed. 1528). On *Sprezzatura* in Leonardo, see Quiviger, *Leonardo* 75–81.

Renaissance is a period of re-connection not only of meaning with form, as long suggested by Panofsky, but also of meaning with materiality and sensoriality.[45]

4 Towards a Conclusion: Imagination, Unconscious and Consciousness of Water

In *L'eau et les rêves* French philosopher Gaston Bachelard examined the various fantasies and images bred by water as part of his project of a psychoanalysis of the four elements. He scoured Western literature to understand water-born imaginations.[46] One famous episode of this history in the Renaissance is the emergence of the nymph, or *ninfa fluida*, a mysterious moving feminine figure draped with fabric evocative of aquatic fluidity and aerial lightness.[47] Another episode of water based imagination is Sandro Botticelli's application of the very same compositional model of the Baptism of Christ to the representation of the *Birth of Venus* (1485–1488, Florence Uffizi), two themes connected by the generative powers of water. Furthermore, baptismal fonts themselves, gendered as female for their association with the *vulva matrix* and *uterus ecclesiae* offered an ideal ground for linking the Baptism of Christ and the Birth of Venus through the life giving power of water.[48] Bachelard's notion of an 'unconscious of water' would also imply a consciousness of water; since we are composed of over 60% of water it could even be argued that we are water with consciousness of which the history, here surveyed through the iconography of the Baptism of Christ, is an instance.

The pre-modern consciousness of water surely differed from that of today. Most conspicuously the modern assumption that water is the most common, basic, and healthy drink dates only from the nineteenth century. From Antiquity to the temperance movement, as urban populations grew alcoholic drinks, beer or wine were the most common means of quenching thirst for mere reasons of hygiene.[49] The well-known medieval dispute of wine and water observes that since water serves to wash it carries filth and infection.[50]

45 Panofsky W., *Renaissance and Renascences in Western Art* (New York: 1969).
46 Bachelard G., *L'eau et les rêves. Essai sur l'imagination de la matière* (Paris: new ed. 1956).
47 Didi-Huberman G., *Ninfa Fluida: essai sur le drapé-désir* (Paris: 2015).
48 Sonne de Torrens, "Fovea peccati et Uterus Ecclesia: The Symbiotic Nature of Female Sexuality on Medieval Baptismal Fonts", in Joynes C.E. – Rowland C.C. (eds.), *From the Margins 2: Women of the New Testament and their Afterlives* (Sheffield: 2009) 334–361, 356.
49 Salzman J., *Drinking Water: A History* (London: 2013).
50 *Le débat du vin et de l'eau*, s.n. (s.l., 1490), fol. 2v. https://gallica.bnf.fr/ark:/12148/bpt6k52409p?rk=21459;2. Last accessed November 3, 2023.

Thus, fertility and hunger, life, and nourishment, rather than thirst and refreshment, are the primary convergence points of the unconscious and the consciousness of water. The theme of water as fertiliser expands in sixteenth-century personifications where, as we have seen, nymphs and river gods merge in a figure of aquatic fertility holding a horn of plenty [figure 5.3]. The fertility associated with water generated an imagery of aquatic materiality, embodiment and nourishment exemplified by Beucklaer's market scenes and Arcimboldo's fish man.

By the sixteenth century, then, two aspects of water had risen to considerable prominence: on the one hand the very bodily understanding of water as a source of nourishment literally represented through the food it produces; on the other hand, the immaterial aspects of transparency, fluidity, and movement as underlying elements of an ideal shaping the imagination and guiding the hand of early modern artists.

For artists images are means of representing; for their audience they are means of apprehending. Thus, the changes in the depiction of waters from medieval to early modern art expose changes of mediation: by 1500 accessing the water-based meaning and symbolism of the Baptism of Christ was mediated through evocation of the tangible properties of water rather than prompted by linear and geometric signs of its presence. In this way increasing visual signalisation of the multi-sensory attributes of water might have further fertilised its power on human imagination.

Bibliography

Bachelard G., *L'eau et les rêves: essai sur l'imagination de la matière* (Paris: new ed. 1956).

Beck I., "The First Mosaics of the Capella Palatina", *Byzantion* 40.1 (1970) 119–164.

Bedard W.M., *The Symbolism of the Baptismal Font in Early Christian Thought*, Studies in Sacred Theology / Catholic University of America; 2nd Ser., No. 45 (Washington, DC: 1951).

Bertola F., *Imago mundi: la rappresentazione del cosmo attraverso i secoli* (Cittadella, Padua: 1997).

Bloch A.R., "Baptism, Movement, and Imagery at the Baptistery of San Giovanni in Florence", in Freni G. – Zchomelidse N. (eds.), *Meaning in Motion: The Semantics of Movement in Medieval Art* (Princeton, NJ: 2011) 131–160.

See Leguay J.P., *L'eau dans la ville au Moyen Age* (Rennes: 2002) loc. 4816 (Kindle edition).

Burnett C., "Perceiving Sound in the Middle Ages", in Smith M.M. (ed.), *Hearing History: A Reader* (Athens, GA: 2004) 75–76.

Carruthers M.J., *The Book of Memory: a Study of Memory in Medieval Culture*, Cambridge Studies in Medieval Literature 10 (Cambridge – New York: 1990).

Carruthers M.J., *The Craft of Thought: Meditation, Rhetoric, and the Making of Images, 400–1200*, Cambridge Studies in Medieval Literature 34 (Cambridge: 1998).

Castiglione B., *The Book of the Courtier* (London – New York: 1974, 1st ed. 1528).

Cheney I.S., "Notes on Jacopino Del Conte", *The Art Bulletin* 52.1 (1970) 32–40.

Cristo S., "The Art of Ravenna in Late Antiquity", *The Classical Journal* 70.3 (1975) 17–29.

Daniélou J. "Le symbolisme de l'eau vive", *Revue des Sciences Religieuses* 32.4 (1958) 335–346.

Daston L. – Lunbeck E., *Histories of Scientific Observation* (Chicago: 2011).

Didi-Huberman G., *Ninfa Fluida: essai sur le drapé-désir* (Paris: 2015).

Gais R.M., "Some Problems of River-God Iconography", *American Journal of Archaeology American Journal of Archaeology* 82.3 (1978) 355–370.

Grabar A., "Les quatre fleuves du Paradis et la source de Castalie" – *Bulletin de La Société Nationale des Antiquaires de France* (1970) 45–53.

Hanford J.H., "The Mediæval Debate Between Wine and Water", *Publications of the Modern Language Association of America* 28.3 (1913) 315–367. https://doi.org/10.1632/457026.

Herrad R.G. – Brown T.J. – Levy K. (eds.), *Hortus deliciarum*, Studies of the Warburg Institute 36 (London: 1979).

Hidiroglou P., *L'eau divine et sa symbolique* (Paris: 1994).

Hyginus C. Julius, Palaephatus, Fabius Planciades Fulgentius, Aratus, Proclus, and Jacobus Micyllus. *C. Ivlii Hygini, Avgvsti liberti, Fabvlarvm liber* [...] (Basil: Apud Ioan. Hervagivm, 1535).

Kemp M., *Leonardo Da Vinci, the Marvellous Works of Nature and Man* (Cambridge, MA: 1981).

Lazzaro C., "River Gods: Personifying Nature in Sixteenth-Century Italy", *Renaissance Studies* 25.1 (2011) 70–94.

Le débat du vin et de l'eau. S.n. S.l., 1490. https://gallica.bnf.fr/ark:/12148/bpt6k52409p?rk=21459;2. Last accessed November 3, 2023.

Leguay J.P., *L'eau dans la ville au Moyen Age* (Rennes: 2002).

Lippincott K., "The Problem with Being a Minor Deity: The Story of Eridanus", in Duits R. – Quiviger F. (eds.), *Images of the Pagan Gods: Papers of a Conference in Memory of Jean Seznec*, Warburg Institute Colloquia 14 (London: 2009) 43–96.

Ninagawa J. – Slachmuylders R., "A Succession of Wishes: The 'Triptych of the Baptism of Christ' by Gerard David", *Zeitschrift für Kunstgeschichte* 76.1 (2013) 13–50.

Nordström F., *Mediaeval Baptismal Fonts: An Iconographical Study*, Acta Universitatis Umensis 6. Umeå (Stockholm: 1984).

Panofsky E., *Renaissance and Renascences in Western Art*, Harper Torchbooks, TB 1447 (New York: 1969).
Park K., "Observation in the Margins, 500–1500", in Daston L. – Lunbeck E. (eds.), *Histories of Scientific Observation* (Chicago – London: 2011) 15–44.
Patton P., "Et Partu Fontis Exceptum: The Typology of Birth and Baptism in an Unusual Spanish Image of Jesus Baptized in a Font", *Gesta* l.33.2 (1994) 79–92.
Pope-Henessy J., *Italian Gothic Sculpture* (London: 1972).
Quiviger F., *Leonardo Da Vinci: Self, Art, and Nature*, Renaissance Lives (London: 2019).
Quiviger F., *The Sensory World of Italian Renaissance Art* (London: 2010).
Réau L., *Iconographie de l'art Chrétien. T. 2, Pt. 2: Iconographie de la Bible. Nouveau Testament* (Paris: 1957).
Salzman J., *Drinking Water: A History* (London: repr. 2013).
Schedel Hartmann, *Liber Chronicarum* (Nuremberg: Anton Koberger, 1493).
Sonne de Torrens H.M. – Torrens M.A (eds.), *The Visual Culture of Baptism in the Middle Ages: Essays on Mediaeval Fonts, Settings and Beliefs* (Farnham, Surrey: 2013).
Sonne de Torrens H.M.,"Fovea Peccati et Uterus Ecclesia: The Symbiotic Nature of Female Sexuality on Medieval Baptismal Fonts", in Joynes C.E. – Rowland C.C. (eds.), *From the Margins 2: Women of the New Testament and Their Afterlives* (Sheffield: 2009) 334–361.
Strzygowski J., *Ikonographie Der Taufe Christi* (Munich: 1885).
Sullivan M.A., "Aertsen's Kitchen and Market Scenes: Audience and Innovation in Northern Art", *The Art Bulletin* 81.2 (June 1999) 236–266.
Vasari G. – Bettarini R. – Barocchi P. (eds.), *Le vite de' più eccellenti pittori scultori e architettori: nelle redazioni del 1550 e 1568* (Florence: 1966).
Wharton J. "Ritual and Reconstructed Meaning: The Neonian Baptistery in Ravenna", *The Art Bulletin* 69.3 (1987) 358–375.

CHAPTER 6

Imagine Oceans: the Many Meanings of the High Sea *c.* 1400–1600

Simon Karstens

Introduction: the One that is Many

On the one hand, the ocean simply is. It has always existed as a spatial reality beyond human comprehension and is much older than mankind itself. It is common knowledge that its resources, winds, currents and the position of its islands shaped human history by making travels easier or harder, guiding routes of expansion and offering or denying maritime resources to different cultures.[1] Yet on the other hand, the historical importance of the ocean cannot be understood without considering the mindsets or mental maps of the humans who looked at it.[2] For example, the thoughts and beliefs the people of a prototypical coastal city held about the ocean at their doorstep could differ and change over time. Their interpretations and perceptions were affected by the technology available to them, as well as by their spiritual beliefs and by the way their epistemic system organised and brought forth new knowledge. But this was not a one-way road. The beliefs and ideas the inhabitants held about the ocean were also influencing the development of new technology, their systems of knowledge or even their society. This is why the ocean is not

1 See for example, Bankhoff G., "Aeolian Empires: The Influence of Winds and Currents on European Maritime Expansion in the Days of Sail", *Environment and History* 23.2 (2017) 163–196.

2 Rau S., *Räume: Konzepte, Wahrnehmungen, Nutzungen* (Frankfurt: 2017); Aulke J., *Spatial Turn(s)*, in Haas S. – Wischermann C. (eds.), *Die Wirklichkeit der Geschichte. Wissenschaftstheoretische, mediale und lebensweltliche Aspekte eines (post) konstruktivistischen Wirklichkeitsbegriffes in den Kulturwissenschaften* (Stuttgart: 2015) 54–62; Dipper C. – Lutz R., "'Raum" in der Europäischen Geschichte; Einleitung", *Journal of Modern European History* 9.1 (2011) 27–41; Langenohl A., "Mental Maps, Raum und Erinnerung Zur kultursoziologischen Erschließung eines transdisziplinären Konzepts", in Damir-Geilsdorf S. (ed.), *Mental Maps – Raum – Erinnerung. Kulturwissenschaftliche Zugänge zum Verhältnis von Raum und Erinnerung* (Münster: 2005) 51–69.

 | DOI:10.1163/9789004696501_008

only a physical entity that shapes human history but is also shaped by the people who thought and wrote about it.[3]

This chapter explores the mindsets of the people who faced the Atlantic Ocean and dared to venture out onto the high seas.[4] It offers insight into different ways of understanding that ocean in the fifteenth and sixteenth centuries. Furthermore, it will show that contradictory perceptions of the Atlantic were an important source of conflict at both an intellectual and physical level.

When looking at earth's oceans, it is easy to forget that the Atlantic was the last one to be conquered by humankind. The Polynesians and the seafarers of the Indian Ocean had established trade routes, built colonies, and raised settlements thousands or at least hundreds of years before the Atlantic was regularly crossed.[5] And yet the exploration of the Atlantic is the best documented and analysed. Firstly, because it was accompanied by the production of written sources still surviving today, and also because it was the steppingstone towards the global connection of all oceans and all people via trade, war and exploitation. The Atlantic was in a way a laboratory of techniques and mentalities which later shaped the world as we know it. This is the reason why this chapter will be focused on the Atlantic, while trying to minimise the inherent eurocentrism of the source material by also including beliefs and actions of African and Native American peoples.

To put this chapter into context, one should start with a sketch of the historiography of the Atlantic as a whole, including circum-, cis- and transatlantic studies.[6] In the last decades, researchers in this field have reconstructed and analysed the whole Atlantic world as one vast space of migration, knowledge, and interaction. Although their points of interest differ broadly, most of these

3 For the later periods one has to consider that mankind had physically changed the oceans and shorelines by overfishing, pollution, global warming and the construction of dams. Although some of these transformations already begun in the time period covered in this chapter, humans had yet barely affected the high seas.

4 This chapter is mainly based on English, French and German sources from the late fifteenth, sixteenth and early seventeenth centuries, including contemporary translations from Spanish and Portuguese material. For more details on the sources and their context, see Karstens S., *Gescheiterte Kolonien, Erträumte Imperien. Eine andere Geschichte der Europäischen Expansion 1492–1615* (Cologne: 2020).

5 Thornton J.K., *A Cultural History of the Atlantic World, 1250–1820* (Cambridge: 2012).

6 On the differentiation between those disciplines, see Armitage D., "Three Concepts of Atlantic History", in Armitage D. (ed.), *The British Atlantic World, 1500–1800* (Basingstoke: 2002) 11–27. For an overview on scholarship on this topic, see Bailyn B., *Atlantic History. Concept and Contours* (Cambridge, MA: 2005); Mancall P.C. (ed.), *The Atlantic World and Virginia 1550–1624* (Chapel Hill, NC: 2007); Canny N. – Morgan P. (eds.), *The Oxford Handbook of the Atlantic World, 1450–1850* (Oxford: 2012).

works share a common approach towards the ocean itself. They mainly focus on the physicality of the ocean with particular emphasis on its current and wind systems. Although the importance of the ocean is universally acknowledged, historians mostly tend to overlook the different perceptions and interpretations of that physical body of water. This is remarkable, since there are hundreds of impressive studies on cultural history that look at different practices, groups, and communities on the Atlantic seaboard. And yet the sea itself mostly remains an empty space between focus points of academic interest.[7]

This is of course only a tendency, since there are a number of works that address questions of perceptions and interpretations of the ocean, and these studies can provide a helpful starting point for this chapter. For instance, Joyce Chaplin's influential article "The Atlantic Ocean and its Contemporary Meanings"[8] is the primary inspiration for this chapter. That being said, it is necessary to emphasise an important difference. Chaplin states that three different perceptions of the ocean have dominated human understanding between 1500 and 1800. In her opinion, the Atlantic after Columbus was at first nothing but a space to cross, before it later became a space of everyday interaction, only to be finally perceived as nothing but the steppingstone to another ocean, the Pacific. Contrary to her interpretation, this chapter will argue that there were more than the three understandings and interpretations of the ocean she identified – some conflicting, some mutually reinforcing – and all were already present even before Columbus first set sail.

This chapter will not only present these perceptions and ideas, but also try to place them within their social context. It will show that some interpretations were more universal, while others were tied to specific social groups. In order to achieve this, this study will have to look beyond concepts of nationality, which are misleading for the early modern era. Instead, it will combine the traditionally separated maritime histories of England, France, Spain, and

7 Bolster J., "Putting the Ocean in Atlantic History: Maritime Communities and Marine Ecology in the Northwest Atlantic, 1500–1800", *The American Historical Review* 113.1 (2008) 19–47.

8 Chaplin J.E., "The Atlantic Ocean and its Contemporary Meanings 1492–1808", in Greene J.P. – Morgan P.D. (eds.), *Atlantic History. A Critical Appraisal* (Oxford: 2009) 35–51. Theories on historical perceptions of the ocean can also be found in well-known works on the history of the sea, like Jourdin M.M., *Europa und das Meer* (Munich: 1993); or Abulafia D., *The Boundless Sea* (London: 2020); and the chapters included in Klein B. (ed.), *Sea Changes: Historicizing the Ocean* (New York: 2003). These works mostly focus on a comparison of the changing perceptions before and after the European expansion. More generally, studies on human perceptions and beliefs of the ocean focus on the time after 1700 due to the higher number of sources available. For an overview see, Mustakeem S.M., *Slavery at Sea: Terror, Sex and Sickness in the Middle Passage* (Urbana, IL: 2016).

other countries. It is an obvious but often neglected fact that fishermen, navy officers, privateers, or merchants of all nations likely shared similar attitudes towards the ocean.

This chapter is divided into five sections. Firstly, the ground will be laid with a look into the tradition of ancient and medieval perceptions of the ocean that were rediscovered or reinvented during the fifteenth century. This intellectual background will be compared with the projects of expansion that preceded the expeditions of Columbus. Each of the following five sections will offer a different interpretation of or perspective on the ocean. All of these rather archetypical perspectives existed simultaneously and were not mutually exclusive, but mostly held in combination with each other throughout the fifteenth and sixteenth centuries. For this reason, the following five categories can also serve as an analytical tool to triangulate any individuals' or groups' attitude towards the ocean. Therefore, this chapter offers not only an insight into the multitude of voices that have been preserved throughout the centuries, but also a steppingstone for further research.

1 Traditions of Knowledge and Knowledge of Practitioners: the Ocean until c. 1500

The Atlantic Ocean was a space of learning for Europeans where they developed their sailing techniques, distinct mentalities and practices that carried them beyond the Mediterranean and all around the globe. It was also the first ocean in which Europeans colonised and exploited societies and landscapes across open water, bringing devastation to indigenous peoples and forcing them to resist or to adapt to a changed world. But how and when did the European exploration of the Atlantic and its consequences begin, and what epistemic systems served as the framework for interpreting and understanding what was out there?

For a long time, the open waters of the western ocean marked the end of the world as it was known to European people. In western European maps from the fifteenth century the ocean was mostly represented as a small strip of water west of Europe and northern Africa, lying beyond the well charted Mediterranean. This perception was passed down from the Middle Ages and backed up by ancient mythology, since Heracles was known to have marked the border of the known world with two stone pillars.[9] This perception also

9 Chaplin, "Atlantic Ocean" 36

correlated with rediscovered ancient geographical knowledge, as can be seen in the work of the geographer and mathematician Claudius Ptolemy.[10] After 1400 his writings circulated in manuscripts and appeared in print around 1470. Maps that included his knowledge still showed the Atlantic as a small strip of water on the western edge of the world. According to Ptolemy, it was part of the world-encompassing, boundless *oceanos*, unlike the Indian Ocean which he considered to be an inland sea like the Mediterranean.

The belief that a great *oceanos* spanned to the end of the world was widespread in the Middle Ages. In Viking mythology, the open ocean was a space of darkness surrounding the whole world, and the Atlantic was nothing but one mighty side sea of that endless ocean, the only one on which the gods allowed men to travel.[11] Further south, the Muslim approach to the Atlantic was quite similar and also based on preserved knowledge of Antiquity. Muslim scholars also understood it to be a sea of darkness that marked the end of the known world. Unlike the well-traveled Indian Ocean, it held little interest for them.[12] The legends of the great oceanic voyages of discovery and conquest by Muslim African kings from Mali are accordingly recorded as vain journeys of no return.

On the other side of the Atlantic, Native American cultures also held similar beliefs. In their peripheral maps made of stones, sand drawings and twigs,[13] shown to the first visiting Europeans, they represented the Atlantic as the end of their known world. Sometimes, for example in Mesoamerica, the Atlantic was also seen as the pathway to another plane of existence, to which only heroes or gods had traveled ages ago. But, just as in Europe and Africa, this quite negative image did not prevent coastal trading, fishing, or travelling to

10 Ibidem; Vogel K.A., "Cosmography", in Park K. – Lindberg D.C. (eds.), *Early Modern Science* (Cambridge: repr. 2008) 469–496. Baumgärtner I., "Neue Karten, neue Welt", *Historische Zeitschrift Supplement* 77 (2019) 274–251; Wilford J.N., *The Mapmakers* (New York: rev. ed. 2000) 29–39; Buisseret D., *The Mapmakers' Quest: Depicting New Worlds in Renaissance Europe* (Oxford: 2003) 14–28; Quinn D.B., "Renaissance Influences in English Colonisation", *Royal Historical Society Transactions* 5.26 (1976) 74–92, 73–76.

11 Johnson D., *Die große Geschichte der Seefahrt / The History of Seafaring. 3000 Jahre Expeditionen, Handel und Navigation* (Hamburg: 2009) 123–134; Jourdin, *Europa und das Meer* 60–68; Borgolte M. – Jaspert N., "Maritimes Zeitalter. Zur Einführung", in Borgolte M. – Jaspert N. (eds.), *Maritimes Mittelalter. Meere als Kommunikationsräume* (Stuttgart: 2016) 9–33, 10–12.

12 Thornton, *A Cultural History*; Feldbauer P., "Die Islamische Welt seit der Jahrtausendwende", in Borgolte M. – Jaspert N. (eds.), *Maritimes Mittelalter. Meere als Kommunikationsräume* (Stuttgart: 2016); Rodger N.A.M., "Atlantic Seafaring", Canny N. – Morgan P. (eds.), *The Oxford Handbook of the Atlantic World, 1450–1850* (Oxford: 2012) 75.

13 Short J.R., *Cartographic Encounters: Indigenous Peoples and the Exploration of the New World* (London: 2009).

islands off the coast to hunt for maritime mammals.[14] The North and Baltic Seas were well travelled – at least from spring to autumn – and well known to people. The same can be seen on the other side of the Atlantic. Interactions between the different islands in the Caribbean and the easy development of trading mechanisms by Native Americans, that Europeans reported after 1500, show that there was coastal trading and traveling as well, albeit on a smaller scale.[15]

Everywhere there were seafarers who knew parts of the ocean. But almost all of them stayed close to the coast and avoided losing sight of the land. The most prominent exceptions were the Norse, who sailed to Iceland, Greenland and, around the year 1000, all the way to Newfoundland.[16] But their expeditions beyond Iceland declined and their colonies in Greenland finally perished in the fifteenth century.

Thus, it can be deduced that the open sea was understood to be a dangerous place. This universal belief influenced all the perceptions and interpretations which will be discussed in the next sections. It seems that everyone knew that people died and ships could be lost out on the open sea, or sometimes even right at the shore. Therefore, in all cultures people living on the coast had to outweigh the risks and gains of seafaring according to their technical abilities. Remarkably, it took the western Europeans longer to decide to take that risk than the north European Vikings or the native peoples of the Arctic and the Caribbean.

But this does not mean that people in Italy, Spain, Portugal, France, England, or Germany did not have positive expectations about what was out there. Throughout the Middle Ages, travel accounts were created and handed down that described the ocean, its islands or the lands beyond it.[17] These included the famous *Navigatio Sancti Brendani Abbatis* (*Voyage of Saint Brendan the Abbot*), but also reports about travels of King Arthur or King Abubakarai Kaita of Mali,

14 Starkey D., "Fish and Fisheries in the Atlantic World", in Coffmann D'M – Leonard A. – O'Reilly W. (eds.), *The Atlantic World* (London: 2015) 61–64.

15 Chaplin, "Atlantic Ocean" 40.

16 Enterline E. – Robert J., *Erikson, Eskimos & Columbus. Medieval European Knowledge of America* (Baltimore, MD: 2002).

17 Gillis J.R., *Islands of the Mind. How the Human Imagination Created the Atlantic World* (New York: 2004). Compare with Egerton and Games, *The Atlantic World: A History, 1400–1888* 51; Lehners J.P., "Anfänge der Portugiesischen Expansion", in Borgolte M. – Jaspert N. (eds.), *Maritimes Mittelalter. Meere als Kommunikationsräume* (Stuttgart: 2016) 165–185; Hancock D., "Atlantic Trade and Commodities", in Canny – Morgan (eds.), *The Oxford Handbook of the Atlantic World, 1450–1850*, 325–326; Allen J.L., "From Cabot to Cartier. The Early Exploration of Eastern North America 1497–1543", *Annals of the Association of American Geographers* 83.3 (1992) 500–521, 504–506 and Jourdin, *Europa und das Meer* 138–140.

or the legend of the seven Spanish bishops who escaped Muslim attacks and fled to the mythical Island of Antilla.[18] All these tales presented the Ocean as a space full of dangers, strange creatures and many islands.

These common perceptions were backed up by the rediscovery of ancient authors like Claudius Ptolemy.[19] Ptolemy's geography already included islands in the western ocean visited by ancient seafarers, and editions after 1470 were enriched with additional islands according to contemporary legends. So, an Island of St. Brendan, the golden Island of Antilla, or the Islands Frizeland and Brezil found a place on the map while the space reserved for the ocean grew.

But the Atlantic islands were not only imaginary places, they were actually explored since the late fourteenth century.[20] The Canaries – believed to be the *fortunate islands* on the maps of Ptolemy and also mentioned by Pliny – were rediscovered even before 1400. The unpopulated Azores, Madeira and Cape Verde islands followed between 1420 and 1460. All these discoveries confirmed the perception of the Ocean as a space full of islands and possibilities.

After those successful expeditions, seafarers from other countries also set out to find islands long imagined.[21] However, most of these voyages had more than one single goal. They were also about whaling and fishing and thus combined nautical curiosity with the quest for profit. The fishermen of Bristol are a good example of this tendency. In the late fifteenth century the Hanseatic League employed a new policy of exclusion and forced all outsiders to look for new fishing grounds. This made the Bristol fishermen venture beyond Iceland. There, they found the vast fishing grounds on the banks of Newfoundland but also kept an eye out for the mystical island of Brezil.[22]

Only a few seafarers asked for the protection of a monarch and an official charter that granted them the permission to discover and conquer certain islands. One early example is that of the Norman noblemen Jean de

18 Allen, "From Cabot to Cartier" 500–521. Loades D.M., *England's Maritime Empire. Seapower, Commerce and Policy, 1490–1690* (Harlow: 2000).

19 Chaplin, "Atlantic Ocean" 36; Vogel, "Cosmography" 469–496. Baumgärtner, "Neue Karten" 274–251. Wilford, *The Mapmakers* 29–39; Buisseret, *The Mapmakers' Quest* 14–28; Quinn, *Renaissance Influences* 73–76.

20 Kicza J.E., "Patterns in Early Spanish Overseas Expansion", *The William and Mary Quarterly* 49.2 (1992) 229–253. Egerton and Games, *The Atlantic World* 68; Fernández-Armesto F., *Before Columbus: Exploration and Colonization from the Mediterranean to the Atlantic, 1229–1492* (Basingstoke: 1987).

21 Rodger, "Atlantic Seafaring" 72.

22 Enterline, *Medieval European Knowledge of America*; and Egerton – Games, "The Atlantic World" 68.

Béthencourt who was sent by the king of Castille in 1402 to conquer the Canaries and started a brutal war of extermination that lasted for decades. Accordingly, the Portuguese kings gave out charters to attract settlers to Madeira and the Azores in the first half of the fifteenth century. Once established, these people and their family networks connected the islands with the coastal trade routes at the European mainland. They planted wine and sugar crop and later abducted people from Africa to work in the first Atlantic plantations. As early as the 1440s, slaves from Sub-Saharan Africa were being brought to Portugal by expeditions that followed the African coastline south to find a way around the continent. So, the knowledge of the ocean expanded, and perceptions began to change in Europe, even though it was still unclear what lay beyond those islands.

The most common answer to this question in the works of the ancient geographers was: Asia. In the late fifteenth century, the fact that the earth was a globe, and that a ship sailing west would reach the most eastern parts of Asia was uncontested. Circumnavigating the world had already been discussed by Strabo in theory, although he considered it impractical due to the world's circumference and suggested instead sailing around Africa. On the other hand, Ptolemy argued that Earth's circumference might be smaller and a voyage towards west was therefore possible, unlike sailing around Africa into the supposedly landlocked Indian Ocean.[23] Therefore, to reach the praised and rich kingdoms of Cathay (China) and Zipangu (Japan) by sea, one had to choose between two contradictory options based on ancient texts.[24]

The first option was to follow the African coastline south and hope to find a passage to the Indian Ocean despite Ptolemy's claim. The Portuguese chose this way and proved Ptolemy wrong in 1488. The second option was to sail west, as pursued by Christopher Columbus. Although his plan was far from revolutionary in theory, he was outstanding in the firmness of his belief that the journey would actually be possible. His major argument was based on a diverging opinion on the earth's circumference in ancient texts.[25] He followed Ptolemy and the later Italian cosmographer Toscanelli, whose calculations placed Japan roughly as far away from Europe as Mexico actually is. But this was still too far for Columbus, who claimed this distance to be even smaller when applying for

23 Parry J.H., *The Discovery of South America* (London: 1979) 45.

24 Wilford, *The Mapmakers* 43–47; Gillis, *Islands of the Mind*; Allen, "From Cabot to Cartier" 502–506; Grafton N., *New Worlds, Ancient Texts. The Power of Tradition and the Shock of Discovery* (Cambridge: 3rd ed. 2000) 70–75.

25 Parry, *Discovery* 48–50; Gewecke F., *Christoph Kolumbus. Leben, Werk, Wirkung* (Frankfurt: 2006) 14–30.

official support. It is no wonder that his plans were rejected by those educated in mathematics and astronomy in Portugal as well as in Castile. They knew that the ancient mathematician Eratosthenes had calculated a much bigger size of the earth's radius. To the majority of the experts, Columbus's plan was doomed to fail because there was no way to sail that far west through open water. Without supply stops, his crew would die long before they would reach their goal. It was a small group of supporters who convinced queen Isabella of Castile to grant Columbus at the very least an official status, which would allow him to attract investors and followers to join him at their own risk. This was the support which made his famous voyage of 1492 possible, on which he found some new islands he proclaimed to be of the coast of Asia.

But if Europeans had already founded colonies in the Atlantic Ocean, conquered and murdered indigenous populations, transported slaves from Africa to plantations on distant islands and even found a pathway to India, what did the journeys of Columbus actually change about the understanding of the ocean? His first remarkable achievement was to uncover the currents and wind patterns of the Atlantic. He found out how easily a westward passage could be made from the Canaries and back eastward to Europe via the Azores. In combination with the North Atlantic route discovered by the English and the Southern Atlantic currents charted by the Portuguese, European sailors unlocked the secrets of travelling open waters within decades. Even more important was the discovery of a mainland much closer to Europe then the ancient texts predicted. New maps were made, and very soon the new lands were described as part of one immense landmass that the cartographer Martin Waldseemüller first named America in 1507.[26] Beyond America there was another ocean, first called the Southern Ocean and later named the Silent or Peaceful Ocean by Magellan. Since the 1530s, the one *oceanos* that once was believed to encompass the world was now understood as just the smaller one of at least two distinct oceans.

This widening of the European geographical horizon has been described as groundbreaking and even as the spark that lit a scientific revolution. But these claims made by Thomas Kuhn or Anthony Grafton, following authors of the sixteenth century who ecstatically celebrated the triumph of their own age over the ancients, have been doubted recently.[27] It has instead been proven

26 Meurer P., "Cartography in the German Lands 1450–1650", in Woodward D. – Harley J.B. (eds.), *The History of Cartography. The Renaissance II* (Chicago: 2007) 1172–1254, esp. 1204–1206.

27 On claims like these and their critique, see Karstens S., "Die Macht der Tradition: Empirie und Muster der Vermittlung transozeanisch erworbenen Wissens ca. 1500–1620", in

that new experiences did not lead to a broad disregard of ancient authorities, but were rather used to enrich mankind's inherited knowledge. For decades, new maps were simply added to old works, like Ptolemy's geography. Proving the ancients wrong could in fact not break the epistemic system of humanism, because that system was never based on blind obedience to the classics, but rather on critical examination and discussion of the sources. For this reason, the inherited knowledge expanded gradually over many decades. Although the ocean was now understood to be much smaller, in people's imagination it remained for a long time a space of limitless possibilities. Accordingly, as they started to exploit these possibilities and to connect the ocean's shores, they were guided by the idea of not one, but many oceans.

2 A Sea of Profit and Violence

Some who sailed the ocean didn't care much about what lands lay beyond the waves. They made their profit on the open water itself or maybe close to distant coasts, avoiding moving inland. Being on the sea was their way of life, and knowing about the winds, currents and dangers of the ocean was their basic requirement to make ends meet. They didn't want to cross the ocean in order to get anywhere, because the ocean itself was the place they needed to be. They were the pirates, privateers, fishermen and whalers of the Atlantic.

Between 1400 and 1600 their activities can be summed up in two major activities: fishing and privateering. The importance of fishing can best be explained with the example of the cod fishery off the coast of Newfoundland.[28] European fishermen had discovered these immensely rich fishing grounds decades before a landfall was made in North America. Soon sailors from France, Spain, Portugal, and England made their way across the ocean to reach places where the fish were so numerous you could allegedly catch them by simply lowering a basket into the water. All these fishermen considered the sea and coastlines to be beyond the control of a European sovereign, a space where they could make their own rules. After St. John's, one of Newfoundland's natural harbors,

---

Burschel P. – Juterczenka S. (eds.), *Das Meer. Maritime Lebenswelten in der frühen Neuzeit* (Vienna: 2020) 117–130. Compare with Grafton, *New Worlds* and Kuhn, *The Structure of Scientific Revolutions*.

28 This followed the tradition of the North and Baltic Sea herring fishery. See Jourdin, *Europa und das Meer* 185–190; Kurlansky M., *Cod. A Biography of the Fish that Changed the World* (London: 1999); Brunelle G.K., *Rouen and the New World: Rouennais Investors in Commerce with North and South America, 1559–1629* (Ph.D Dissertation, Emory University: 1988) 95–103.

had become a base for repairs and supplies, the multinational community of fishermen chose a new Harbour-Admiral each year who would settle potential disputes. Usually, they chose the Master who had made the first crossing of the year and thus had proven his maritime bravery. Sources indicate that hundreds of ships and thousands of sailors came each year, but compared to the expeditions of discovery, they left almost no reports of their lives and journeys.[29]

Only statistical information from harbour registers tells the story of how those sailors built a vast transatlantic network, connecting England, France, Spain, and Newfoundland with markets in Germany and Italy. They created what Joan Chaplin has called the everyday Atlantic long before France or England even had colonies and did so while other sailors were already searching for a passage into the next ocean.[30] However, due to the lack of written sources, historians are forced to deduce these people's attitude towards the ocean in a more indirect way. For this they can look at the fishermen's actions and also compare the sources belonging to another group that made its living out on the ocean – pirates and privateers.

The basic difference between the last two was that piracy was robbery at sea and considered to be illegal everywhere.[31] Pirates were 'enemies of all nations' and could not legally collect funding from investors or use the harbours of a country as official safe havens. All the European monarchs had over time tried to fight piracy in their home or even colonial waters, and publicly executed those they captured.

Privateers, on the other hand, held an official letter of marque by a European sovereign that allowed them to use their harbours as safe havens, and to legally approach investors and trade with their bounty. In exchange, they were allowed to attack only ships flying the flag of certain enemies. They were further obliged to follow rules of engagement, and faced legal reprimands if they used excessive violence or attacked unjustly – at least in theory. The privateers themselves were considered to be part of a sovereign's fighting force at sea and

29 For the surviving sources, see Quinn's edition, *North American World*.

30 Chaplin, "Atlantic Ocean" 41.

31 Childs D., *Pirate Nation, Elizabeth I and her Royal Sea Rovers* (Barnsley: 2014) 2–4; Loades, *England's Maritime Empire* 42–43; and Moreau J.P., *Pirates. Flibuste et piraterie dans la Caraïbe et les mers du sud, 1522–1725* (Paris: 2006) 35–36, and 211, Moreau states that Jean Ango became the first official French privateer in the Atlantic in 1522. The following argue for an older tradition dating back to the hundred years war: Villiers P., *La France sur mer. De Louis XIII à Napoléon* (Paris: 2015) 14–18 and Rodger, *The Safeguard of the Sea* 198–202, 238–297.

could therefore muster broad support, build up entire fleets and even become rich or famous.[32]

But the difference between pirates and privateers often remained theoretical. A sovereign's privateers were nothing but pirates to his enemies and vice-versa. Furthermore, a lot of privateers, and even fishermen, took every shot they got at capturing prizes, even if it was officially off-limits.[33] But compared to pirates or the fishermen, privateers left more sources about their lives, thereby granting insight into their perceptions of the ocean.

For privateers, the ocean was a place to make profit by taking from others through semi-legalised force. For them, the sea was – at least partially – separated from the European sphere of laws and justice. During the fifteenth and sixteenth centuries, they often continued perpetrating violence in times of peace. This course of action was at least tolerated by the monarchs and admirals in France and England, who received 20% of the bounty. The most successful privateers were able to build a profitable business enterprise. Men like Jean Ango in early sixteenth-century Dieppe commanded a whole fleet backed by Italian bankers.[34] He saw the Ocean as a private field of war and planned coordinated raids all over the North and South Atlantic, including a siege of Lisbon.

But what was life like for the individual sailor on board a ship setting out to make a profit on the high seas – be it by fishing, whaling, or privateering?[35] For them, the ship was not merely a medium of transportation, but their part-time home. Unlike passengers, they were part of a crew and formed a community that had to establish and enforce rules, settle differences, and unite to achieve common goals. Naturally, there was a hierarchy in place that was enforced brutally when challenged, but in the fifteenth and sixteenth centuries masters of a ship could not simply impose their will on their subordinates arbitrarily. They had to accept maritime tradition to avoid mutinies. This included a certain

32 On England, see Loades, *England's Maritime Empire* 122–131; on France, Augeron M., "Coligny et les Espagnols à travers la course (c.1560–1572) une politique maritime au service de la cause protestante", in Acerra M. – Martinière G. (eds.), *Coligny, les protestants et la mer* (Paris: 1999) 155–176.

33 Childs, *Pirate Nation* 8; Augeron, Coligny, "Coligny et les Espagnols" 55–176.

34 Guénin E., *Ango et ses pilotes, d'après des documents inédits tirés des archives de France, de Portugal et d'Espagne* (Paris: 1901).

35 On this topic compare Childs, *Pirate Nation*; Fury C.A., *Tides in the Affairs of Men. The Social History of Elizabethan Seamen, 1580–1603* (MA dissertation, McMaster University Canada: 1998).
Andrews K.R., "The Elizabethan Seaman", *The Mariner's Mirror* 68 (1982) 245–262. James A., *The Navy and Government in Early Modern France* (Ph.D dissertation, University of Manchester: 1997); Loades, *England's Maritime Empire*; Moreau, *Pirates*; Rodger, *Safeguard*; Villiers, *La France sur mer* 14–34.

level of participation in the decision making. If there was no imminent crisis, they were expected to summon a ship's council to collectively decide on punishments, rations, or other things. This helped sailors to build a community. Sailors from all backgrounds shared similar knowledge, experiences, rituals, behaviour, language and even walked in a recognisable way. This international community was also open to non-Europeans.[36] Seafarers, whalers and privateers could in fact be of African, Native American, or mixed background and it is only plausible that all of them adapted very similar attitudes towards the ocean.[37]

In the centuries before press-gangs and modern navies, after being trained on board by their seniors for several years, the sailors had a certain level of freedom due to their skills and knowledge. Their loyalty was accordingly somewhat flexible. They mostly followed a certain patron or their current employer rather than a monarch and his flag, as was the case with mercenaries in the European wars of this age. This is why French protestants and Englishmen could work together closely and share a common labour market during the sixteenth century.[38] The vast majority of these people were motivated by shared perceptions and understandings of the sea. The ocean was a place full of resources waiting to be exploited by those who had the knowledge and courage to seize them. As part of a community of seamen, they accepted the ocean as a space of danger and violence and as their temporary home.[39]

3 A Void to Cross

Not everybody who organised or took part in a voyage was looking to find wealth or fame or to make a living on the sea itself. For many people, the Atlantic Ocean was nothing but a void, an emptiness to be crossed quickly and with as little distraction as possible. This perspective was shared by four groups, three of which were until 1600 mostly of European origin: merchants, empire-builders, and settlers; the fourth group came from Africa and its members had

36 Jourdin, *Europa und das Meer* 137; Fury, *Tides in the Affairs of Men* 198–215; Rodger, *Safeguard* 319–320.

37 Chaplin, "Atlantic Ocean" 41

38 Fury, *Tides in the Affairs of Men* 79–87. Tyacke S., "Chartmaking in England", in Woodward D. – Harley J.B. (eds.), *The History of Cartography. The Renaissance II* (Chicago: 2007) 1722–1753, 1726–1729.

39 Jourdin, *Europa und das Meer* 191–200, Schaub J.F., "Violence in the Atlantic, Sixteenth and Seventeenth Century", in Canny – Morgan (eds.), *The Oxford Handbook of the Atlantic World: c.1450–c.1850*.

their very own, distinct perceptions of the ocean since they were forced to cross it after being enslaved.

For merchants, it was of highest importance to move their goods from one harbour to another with little risk and within a certain timeframe.[40] Their wish to minimise risks brought forth several innovations in the late medieval and early modern era that also changed the perception of the ocean's dangers. Italian seafarers were the first to develop an insurance system with risk calculation and differentiated tariffs while the Dutch and English formed semi-sovereign trading companies that put the risk on many shoulders.[41] For those involved or investing in the trade or working in insurances, the ocean was an empty, yet dangerous space.

During the fifteenth and sixteenth centuries, European merchants massively expanded their trade routes. Many expeditions set out to find passages beyond the Atlantic.[42] In England, this was actually of much more importance than any project of colonisation until the seventeenth century. The Atlantic was in this perspective just a steppingstone to the riches of Asia, or even new worlds to be discovered, like a mystical 'Third World' in the Pacific, as imagined by the French Author La Popeliniére as a space for colonial expansion.[43]

This idea – based on ancient beliefs about a *terra australis* on the opposite side of the world – leads to the perspective of those who organised projects of colonisation and imagined transoceanic empires.[44] For them, the ocean was mostly an empty space between the motherland and territories to acquire. They needed settlers to be transported and troops to be deployed as quickly as possible to avoid sickness and starvation. They also wanted to build new trade networks between the motherland and its colonies that would create revenue through customs and taxes. For this, those who imagined colonial empires had a certain mercantilist economic system in mind.[45] They wanted raw materials

---

40 Johnson, *Geschichte der Seefahrt* 103–105; Scheller B., "Kaufen, Verkaufen und Verstehen in die Atlantikexpansion der Europäer, die Fernhändler und die neue Erfahrung des Fremden im 14. und 15. Jahrhundert", in Borgolte – Jaspert (eds.), *Maritimes Mittelalter. Meere als Kommunikationsräume*; Lehners, *Anfänge* 177; Jourdin, *Europa und das Meer* 68–69 and 204–205.

41 Loades, *England's Maritime Empire* 117.

42 Quinn D.B., "The Northwest Passage in Theory and Practice", in Allen J.L. (ed.), *North American Exploration* (Lincoln, NE: 1997) 292–343, and Karstens, *Gescheiterte Kolonien*, both contradicting the idea that the desire to reach another ocean only came into play much later as presented by Chaplin, "Atlantic Ocean" 46–47.

43 For a modern edition, see Voisin de La Popelinière, *Les trois mondes de La Popelinière*, ed. A.M. Beaulieu (Geneva: 1997).

44 Karstens, *Gescheiterte Kolonien*.

45 Hancock, "Atlantic Trade" 326.

from distant colonies to be transported by sea to the motherland, where refined products would be made and sent back to the colonies. This cycle required regular and undisturbed ocean crossings to work. In the circles of those who imagined and promoted colonial empires, there were also cosmographers and historians of the New World to be found, who wrote about cultural encounters, natural resources of the Americas and conquest. For many of those writers, the ocean was also of little interest. It was just the space between themselves and the people, landscapes, and resources they wished to tell their readers about.[46]

While the organisers and promoters of colonial enterprises themselves often stayed in Europe, their plans nonetheless required many people to cross the ocean. Thousands of colonisers set sail. Their hopes of fame, treasure or at least a better way of living was bound to the land across the ocean, not to the sea. Even more so, since one of the key objectives for many colonising enterprises was to spread Christianity, a feat impossible to be accomplished on open water. So, the colonists' perception of the ocean was simply opposite to that of the seamen's, which could lead to an estrangement between the two groups.[47]

Despite their differences, sailors, colonists, and merchants mostly chose seafaring voluntarily. This was completely different from the millions of African people that were enslaved and shipped against their will as human cargo.[48] Not only for their numbers, but even more for the importance their brutal exploitation had for the building of Atlantic Worlds, their perspective needs to be included in this chapter. For enslaved people, leaving the shores of Africa marked the end of hope to return to their former lives. Along with the shoreline, there also disappeared their last chance of physical escape, which is why the departure was often scheduled at night or accompanied by extra strict security measures.[49]

After the departure of their floating prisons, when the slaves were brought on deck in small groups to exercise or eat, they saw nothing but emptiness around them. Inexperienced in sailing or at least unaware of their nautical position or the distance to the end of their journey, the ocean itself became part of their imprisonment. What remained for the people of some African cultures – albeit not for all – was the hope of a spiritual escape. Some believed that by ending one's own life they would be able to enter a spiritual realm and

46 This includes Spanish as well as French cosmographers and historians such as André Thevet, Francois Belleforest, Petrus Martyr de Anghiera or José de Acosta.

47 See Karstens, *Gescheiterte Kolonien* 457–476, 498–512.

48 See Rediker M., *The Slave Ship: A Human History* (New York: 2007); Mustakeem, *Slavery at the Sea*.

49 Mustakeem, *Slavery at the Sea* 115 and 125.

meet friends and relatives that had already passed away.[50] To achieve this, the enslaved could starve themselves to death or jump overboard. The latter method was favourable at least to people of some African cultures, who considered the water as a pathway to the spirit world. This form of spiritual escape was often understood as suicide by the Europeans, who considered this to be a sin. They tried to prevent this – obviously also because of financial reasons – by installing nets or walls. For those who broke the barriers and jumped, as well as for many more, the ocean became the final resting place. Dead and sometimes even sick people were thrown overboard and sometimes mutilated beforehand, to show others that their bodies would not enter the spiritual world as a whole. Estimates may vary, but approximately 15% of the enslaved did not survive the crossing, with the death toll being close to 20% in the beginning and decreasing to less than 10% over time.[51]

While being on the ship in the middle of the emptiness, the enslaved were subdued to what Sowande Mustakeem has called 'a continuum of torture, abuse and survival'.[52] She and other historians have argued that the crossing itself played an important part in breaking the captured people and turning them into slaves. But this cruel practice failed to succeed on more than one occasion. Instead, the daily suffering made several of the enslaved resist their captors in more direct and violent ways and sparked rebellion. The fear of such uprisings, on the other hand, made the vastly outnumbered crew constantly feel a sense of danger on their own ship. The empty ocean was for them surely adding to this feeling, because no reinforcements would come. This is why they took many precautions. But despite guns, barriers, and chains more than one report tells of an uprising. If the enslaved people failed to seize the vessel, they often tried to set fire to the ship and thereby kill themselves and their captors. But even when they were successful, they often had no means to navigate, and so sometimes surrendered to the emptiness around them by setting fire to the ship anyway. Only in a few cases, they successfully sailed the ship to a coast unknown to them where they could escape.

4 A Nautical, Mathematical or Cosmographical Challenge

The previous sections have shown that the ocean was indeed a dangerous place. It is therefore obvious that people tried to minimise its risks through a better

---

50 Ibidem 117 and 128.

51 Priesching N., *Sklaverei in der Neuzeit* (Darmstadt: 2014) 57–64.

52 Ibidem Introduction.

understanding of it. Learning about its patterns, currents and coastlines and structuring this data into processed knowledge became of higher importance as sailing intensified. From this basic interest came two different approaches to taming the ocean, shared by social groups that considered themselves to be much different from each other and that also had different ideas of the ocean. The first one is nautical, a hands-on approach by those who sailed and risked their own lives. The second one is more abstract. It was the attempt to capture the unstable oceans on maps by mathematics and astronomy or even to harmonise them into a complete cosmographic description of God's creation.

In the fifteenth and sixteenth centuries, navigators made a living out of their abilities. They were educated on board and learned their trade over many years from different masters, always expanding their knowledge with first-hand experience.[53] Much of their knowledge was stored in their minds, supported by handbooks with sketches of coastlines and harbours, as well as notes on reefs or shallows.[54] Masters copied parts of their notes and traded them, while also guarding some of their knowledge to have an advantage. Only after more than one hundred years of travel to distant islands did the first European rutters or handbooks appear in print. Finally, at the end of the sixteenth century, compilations of these works appeared, depicting all major harbors and waterways of Europe and other continents.[55]

These navigational experts had a special relationship with the ocean. They not only made a living on it like other seamen, but they also had a deeper practical understanding of it. It is important to consider that these abilities did not require any European education or skills such as writing or drawing. The history of discovery is full of famously skilled indigenous pilots, who guided voyagers such as Vasco Da Gama or James Cook.[56]

In the sixteenth century, the monarchs of Portugal and Castile ordered the collection of all the knowledge of their navigators and seafarers in specialised institutions. It was supposed to be processed, kept secret and put into order for future use – a goal that never could be fully achieved.[57] But nonetheless, an immense amount of data was collected, and some of the pieces of information were transferred into master-maps depicting all lands and islands known.

---

53 Tyacke, *Chartmaking* 1722–1753; Compare with Ash E.H., "Navigation Techniques and Practice in the Renaissance", in Woodward D. – Harley J.B. (eds.), *The History of Cartography, The Renaissance I* (Chicago: 2007) 509–528; Brendecke, *Imperium und Empirie*; Edney, "Knowledge and Cartography in the Early Atlantic".

54 Jourdin, *Europa und das Meer* 125.

55 The most important was Lucas Janszoon Waghenaer's *Spieghel der Zeevaertd* from 1585

56 Chaplin, "The Atlantic Ocean" 46–47. Short, *Encounters*.

57 Jourdin, *Europa und das Meer* 274; Chaplin, "Atlantic Ocean" 44–45; Brendecke, *Imperium*.

So, the ocean became a space comprised of measured and collected data and growing knowledge.

There were also plans to make this go full circle by instructing the masters to collect data according to common guidelines. Similar ideas were also discussed in England and France, but there the collection of knowledge was left to certain groups of investors and scholars who lobbied for colonial expansion and trade. In England, it was colonial promoter and clergyman Richard Hakluyt who argued for official lectures in navigation and collected travel reports, while in France the privateer organiser Jean Ango invited cartographers and navigators to Dieppe, where he created a knowledge centre to improve his activities.

The creation of complex maps led to the question of how the ocean was perceived on a more academic level by mathematicians and cosmographers.[58] For them, who like the famous German cosmographers of the Renaissance often worked far away from the sea, the ocean was primarily a sum of numbers and measurements and its correct representation on a map a puzzle to be solved. While this was true for those who followed a more mathematical approach, others who were cosmographers in the narrow sense aimed less at a precise depiction and more at harmonising the numbers with the traditional worldview and creating a unified description of God's creation. Cosmographers would even add or omit an island to make their point. At the end of the sixteenth century, the more mathematical approach became predominant and the more literary cosmography vanished.[59] To achieve their respective goals, cartographers and cosmographers separated oceans like the Atlantic into a separate Northern and Southern ocean, the term Atlantic was indeed uncommon before 1600.[60] They all saw their work as enrichment, rather than a refutation of the ancient knowledge of Pliny, Ptolemy and others, whose works were the foundation of their own epistemic framework. Nonetheless, their effort brought forth new worldviews, for example when they changed England's position from the edge of Europe to the centre of an Atlantic world.[61]

When the nautical and the two more abstract mathematical or cosmographic approaches are considered together, it becomes evident that they

58 Lestringant F., *André Thevet: cosmographe des derniers Valois* (Geneva: 1991); Mosley A., "The Cosmographer's Role in the Sixteenth Century", in Dupré S. – Hallyn F. (eds.), *Early Modern Cosmography. Proceedings of a Conference at the Centre for History of Science, Ghent University, Ghent and Louvain, 28–30 May 2008* (Turnhout: 2009) 423–440; Buisseret, *The Mapmakers' Quest*; Lestringant, "Fictions cosmographiques à la renaissance".

59 Lestringant, "Fictions cosmographiques à la renaissance" 293–318.

60 Edney, "Knowledge and Cartography" 98–99.

61 Cormack L.B., *Charting an Empire. Geography at the English Universities, 1580–1620* (Chicago: 1997) 8–9.

were not as exclusive as the contemporaries themselves described them to be.[62] Although authors proclaimed a difference, leading to a battle of practitioners and theorists, this was often a part of personal or institutional rivalries. There was not much reason for conflict since 'experience' was understood as a mixture of traditional knowledge and new practical experiences, both presented and structured according to traditional epistemic rules.[63] This connection can best be seen in the work of Gerhard Mercator, who solved the puzzle of making a conformal projection of the world and its oceans on two-dimensional maps.[64] With his innovation a straight line on the map was now equal to a constant course at sea. This came at a price though, since this depiction inflated objects away from the equator. Nonetheless, with the simultaneous invention of conformal projection and the modern atlas, Mercator gave navigation the most powerful tools until the invention of the GPS.

But still, his maps visualise ancient theories of an ice-free North Pole as well as mystical islands from medieval legends.

5 Where the Devil Lurks and Spirits Roam

The previous section has shown how thoroughly people in the fifteenth and sixteenth centuries analysed their environment and tried to minimise its risks. Many of their methods such as insurances, better map making, or a practical onboard training seem familiar and rational from a modern point of view. But this is just one side of the coin. Obviously, people of the time also imagined spiritual and other supernatural forces at work on the open waters. These were understood in two ways: as the reason for maritime dangers, but also as the best protection against them.

One major source for dangers was the great antagonist himself – the devil. Shakespeare puts this idea in the words of Prince Ferdinand in *The Tempest* when he jumps from his burning ship into a raging sea: 'Hell is empty and all the devils are here'.[65] This line is rooted in contemporary belief that the devil is lurking out on the ocean, waiting to attack either those who are no longer protected by God because of their sins, or those who disturb the devil's plans

62 Mosley, *The Cosmographer* 432–440; Grafton, *New World*. See also Karstens, "Die Macht der Tradition".

63 Dear, "The Meanings of Experience".

64 Ash, "Navigation Techniques and Practice" 519–523.

65 Ferdinand's words are related to Prospero through the Spirit-servant Ariel who witnessed the events, see Jourdin, *Europa und das Meer* 251–257.

by doing God's work beyond the sea. The last idea offers of course immense narrative possibilities for promotional literature. Authors could describe their project as godly and use the storms they overcame as proof of the devil's failed attempts to stop them from doing God's will.[66]

Beyond the devil himself, sea monsters or witches were also believed to be endangering ships in many Atlantic cultures.[67] These beliefs were so strong that they even shaped the way sailors perceived the maritime worlds they encountered. Maritime life forms were seen as mermaids, and Martin Frobisher even presented Elizabeth I with the tooth of a narwhal believed to be the magical horn of a sea unicorn.[68]

But a closer look at travel narratives shows that the most important supernatural reason for storms and wreckage was God himself.[69] It was most common to describe maritime dangers like storms as a test of faith or as a divine punishment for wrongdoing. This line of interpretation also allowed one to avoid any questions about the devil's spiritual role and his actual powers or the nature of ghosts and monsters.

To overcome the multitude of supernatural maritime dangers, sailors could apply several appropriate countermeasures. The most important one was staying calm and having faith in God when danger was on the horizon. This carried a strong biblical meaning by referring to the story of Jesus calming the storm in the Gospel of Mark, Chapter 4. To receive God's protection Catholic seafarers prayed to saints, and especially also the Virgin Mary, who was called *stella maris*. Protestants instead sung certain psalms and saw their survival more as an act of God's direct will or even as proof that they had already been predetermined to be saved long ago. In all cases, God was described as the 'heavenly pylot'[70] who could make a passage safe for those he protected.

But there was more to do than just to act appropriately when danger was imminent. The masters were obliged and expected to enforce obedience to the Ten Commandments and punish any behaviour which was thought of being offensive to God. Furthermore, the crews held their own religions ceremonies by reading from the Bible or singing songs on Sundays. But this was

66 This is a trope more used by Catholic than Protestant writers, Karstens, *Gescheiterte Kolonien* 462–475.

67 Jourdin, *Europa und das Meer* 76–77; 210–213; 241–242; 247.

68 Best G., *A True Discourse of the Three Voyages of Discoverie for the finding of a passage to Cathaya, by the Northvveast, vnder the conduct of Martin Frobisher Generall deuided into three bookes* (London: Henry Bynnyman, 1578) 15.

69 Karstens, *Gescheiterte Kolonien* 462–475.

70 Whitaker A., *Good Newes from Virginia sent to the Counsell and Company of Virginia, resident in England from Alexander Whitaker* (London: Welby, 1613) 7.

only the more official part of the spiritual and supernatural protection. Magic, in form of rituals, lucky charms, tattoos, scarrings, talismans or certain rules and taboos, was a global phenomenon and could be found in any seafaring culture.[71] A rather distinctive European practice was to hang ship models in churches to thank for God's protection and ensure his goodwill. But not only people on board could be shielded spiritually. The ship itself could be as well. In Europe, the ships were christened and blessed from bow to stern. The mast, as the spiritual centre, carried highest importance. In Christian countries, a coin was put underneath it, while some reports on northern European pagan societies indicate animal sacrifices when it was erected. So, the sailors faced the dangers of the deep armed with faith in God or other higher powers.

The supernatural was also a part of the stories people told about the ocean. Overcoming dangers was seen as a triumph of God over the devil or as God's blessing for the voyage itself.[72] If a ship did succumb to the dangers of the sea, it was interpreted as a punishment, or even a sacrifice, to restore God's benevolence. Mary C. Fuller has shown this masterfully in her analysis of the reports on the wreckage of Humphrey Gilbert's 1583 expedition.[73] It was a different matter, though, if ships of colonial rivals perished. Promoters jumped to the occasion and presented this as a manifestation of God's will to reserve parts of the new world for themselves: 'strongly defended on our behalf by the almighty hand of God withstanding the enterprises of other nations'.[74] They presented the open sea as the place where God smites the enemies of his people.

6 Oceans between Freedom and Possession

The final perspective to be discussed here is one shared by European monarchs, members of their centralised institutions, powerful trading companies and pro-colonial pressure groups made up of investors, scholars, and courtiers. It is the idea that the ocean itself can be owned exclusively. This perception

71 Jourdin, *Europa und das Meer* 76–77; 210–213; 241–242 and 247.

72 Burghartz S., "Erfolg durch Scheitern? Zur Konstruktion von Überlegenheit im kolonialen Diskurs um 1600", in Dürr R. (ed.), *Expansionen in der frühen Neuzeit* (Berlin: 2005) 307–324. These interpretations were based on medieval pilgrimage reports discussed by Jahn B., *Raumkonzepte in der frühen Neuzeit. Zur Konstruktion von Wirklichkeit in Pilgerberichten, Amerikareisebeschreibungen und Prosaerzählungen* (Frankfurt: 1993).

73 Fuller M.C., *Voyages in Print. English Travel to America, 1576–1624* (Cambridge: 1995).

74 See E. Hayes's report on Gilbert's final expedition "A report of the Voyage and the Success thereof", in Archer I.W. – Price D. (eds.), *English Historical Documents, 1558–1603* (London: 2011) 602.

directly leads to the question of how this should be done practically. After all, the sea is always moving and eludes all attempts of physically taking possession of it.[75] Several ways to deal with this problem were developed in Europe in Antiquity and throughout the Middle Ages. The basic idea was to use land to control the sea. Fortifying straights, building ships to patrol coastal waters, setting up watch-posts and putting garrisons in harbour cities were most common, all superseded by establishing monopolies on trade. But these attempts to control maritime resource extraction and trading were never fully successful.[76] Oftentimes they were rather symbolic, just like claiming by naming which describes the global phenomenon that sea powers named their spheres of influence.

While medieval attempts at controlling the sea were focused on land, the early modern era changed that at least in theory. In the years 1479, 1493 and 1529 Castile and Portugal signed several treaties that divided the world's oceans into spheres of influence.[77] Their zones were always considered to be exclusive and were separated by a simple straight line on maps crossing over open waters. Although the treaties were primarily about land to be discovered and the right to transport slaves out of Africa, they still showed the perception that oceans can be owned.[78] Both powers forbid outsiders from travelling and threatened them with being destroyed or having their cargo confiscated.

This partition of the world was contested right away, especially by the kings of France and England and their subjects.[79] They stated that no one could possess the oceans or could claim coastlines just by passing by or raising a flag. One needed to actually subdue the country to call it one's own according to Genesis 1:28. Remarkably, both kingdoms already had their own history of trying to control the seas close to their own countries.[80] They made a differentiation, and considered the open ocean as well as islands and coastlines not yet in

75 Rüdiger J., "Kann man zur See herrschen? Zur Frage Mittelalterlicher Thalassokratien", in Borgolte – Jaspert (eds.), *Maritimes Mittelalter Meere als Kommunikationsräume* 35–58.

76 Borgolte – Jaspert, *Maritimes Zeitalter* 16 and 26–27; Cormack, *Charting* 12; MacMillan K., "Centers and Peripheries in English Maps of America 1590–1685", in Brückner M. (ed.), *Early American Cartographies* (Chapel Hill, NC: 2011) 67–92, 78–79.

77 Rüdiger, *Kann man zur See herrschen?* 25–26; Alves G.J., "De la Mediteranée a l'océan. Nouveaux problèmes, nouvelle solutions", in Hofmann C. (ed.), *L'âge d'or des cartes marines. Quand l'Europe découvrait le monde* (Paris: 2012) 126–135, 132–133.

78 Lehners, *Anfänge* 168; Bethencourt F. – Ramada Curto D., *Portuguese Oceanic Expansion, 1400–1800* (Cambridge: 2007); Gaspard, "De la Mediteranée a l'océan" 132–133.

79 Villiers, *La France sur mer* 18.

80 Armitage, *The ideological Origins of the British Empire*; Jourdin, *Europa und das Meer* 142–154. Hancock, "Atlantic Trade, 325–327"; Rüdiger, *Kann man zur See herrschen.*

real possession of a Christian monarch as fair game while their own territories were off limits. So, the basic European mindset of colonialism was formed.

Naturally, colonialism led to colonial rivalry. This also included a reflection on the nature of the ocean, and the juridical question if it could actually be owned. The most famous analysis on this topic was published in 1609 as an extract from a study on privateering. It was written by the Dutchman Hugo Grotius who declared the open sea – beyond a certain coastal zone – to be a *mare liberum*, a space of free interaction that God had given to all mankind.[81] His position became famous, but it should be remembered that it was already contested in his time. A response called *mare clausum* followed from the Englishman Edward Selden, and some years after Grotius's treatise, the Dutch trading companies nevertheless enforced monopolies and exclusive sailing rights themselves. So about one hundred and twenty years after Columbus, the contrary perceptions of the open waters as a space of freedom or possession were formed. They derived from conflict and are still the two major rivaling perceptions that inspire conflict today.

7 Imagine Oceans – Conclusion

Finally, after sketching different interpretations and perceptions of the ocean, it is necessary to go back to a point made in the introduction to this chapter. The different perspectives discussed here were not clearly distinguished in their time and while some were incompatible, others could be combined and be held by one and the same individual simultaneously. Therefore, these distinctive imagined oceans are more to be understood as an analytical tool to triangulate an individual's or a group's attitude towards the ocean. This could allow for broader, methodological comparisons of all the different meanings given to this vast amount of water. The sections above have shown that the ocean was – at the same time – the unknown end of the world and a space limited on each side by continents. It was an empty void to cross, and a space filled with opportunities to make a living by fishing or violence. It was a complicated mathematical puzzle and the place where the devil lurked and also where one's spirit could find freedom even if the body was chained. It was free and

81 Armitage, *The ideological Origins* 103–104. For a modern edition of a contemporary translation of Hugo Grotius's work with extensive comments, see Armitage, Hakluyt and Welwood, *The Free Sea*.

exclusively owned and it was full of hope, and even more of despair. The only thing everybody agreed upon was – the ocean could be a dangerous place.

The fact that these perceptions are ex post facto analytical categories does not mean that the ideas encompassed by them did not lead to conflict. The most obvious example for this is the contradictory understanding of the sea as a free space or as an exclusive zone of influence followed by mutually exclusive claims of dominance, be it mercantile or political. There are also fundamental lines of conflict between the interpretation of the ocean as a space of violence with resources for the taking and as a space for safe and peaceful trade. Those who followed the former idea encouraged war while their counterparts opted for peace.

But arguably of highest importance and yet mostly neglected in history is the everyday conflict of interpretations that arose between those sailing the ocean to make a living and those who just wanted to cross it as quickly as possible. Almost all colonial projects undertaken under the flags of France and England suffered massive problems that can be tracked down to this opposition.[82] Furthermore, this very conflict also held vast consequences for the historiography of this period. The great majority of the people who wrote about sea voyages in the fifteenth and sixteenth centuries were not seamen. Rather the opposite, they openly mistrusted and despised seamen and doubted their competence as well as their intentions because of basic conflicts of interest and their distinctive group identity. This bias shaped the vast majority of the sources that survived until today and therefore still affects our understanding of this era.

The mentioning of neglected perspectives leads to another important, final aspect. There still is comparatively little research available on non-European cultural approaches to seafaring and to the Atlantic Ocean in this time. Due to the lack of written evidence, investigations have mostly been done by archaeologists who aimed at reconstructing trade connections or settlement patterns. While this is an important step that indicates immense achievements now long forgotten, there is still much to find out about the individual perceptions and interpretations of the ocean held by all those cultures who lived on its shores.

---

82 Karstens, *Gescheiterte Kolonien* 371–386 with several examples for the conflicts mentioned, 457–476, 498–512.

Bibliography

Primary Sources

Best G., *A true discourse of the three voyages of discoverie for the finding of a passage to Cathaya, by the Northvveast, vnder the conduct of Martin Frobisher Generall deuided into three bookes* (London: Henry Bynnyman, 1578).

Grotius H., *The Free Sea. Translated by Richard Hakluyt with William Welwod's Critique and Grotius' Reply*, ed. D. Armitage (Indianapolis, IN: 2004).

Beaulieu A.M. (ed.), *Les trois mondes de La Popelinière* (Geneva: 1997).

Hayes E., "A report of the Voyage and the Success thereof", in Archer I. – Price F.D. (eds.), *English Historical Documents 1558–1603* (Milton Park, Oxfordshire: 2011).

Whitaker A., *Good newes from Virginia sent to the Counsell and Company of Virginia, resident in England from Alexander Whitaker* (London: Welby, 1613).

Secondary Sources

Abulafia D., *The Boundless Sea* (London: 2020).

Allen J.L., "From Cabot to Cartier: The Early Exploration of Eastern North America 1497–1543", *Annals of the Association of American Geographers* 83.3 (1992) 500–521.

Alves Gaspard J., "De la Mediteranée a l'océan. Nouveaux problèmes, nouvelle solutions", in Hofmann C. (ed.), *L'âge d'or des cartes marines. Quand l'Europe découvrait le monde* (Paris: 2012) 126–135.

Andrews K.R., "The Elisabethan Seaman", *The Mariner's Mirror* 68 (1982) 245–262.

Armitage D., *The Ideological Origins of the British Empire* (Cambridge: 2009).

Armitage D., "Three Concepts of Atlantic History", in Armitage D. (ed.), *The British Atlantic World, 1500 – 1800* (Basingstoke: 2002) 11–27.

Ash E.H., "Navigation Techniques and Practice in the Renaissance", in Woodward D. – Harley J.B. (eds.), *The History of Cartography, The Renaissance I* (Chicago: 2007) 509–528.

Aulke J., "Spatial Turn(s). Die Wirklichkeit des Raumes in der neuen Kulturgeschichte", in Haas S. – Wischermann C. (eds.), *Die Wirklichkeit der Geschichte. Wissenschaftstheoretische, mediale und lebensweltliche Aspekte eines (post-)konstruktivistischen Wirklichkeitsbegriffes in den Kulturwissenschaften* (Stuttgart: 2015) 54–62.

Augeron M. "Coligny et les Espagnols à travers la course (c.1560–1572) une politique maritime au service de la cause protestante", in Acerra M. – Martinière G. (eds.), *Coligny, les protestants et la mer* (Paris: 1999) 155–176.

Bailyn B., *Atlantic History: Concept and Contours* (Cambridge, MA: 2005).

Bankhoff G., "Aeolian Empires: The Influence of Winds and Currents on European Maritime Expansion in The Days of Sail", *Environment and History* 23.2 (2017) 163–196.

Baumgärtner I., "Neue Karten, neue Welt", *Historische Zeitschrift Supplement* 77 (2019) 274–251.

Bethencourt F. – Ramada Curto D., *Portuguese Oceanic Expansion, 1400–1800* (Cambridge: 2007).

Bolster J., "Putting the Ocean in Atlantic History: Maritime Communities and Marine Ecology in the Northwest Atlantic, 1500–1800", *The American Historical Review* 113.1 (2008) 19–47.

Borgolte M. – Jaspert N., "Maritimes Zeitalter. Zur Einführung", in Borgolte M. – Jaspert N. (eds.), *Maritimes Mittelalter. Meere als Kommunikationsräume* (Stuttgart: 2016) 9–33.

Brendecke A., *Imperium und Empirie. Funktionen des Wissens in der spanischen Kolonialherrschaft* (Köln: 2009).

Brunelle G.K., *Rouen and the New World: Rouennais Investors in Commerce with North and South America, 1559–1629* (Ph.D dissertation, Emory University: 1988).

Buisseret D., *The Mapmakers' Quest: Depicting New Worlds in Renaissance Europe* (Oxford: 2003).

Burghartz S., "Erfolg durch Scheitern? Zur Konstruktion von Überlegenheit im kolonialen Diskurs um 1600", in Dürr R. (ed.), *Expansionen in der frühen Neuzeit* (Berlin: 2005) 307–324.

Canny N. – Morgan P. (eds.), *The Oxford Handbook of the Atlantic World, 1450–1850* (Oxford: 2012).

Chaplin J.E., "The Atlantic Ocean and its Contemporary Meanings 1492–1808", in Greene J.P. – Morgan P.D. (eds.), *Atlantic History. A Critical Appraisal* (Oxford: 2009) 35–51.

Childs D., *Pirate Nation, Elizabeth I and her Royal Sea Rovers* (Barnsley: 2014).

Cormack L.B., *Charting an Empire. Geography at the English Universities, 1580–1620* (Chicago: 1997).

Dear P., "The Meanings of Experience", in Park K. – Lindberg D.C. (eds.), *Early Modern Science* (Cambridge: 2008) 106–130.

Dipper C. – Lutz R., "Raum in der Europäischen Geschichte. Einleitung", *Journal of Modern European History* 9.1 (2011) 27–41.

Edney M.H., "Knowledge and Cartography in the Early Atlantic", in Canny N. – Morgan P. (eds.), *The Oxford Handbook of the Atlantic World, 1450–1850* (Oxford: 2012) 87–112.

Egerton D.R. – Games A., *The Atlantic World: A History, 1400–1888* (Wheeling, ILL: 2007).

Enterline E. – Robert J., *Erikson, Eskimos & Columbus. Medieval European Knowledge of America* (Baltimore, MD: 2002).

Feldbauer P., "Die Islamische Welt seit der Jahrtausendwende", in Borgolte M. – Jaspert N. (eds.), *Maritimes Mittelalter. Meere als Kommunikationsräume* (Stuttgart: 2016) 139–163.

Fernández-Armesto F., *Before Columbus. Exploration and Colonization from the Mediterranean to the Atlantic, 1229–1492* (Basingstoke: 1987).

Fuller M.C., *Voyages in Print. English Travel to America, 1576–1624* (Cambridge: 1995).

Fury C.A., *Tides in the Affairs of Men: The Social History of Elizabethan Seamen, 1580–1603* (Ph.D dissertation, McMaster University Canada: 1998).

Gewecke F., *Christoph Kolumbus. Leben, Werk, Wirkung* (Frankfurt: 2006).

Gillis J.R., *Islands of the Mind: How the Human Imagination Created the Atlantic World* (New York: 2004).

Grafton A., *New Worlds, Ancient Texts. The Power of Tradition and the Shock of Discovery* (Cambridge: 3rd ed. 2000).

Guénin E., *Ango et ses pilotes, d'après des documents inédits tirés des archives de France, de Portugal et d'Espagne* (Paris: 1901).

Hancock D., "Atlantic Trade and Commodities", in Canny N. – Morgan P. (eds.), *The Oxford Handbook of the Atlantic World, 1450–1850* (Oxford: 2012) 324–340.

Kuhn T., *The Structure of Scientific Revolutions* (Chicago: 3rd ed. 2008).

Jahn B., *Raumkonzepte in der frühen Neuzeit. Zur Konstruktion von Wirklichkeit in Pilgerberichten, Amerikareisebeschreibungen und Prosaerzählungen* (Frankfurt: 1993).

Johnson D., *Die große Geschichte der Seefahrt / The History of Seafaring. 3000 Jahre Expeditionen, Handel und Navigation* (Hamburg: 2009).

Karstens S., "Die Macht der Tradition. Empirie und Muster der Vermittlung transozeanisch erworbenen Wissens ca. 1500–1620", in Burschel P. – Juterczenka S. (eds.), *Das Meer. Maritime Lebenswelten in der frühen Neuzeit* (Vienna: 2020) 117–130.

Karstens S., *Gescheiterte Kolonien, Erträumte Imperien. Eine andere Geschichte der Europäischen Expansion 1492–1615* (Cologne: 2020).

Klein B. (ed.), *Sea Changes: Historicizing the Ocean* (New York: 2003).

Kicza J.E., "Patterns in Early Spanish Overseas Expansion", *The William and Mary Quarterly* 49.2 (1992) 229–253.

Kurlansky M., *Cod. A Biography of the Fish that Changed the World* (London: 1999).

James A., *The Navy and Government in Early Modern France. 1572–1661* (Ph.D dissertation, University of Manchester: 1997).

Jourdin M.M., *Europa und das Meer* (Munich: 1993).

Langenohl A., "Mental Maps, Raum und Erinnerung. Zur kultursoziologischen Erschließung eines transdisziplinären Konzepts", in Damir-Geilsdorf S. (ed.), *Mental Maps – Raum – Erinnerung. Kulturwissenschaftliche Zugänge zum Verhältnis von Raum und Erinnerung* (Münster: 2005) 51–69.

Lehners J.P., "Anfänge der Portugiesischen Expansion", in Borgolte M. – Jaspert N. (eds.), *Maritimes Mittelalter. Meere als Kommunikationsräume* (Stuttgart: 2016) 165–185.

Lestringant F., *André Thevet: cosmographe des derniers Valois* (Geneva: 1991).

Lestringant F., "Fictions cosmographiques à la renaissance", in Lestringant F. (ed.), Écrire le monde à la renaissance. Quinze études sur Rabelais, Postel, Bodin et la littérature géographique (Caen: 1994) 293–318.

Loades D.M., *England's Maritime Empire: Seapower, Commerce and Policy, 1490–1690* (Harlow: 2000).

MacMillan K., "Centers and Peripheries in English Maps of America 1590–1685", in Brückner M. (ed.), *Early American Cartographies* (Chapel Hill, NC: 2011) 67–92.

Mancall P.C. (ed.), *The Atlantic World and Virginia, 1550–1624* (Chapel Hill, NC: 2007).

Meurer, P., "Cartography in the German Lands 1450–1650", in Woodward D. – Harley J.B. (eds.), *The History of Cartography. The Renaissance II* (Chicago: 2007) 1172–1254.

Moreau J.P., *Pirates. Flibuste et piraterie dans la Caraïbe et les mers du sud, 1522–1725* (Paris: 2006).

Mosley Adam., "The Cosmographer's Role in the Sixteenth Century", in Dupré S. – Hallyn F. (eds.), *Early Modern Cosmography. Proceedings of a Conference at the Centre for History of Science, Ghent University, Ghent and Louvain, 28–30 May 2008* (Turnhout: 2009) 423–440.

Mustakeem S.M., *Slavery at Sea: Terror, Sex and Sickness in the Middle Passage* (Urbana: 2016).

Parry J.H., *The Discovery of South America* (London: 1979).

Priesching N., *Sklaverei in der Neuzeit* (Darmstadt: 2014) 64–57.

Quinn D.B., *North American World: A Documentary History of North America to 1612*, Vol. 1 (London: 1979).

Quinn D.B., "The Northwest Passage in Theory and Practice", in Allen J.L. (ed.), *North American Exploration* (Lincoln, NE: 1997) 292–343.

Quinn D.B., "Renaissance Influences in English Colonisation", *Royal Historical Society Transactions* 5 Ser. 26 (1976) 74–92.

Rau S., *Räume. Konzepte, Wahrnehmungen, Nutzungen* (Frankfurt: 2017).

Rediker M., *The Slave Ship: A Human History* (New York: 2007).

Rodger N.A.M., "Atlantic Seafaring", in Canny N. – Morgan P. (eds.), *The Oxford Handbook of the Atlantic World, 1450–1850* (Oxford: 2012) 71–86.

Rodger N.A.M., *The Safeguard of the Sea 660–1649* (London: 2004).

Rüdiger J., "Kann man zur See herrschen? Zur Frage Mittelalterlicher Thalassokratien", in Borgolte M. – Jaspert N. (eds.), *Maritimes Mittelalter Meere als Kommunikationsräume* (Stuttgart: 2016) 35–58.

Schaub J.F., "Violence in the Atlantic, Sixteenth and Seventeenth Century", in Canny N. – Morgan P. (eds.), *The Oxford Handbook of the Atlantic World, 1450–1850* (Oxford: 2012) 113–129.

Scheller B., "Kaufen, Verkaufen und Verstehen. Die Atlantikexpansion der Europäer, die Fernhändler und die neue Erfahrung des Fremden im 14. und 15. Jahrhundert", in Borgolte M. – Jaspert N. (eds.), *Maritimes Mittelalter. Meere als Kommunikationsräume* (Stuttgart: 2016) 233–259.

Short J.R., *Cartographic Encounters: Indigenous Peoples and the Exploration of the New World* (London: 2009).

Starkey D., "Fish and Fisheries in the Atlantic World", in Coffmann D'M – Leonard A. – O'Reilly W. (eds.), *The Atlantic World* (London: 2015) 61–64.

Thornton J.K., *A Cultural History of the Atlantic World, 1250–1820* (Cambridge: 2012).

Tyacke S., "Chartmaking in England and its Context 1500–1660", in Woodward D. – Harley J.B. (eds.), *The History of Cartography. The Renaissance II* (Chicago: 2007) 1722–1753.

Villiers P., *La France sur mer. De Louis XIII à Napoléon Ier* (Paris: 2015).

Vogel K.A., "Cosmography", in Park K. – Lindberg D.C. (eds.), *Early Modern Science* (Cambridge: repr. 2008) 469–496.

Wilford J.N., *The Mapmakers* (New York: rev. ed. 2000).

PART 3

Water in Thought and Practice

CHAPTER 7

Fluid Dynamics: Aquatic Agency in the *Poetic Edda* of the Codex Regius

Michael Bintley

Taking the Codex Regius of the *Poetic Edda* as its focus, the primary source for the majority of surviving Old Norse Eddic poetry, this chapter contends that this manuscript preserving mythology and wisdom from medieval Iceland is not merely a record of thought and experience from a former age. Using the cognitive archaeological framework of Material Engagement Theory (MET), it proposes that the Codex Regius is not the sign of something achieved by a human mind in the distant past, but a residual component of a mind in action.[1] A 'mind', conceived in these terms, is constituted not only by the brain, the central nervous system, and the body, but also the objects and environment encountered and engaged with by the human organism. This includes objects which have been described as *exograms* – external memory records of ideas analogous to the biological 'engrams' of the brain. In the contexts of medieval Iceland these exograms might include objects such as manuscripts written largely in Roman script and other objects inscribed with runes. Memory records such as these formed parts of a continuum that reflected and affected human interactions with the environment. This chapter presents an aqueous *exogrammar* of the Codex Regius, a body of evidence reflecting the appearances of water in all its forms throughout this single manuscript, considering what this one text-object reveals about interactions with water in the contexts of thirteenth-century Iceland, and how it would, in turn, have contributed to the way in which its readers thought about and experienced waters of all kinds.

Employing the hydrological cycle as its organising principle, the chapter follows rainfall and dewfall into rivers and streams, before moving out into the open seas, and finally considering water in its solid form – as ice and hail. Towards the end of the chapter, I discuss the various ways in which the Codex Regius can be understood as a residual component of minds in medieval Iceland, and, importantly, as a component of modern minds affecting human interactions with the environment in the present day. Though few experience

1 On medieval minds and mentalities in this volume see also discussion by Karstens 196–224; and on embodied minds and metaphors Borghetti 317–343.

 | DOI:10.1163/9789004696501_009

the Codex Regius directly in its 'original' form, every edition and translation of its texts has the potential to serve a similar function to the manuscript for the modern reader, and in some way, however small, to play a determining role in their experience of water. I conclude by suggesting that approaching medieval text-objects in this way could offer more practicable ways of considering them as part of material-textual continua, and that it might also serve as a means of bridging the aesthetic separation between texts and environment which has long been seen as a stumbling block by ecotheorists and ecocritics.

1 Extended Minds and Material Engagement

Extended Mind Theory (EMT) and Material Engagement Theory (MET) propose a model of the mind which is not limited to the confines of the brain, the central nervous system or the human body, but extends beyond the limits of the organism into the surrounding environment.[2] This model proceeds from Maurice Merleau-Ponty's thought experiment of the 'blind man's stick', expressed in his *Phenomenology of Perception*, which questions where the 'mind' begins and ends if an object external to the human body is essential to the workings of the 'mind'.[3] On these grounds, cognitive processing is not something that takes place only within the brain, but in the interactions between the brain, body, and things external to the body. This includes, but is not limited to, interactions with artefacts, other humans and non-human animals, spaces, aspects of environment and so on.[4] In this respect, MET builds on the entanglement theory of Ian Hodder, which in turn builds on Bruno Latour's Actor-Network Theory; the former emphasising the agency of connections and entanglements between things, and the latter the agency of the things themselves.[5] Lambros Malafouris, principal proponent of MET, has demonstrated this theory in

---

2 Clark A. – Chalmers D., "The Extended Mind", *Analysis* 58.1 (1998) 7–19; Malafouris L., *How Things Shape the Mind: A Theory of Material Engagement* (Cambridge, MA: 2013); Malafouris L. – Renfrew C., "The Cognitive Life of Things: Archaeology, Material Engagement and the Extended Mind", in Malafouris L. – Renfrew C. (eds.), *The Cognitive Life of Things: Recasting the Boundaries of the Mind* (Cambridge: 2010) 1–12.

3 Merleau-Ponty M., *Phenomenology of Perception*, trans. D.A. Landes (Abingdon: 2012) 165–166, 175–77.

4 Malafouris L., "Metaplasticity and the Primacy of Material Engagement", *Time and Mind* 8.4 (2015) 351–357; also, Malafouris L., "The Cognitive Basis of Material Engagement: Where Brain, Body and Culture Conflate", in DeMarrais E. – Gosden C. – Renfrew C. (eds.), *Rethinking Materiality: The Engagement of Mind with the Material World* (Cambridge: 2004) 53–62.

5 Hodder's Entanglement theory is outlined in Hodder I., *Entangled: An Archaeology of the Relationships Between Humans and Things* (Malden, MA – Oxford: 2012), and revisited in

practice using the example of a potter making a pot on a wheel. Within this act of making, the creation of the 'finished' pot is not something enacted by the potter's brain on a piece of clay but depends on all of the actants involved in this process. This conception of the extended mind-in-action includes not only the potter's brain, but also the potter's senses, their bones and body, the operations of the wheel, and the nature of the clay shaped by their hands. Importantly, other determining factors would arise from the cultural contexts in which the pot is manufactured, such as styles or types of pot familiar to the potter, the projected contents of such a pot, the training the potter has received, and the timeframe and associated social and economic contexts in which the pot is manufactured. In this respect, the pot is not something achieved by a mind in a brain, but a sign of a mind in process constituted by all of these elements and more.[6] Objects inscribed with writing have an interesting role within this conception of mind. Malafouris draws on Merlin Donald's concept of the *exogram* to explain their function. For Donald, an exogram is an 'external memory record of an idea', analogous to Karl Lashley's 'engram', the latter being a 'single entry in the biological memory system'; the engram, to put it more crudely, is a memory preserved within the brain, whilst exograms are arrangements of objects *external* to the brain.[7] These exograms, like any other object, can act as components of the extended mind, but are different in so far as they enable the *storage* of information outside of the biological human brain. Such objects can, in turn, become part of other minds through material transmission to other humans. Exograms of this kind can be rewritten, reinscribed, and copied in ways that are not possible for the engrams of biological memory.

For medieval scholars, who rely on a more limited range of written evidence than researchers working on more recent history, MET (though developed by prehistorians) may offer a useful framework for the integration of evidence from sources often treated as the territory of different disciplines, and described as 'textual' or 'material'. The concept of the *exogram* offers the

---

Where Are We Heading? The Evolution of Humans and Things (New Haven, CT – London: 2018).

6 Malafouris, *How Things Shape the Mind* 209–213; also Malafouris L., "At the Potter's Wheel: An Argument *for* Material Agency," in Knappet C. – Malafouris L. (eds.), *Material Agency: Towards a Non-Anthropocentric Approach* (Boston, MA.: 2008) 19–36; Malafouris, "Metaplasticity"; Malafouris, "The Cognitive Basis of Material Engagement".

7 Donald M., *Origins of the Modern Mind: Three Stages in the Evolution of Culture and Cognition* (Cambridge, MA – London: 1991) 314; Lashley K., "In Search of the Engram", in Beach F.A. – Hebb D.O. – Morgan C. – Nissen H. (eds.), *The Neuropsychology of Lashley* (New York: 1950) 2–31.

opportunity to consider texts as material objects that are elements of the extended mind. Taking this a step further, in recent work I have suggested ways in which we might seek to implement this understanding practically in the contexts of interdisciplinary research, using the example of vernacular charter boundary clauses from early medieval England.[8] These textual cartographies lead the reader around the boundaries of an estate, defining this boundary by identifying features such as stones, hedges, bodies of water, trees, areas of woodland and enclosures, and in this respect are some of the most materially-engaged texts to have survived from the Middle Ages.[9] Charter bounds, as exograms, define and are defined by the landscape they describe. As elements of extended minds, they formed part of a set of ongoing feedback loops between the early medieval humans who created them and wrote them down, and the environments in which they lived.

I have suggested describing a set of ideas distributed across one or more exograms as an *exogrammar*, reflecting and affecting human/environmental interactions as components of extended minds. An exogrammar of this kind might be assembled from one hundred manuscripts, one hundred folios within a manuscript, one hundred lines on a folio, or one hundred letters on a line, depending on the subject of the exogrammar under investigation by the researcher. An exogrammar is limited only by the availability of text-objects and other decisions made as a matter of necessity by a researcher framing their research with consideration for other factors such as temporal and linguistic contexts. Importantly, an exogrammar of these text-object/s should endeavour to be as comprehensive as possible within these parameters in order to avoid the kind of selective approaches to texts which can obscure information which, while it may not appear especially significant to a modern reader in isolation, nevertheless forms part of the broader exogrammar *whether or not it may immediately seem to be important*. Assembling an aqueous exogrammar of *all* extant Old Norse texts-objects, for this reason, would be a huge task, and various decisions would need to be made about the parameters of the exogrammar, such whether or not to include runic inscriptions, which manuscripts to include and exclude, whether or not to include later copies of earlier

8 This argument and theoretical approach are advanced in Bintley M., "Reading Early Medieval Landscape and Environment: Materially Engaged Approaches to Documentary Sources", in Willard T. (ed.), *Reading the Natural World: Perceptions of the Natural World and Environment in the Global Middle Ages and Renaissance* (Turnhout: 2020) 3–20.

9 For a general discussion of the emergence of early English boundary clauses, see Lowe K.A., "The Development of the Anglo-Saxon Boundary Clause", *Nomina* 21 (1998) 63–100.

manuscripts, and whether or not to include words or terms borrowed from other languages (e.g. Old English and Latin).

2 Codex Regius as Exogram

An exogrammar need not be so wide ranging, however, and can be assembled from a single text-object such as the Codex Regius of the *Poetic Edda*, GKS 2365 4°, itself an assemblage of text-objects in the form of 45 folios numbered 1–90, measuring 194 mm × 134 mm, which are inked with the majority of Old Norse Eddic poetry surviving from the Middle Ages.[10] The Codex Regius was written c. 1270 by a single unknown scribe, and later obtained in 1643 (in unknown circumstances) by Brynjólfur Sveinsson, bishop of Skáholt, who erroneously attributed it to Sæmundr fróði.[11] Though not an inexpensive creation, it is small and unostentatious in comparison with many contemporary medieval manuscripts. The scribe made good use of the skin, writing around a tear across most of a page, and elsewhere skirting the edge of holes in the parchment, one of which is fairly substantial.[12] It is not my intention to become entangled in the wide-ranging discussion of what the scribe and/or patron of the manuscript may or may not have intended in gathering together these works, though they are divided into fourteen *Goðakvæði* and twenty-four *Hetjukvæði* in the 2014 two-volume Íslenzk Fornrit edition produced by Jónas Kristjánsson and Vésteinn Ólason, updating the Neckel-Kuhn edition that had been the standard scholarly text for almost a century.[13] A number of these texts are thought to be significantly older than the manuscript they occupy, and draw on narratives and mythology thought (or known, in some instances) to be older still, including legendary events with a historical basis in the fifth century CE.[14] The thorny issue of dating these works is of no concern to this essay beyond the

10 https://handrit.is/en/manuscript/view/is/GKS04-2365. The manuscript may here be viewed in its entirety. Accessed May 15, 2024.

11 Clunies Ross M., "The Transmission and Preservation of Eddic Poetry", in Larrington C. – Quinn J. – Schorn B. (eds.), *A Handbook to Eddic Poetry: Myths and Legends of Early Scandinavia* (Cambridge: 2016) 12–32, 23.

12 See folios 28, and (for example) 35/41.

13 All references to poems of the CR from *Eddukvæði*, eds. J. Kristjánsson – V. Ólason, 2 vols, Íslenzk fornrit (Reykjavík: 2014). The numbering of stanzas in this article follows this edition.

14 See, e.g. discussion in Gunnell T., "Eddic Poetry", in McTurk R. (ed.), *A Companion to Old Norse-Icelandic Literature and Culture* (Oxford: 2005) 82–100, 93–94; also Kristjánsson J., *Eddas and Sagas: Iceland's Medieval Literature*, trans. P. Foote (Reykjavík: 1997) 28–29; Thorvaldsen B.O., "The Dating of Eddic Poetry", in Larrington C. – Quinn J. – Schorn B.

dating of the manuscript. It is, nonetheless, worth considering that representations of aquatic environments preserved in these texts in all likelihood stem from much longer-term perspectives than the thirteenth century contexts of the the manuscript. They may predate the writing of even the earliest poems, some of which may already be centuries older. These are likely to reflect the material conditions of life, and in this case the experience of water, across the broader geographical sphere in which Old Norse literary culture developed, which was connected to a significant extent by this element, with Norse speakers at one time inhabiting regions as far west as North America and Greenland, and as far south as the Mediterranean, as well as a more substantial core in western Russia, and across Scandinavia, the north-eastern Atlantic archipelago and the North Sea zone.[15] Practically speaking, assembling an exogrammar involves the same detailed fine-tooth-combing of texts that forms part of the usual process by which literary scholars and historians gather evidence, with the proviso that this should endeavour to be more exhaustive than selective; it should represent the exogram as a whole, rather than choosing examples which appear – in the critic or historian's view – to be most representative of the subject of study. This is not to say that some examples from the exogrammar may not reveal more than others (as the reader will see from what follows), but that all this evidence should be represented as fully as possible.

3 Rainfall, Dewfall

Several of the poems in this manuscript offer insight into the mythic structure of the cosmos, whose organisation lends itself to a systematic appraisal of the waters within their text-worlds. Our impression of rain in the Codex Regius is formed mainly by *Alvissmál*. This world is a watery world, as the dwarves seem to know best, and much of this water drips in through the roof over our heads or may even emanate *from* it. In stanza 12 we hear about the names of the skies – *himinn* ('heaven') for men, *hlýrnir* ('the heights') for the gods, a weaver of wind for the Vánir (*vindófni*), *uppheim* ('up world') for the giants, and *fagraræfr* ('fair roof') for the elves. For the dwarves, however, who are better acquainted with the earth into and through which the waters flow, the heavens are a roofed

(eds.), *A Handbook to Eddic Poetry: Myths and Legends of Early Scandinavia* (Cambridge: 2016).

15 Clunies Ross M., *A History of Old Norse Poetry and Poetics* (Cambridge: 2005) 114–115. On water in Norse mythology more broadly, see Birkett below 406–429.

hall, an element of cosmic structure found elsewhere in Old Norse and early English texts.[16] The dwarf Alvíss, Þórr's interlocutor in this poem, is especially well positioned to discuss dwarven understanding of the organisation of the cosmos. Our world – 'the dripping hall', as the dwarves describe it, is watered by the clouds of the heavens. Þórr asks Alvíss how the beings that inhabit the cosmos describe the clouds and discovers how prosaic human beings can be. According to our species, the heavens are the heavens, and the clouds are the clouds.[17] In *Alvissmál* 17, Þórr asks Alvíss *hvé þau ský heita er skúrum blandask* ('what the clouds are called that are blended with showers'), suggesting that the two are both closely related, but also distinct from one another. In 18, Alvíss explains that whilst the Vánir may call them *vindflot* ('wind-floaters'), the elves *veðrmegin* ('weather-might'), and the denizens of Hel *hjálm huliðs* ('the helm of obscurity'), their origins and constitution are clear to the Æsir and the giants. The clouds bring showers of rain, and thus for the Æsir they are *skúrván* ('the expectation of showers'), and for the giants *úrván* ('the expectation of drizzle').

The waters of the earth are better known to those who inhabit it, or in the case of the dwarves, those who live *in it*, whilst the elevation of the gods and the stature of the giants gives them a more prognostic sense of what gathering clouds portend. None of these terms, notably, has much to say about rain as a source of life-giving waters; their wetness is the thing-in-itself, and they show less interest in its potential for conjuring growth or new life. A similarly bald sense of water's presence in the landscape can be found in descriptions of other waters that have their origins in the air – the dews that fall in the valleys. The source of these waters in *Vǫluspá* (19) is the world tree Yggdrasill, from whose branches this dew descends. The seeress says,

> Ask veit ek standa,
> heitir Yggdrasill,
> hár baðmr, ausinn
> hvíta auri;
> þaðan koma dǫggvar
> þærs í dala falla,
> stendr æ yfir grœnn
> Urðarbrunni.

16 See discussion of the roofed heavens in Bintley M., *Settlements and Strongholds in Early Medieval England: Texts, Landscapes, and Material Culture* (Turnhout: 2020) 87–90.

17 On Þórr's relationship with water, see also Birkett 406–429.

I know an ash-tree stands,
named Yggdrasill,
a high tree, drenched
with shining mud;
from there come the dews
that fall in the dales,
it stands ever green
over the well of Urðr.

This tree, as *axis mundi*, offers a parallel with the cross-as-axis-mundi in the Christian tradition, which the visionary of *The Dream of the Rood* sees drenched in blood and water flowing from the wound in Christ's side.[18] Ultimately, as these lines suggest, the roots of this tree draw sustenance from the well of Urðr, whose waters keep it *yfir grœn*. This tree may also be the source of the waters that drench the cosmic stag *dǫggo slunginn* ('dripping with dew'), who is compared with Helgi in *Helgakviða Hundinsgbana II* (38). In *Vǫluspá* the tree rises once again following Ragnarǫk, and the same seems to be true in *Vafþrúðnismál* (45), in which these waters nourish not only the earth, but also the two survivors of the final conflagration; hiding in Hoddmimír's wood, *Líf* ('Life') and *Lífþrasir* ('Life-eager') derive their sustenance from *morgindǫggvar* ('morning dew') in terms that evoke Mosaic manna in the book of Exodus (ch. 16). This seems to conflict with the origins of the waters described elsewhere in the Codex Regius. *Vafþrúðnismál* (14) attributes their origin to one of the celestial horses who champs at its bit, releasing foam that falls as *dǫgg um dala* ('dew in the dales'), an image no doubt familiar to those responsible for equine husbandry in medieval Iceland.

Other everyday images of water evoke the quotidian experience of life in this environment. In *Skírnismál* and *Helgakviða Hundinsgbana I*, difficult journeys are made across *úrig fjǫll* ('damp fells' *Skn* 10) and *dala dǫggótta* ('dew-wet dales' *HH I* 47) – perilous travel for those on foot or on horseback who find themselves at the mercy of the elements. *Guðrúnarkviða II* (23) offers the image of ale brewed with all manner of unpleasantness, including *umdǫgg arins* ('hearth-dew') – a domestic image still familiar today for those whose homes continue the winter season's battle between the warmth of burning wood and the pervasiveness of cold and damp.

18 On the tree and cross as *axis mundi*, see Cusack C.M., *The Sacred Tree: Ancient and Medieval Manifestations* (Newcastle Upon Tyne: 2011).

4 River

Whatever their origins, the worlds of the Codex Regius are essentially saturated with these waters, even before they begin to coalesce as rain, or to channel their way into streams and rivers. Reflecting the landscapes of Iceland and much of northern Europe, rivers find their way through many of the works in this manuscript. The source of these rivers, in some cases, is obscure. Like the dews that settle in the valleys, *Grímnismál* connects their origins with a cosmic beast, the stag Eikþyrnir, who feeds on the branches of the cosmic tree Læraðr, and from whose antlers dew falls into Hvergelmir, *þaðan eigu vǫtn ǫll vega* ('from whence rise all the waters', 26).[19] *Grímnismál* lists these rivers, some of which Þórr wades through on his way to the judgement seats of Yggdrasill on a daily basis (29). Earlier in the poem, the river Þund is a barrier to be overcome in a similar fashion by those approaching Valhǫll (21).[20] Wading through rivers is seen elsewhere in *Vǫluspá* (38), where the seeress recalls murderers and traitors wading through *þunga strauma* ('powerful rivers') similar to the waters of the river Vaðgelmír in which liars wade in *Reginsmál* (4).[21]

Elsewhere, flowing waters present a more substantial and perilous barrier. In *Vafþrúðnismál* (15–16), during the conversation between Óðinn and the giant Vafþúðnir, the latter asks the former:

hvé sú á heitir,
er deilir með jǫtna sonum
grund ok með goðum.

19 *Grímnismál* 26–29. This presumably includes the *heilǫg vǫtn* ('holy streams') that fall *af Himinfjǫllum* ('from the heaven-cliffs') in *Helgakviða Hundingsbana I* (1), but not the river *Ván* ('Hope') that flows from the slathering jaws of the chained Fenrir in *Lokasenna* 41. Læraðr may or may not be the same tree as Yggdrasill; Snorri Sturluson, in *Gylfaginning*, puts Hvergelmir's location beneath Yggdrasill, though this may be his attempt to make sense of conflicting elements in earlier poems which did not have a single, doctrinal mythology. Equally obscurely, in *Vǫluspá* (27) the horn of Heimdallr pours water on Yggdrasill.

20 Christopher Hale identifies forty-three named rivers in Eddic poetry; forty of these appear in *Grímnismál*. See Hale C.S., "The River Names in *Grímnismál* 27–29", in Glendinning R.J. – Bessason H. (eds.), *Edda: A Collection of Essays*, University of Manitoba Icelandic Studies 4, 165–186 (Winnipeg: 1983) 165.

21 Throughout *Reginsmál* the dwarf Andvari lives in a waterfall in the form of a pike, transformed into this shape by one of the norns (2). Perhaps this punishment of liars, murderers, and traitors has something to do with the waters of the river Leipt (one of those named in *Grímnismál*) and the *úrsvǫlum Unnarsteini* ('ice-cold stone of Urðr') by which the Valkyrie Sigrún curses her brother Dagr for killing her lover Helgi, to whom the former had previously sworn allegiance (*HH II* 31).

what is the name of the river
that divides the land
between the sons of giants, and the gods?

He is told that

Ífing heitir á,
er deilir með jǫtna sonum
grund ok með goðum;
opin renna
hon skal of aldrdaga;
verðr-at íss á á.

Ífingr is its name
that divides the land
between the sons of giants and the gods;
it shall run freely
for all time,
and ice will never form upon the river.

The surging waters of the Ífingr, useful in *Vafþrúðnismál* as a means of separating the mythological spaces of the giants and the gods, could be the same ones Þórr seeks to cross in *Hárbarðsljóð*, when he is refused passage by the ferryman Hárbarðr; Þórr here explains that he has been guarding the banks of a river against the giants.[22]

An even more dangerous river appears in *Vǫluspá*, a surging torrent of *sǫxum ok sverðum* ('seaxes and swords', 35) named *Slíðr* ('cutting', 36). This is not the only poem in Codex Regius to evoke the relationship between rivers and metal, and at least two others refer to the Rhine gold.[23] In *Sigurðarkviða in skamma*, whilst in conversation with Hǫgni over the betrayal of Sigurðr, Gunnarr reflects that *gott er at ráða Rínar málmi* ('it is good to rule the Rhine's metal', 16). In *Guðrúnarkviða II*, reflecting on the fate of Sigurðr, Gunnarr recalls that Hǫgni told him: *Liggr of hǫggvinn fyr handan ver Gothorms bani, of gefinn úlfum*

22 On the significance of these places of crossing in texts and material culture, see Lund J., "Thresholds and Passages: The Meanings of Bridges and Crossings in the Viking Age and Early Middle Ages", *Viking and Medieval Scandinavia* 1 (2005) 109–135; Lund J., "Banks, Borders and Bodies of Water in a Viking Age Mentality", *Journal of Wetland Archaeology* 8 (2008) 53–72.

23 Abram C., "New Light on the Illumination of Grendel's Mere", *Journal of English and Germanic Philology* 109.2 (2010) 198–216.

('he lies struck down on the bank of the river, Guthorm's bane, given to wolves', 7). The identity of this river is unclear, though on the basis of the imagined location of Gunnarr's court, it could be the Rhine. *Sigurðarkviða in skamma* (*SiS*) is preceded by *Guðrúnarkviða I*, and followed by *Helreið Brynhildar*, before *Guðrúnarkviða II*. If read in the light of *SiS*, this line in *GII* could be taken to connect Sigurðr's death with the source of the Rhine's gold.

The source of this river is divine, as Gunnarr later recalls in *Atlakviða* (28), refusing to trade gold for his life even when the heart of Hǫgni is presented to him, cut from his breast at the command of king Atli. Gunnarr's words seem to play on the same circularity which may be present in *Guðrunarkiða II* surrounding the death of Sigurðr – of gold, water, and the fates of humans. In Gunnarr's case, he alone knows the location of the Rhine gold. Here in *Atlakviða*, he says that *Rín skal ráða rógmálmi skatna* ('the Rhine must govern the strife-metal of men', 27), and that *í veltanda vatni lýsask valbaugar, heldr en á hǫndom gull skíni Húna bǫrnom* ('in the welling waters splendid rings will shine, before gold shines on the hands of the children of Huns', 27). Fittingly, one of the many images that appears in Guðrún's visions portending the death of Atli and his sons in *Atlamál* (26) is that of a great river bursting through his hall:

> Á hugða ek hér inn renna
> at endilǫngu húsi,
> þyti af þjósti,
> þeystisk of bekki,
> bryti fœtr ykkra
> brœðra hér tveggja,
> gerðit vatn vægja;
> vera mun þat fyr nekkvi.
>
> I thought a river ran in here
> throughout the length of the house
> that thundered in its course,
> rushed over the benches,
> broke the legs of
> both your brothers here,
> nothing could stop the water;
> that must mean something.

Here, in full spate, the flood catches up with Atli, a torrent of such power in Guðrún's vision that it splinters the legs of his children and overwhelms the benches: a clear metonym for the destruction of his hall *and* his family.

5 Sea

This image of the hall, as an emblem of civilisation, kingship and kinship, powerless in the face of this inundation, bears us out into open water. According to *Vǫluspá*, unlike the primordial waters of Genesis, there was nothing in the beginning when the cosmic giant Ymir made a home for himself: *vara sandr né sær né svalar unnir, jǫrð fannsk æva né upphiminn* ('there was no sand nor sea, nor cooling waves, no earth to be found, nor heavens above', 3). Girded by the world serpent, the oceans make no further appearance until the end of the poem, at Ragnarǫk, when its coils strike the waves, and Naglfar breaks its moorings (48), making its way from the east (also the source of the river Slíðr), piloted by Loki and bearing the sons of Muspell (49).[24] And now

Sól tér sortna,
sígr fold í mar,
hverfa af himni
heiðar stjǫrnur;
geisar eimi
við aldrnara,
leikr hár hiti
við himin sjálfan.

The sun grows black,
the land sinks into the sea,
the bright stars disappear
from the heavens;
fire rages
against the life-nourisher [?Yggdrasill],
fire flies high
against heaven itself. (55)

The new earth will rise from these waters once again, *jǫrð ór ægi iðiagrœna* ('the earth from the sea, eternally green', 57).[25] In both *Vafþrúðnismál* (21)

24 Þórr's fishing expedition for the world serpent, the *brimsvín jǫtuns* ('giant's sea-swine'), can be found in *Hymiskviða* 28.

25 This may be Yggdrasill or another cosmic tree or grove. See Tolley C., *Shamanism in Norse Myth and Magic* (Helsinki: 2009) 351; Bintley M.D.J., *Trees in the Religions of Early Medieval England* (Woodbridge: 2015) 81.

and *Grímnismál* (40) the sea is formed from the blood of Ymir himself.[26] The gods engage with it in a characteristically oblique fashion. The seahorses of humans pass over it in Old Norse, Old English and other related literatures, but in *Fáfnismál* the steeds of the gods swim in the waters as they spar with their spears (15). According to *Grímnismál,* two of the gods have homes with aquatic associations – for obvious reasons in the case of the sea-ván Njǫrðr, whose hall bears the name *Nóatún* ('naval yard', 16), though less in the case of Óðinn and Sága's drinking hall, *Søkkvabekkr* ('sunken bank'), where they drink from golden cups, *en þar svalar knegu unnir yfir glymia* ('and where cool waves resound over it', 7). The fundamental opposition between waters and places of human habitation, as expressed by Guðrún's dream of the flood in the hall, does not seem to pose the same problems for these Æsir. A gnomic phrase from *Hávamál* expresses this anxiety more keenly, amidst a smattering of advice to Loddfáfnir from the High One on the amuletic properties of different elements used in sympathetic magic: *fold skal við flóði taka* ('earth must stand against water' 137). Why? Because it must, until the end of time, when as Judy Quinn writes, the sea 'exceeds land in power and expanse, rising to cover it entirely', only to recede once more 'having recreated it as beautiful and productive terrain for the new generation of gods to live in'.[27]

Like the names of clouds, those given to the sea by different beings in *Alvíssmál* are a matter of perspective:

Sær heitir með mǫnnum,
enn sílægja með goðum,
kalla vág vanir,
álheim jǫtnar,
álfar lagastaf,
kalla dvergar djúpan mar.

Men call it 'sea',
but it's 'ever-lying' amongst the gods,
the Vanir call it 'the wave':

26 On potential connections between Ymir's blood and the 'origin of wells beneath the surface of the earth', see Quinn J., "Liquid Knowledge: Traditional Conceptualisations of Learning in Eddic Poetry", in Ranković S. – Melve L. – Mundal E. (eds.), *Along the Oral-Written Continuum: Types of Texts, Relations and Their Implications* (Turnhout: 2010) 183–226, 207.

27 Quinn J., "Mythologizing the Sea: the Nordic Sea-Deity Rán", in Tangherlini T. (ed.), *Nordic Mythologies: Interpretations, Institutions, Intersections*, The Wildcat Canyon Advanced Seminars: Mythology 1, 71–97 (Berkeley – Los Angeles: CA.: 2014) 73.

the giants 'eel-home', the elves 'lagastaf',[28]
and the dwarves call it 'deep sea'. (24)

On the whole, the sea appears in Codex Regius as a kind of terrain, comparable to mountains and plains; the same sun shines on earth and water in *Grímnismál* (38), and in *Skírnismál*, when Freyr sees Gerðr, the daughter of the giant Gymir, *allt lopt ok lǫgr* ('all the sea and sky shone', 6), illumined by the brightness of her arms. It has some inhabitants, such as whales,[29] but for the most part it is a barrier to be overcome. *Hávamál* offers the image of an eagle soaring above the waves: *Snapir ok gnapir, er til sævar kømr, ǫrn á aldinn mar* ('it snaps and cranes its neck when it comes to the sea, an eagle over the ancient ocean', 62), just as the Valkyries pass over the sea in *Vafþrúðnismál* (48), and the wind *ferr vág yfir* ('passes over the waves', 21). We have already seen how rivers form a barrier for Þórr in *Hárbardsljóð*; fjords do much the same elsewhere in *Atlamál* (31) and *Hávamál* (116). Overall, however, humans overcome the seas by sailing. In addition to the journey of Loki and the sons of Muspell (considered above in *Vǫluspá* (49)), *Hávamál* advises sea travel in good weather (82), *Hymiskviða* shows us Þórr's sea-fishing (17, 20, 27), demonstrating temporary mastery over the sea according to Preben Meulengracht Sørensen, and intrepid seafaring appears repeatedly as a backdrop to a number of the 'heroic' poems in the latter half of the manuscript.[30] Especially dramatic examples of this are found in *Reginsmál* (16–17) when Hníkar (Óðinn in disguise) asks,

Hverir ríða þar
Ræfils hestum
hávar unnir,
haf glymjanda?

28 Larrington writes that 'the meaning of this word is unknown', in her translation and notes on *The Poetic Edda* 278. Orchard translates it as 'brew-stuff', writing that 'the same word (*lagastaf*) is used in 32, and so is part of the elves' vocabulary for both 'ocean' and 'barley': brew-stuff is a compromise, since ale is composed of both barley and water'; see Orchard A. (trans.), *The Elder Edda: Myths, Gods and Heroes from the Viking World* (London: 2011) 310.

29 As elsewhere in Germanic-language literature, it is the whale's home in *Hymiskviða* 26.

30 See, for example, in *Helgakviða Hundingsbana I* (29, 31, 50), *Helgakviða Hjorvarðsonnar* (12, 18, 26), *Helgakviða Hundingsbana II* (5, 6, 13), *Guðrúnarkviða I* (7), *Guðrúnarkviða II* (16, 35), *Oddrúnargrátr* (31), and *Atlamál* (3). Preben Meulengracht Sørensen, "Þorr's Fishing Expedition" [*Hymiskviða*] [trans. K. Williams], in *The Poetic Edda: Essays on Old Norse Mythology* 129.

> Seglvigg eru
> sveita stokkin,
> munat vágmarar
> vind um standask.
>
> Who rides there
> the horses of Rævil,
> on the towering waves,
> the roaring ocean?
> The sail-steeds are
> dripping with sweat
> the seahorses won't
> withstand the wind.

Reginn responds:

> Hér ero vér Sigurðr á sætriám,
> er oss byrr gefinn við bana siálfan;
> fellr brattr breki brǫndom hæri,
> hlunnvigg hrapa; hverr spyrr at því?
>
> Here are Sigurd and I on sea-trees
> and we've been given a wind that'll be the death of us;
> the steep breakers are falling higher than the prow of the ship,
> the roller-steeds are rushing headlong; who's asking about that?

One of the ever-present perils of the seas in these stanzas, as elsewhere in Old Norse literature, is their treacherousness. In *Hávamál*, similarly, the seas appear repeatedly in lists of things that are untrustworthy. In stanzas 85–6, readers are warned against *vaxanda vági* ('a waxing wave') and a *fallandi báru* ('a falling tide'). Equally perilous is seafaring when one has lost control of one's vessel. Much like love in Petrarch's *S'amor non è*, co-opted by Chaucer for *Troilus and Criseyde*, *Hávamal* compares the love of a treacherous woman with *í byr óðum beiti stjórnlausu* ('steering a rudderless boat in a gale', 90).[31] Fortunately, there are ways in which these dangers can be overcome. One of the songs known by

31 As Troilus says: 'Thus possed to and fro, Al sterelees withinne a boot am I, Amydde the see, bitwixen wyndes two, That in contrarie stonden ever mo' (I, 415–18); *Troilus and Criseyde*, in Benson L.D. (ed.), *The Riverside Chaucer* (Oxford: 3rd ed. 2008) 471–585.

Óðinn, to which he alludes in *Hávamál* (154), grants him Christ-like power to calm tempestuous waters:

> Þat kann ek iþ níunda,
> ef mik nauðr um stendr
> at bjarga fari mínu á floti,
> vind ek kyrri
> vági á
> ok svæfik allan sæ.
>
> I know a ninth if I need it,
> if I am required
> to protect my ship
> the wind I quell
> on the wave
> and put the sea to slumber.

Advice for those who lack this knowledge is offered in *Sigrdrífumál* (11), which presents a magical, runic solution for those looking to keep safe at sea:

> Brimrúnar skaltu rísta,
> ef þú vilt borgit hafa
> á sundi seglmǫrum;
> á stafni skal rísta
> ok á stjórnar blaði
> ok leggja eld í ár;
> era svá brattr breki
> né svá blár unnir,
> þó kømstu heill af hafi.
>
> You must carve sea-runes,
> if you want to have protection,
> to sail seahorses on the waters;
> you must carve them on the prow
> and on the blade of the rudder
> and burned into the oars with fire;
> there is no breaker so steep,
> nor wave so dark,
> that you will not return home from the sea.

But death by water is never far away in the Codex Regius, as elsewhere in Old Norse poetry.[32] *Sigrdrífumál* (33) and *Fáfnismál* (11) draw attention to this risk, as does *Helgakviða Hjǫrvarðssonar* (19), recalling the drowning Hlöðvarð's sons.[33] *Guðrúnarhvǫt* (13) is exceptional, however, as here Guðrún recounts her inability to drown herself and thus thwart the designs of the Norns:

Gekk ek til strandar,
grǫm vark nornum,
vilda ek hrinda
stríð grið þeira;
hófu mik, né drekkðu,
hávar bárur,
því ek land um sték,
at lifa skyldak.

I went to the strand,
angry with the Norns,
I wished to cast off
their grievous truce;
but the high waves
raised me up, and did not drown me,
and so I stood upon the land,
and had to live.

The sea has a peculiar agency in these lines, as Pete Sandberg has suggested. Perhaps, treacherous to the end, it here rejects Guðrún's attempt to exercise agency over either her fate *or* the open seas, both of which must – to her – seem equally difficult to fathom.[34]

---

32 An early example of this is found in Egill Skallagrímsson's *Sonatorrek*, in which the skaldic warrior poet (c. 904–955) mourns the death of his sons, one of whom (Bǫðvarr) was drowned in a shipwreck. Quinn, "Mythologizing the Sea" 84, 97; see also Jesch J., "The Threatening Wave: Norse Poetry and the Scottish Isles", in Barrett H. – Gibbon S.J. (eds.), *Maritime Societies of the Viking and Medieval World* (London: 2015) 320–332.

33 See also stanza 38. On drowning see also, in the present volume, Hayden 90–93, Borysławski 361, 368, Magennis 380–382, Birkett 421.

34 Personal communication.

6 Ice

Predictably, for the literature of northern Europe, Codex Regius has much to say about water in its solid state. Ice forms in various places and has many of the same properties as liquid water. It will not form on the river Ífing, fortunately for those threatened by the giants on its far bank (*Vafþrúðnismál* 16), but in many other cases it is an especially treacherous form of water. *Hávamál* has still more to say about this; the ocean, perilous to navigate in a rudderless boat, is as dangerous as riding unshod over ice (90). *Ísi einnættum* ('one-night ice'), ice formed overnight, is not to be trusted (*Hávamál*, 86), or at the very least should only be praised once overcome (*Hávamál*, 81). Ideally, this poem seems to suggest, ice is best crossed on skates (*Hávamál* 83). As far as I can discern, ice appears in four further instances, one of which is as icicles hanging from the frozen beard of the giant Hymir in *Hymiskviða* (10). Ice appears as hail caught up in the hair of Helgi when he briefly returns from Valhǫll in *Helgakviða Hundingsbana II* (44); when he meets with Sigrún, she finds his hands, too, to be *úrsvalar* ('cold and wet'). This, as he explains in the following stanza (45), is a consequence of the tears she has shed for him as he lies dead and buried:

Ein veldr þú, Sigrún
frá Sevafjǫllum,
er Helgi er
harmdǫgg sleginn.
Grætr þú, gullvarið,
grimmum tárum,
sólbjǫrt, suðrœn,
áðr þú sofa gangir;
hvert fellr blóðugt
á brjóst grami,
úrsvalt, innfjálgt,
ekka þrungit.

You alone, Sigrún
from Sefafells,
make Helgi
soaked in sorrow-dew.
You grieve, golden girl,
with grim tears,
sun-bright, southern,
before you go to sleep;

each one falls like blood
on the breast of the prince,
rain-cold, burning-hot,
weighted with sorrow.

This is an extraordinary image, extending the grief of Sigrún to the point where tears shed in one place quite literally saturate the lifeless corpse of Helgi, in another. A similar emotional state is expressed using icy imagery in *Sigurðarkviða in skamma* (8), to describe Brynhildr's reaction to the relationship between Guðrún and Sigurðr: *opt gengr hon innan, ills um fylld, ísa ok jǫkla, aptan hvern* ('she always goes out filled with bitterness, with ice-floes and glaciers every evening', 8).

7 The Properties of Water in the Codex Regius

A perhaps surprising absence from this overview of water in its many forms is standing water, despite there being plenty of this in Iceland, Scandinavia, and throughout most (though not all) of the islands in the north-east Atlantic archipelago. The 'wells' that nourish the roots of the world tree are an obvious exception, and still waters appear in some placenames, but as this overview has shown, these appearances are relative anomalies as far as the Codex Regius is concerned. Why might this be the case? Water is rarely passive or static when it is found in these poems. There is a consistent emphasis on its agency, and the way in which it interacts with humans, non-human animals, and other elements of environment. Far less common is the sense of bodies that are suspended, or inert, within or upon water. The structures of the earth itself are wet through; water falls from above and rises from below. There is clear understanding that the waters which fall from cosmic heights, either as rain or as the dews that settle in the valleys, are essentially connected with the vitality of plant-life and the animals (human and non-human) which depend upon it for survival.

More often than not, however, water is a dangerous impediment to human activity. Rivers present barriers to human (and divine) movement, and partially for this reason they are places where gods and people meet, and sometimes boundaries that require defence. The (violent) poems relating to the Vǫlsungs, interestingly, seem to allude to several of these qualities. They include the cyclicality of rivers within the hydrological cycle, the celestial origin of the Rhine's gold and *all* rivers, and the part that both of these play in the various conflicts and intrigues that take place near the banks of the Rhine, concluding with Guðrún's vision of the torrent that destroys Atli's hall. The sea, like rainfall and rivers, has an essential primordial relationship with the land, but also imposes on the world of landlubbers in more practical, quotidian contexts. Like rivers, the sea is a barrier, though more difficult to overcome. Waters such as these, as *Hávamál* makes plain on several occasions, are not to be trusted, and the threat of drowning is never far away. Intriguingly, water does not seem to shed many of these perilous properties when it appears in solid form, as ice. It remains just as treacherous to cross.

For its medieval Icelandic readers, how then did the aqueous exogrammar of the Codex Regius, as an exogram, and thus a constituent element of mind within the model proposed by Extended Mind Theory, play a part in determining relationships between the humans who read it and the environment in which they lived? Within their thirteenth-century Christian worldview, shaped by the environments which had formed their literary culture (geographical, literary, social, and so on), various aspects of water as they appear in the Codex Regius would not have been understood literally, just as Snorri's repetition of the mythic creation of the first humans (Askr and Embla) would not have superseded what his readers took to be the literal truth of the creation of Adam and Eve recounted in Genesis. With this in mind, what effect might this aqueous exogrammar have had on understandings of, and interactions with, the waters of the air, land and seas? In part, this has already been explored in detail by Judy Quinn in her analysis of the deity Rán, in which Quinn demonstrates how this goddess reflects contemporary awareness of the hydrological cycle.[35] Taken as a whole, the exogrammar of Codex Regius operates in a broadly similar fashion. Water is a fundamental element, permeating all things, and it has an integral relationship with the earth it traverses, descending from above as dew or rain, and passing through or over the earth in various forms and via many channels. The ways in which it is transformed, accordingly, affect everything it touches; there is nothing from which it is divisible. It is essential, consequently, to all of the life that takes place within the world. It sustains plant-life, and all that feeds upon vegetation, but is also *depends* upon these things, and is entangled with them in various ways.

Water may be a necessity, but more often than not it appears in contexts where it creates problems for humans. Water is slippery. It trips us up, and the horses we ride in on. It bars our way, and sometimes we drown in it. On occasion, we can demonstrate our skills and temporary mastery of it, but this is a risky business, and there is every likelihood that it will rear up of its own

35 Quinn, "Liquid Knowledge".

accord, demonstrating its awesome power by dashing our ships to pieces, or racing inland to inundate and engulf our temporary enclosures on dry land. How, then, did the aqueous exogrammar of the Codex Regius function for readers in medieval Iceland? Water surrounded and shaped their habitats. It erupted from the earth and poured through it with extraordinary power. It was then, as now, both glorious and terrifying in its tremendous force, but it also represented a more mundane and everyday challenge to be overcome. The appearance of water in this gathering of skins was a reminder of the flimsy grasp of humans on the land. No matter how accomplished medieval Icelanders may have been as seafarers, like the rest of our species they were terrestrial: reliant and dependent upon the solidity of the earth, and engaged on a daily basis with the same struggle that confronts humans today in our global engagements with our environmental commons. *Earth must stand against water.*

When the Codex Regius was created in the thirteenth century, as an exogram, intentionally or otherwise, it carried a set of information which, within the model of mind proposed by EMT/MET, constituted a component of the extended minds of both its scribes and readers. The customary approach would have been to conclude this essay at the end of the preceding paragraph, having made a series of observations about the way in which the texts of the Codex Regius reflect engagements with water in medieval Iceland and in other environments which gave rise to the texts preserved within in it. But this would be to overlook the active role of the Codex Regius and the various exograms which have borne its imprint in the forms of editions and translations, including the editions from which I have worked in writing this chapter, and the notes and translations that appear both in the pages you are reading, and locked in electronic code (beyond my comprehension) on my laptop. The Codex Regius formed part of medieval minds, just as it – and its many offshoots – form part of modern minds, themselves shaped by the centuries of material change which have passed, including the production of countless exograms which have shaped human interactions with environment and the production of other exograms within it. The poems of the Codex Regius are not, then, relics of a former age locked in letters and bound in hides. They are material components of the minds that inked them eight centuries ago, but they are always, also, potential components of extended human minds *in the present moment of each and every reader.*

As a component of contemporary Extended Minds, what role might the aqueous exogrammar of Codex Regius play in determining both our understanding of the minds of those who created it in the past, and those who encounter it in the present? In certain respects, little has changed in our experience of water over the course of these seven centuries. From an ecocritical standpoint, this truism could precede an extended sermon on the necessity of water for the life of everything on this planet, but thinking about 'Nature' in these terms, as Timothy Morton has shown, is part of what creates distance between our species and the environment of which we are part, and which is part of us.[36] As I have noted, Codex Regius directly reconnects our extended minds with those of people whose lives were more immediately threatened by water than is likely to be the case for those who will read this essay. They depended on water in forms uncontrolled by modern engineering, for travel and sustenance. Their lives depended on the things that swam in it, and upon the craft they could pilot on its surface, and they had direct experience of the perils of water because of this proximity. Many people today are no less threatened by these same dangers, and by the displacement of peoples that has already begun to take place as a consequence of the deep ripples of the unfurling ecological crisis.

The world at the time of the European Middle Ages was in many ways different from the world of the present, though as many commentators have observed, it has also been shaped by ideas that developed in this period.[37] The past two centuries years in particular, with the advent of the Anthropocene, and climate change vastly accelerated by human action (and inaction), have seen unprecedented changes to the rhythms of life on this planet. With this context in mind, what use for ecological thought can there be in scraping the skins of a manuscript predating the internal combustion engine by several hundred years, inked with poems still older? As Timothy Morton has observed, the earth is not on the brink of climate crisis, but deep within it. The prevailing attitude towards Nature, with a capital 'N', has been to treat our environmental commons as though it is something from which humans are distinct – as though an 'ecological criticism' could in any way be separated from 'criticism'. This misconception, of the separation between the human and the environment is not, as Chris Abram and others have considered, the product of any one tradition of thought or experience, and is as visible in Old Norse literature as anywhere else.[38] As is the case in other ancient world literatures, whilst understanding of the human relationship with the environment runs deep,

36 See Morton T., *Ecology Without Nature* (Cambridge, MA: 2007); and *The Ecological Thought* (Cambridge, MA: 2010).

37 See (e.g.) Heng G., *Empire of Magic: Medieval Romance and the Politics of Cultural Fantasy* (New York: 2003); Estes H., *Anglo-Saxon Literary Landscapes* (Amsterdam: 2018).

38 Phelpstead C., "Ecocriticism and *Eyrbyggja Saga*", *Leeds Studies in English* n.s. 45 (2014) 1–18, also discussion in Abram C., *Evergreen Ash: Ecology and Catastrophe in Old Norse Myth and Literature* (Charlottesville, VA: 2019).

there is nevertheless a fundamental anxiety about our relationship with it, accentuated by our ability – unlike any other species that has ever lived on this planet – to affect the material conditions of our environment with a significant degree of intentionality. The written text, then as now, occupies an odd third position within the relationship between humans and the environment, both articulating this gap and, in doing so, maintaining, modifying, and mediating it. Reading, the human understands the text as material object, produced from animal skin or tree pulp, but the text as a communiqué from author to reader is like Graham Harman's 'third object', constantly retreating from view as one attempts to pin down its position in the reader-text-environment triangle.[39] One solution, superseding even the desire to read anachronically,[40] is to read the Codex Regius in a way that has something in common with deep historical approaches,[41] recognising that this manuscript, through its derivatives (editions, translations, and the critical literature it has produced) is not a 'medieval' book in any narrow sense. Most of its existence as an object has been spent in a post-medieval world, and it has had vastly more readers in the twentieth and twenty-first centuries than at any other time in its history. In both its 'medieval' and 'modern' incarnations, it has been part of the material continuum of interaction between humans and environment. A move towards recognising texts and environment as elements of extended minds could be a useful way of understanding literature and other forms of written texts within a material continuum, and in doing so, offer a framework in which to view texts like this one as active components of the extended minds of humans – medieval *and* modern.

Bibliography

Primary Sources

Benson L.D. (ed.), *The Riverside Chaucer* (Oxford: 3rd ed. 2008).

Jónas K., *Eddas and Sagas: Iceland's Medieval Literature*, trans. P. Foote (Reykjavík: 1997).

39 Harman G., *The Quadruple Object* (Winchester: 2011).

40 Following the example of Nagel and Wood in: Nagel A. – Wood C.S., *Anachronic Renaissance* (New York, NY: 2010); and Nagel A., *Medieval Modern Art Out of Time* (London: 2012).

41 E.g., those found in Smail D.L., *On Deep History and the Brain* (Berkeley, CA: 2008); and Shryock A. – Smail D.L., (eds.), *Deep History: The Architecture of Past and Present* (Berkeley, CA: 2011).

Jónas K. – Ólason V. (eds.), *Eddukvæði*, 2 vols, Íslenzk fornrit (Reykjavík: 2014)
Larrington C. (trans.), *The Poetic Edda* (Oxford: 1996).
Orchard A. (trans.), *The Elder Edda: Myths, Gods and Heroes from the Viking World* (London: 2011).

Secondary Sources

Abram C., "New Light on the Illumination of Grendel's Mere", *Journal of English and Germanic Philology* 109.2 (2010) 198–216.
Abram C., *Evergreen Ash: Ecology and Catastrophe in Old Norse Myth and Literature* (Charlottesville, VA: 2019).
Bintley M.D.J., *Trees in the Religions of Early Medieval England* (Woodbridge: 2015).
Bintley M.D.J., "Reading Early Medieval Landscape and Environment: Materially Engaged Approaches to Documentary Sources", in Willard T. (ed.), *Reading the Natural World: Perceptions of the Natural World and Environment in the Global Middle Ages and Renaissance* (Turnhout: 2020) 3–20.
Bintley M.D.J., *Settlements and Strongholds in Early Medieval England: Texts, Landscapes, and Material Culture* (Turnhout: 2020).
Clark A. – Chalmers D., "The Extended Mind", *Analysis* 58.1 (1998) 7–19.
Clunies Ross M., *A History of Old Norse Poetry and Poetics* (Cambridge: 2005).
Clunies Ross M., "The Transmission and Preservation of Eddic Poetry", in Larrington C. – Quinn J. – Schorn B. (eds.), *A Handbook to Eddic Poetry: Myths and Legends of Early Scandinavia* (Cambridge: 2016) 12–32.
Cusack C.M., *The Sacred Tree: Ancient and Medieval Manifestations* (Newcastle Upon Tyne: 2011).
Donald M., *Origins of the Modern Mind: Three Stages in the Evolution of Culture and Cognition* (Cambridge, MA – London: 1991).
Estes H., *Anglo-Saxon Literary Landscapes* (Amsterdam: 2018).
Gunnell T., "Eddic Poetry", in McTurk R. (ed.), *A Companion to Old Norse-Icelandic Literature and Culture* (Oxford: 2005) 82–100.
Hale C.S., "The River Names in *Grímnismál* 27–29", in Glendinning R.J. – Bessason H. (eds.), *Edda: A Collection of Essays*, University of Manitoba Icelandic Studies 4 (Winnipeg: 1983) 165–186.
Harman G., *The Quadruple Object* (Winchester: 2011).
Heng G., *Empire of Magic: Medieval Romance and the Politics of Cultural Fantasy* (New York: 2003).
Hodder I., *Entangled: An Archaeology of the Relationships Between Humans and Things* (Malden, MA – Oxford: 2012).
Hodder I., *Where Are We Heading? The Evolution of Humans and Things* (New Haven, CT – London: 2018). https://handrit.is/en/manuscript/view/is/GKS04-2365.

Jesch J., "The Threatening Wave: Norse Poetry and the Scottish Isles", in Barrett J.H – Gibbon S.J. (eds.), *Maritime Societies of the Viking and Medieval World* (London: 2015) 320–332.

Lashley K., "In Search of the Engram", in Beach F.A. – Hebb D. O. – Morgan C. – Nissen H. (eds.), *The Neuropsychology of Lashley* (New York: 1950) 2–31.

Lowe K.A., "The Development of the Anglo-Saxon Boundary Clause", *Nomina* 21 (1998) 63–100.

Lund J., "Thresholds and Passages: The Meanings of Bridges and Crossings in the Viking Age and Early Middle Ages", *Viking and Medieval Scandinavia* 1 (2005) 109–135.

Lund J., "Banks, Borders and Bodies of Water in a Viking Age Mentality", *Journal of Wetland Archaeology* 8 (2008) 53–72.

Malafouris L., "The Cognitive Basis of Material Engagement: Where Brain, Body and Culture Conflate", in DeMarrais E. – Gosden C. – Renfrew C. (eds.), *Rethinking Materiality: The Engagement of Mind with the Material World* (Cambridge: 2004) 53–62.

Malafouris L. – Renfrew C., "The Cognitive Life of Things: Archaeology, Material Engagement and the Extended Mind", in Malafouris L. – Renfrew C. (eds.), *The Cognitive Life of Things: Recasting the Boundaries of the Mind* (Cambridge: 2010) 1–12.

Malafouris L., *How Things Shape the Mind: A Theory of Material Engagement* (Cambridge, MA: 2013).

Malafouris L., "Metaplasticity and the Primacy of Material Engagement", *Time and Mind* 8.4 (2015) 351–357.

Malafouris L., "At the Potter's Wheel: An Argument *for* Material Agency", in Knappet C. – Malafouris L. (eds.), *Material Agency: Towards a Non-Anthropocentric Approach* (Boston, MA: 2008) 19–36.

Merleau-Ponty M., *Phenomenology of Perception*, trans. D.A. Landes (Abingdon: 2012).

Morton T., *Ecology Without Nature* (Cambridge, MA: 2007).

Morton T., *The Ecological Thought* (Cambridge, MA: 2010).

Nagel A. – Wood C.S., *Anachronic Renaissance* (New York, NY: 2010).

Nagel A., *Medieval Modern Art Out of Time* (London: 2012).

Phelpstead C., "Ecocriticism and *Eyrbyggja Saga*", *Leeds Studies in English* n.s. 45 (2014) 1–18.

Quinn J., "Liquid Knowledge: Traditional Conceptualisations of Learning in Eddic Poetry", in Ranković S. – Melve L. – Mundal E. (eds.), *Along the Oral-Written Continuum: Types of Texts, Relations and Their Implications* (Turnhout: 2010) 183–226.

Quinn J., "Mythologizing the Sea: The Nordic Sea-Deity Rán", in Tangherlini T. (ed.), *Nordic Mythologies: Interpretations, Institutions, Intersections*, The Wildcat Canyon Advanced Seminars: Mythology 1 (Berkeley and Los Angeles, CA: 2014) 71–97.

Shryock A. – Smail D.L. (eds.), *Deep History: The Architecture of Past and Present* (Berkeley, CA: 2011).

Smail D.L., *On Deep History and the Brain* (Berkeley, CA: 2008).

Sørensen P.M., "Þorr's Fishing Expedition [*Hymiskviða*]", trans. K. Williams, in Acker P. – Larrington C. (eds.), *The Poetic Edda: Essays on Old Norse Mythology* (New York – London: 2002) 119–137.

Thorvaldsen B.Ø., "The Dating of Eddic Poetry", in Larrington C. – Quinn J. – Schorn B. (eds.), *A Handbook to Eddic Poetry: Myths and Legends of Early Scandinavia* (Cambridge: 2016) 72–91.

Tolley C., *Shamanism in Norse Myth and Magic*, 2 vols. (Helsinki: 2009).

CHAPTER 8

The Semantic Field of 'Water' in the Gothic Bible: the Letter and the Spirit

Marusca Francini

In fourth-century Moesia Inferior (modern northern Bulgaria and Romanian Dobrudja) the Visigothic bishop Wulfila translated the whole Bible from Greek into Gothic, except for the Book of Kings. The Gothic version is one of the first translations of the Bible in the Christian world and the earliest extensive document written in a Germanic language. A large part of the New Testament (the gospels and the Pauline epistles) has survived, transmitted in manuscripts written in the Ostrogothic kingdom of Italy, in the sixth century. This chapter deals with terms belonging to the semantic field of 'water' in Wulfila's vocabulary, with a view to assessing the linguistic and stylistic strategies by which the translation managed to convey the metaphorical overtones of 'water' and related terms in the New Testament. 'Water' is a keyword of the Fourth Gospel, where the concept of 'living water' bears both a literal and a spiritual meaning (as a symbol of Christ and of the Spirit), while, in the synoptic gospels, water (especially the sea) is above all a force of chaos, tamed by the word of God. The present study will be carried out using philological evidence, investigating linguistic aspects such as loanwords and synonyms and stylistic aspects such as alliteration.

To place the study in context, some initial attention will be given to theoretical views from the fields of translation studies and the geography of religion in a brief sketch of various strands and themes concerning the question of translation and of the relationship between religion, landscape, and language. Discussion of such theoretical considerations in approaching the Gothic New Testament will be followed by an overview of the manuscripts of the translation and then an examination of the references to water and water-words as presented in significant passages of the Gothic text, the focus throughout being on the textual evidence. The intention is to explore biblical Gothic vocabulary, analysing the contextual realisation of a particular family of words in a way that makes clear the translator's distinctive rhetorical strategies and draws attention to the complex question of the relationship between target text and source text. The rendering of the 'water' terms provides a case study of how symbolic meanings are conveyed by the Gothic translator, thus throwing

 | DOI:10.1163/9789004696501_010

new light on an aspect of the Gothic language. Finally, the chapter will end with a list of the terms attested in the Gothic Bible pertaining to the semantic field of 'water', identifying the water vocabulary, and discussing etymology and semantics.

1 Religion, Landscape, and Language

In order to tease out from the samples of texts discussed below how form and content interact in the process of transfer from the language of the Greek *Vorlage* to biblical Gothic, a word must be said about theoretical underpinning with reference to the interaction of religion, landscape and language. When discussing aspects of the Gothic rendering of the terms related to the semantic field of 'water', translation studies, the geography of religion and ecotheology can provide a frame of reference. Bringing together principles from these fields proves useful for the purpose of gaining insight into the processes of the Gothic translation, allowing us to draw upon an understanding of the nature of the translation itself and a recognition of the relationship between landscape features, religion, and language in the translated Greek New Testament texts.

A number of theories relevant to our topic have been developed in the pursuit of the understanding of translation, touching on cultural studies and linguistics (especially contrastive linguistics and semantics). Theories of translation have emerged in the last few decades from the pens of scholars such as Nida, Lefevere, Toury, Reiss and Vermeer, Venuti and Berman, among others. Nida[1] distinguishes between formal equivalence (author-oriented) and dynamic equivalence (or equivalence of effect: reader-oriented) in translation. Visibility and invisibility are the key notions in the theory of Venuti, linked to the concepts of foreignisation and domestication.[2] According to Lefevere, the context and the situation in which the translation was produced must be taken into account.[3] Culturally-oriented translation studies focus on issues of intercultural communication, on the cultural contexts in which the source and target texts are embedded and on the ideology that dominates a particular culture.

1 Nida E., *Towards a Science of Translating, With Special Reference to Principles and Procedures Involved in Bible Translating* (Leiden: 1964).
2 Venuti L., *The Translator's Invisibility: A History of Translation* (London – New York: 1995).
3 Lefevere A., *Translation, Rewriting, and the Manipulation of Literary Fame* (London: 1992).

In the approach referred to as Descriptive Translation Studies, Gideon Toury[4] suggests an analysis of translation based on the combination of linguistic comparison of the source text and target text with study of the cultural framework of the target text, in order to identify the norms at work during the translation process. Antoine Berman, furthermore, suggests that translation involves the elaboration of a system of choices; in Berman's 'circular approach', translation is considered as the construction of identity, an operation through which a culture becomes aware of itself by means of a deliberate relation to the Other, to the foreign.[5] This approach is particularly relevant with reference to the Gothic New Testament. As shown below, the cultural context in which the work of translation was carried out was characterised by a religious-linguistic system in which the key principle at work was close adherence to the Greek source, fostering an 'overt' and 'foreignising' type of translation.

Also relevant to the present chapter is the study of geography and religion. Geography and religion have been related since ancient Greek times, when the mapmaker Anaximander saw the world as the manifestation of a religious principle, and they continue to be so in modern scholarship. Jordan-Bychkov, Damosh and Rowntree, for example, consider geography and religion together,[6] while Ellen Semple claimed that the imagery and symbolism of a religion is affected by its place of origin.[7] The role that sea, lakes, springs and wells had in the waterscape of Palestine as reflected in the New Testament prompts reflection on their treatment in the Gothic translation. The figurative meaning of water imagery in the New Testament is an aspect of the perception of the world and its natural features in the religious mind, building up a symbolic biogeography.

Geography of religion is a branch of cultural geography, which focuses on the human construction and experience of natural landscape and examines the intersection of cultural practices and beliefs with the concepts of place and space and their symbolism.[8] It is the so-called 'new' cultural geography in

---

4 Toury G., "The Nature and Role of Norms in Translation", in Toury G. (ed.), *Descriptive Translation Studies and Beyond* (Amsterdam: 1995) 53–69.

5 Berman A., *The Experience of the Foreign: Culture and Translation in Romantic Germany* (Albany, NY: 1992) 32.

6 Jordan-Bychkov T.G. – Domosh M. – Rowntree L., *The Human Mosaic: A Thematic Introduction to Cultural Geography* (New York: 1994).

7 Semple E., *Influences of Geographic Environment: On the Basis of Ratzel's System of Anthropo-Geography* (New York – London: 1911).

8 For an overview of the geography of religion, see Sopher D.E., *Geography of Religions* (Englewood Cliffs, NJ: 1967).

particular that deals with such symbolism,[9] striving to show how landscape might affect religion, and vice versa, through the interaction of natural features and spirituality. Relationships between religion and environment are mutual: environment affects religion, which in turn affects people and landscape.[10] According to Lily Kong, geography of religion should investigate 'pertinent and crucial themes, such as imagery and symbolism, since religion is one aspect of life that does give itself especially well to symbolism and meaning'.[11]

Ecotheology can also provide a standpoint from which to examine the relation between landscape, language and culture. This branch of scholarly research addresses ecological issues from a religious point of view and considers questions about humanity's relationship with the natural world such as the creation of God.[12] An ecotheological analysis of the environmental references in the texts of the four gospels has been performed using descriptive statistics by Zoe Gareiou and Ethimios Zervas; their study reveals that the Gospel of John 'deals more with the spiritual than the anthropogenic items, as compared to the other three gospels'.[13]

Before surveying the manuscripts of the translation and then going on to examine the Gothic translation in the light of some of the above-mentioned theoretical approaches, it is relevant to refer, however briefly, to the critical literature on the translation that has built up over time. Scholars have long studied the Gothic lexis, and the accumulation of small-scale lexical studies amounts to a very large body of scholarship. Since the literature on lexicography is so extensive, suffice it here to mention only some of the most recent studies, Erik Carlson's examination of the vocabulary of fear, Artūras Ratkus's article on the rendering of Greek ἀρχιερεύς, and Brendan Wolfe's on semiticisms.[14]

---

9 Kong L., "Geography and Religion: Trends and Prospects", *Progress in Human Geography* 14 (1990) 355–371, 362.

10 Park C., *Sacred Worlds. An Introduction to Geography and Religion* (London – New York: 1994) 17. On geography and religion, see Fickler F., "Fundamental Questions in the Geography of Religion", in Wagner P.L. – Mikesell W.M. (eds.), *Readings in Cultural Geography* (Chicago: 1962) 121–144 and Cooper A., "New Directions in the Geography of Religion", *Area* 24 (1992) 123–129.

11 Kong, "Geography and Religion: Trends and Prospects" 361.

12 See Dean-Drummond C., *Eco-Theology* (Winona, MN: 2008).

13 Gareiou Z., "Analysis of Environmental References in the Texts of the Four Gospels Using Descriptive Statistics", *Religions* 9 (2018) 1–14, see also Hultkrantz A., "An Ecological Approach to Religion", *Ethnos* 31 (1966) 131–150.

14 Carlson E., "The Gothic Vocabulary of Fear", *Journal of English and Germanic Philology* 111 (2012) 285–303, Ratkus A.,"Greek ἀρχιερεύς in Gothic Translation", NOWELE 71 (2018) 3–34

2 The Letter: the Evidence of the Gothic Manuscripts

Of all the East Germanic dialects, which include the languages of the Goths, Burgundians, Gepides, Rugians, Scirii and Vandals, only texts in Gothic are attested, comprising the translation of the Bible from Greek.[15] It is commonly agreed that the translation dates from the second half of the fourth century in the Balkans by the Visigothic bishop Wulfila (ca. 311–383),[16] who also created the Gothic alphabet, a writing system based on the Greek alphabet with the addition of selected Latin and runic characters.[17] It is thought that the translation was carried out by Wulfila with the aid of a team of scholars, which he supervised.[18]

The extant manuscripts of the translation were written later, in the early-sixth-century Ostrogothic kingdom of Italy (493–553) during the reign of Theoderic the Great (493–526), a period in which the ruling Ostrogoths fostered a lively scribal tradition.[19] The most famous Gothic manuscript is the *Codex Argenteus*, a *deluxe* gospel book written in Ravenna in silver and golden

and Wolfe B.N., "The Relevance of Certain Semiticisms in the Gothic New Testament", *NOWELE* 71 (2018) 249–256.

15 A broader discussion of the Greek *Vorlage* is given in Francini F., "L'edizione della Bibbia gotica tra testo vivente e tradizione quiescente", in Ferrari F. – Bampi M. (eds.), *Storicità del testo, storicità dell'edizione* (Trento: 2009) 313–340. See also Falluomini C., *The Gothic Version of the Gospels and Pauline Epistles* (Berlin – Boston, MA: 2015).

16 The sources on the bishop and his translation are represented by the fifth-century Church historians Philostorgius, *Historia Ecclesiastica* II, 5, eds. J. Bidez – F. Winkelmann (Berlin New York: 1981); Socrates, *Historia Ecclesiastica* IV, 33, in Hansen G.C. – Širinjan M. (eds.), *Sokrates. Kirchengeschichte* (Berlin: 1995); Sozomen, *Historia Ecclesiastica* VI, 37, in Bidez J.- Hansen G.C. (eds.), *Sozomenus, Kirchengeschichte* (Berlin: 1995).

17 The Gothic texts, with the exception of the fragments of the *Codex Bononiensis*, are edited by Streitberg W., *Die gotische Bibel, I. Der gotische Text und seine griechische Vorlage Mit Einleitung, Lesarten und Quellennachweisen sowie den kleineren Denkmälern als Anhang. 7. Aufl. Mit einem Nachtrag von Piergiuseppe Scardigli* (Heidelberg: 2000); the *Codex Bononiensis* is edited by Finazzi R.B. – Tornaghi P., "*Gothica Bononiensia*. Analisi linguistica e filologica di un nuovo documento", *Aevum* 87 (2013) 113–153. The Gothic text, the Greek text and the English translations in this chapter are taken from the database http://wulfila.be/(last accessed November 3, 2023) based on the Streitberg edition.

18 See Jellinek M.H., *Geschichte der gotischen Sprache* (Berlin: 1926) 10–11; Friedrichsen G.W.S., *Gothic Studies* (Oxford: 1961) 104.

19 Metzger B.M., *The Early Versions of the New Testament: Their Origin, Transmission, and Limitations* (Oxford: 1977) 377.

ink on purple parchment.[20] *Ambrosianus C*[21] and *Codex Gissensis*[22] contain small portions of the Gospel of Matthew and the Gospel of Luke, respectively, while *Ambrosianus A*,[23] *Ambrosianus B*[24] and the *Codex Carolinus*[25] transmit the Pauline Epistles. *Ambrosianus E*[26] and *Codex Vaticanus Latinus* 5750[27] were originally a single manuscript preserving fragments of the *Skeireins*, a commentary on John which contains citations from the gospels,[28] while the now lost Tablet of Hács-Béndekpuszta contained John 17:11–12. In the year 2010, a fragment of a brief Gothic text (a sermon?) was discovered in the Bologna manuscript Archivio della Fabbriceria della Basilica di San Petronio Cart. 716/1, n° 1; it contains citations from the Old and the New Testament. *Codex LI* (49) in the Biblioteca Capitolare in Verona comprises a collection of Latin homilies and marginal notes in Gothic (called *Gotica Veronensia*) that are biblical quotations. *Codex Vindobonensis 795* (Österreichische Nationalbibliothek, Vienna) contains Gothic alphabets and quotations from the Gothic version of Luke (called *Gotica Vindobonensia*) and Paris, BNF, MS Lat. 528 contains Biblical names in Gothic (called *Gotica Parisina*).[29]

Our survey of the manuscript tradition enables us to state that the Gothic version is unlikely to have been transmitted in the form in which it was originally written. Since the extant manuscripts postdate the original translation by about two centuries, some features of their texts are probably scribal rather

20 Uppsala, Universitetsbiblioteket, MS D G 1, 187 leaves (originally 336), containing the four gospels in the so-called 'Western' order Matthew-John-Luke-Mark; see von Friesen O. – Grape A. (eds.), *Codex Argenteus* (Uppsala: 1928).

21 Milan, Biblioteca Ambrosiana, MS I 61 sup., fols. 90–91, a fragmentary palimpsest containing Matthew 25:38–46; 26: 1–3; 26: 65–75; 27:1.

22 Giessen, Universitätsbibliothek, MS 651/20, a now lost Gothic-Latin bilingual bifolio from the first half of the sixth century, containing Luke 23:11–14; 24: 13–17.

23 Milan, Biblioteca Ambrosiana, MS S 36 sup. (102 leaves) + Torino, Biblioteca Nazionale Universitaria, F. IV. 1 Fasc. 10 (four leaves), a palimpsest.

24 Milan, Biblioteca Ambrosiana, MS S 45 sup. (78 leaves), palimpsest.

25 Wolfenbüttel, Herzog August Bibliothek, MS Guelf. 64 Wiess. (ff. 255–256, 277, 280), a Gothic-Latin bilingual dating to the first two decades of the sixth century. It preserves chapters 11–15 of the *Epistle to the Romans*.

26 Milan, Biblioteca Ambrosiana, MS E 147 sup.

27 Rome, Biblioteca Apostolica Vaticana.

28 Bennett W.H. (ed.), *The Gothic Commentary on the Gospel of John. Skeireinsaiwaggeljons þairh Iohannen: A Decipherment, Edition and Translation* (New York: 1960).

29 Only a fragment of the translation of the Old Testament survives (a small portion of the Book of Nehemiah), handed down in *Ambrosianus D* (Milan, Biblioteca Ambrosiana, MS G 82 sup., fols. 209–210, 451–452, 461–462). As well as the biblical translations, the manuscripts also preserve part of a Gothic Calendar (*Ambrosianus A*) and the Deeds of Naples and Arezzo, charters from sixth-century Italy: see Tjäder J.O., *Die Nichtliterarischen Papyri Italiens aus der Zeit 445–700* (Lund: 1954).

than authorial. In the span of time that elapsed between the translation and the date of its extant manuscripts, the text would have undergone revisions and rewording; for example, it has been shown that the extant manuscripts show influence from the Latin versions.[30] The *Codex Argenteus* (along with the other manuscripts) represents only a single stage in a sequence of manuscripts now lost to us, extending from the fourth to the sixth century, crossing geographical boundaries, from Moesia to Italy, and dialectal boundaries, from the Visigothic of Wulfila to the Ostrogothic dialect. Even though scribes seem to have taken considerable care to reproduce the text as faithfully as possible, being the Sacred Text, the Wulfilian original would have been modified during two centuries of textual transmission.

3 The Spirit: the Gothic Rendering of the Symbolic Water

In the Gothic New Testament as preserved in the manuscripts, we come across the theme of water many times: the storm on the Sea of Galilee, the concept of Jesus as 'living water', baptism, and the use of water for ritual cleaning in the practice of foot-washing. The prominence of water, real and metaphorical, becomes clear because 'water is the essential ingredient of life, and water supply is the essential ingredient of civilization'.[31] Water is a powerful metaphor. In the Gospel of John, Jesus himself is represented as a life-giving spring of water. Symbolism lies at the core of the Johannine gospel, where water is a recurring symbol, clearly referring to the Paraclete (the Spirit); the quality and characteristics of water serve as a vehicle to portray Jesus and the Spirit in the Fourth Gospel. Water is salvation and redemption amidst the hardships of human life, but the deadly force that this element can exert is also represented in the New Testament, where water, in the synoptic gospels and in the Pauline epistles, is portrayed as the threat of chaos, and a metaphor of the tribulations of human life.

Water terms in the New Testament sometimes denote the elements 'sea', 'lake', 'rain' in a general sense, but in certain cases they refer to specific features of the Palestinian landscape, such as the Jordan river, or the Lake of Tiberias. This lake went by different names in the New Testament: Lake of Gennaseret (Luke 5:1), Sea of Galilee, and Sea of Tiberias (John 6:1). The Gothic text uses *saíws* and the compound *marisaíws* to translate the Greek word for 'lake'(λίμνη), but

30 See Friedrichsen G.S.W., *The Gothic Version of the Gospels A Study of its Style and Textual History* (London: 1926; rev. ed. 1939).

31 Sedlak D., *Water 4.0: The Past, Present and Future of the World's Most Vital Resource* (New Haven, CT – London: 2014) i.

marei to translate θάλασσα 'sea' when the lake, which is the largest freshwater lake in the country, is described in the traditional Palestinian fashion as a 'sea'. The Gothic word *marei* is also recorded in the *Deed of Naples*, a charter from sixth century Italy, where it means 'marshland', denoting the lagoon environment around Ravenna. The distribution of the terms reveals that *marei* refers to the sea, whereas *saiws* refers to freshwater and marshland. As for geographical sensibility as expressed by language, the occurrences of *saiws* in biblical Gothic on the one hand and the usage in the charter from Italy on the other hand show that the term could be employed to denote different landscape features.

The Sea of Galilee is known for its sudden violent storms; it is potentially hazardous, with unpredictable winds and waves. The wind produced by cold-air drainage from the tops of the mountains west of the Sea of Galilee (crests 2000 feet above the sea level) can create waves one foot high or higher. The three synoptic gospels (Matthew 8:24–26; Luke 8:24; Mark 4:37) tell how Jesus, crossing this sea with his disciples in a vessel, stills a tempest with a simple word. The stilling of the storm on the sea is to be understood as a revelation of Jesus's nature; he is the Lord who can calm the roughest sea, the Deliverer who saves those overwhelmed by the afflictions of life, which are symbolised by the chaos brought by wind and rough water. The miracle shows Jesus's authority over nature, as wind and water obey him, as well as his power to overcome evil, symbolised by the threatening chaos of the sea. The gospels use the antithetic pair 'storm' and then 'calm of the sea', after the miracle has been performed. The Gothic version manages to convey the antithesis and endeavors to place emphasis on the symbolic meaning of the miracle by means of an alliteration, using the alliterating pair *wegs* (for 'storm') and *wis* (for 'calm of the sea').

In Luke 8:24 the alliteration involves *wind* ('wind'), *wega watins* (κλύδωνι τοῦ ὕδατος, *tempestatem aquae* in the Latin version, 'raging of the water') and *warþ wis*:

> Ποσελθόντες δὲ διήγειραν αὐτὸν λέγοντες, ἐπιστάτα ἐπιστάτα, ἀπολλύμεθα. ὁ δὲ διεγερθεὶς ἐπετίμησεν τῷ ἀνέμῳ καὶ τῷ κλύδωνι τοῦ ὕδατος: καὶ ἐπαύσαντο, καὶ ἐγένετο γαλήνη.
>
> Duatgaggandans þan urraisidedun ina qiþandans: talzjand, fraqistnam! Iþ is urreisands gasok wind jah þamma wega watins jah anaslawaidedun jah warþ wis.
>
> And they came to him, and awoke him, saying, Master, master, we perish. Then he arose, and rebuked the wind and the raging of the water: and they ceased, and there was calm.

In Mark 4:37 the *w*- alliteration runs through the whole clause *jah warþ skura windis mikila jah wegos waltidedun in ski*p ('And there arose a great storm of wind, and the waves beat into the ship [...], Greek καὶ γίνεται λαῖλαψ μεγάλη ἀνέμου, καὶ τὰ κύματα ἐπέβαλλεν εἰς τὸ πλοῖον, ὥστε ἤδη γεμίζεσθαι τὸ πλοῖον), where *weg* alliterates with *wind;* and in the following Mark 4:39 *wind* alliterates with *wis*:

> Καὶ διεγερθεὶς ἐπετίμησεν τῷ ἀνέμῳ καὶ εἶπεν τῇ θαλάσσῃ, σιώντα, πεφίμωσο, καὶ ἐκόπασεν ὁ ἄνεμος καὶ ἐγένετο γαλήνη μεγάλη.
>
> Jah urreisands gasok winda jah qaþ du marein; gaslawai, afdumbn! Jah anasilaida sa winds jah warþ wis mikil.
>
> And he arose, and rebuked the wind, and said unto the sea, Peace, be still. And the wind ceased, and there was a great calm.

Gothic *wegs* serves to translate three different Greek words: κῦμα (in the plural) 'wave' (in Matthew and Mark), κλύδιον 'surging (sea)' (in Luke), and σεισμός 'quake' (in Matthew). In the Gospel of Matthew (8:24) *wegs* serves to translate both σεισμός 'quake' and (in the plural) κῦμα 'wave'; the alliteration emphasises *wegs* and *warþ* in the first clause, *wairþan* and *wegim* in the second:

> Καὶ ἰδοὺ σεισμὸς μέγας ἐγένετο ἐν τῇ θαλάσσῃ, ὥστε τὸ πλοῖον καλύπτσθαι ὑπὸ τῶν κυμάτων. Αὐτὸς δὲ ἐκάθευδεν.
>
> Jah sai, wegs mikils warþ in marein, swaswe þata skip gahuliþ wairþan fram wegim; iþ is saislep.
>
> And behold, there arose a great tempest in the sea, insomuch that the ship was covered with the waves; but he was asleep.

In Matthew 8:26 *wis* alliterates with *windam*:

> [...] ἐγερθεὶς ἐπετίμησεν τοῖς ἀνέμοις καὶ τῇ θαλάσσῃ καὶ ἐγένετο γαλήνη μεγάλη.
>
> [...] þanuh urreisands gasok windam marein jah warþ wis mikil.
>
> [...] then he arose, and rebuked the winds and the sea; and there was a great calm.

Chapters 8–9 of the Gospel of Matthew string together ten miracle-stories, presenting an illustration of the miraculous power of Jesus.[32] The account given by Matthew in chapter 8 intensifies the force of the storm on the sea by using, instead of the 'great wind storm' of Mark 4:37, the more cosmically dramatic 'great quake in the sea'. The word σεισμός used by Matthew literally means 'earthquake'. In the same gospel (Matthew 27:51), it is reported that by an earthquake the natural world will witness to the death of Jesus: ἡ γῆ ἐσείσθη ('the earth quaked, shook') where the Greek verb has the same root as σεισμός found in Matthew 8:24. The Gothic rendering of this passage is *airþa inreirada*. By means of the use of one Gothic word *wegs* to render two different Greek terms, σεισμός 'quake' and κῦμα 'wave', the parallelism between Jesus's stilling of the storm and his death on the cross is lost, though alliteration is provided.

The sudden 'great calm' (γαλήνη μεγάλη) that ensues after Jesus has called for the sea to be still, underlines Jesus's authority over the cosmos. The stilling of wind and water is a nature miracle that seen symbolically attests that Jesus is the Deliverer who saves those overwhelmed by chaos, his identity as the Son of God.[33] In the Old Testament God is shown to have power over the waters. Psalm 107:23–32 tells that when people cried out to God in distress, 'he made the storm be still, and the waves of the sea were hushed'. God alone stretched out the heavens and trampled the waves of the sea (Job 9:8). Control of the waters, the forces of primeval chaos, is the prerogative of God alone; Jesus does what only God can do.[34] It is worth noting that Gothic *wis*, translating γαλήνη ('calm') and probably related to Gothic *wisan* 'to be', and *wists* 'nature', has no equivalents in the other Germanic languages.

With regard to the presence of alliteration (which was the structuring metrical principle of Germanic poetry), Harry Velten noted that 'some imagined to recognise the ring of Old Germanic poetry in Gothic Bible passages'.[35] As the geographical world is interpreted by religion, alliteration is one response to the religious experience with landscape features and the ensuing dialectic between water and symbolism in the gospels, since the rhetorical effect

32 Klostermann E., *Das Matthäusevangelium* (Tübingen: 1927) 72.

33 Feiler P.F., "The Stilling of the Storm in Matthew: A Response to Gunther Borkamm", *Journal of Theological Studies* 26 (1983) 399–406, 406.

34 Brown S., "Water Imagery and the Power and Presence of God in the Gospel of John", *Theology Today* 72 (2015) 289–298, 290.

35 Velten H.V., "Studies in Gothic Vocabulary with Especial Reference to Greek and Latin Models and Analogues", *Journal of English and Germanic Philology* 29 (1930) 489–509, 507. On the use of alliteration in the Gothic translation, see Stolzenburg H., *Zur Übersetzungstechnik des Wulfila* (Halle: 1905) 155; see also Stutz E., *Gotische Literaturdenkmäler* (Stuttgart: 1966) 53–54.

produced makes the meaning of these passages more relevant. It is worth noting that, although the Gothic translator as a rule employs a 'foreignising' or 'author oriented' method, the use of the stylistic device common to Germanic poetry points to a 'domesticating' or 'reader oriented' technique of translation.

Another alliteration involving the term for 'sea' can be found in the expression *malma mareins* for ἄμμος τῆς θαλάσσης ('sand of the sea', said of the children of Israel) in the Pauline *Epistle to the Romans* 9:27.

Paul describes the dangers of water in 2 Corinthians 11:25–26, where shipwreck, rivers and the sea are listed among the perils and sufferings he has been through:

> Τρὶς ἐρραβδίσθην, ἅπαξ ἐλιθάσθην, τρὶς ἐναυάγγησα, νυχθήμεραν ἐν τῷ βυθῷ πεποίηκα. Ὁδοιπορίαις πολλάκις, κινδύνοις ποταμῶν, κινδύνοις λῃστῶν, κινδύνοις ἐκ γένους, κινδύνοις ἐξ ἐθνῶν, κινδύνοις ἐν πόλει, κινδύνοις ἐν ἐρημίᾳ, κινδύνοις ἐν θαλάσσῃ, κινδύνοις ἐν ψευδαδέλφοις.
>
> Þrim sinþam wandum usbluggwans was, ainamma sinþa stainiþ was; þrim sinþam usfarþon gatawida us skipa; naht jah dag in diupiþai was mareins. Wratoþun ufta, bireikeim aƕo, bireikeim waidedjane, bireikeim us kunja, bireikeim us þiudom, bireikeim in baurg, bireikeim in auþidai, bireikeim in marein, bireikeim in galiugabroþrum.
>
> Thrice I was beaten with the rods, once I was stoned, thrice I suffered shipwreck, a night, and a day I have been in the deep. In journeyings often, in perils of waters, in perils of robbers, in perils by mine own countrymen, in perils by the heathen, in perils in the city, in perils in the wilderness, in perils in the sea, in perils among false brethren.

In diupiþai mareins translates ἐν τῷ βυθῷ, where βυθός has the meaning 'the depth', especially of the sea. The Gothic translation specifies *mareins* ('of the sea'), as does the *Vulgate* Latin version, which reads *in profundo maris*. As for shipwreck and sea, in the new Heaven announced in Revelation (21:1) there is no longer any sea: 'And I saw a new heaven and a new earth; for the first heaven and the first earth were passed away; and there was no more sea', where 'no more sea' is a metaphor for 'no more evil'. The sea in the Bible often represents evil: overflowing with rain, the sea caused the Flood (Genesis 7:11); the Leviathan is at home in the sea (Psalm 104:25–26), the wicked are compared to a troubled sea (Isaiah 57:20), and the Antichrist himself arose from the sea (Revelation 13:1). As for rivers, they were dangerous since in the times of Paul there were not many bridges.

The parallel passages Matthew 7:25–27 (the concluding section of the Sermon of the Mount) and Luke 6:48–49 (the concluding section of the Sermon of the Plain) contain the parable of the Wise and the Foolish Builders, where rain and flood are described in their power of destruction. The rainstorm, and the flood, are a symbol of trials and tribulations.[36] In this parable, Jesus compares wise and foolish builders. In Matthew's account, the wise one builds a house on the rock, but the foolish one builds his own on the sand. When the rain descends and the wind blows, and the streams rise, the house whose foundations are built on the rock does not fall, but when the same elements beat against the house built on the sand, they dislodge the structure, and the house collapses. In Luke's parallel account, the wise builder digs deep and lays the foundation on the rock, while the foolish builder builds his house on the surface. When the nearby river overflows its banks and floods the area, the house of the wise builder is not shaken, but the house of the foolish one falls.

Matthew 7:27 gives a dramatic description of a rainstorm:

> Καὶ κατέβη ἡ βροχὴ καὶ ἦλθον οἱ ποταμοὶ καὶ ἔπνευσαν οἱ ἄνεμοι καὶ προσέκουψαν τῇ οἰκίᾳ ἐκείνῃ, καὶ ἔπεσεν, καὶ ἦν ἡ πτῶσις αὐτῆς μεγάλη.
>
> Jah atiddja dalaþ rign jah qemun aƕos jah waiwoun windos jah bistugqun bi jainamma razna, jah gadraus, jah was drus mikill.
>
> And the rain descended, and the floods came, and the winds blew, and beat upon the house; and it fell: and great was the fall of it.

Luke 6:48–49 is more interested in the dramatic effects of a flood:

> Ὁμοιός ἐστιν ἀνθρώπῳ οἰκοδομοῦντι οἰκίαν ὃς ἔσκαψεν καὶ ἐβάθυνεν καὶ ἔθηκεν θεμέλιον ἐπὶ τὴν πέτραν: πλημμύρης δὲ γενομένης προσέρηξεν ὁ ποταμὸς τῇ οἰκίᾳ ἐκείνῃ, καὶ οὐk ἴσχυσεν σαλεῦσαι αὐτὴν διὰ τὸ καλῶς οἰκοδομῆσθαι αὐτήν.Ὁ δὲ ἀκούσας καὶ μὴ ποιήσας οἰκοδομῆσαντι οἰκίαν ἐπὶ τὴν γῆν χορὶς θεμελίου, ᾗ προσέρηξεν ὁ ποταμός, καὶ εὐθὺς συνέπεσεν, καὶ ἐγένετο τὸ ῥῆγμα τῆς οἰκίας ἐκείνης μέγα.
>
> Galeiks ist mann timrjandin razn, saei grob jah gadiupida grunduwaddjau ana staina, at garunjon þan waurþanai bistagq aƕa bi jainamma razna jah ni mahta gawagjan ita; gasuliþ auk was ana þamma staina. Iþ sa

36 Harrington D.J., *The Gospel of Matthew* (Collegeville, MN: 1991) 109.

> hausjands jah ni taujands galeiks ist mann timjandin razn ana airþai inuh grunduwaddju, þatei bistagq flodus [aƕa] jah suns gadraus, jah warþ so uswalþeins þis raznis mikila.
>
> He is like a man which built a house, and digged deep, and laid the foundation on a rock; and when the flood arose, the stream beat vehemently upon that house, and could not shake it: for it was founded upon a rock. But he that heareth, and doeth not, is like a man that without a foundation built a house upon the earth, against which the stream did beat vehemently, and immediately it fell; and the ruin of that house was great.

Flodus is a *hapax* which translates ποταμὸς in Luke 6:49, but the same hand that wrote the text has inserted the gloss *aƕa*, bringing it in line not only with Luke 6:48 where ποταμὸς is translated *aƕa*, the usual rendering for the Greek word, but also with the parallel passage in the Gospel of Matthew 7:27, where *aƕa* is used. According to Friedrichsen, the gloss may be the result of the substitution of a word more familiar to the reviser or scribe than the reading of the Wulfilian text.[37] The reason behind the rendering *flodus* may be stylistic diversification on the part of the translator, in a context where the Greek word occurs three times in close succession, twice translated with *aƕa*. But the term and its gloss demand one further reflection: we may have an example of lexical diversification, aimed, perhaps, at rhetorical effects; or the presence of *flodus* may be due to the influence of parallel passages. But here ποταμὸς is not simply a 'river', but a river that overruns its banks and floods the area. *Flodus* and *aƕa* are probably not synonyms, since, prompted by the context, the lexical choice of *flodus*, maybe meaning 'flood' rather than 'river', emphasises the image of a devastating river flood. *Flodus* (taking 'flood' as its meaning) is of particular interest here because it shows that, in the process of interpretation of the natural phenomenon caused by water as described in the source text, the translator had been able to use a term he found more appropriate. The *Codex Argenteus* contains fifteen interlinear or marginal glosses, which are, usually, synonyms of the readings in the text; the marginal note *aƕa* to *flodus* in the *Codex Argenteus* bears witness to the fact that the concrete texts (the manuscripts) which actualise the Gothic version present an interplay of variant readings.

The impress of religion on landscape results in the symbolism with which water terms are imbued in the Fourth Gospel. As we will see, the linguistic transfer and incorporation of culture-specific aspects of the interface between

37 Friedrichsen, *The Gothic Version of the Gospels* 119.

geographical features and Christian symbols involves above all the creation of calques in biblical Gothic.

Water imagery permeates the Gospel of John, having a spiritual sense, in which the Logos (Christ) comes to man as the bestower of light, life, truth, joy, bread and water. The 'living water' (as it is called in the Fourth Gospel) is a fundamental element that makes existence possible and, in this case, the Gothic word for 'water', *wato*, denotes a Christian abstract idea. The Gospel of John is well known for its wealth and depth of figurative language, metaphors, and symbols; symbolic expression is part and parcel of its style.[38] The Johannine concept of 'living water' metaphorically refers to divine activity, as in the Old Testament Yahweh is the fountain of living water (Jeremiah 2:13). The Fourth Gospel is the only New Testament writing to mention living water. The Johannine Jesus portrays himself as the source of living water in John 7:38, which is the only passage of what remains of the Gothic translation (handed down in the fragmentary *Codex Argenteus*) where living water is mentioned. It relates the discourse of Jesus at the Feast of Tabernacles:

> ὁ πιστεύων εἰς ἐμέ, κατὼς εἶπεν ἡ γραφή, ποταμοὶ ἐκ τῆς κοιλίας αὐτοῦ ρεύσουσιν ὕδατος ζῶντος.
>
> Saei galaubeiþ du mis, swaswe qaþ gameleins, aƕos us wambai is rinnand watins libandins.
>
> He that believeth on me, as the scripture hath said, out of his belly shall flow rivers of living water.

The Gothic for 'living water' is *watins libandins*; the Gothic participle *libandins* here translates a Greek participle (ζῶντος). The Greek syntactic structure of noun and postposed present participle is matched by the same arrangement of elements in the Gothic rendering. The noun (*watins*) followed by its determinant mirrors the Greek arrangement of elements, not the Germanic placement of a present participle with respect to its head noun; the Gothic translation, choosing a postnominal modifier, has shaped the word order according to Greek patterns. The word order reveals Greek influence on syntactic patterns

38 See Joubert J., "Johannine Metaphors/Symbols Linked to the Paraclete-Spirit and their Theological Implications", *Acta Theologica* 27 (2007) 83–103, 83 and 86. On John's use of symbols, imagery, and metaphors, see especially Dodd C.H., *The Interpretation of the Fourth Gospel* (Cambridge: 1953) and Schneiders S., *Written That You May Believe: Encountering Jesus in the Fourth Gospel* (New York: 2003) 65–69.

in Gothic writings; the underlying impetus for the word-for-word translation system stemmed from Christian views. The Gothic rendering adheres so strictly to the wording of the Greek text because, according to early Christian philosophy, the spirit of God dwelt literally in each word, and in the very word order of the Holy Writ. Since conservatising pressures grow up within a religious tradition, dependency of Gothic on Greek word order is due to the specific nature of the source text, which is the Sacred Text. Fidelity in translation of the Bible is common in late Antiquity, motivated by the need to reproduce the words of the Sacred Text, in precisely the same order. The language of the Greek New Testament had a tremendous impact on the early translations of the Bible, acting as conduit for linguistic innovation, providing a new lexicon, a special syntax, and a style of its own. Linguistic patterns of the Greek text became a 'sacral stamp',[39] imbued as they were with religious and social value to connote a 'linguistic emblem' of membership in the Christian community, distinguishing insiders from outsiders, and were therefore imitated by the translators.[40]

The character of the language in John differs significantly from that of the synoptics. In the synoptic gospels Jesus's sayings are relatively brief, for he often teaches with parables. In the Fourth Gospel the style of Jesus's language is different, for he uses longer discourses with expansions on theological themes. In the Johannine gospel water plays a symbolic role, but the water imagery is multivalent, both real and symbolic. Culpepper states that light, bread, and water are the three core symbols in the Fourth Gospel, where the term for 'water' occurs 21 times[41] (in passages for the most part lost to the Gothic version). The three core symbols communicate the transcendent in the natural world and bridge 'the chasm between what is "from above"' and what is "from below" without collapsing the distinction'.[42] Though in the world, Jesus is not of the world: the concept of the vertical dimension of existence underlies the dual levels of meaning, the presence of both the literal and symbolic meanings for earthly features. Earthly images are used to bear witness to heavenly realities;

39 Psaltes S.B., *Grammatik der Byzantinischen Chroniken* (Göttingen: 1913) VIII. Psaltes talks of a 'stamp of sacrality' referring to periphrastic constructions.

40 Drinka B., "The Sacral Stamp of Greek: Periphrastic Constructions in New Testament Translations of Latin, Gothic, and Old Church Slavonic", in Welo E. (ed.), *Indo-European Syntax and Pragmatics: Contrastive Approaches* (Oslo: 2011) 41–73, 41.

41 Culpepper R.A., *Anatomy of the Fourth Gospel: A Study in Literary Design* (Philadelphia: 1983) 189–198.

42 Koester C., *Symbolism in the Fourth Gospel: Meaning, Mystery, Community* (Minneapolis, MN: 2003) 4.

through analogy between the two realities, the Gospel of John creates a figurative world, which metaphorises aspects of the earthly world.[43]

In John 3:23–26 a geographical feature is used for a symbolic act: Jesus and his disciples engage in baptism in Judea, in the area of the river Jordan, where water is plentiful.[44] In the fourth evangelist's use of water imagery, water has a symbolic function in baptising; John the Baptist in the Jordan river physically baptised in water and metaphorically immersed in the living water that is Christ. Baptism establishes a connection between water and the Spirit, so that the symbolism of baptism in water represents the birth of a new being. The Gothic verb *daupjan* translates βαπτίζειν 'baptise'; the prefixed form *ufdaupjan* translates βαπτίζαι 'baptise', but also ἐμβάψαι 'dip in' (once in John 13:26, said of bread in the dish, during the Last Supper), which is the original meaning of *daupjan*. *Daupjan* is a case of lexical influence, where the Gothic word has taken on an additional sense from its Greek counterpart.[45] The sense 'to baptise' is a loan meaning modelled on the Greek βαπτίζειν and βαπτίζαι; both Gothic verbs meant 'dip in' and acquired the Christian sense 'baptise', since baptism was originally performed by immersion. The Gothic translator created Christian cult-words above all by coining calques out of native speech material (as is the case with the verb *daupjan*), rather than borrowing. According to Jordanes's *Gothic History*, the Goths preached the gospel to other Germanic peoples and a possible Gothic lexical influence on Old High German has been suggested;[46] among German religious terms, Gothic influence has been detected in the meaning of Old High German *touffan* (New German *taufen*) whose meaning 'to baptise' would be derived, according to this theory, from the Christian sense of Gothic *daupjan*.[47]

In the sections of John that involve water, the element serves primarily as a symbol of the Spirit and of Christ, but it functions also as a symbol of cleansing and rebirth, as in the episode of the healing of the man born blind who must wash his eyes in the pool to be able to see (John 9:7). The Gothic verb for 'wash'

43 Van der Watt J., *Family of the King. Dynamics of Metaphor in the Gospel According to John* (Leiden – Boston – Cologne: 2000) 310.

44 Bruce F.F., *The Gospel of John. Introduction, Exposition, and Notes* (Basingstoke: 1983) 93.

45 But Greek βαπτίζειν also, like its Aramaic equivalent, originally meant 'to dip, wash (by immersing)'.

46 Green D.J., *Language and History in the Early Germanic World* (Cambridge: 1998) 236–253; idem, "Linguistic and Literary Traces of the Ostrogoths", in Barnish S.J. – Marazzi F. (eds.), *The Ostrogoths. From the Migration Period to the Sixth Century. An Ethnographic Perspective* (Woodbridge: 2007) 387–416, 145, 158–159. For Jordanes's *Gothic History*, see Mommsen T. (ed.), *Iordanis Romana et Getica* (Berlin: 1882).

47 Kluge F., "Gotische Lehnwörter in Althochdeutschen", *Beiträge zur Geschichte der deutschen Sprache und Literatur* 35 (1909) 124–160.

is here *þwahan*, translating Greek νίπτειν, which in the Gospel of John is also used when Jesus washes his disciples' feet (John 13:12–13). A case of multiple rendering of one Greek word is represented by the translation of this verb, νίπτω ('wash the hands or feet, purge, cleanse'), which is translated as *þwahan* and its prefixed forms *afþwahan*, *biþwahan* and *usþwahan*. Like other Indo-European languages, Gothic shows a capacity to form new verbs by means of elements termed preverbs or verbal prefixes, which express diversified semantic and grammatical features; the less grammaticalised layer consists of prefixes that have prepositional counterparts,[48] as is the case of *afþwahan*, *biþwahan* and *usþwahan*. Over half of the attested Gothic verbs are prefix formations of some kind.[49] In chapter 9 of the Gospel of John (the blind man must wash his face in the pool to be healed) the Greek verb νίπτω is translated as *þwahan*, *afþwahan*, *biþwahan*. *Þwahan* (in the infinitive form) and *afþwahan* (in the past tense) are used in John 9:7:

> Καὶ εἶπεν αὐτῷ, ὕπαγε νίψαι εἰς τὴν κολυμβήθραν τοῦ σιλωάμ [...] ἀπῆλθεν οὖν καὶ ἐνιψάτο, καὶ ἦλθεν βλέπων.
>
> Jah qaþ þu imma: gagg þwahan in swumsl Siloamis [...] Galaiþ jah afþwoh jah qam saiƕands.
>
> And said unto him: Go, wash in the pool of Siloam [...] He went his way therefore, and washed, and came seeing.

The basic semantic function of the prepositional prefix *af* 'from, by, out of' is spatial and connotes motion away from; *afþwahan* can impart the meaning 'to wash (the disease) off' but the prefix can also have a perfective function. The prefixed form of the verb appears in John 9:7 (see above), and John 9:15:

> ὁ δὲ εἶπεν αὐτοῖς πηλὸν ἐπεθηκὲν μου ἐπὶ τοὺς ὀφθαλμούς καὶ ἐνιψάμην καὶ βλέπω.
>
> Iþ is qaþ jah þaim: fani galagida mis ana augona jah afþwoh, jah saiƕa.
>
> He said unto them, he put clay upon mine eyes, and I washed, and I do see.

48 Goetting L., "Greek Textual Influence on Gothic Complex Verbs with Pleonastic Prepositions", *Journal of Germanic Linguistics* 19 (2007) 303–347, 314.

49 West J., "The Semantics of Preverbs in Gothic", *Indogermanische Forschungen* 87 (1982) 138–165, 138–139.

The preverb *bi-* in *biþwahan* is the preposition 'by, around, at, near' and can impart a perfective function to the verb, as is the case in John 9:11 where the participle *biþwahands* translates νιψάμενος:

> Καὶ εἶπεν μοι ὅτι ὕπαγε νίψαι εἰς τὸν σιλωάμ καὶ νίψαι: ἀπῆλθῶν οὖν καὶ νιψάμενος ἀνέβλεψα.
>
> Jah qaþ mis: gagg afþwahan in þata swumfsl Siloamis; iþ ik galaiþ jah biþwahands ussaƕ.
>
> And he said unto me, go to the pool of Siloamm, and wash: and I went and washed, and I received sight.

As already stated, imitation of Greek syntactic patterns was common in late antique translations of the Christian Sacred Word. In particular instances (involving aspect, for example) the Gothic rendering seems based more on the exigencies of the context than on the Greek *Vorlage*, for example in the use of aspectually charged prefixes that imply perfectivity.

The simplex *þwahan* is also used also in Mark 7:3 (*þwahand handuns* 'they wash their hands'), Matthew 6:17 (*ludja þeina þwah* 'wash thy face'), and I Timothy 5:10 (*weihaim fotuns þwohi* 'if she has washed the saints' feet'). In John 13:12 and John 13:14 Jesus washes his disciples' feet; in these Johannine passages νίπτειν is translated as *þwahan,* but, in the perfective forms, the Greek verb is rendered as *usþwahan.* The preverb *us-* in *usþwahan* is the preposition 'by, around, at, near'; it has also a perfective aspect and a resultative function, as in John 13:12:

> ὅτε οὖν ἔνιψεν τοὺς πόδας αὐτῶν [...]
>
> Biþeh þan usþwoh fotuns ize [...]
>
> So after he had washed their feet [...]

And John 13:14:

> Εἰ οὖν ἐγὼ ἔνιψα ὑμῶν τοὺς πόδας, ὁ κύριος καὶ ὁ διδάσκαλος, καὶ ὑμεῖς ὀφείλετε ἀλλήλων νίπτειν τοὺς πόδας,
>
> Jabai nu <ik> usþwoh izwis fotuns, frauja jah laisareis, jah jus skuluþ izwis misso þwahan fotuns.

> If then, your Lord and Master, have washed your feet, ye also ought to wash one another's feet.

On the other hand, Gothic *usþwahan* translates two different verbs: νίπτω/νίζω in John 13:12 and John 13:14 (Jesus washes his disciples' feet) and πλύνω (the sailors are washing the fishing nets) in Luke 5:2. While the Greek verb νίπτω/νίζω means 'wash the hands or feet, purge, cleanse', Greek πλύνω is used for the washing of clothes (or fishing nets).

In the Gospel of Luke Jesus proclaims his message of grace and salvation for sinners. In Luke 7:38–44, Jesus accepts the foot-washing of a prostitute, symbolising the salvation he was extending to all mankind. When Jesus praises the woman who, in the house of Simon, washes his feet with her tears, the Greek verb is βρέχειν 'moisten' rendered by the inchoative verb *natjan* when the woman is beginning to perform the foot-washing (*dugann natjan fotuns is tagram* 'she began to wet his feet with her tears'), but rendered by the perfective verb *ganatjan* after the action has been accomplished (*si tagram seinaim ganatida meinans fotuns* 'she wet my feet with her tears'). Foot-washing was a form of hospitality and a form of servitude. The host furnished the water, and the guests washed their own feet, but, in the richer houses the washing was done by a slave. Jesus contrasts the host's neglect in not even giving him water for his feet to the woman's washing them with her tears and wiping them with her hair. The Gospel of John focuses on Jesus's self-revelation as the eternal Son of God and portrays Jesus in his divinity as the 'living water', which is the source of life and salvation. The theme of universal salvation is at the heart of the Gospel of Luke, which portrays Jesus as the Messiah who is sovereign Lord and redeeming Saviour,[50] and pays particular attention to the outcasts of Jewish society, including women. While John's is the gospel of the glory of Christ, Luke's emphasis is on Jesus's solidarity with the marginalised and outcast and on his authority to forgive, symbolised by the tears of the woman who threw herself to his feet, washed them with her tears, and thereby found forgiveness.

As we have seen, the verb βρέχειν 'moisten' has been translated with *natjan* (in the infinitive) in Luke 7:38 and *ganatjan* (in the past form *ganatida*) in the following Luke 7:44:

Luke 7:38

50 Squires J.T., "The Gospel according to Luke", in Barton S.C. (ed.), *The Cambridge Companion to the Gospels* (Cambridge: 2021) 158–188, 170.

> Καὶ στᾶσα ὀπίσω παρὰ τοὺς πόδας αὐτοῦ κλαίουσα, τοῖς δάκρουσιν ἤρξατο βρέχιν τοὺς πόδας αὐτοῦ καὶ ταῖς θριξὶν τῆς κεφαλῆς αὐτῆς ἐξέμασσεν, καὶ κατεφίλει τοὺς πόδας αὐτοῦ καὶ ἤλειφεν τῷ μύρῳ.
>
> Jah standandei faura fotum is aftaro greitandei, dugann natjan fotuns is tagram jah skufta haubidis seinis biswarb jah kukida fotum is jah gasalboda þamma balsana.
>
> And stood at his feet behind him weeping, and began to wash his feet with tears, and did wipe them with the hairs of her head, and kissed his feet, and anointed them with the ointment.

Luke 7:44

> [...] εἰσῆλθόν σου εἰς τὴν οἰκιάν, ὕδωρ μοι ἐπὶ πόδας οὐκ ἔδωκας: αὕτη δὲ τοῖς δάκουσιν ἔβρεξέν μου τοὺς πόδας καὶ ταῖς θριξὶν αὐτῆς ἐξεμάζεν.
>
> [...] Ataggandin in gard þeinana wato mis ana fotuns meinans ni gaft; iþ si tagram seinam ganatida meinans fotuns jah skufta seinamma biswarb.
>
> [...] I entered into thine house, thou gavest me no water for my feet; but she haths washed my feet with tears, and wiped them with her hair.

Given the perfectivising function of the preverb *ga-*, the formal shift from simplex to *ga-* bears a relation to context and verbal set, expressing an aspectual contrast. *Natjan* in Luke 7:38 is the prior or governing form in a two-member sequence where *ganatjan* is combinative. The preverb *ga-* in *ganatjan* functions as a marker of temporal differentiation of the tense incidence of the combinative in relation to its prior form; the incidence of the *ga*-form is posterior to that of the prior (or governing) form. The use of simplex and *ga*-form in this sequence reveals an aspectual sophistication on the part of the Gothic translator.[51]

4 Concluding Remarks

Building upon textual evidence, this chapter provides but a glimpse of the ways in which water and religious symbolism came to be linked in the Greek

51 Scherer P., "The Theory of the Function of the Gothic Preverb *Ga-*", *Word* 20 (1964) 222–245, 231.

New Testament, and the ways in which the Gothic version, in turn, manages to fuse the linguistic, the metaphorical and the symbolic perspectives found in the source text.

The texts under discussion are theological texts, where the discourse often follows parallels, symbolism and allegories; water is an archetypal symbol. The various New Testament texts reveal different environmental imaginations. The synoptic gospels encourage their readers to perceive water (and the wind) as part of a landscape full of dangers, depicting floods, heavy rain, sea storms; and the sea and rivers of Paul were threats that travelers would fear. In the synoptic gospels, water is an element of nature over which Jesus reveals his mastery, as it illustrates his control over the chaotic elements of nature; a number of passages in the synoptics and in the Pauline epistles are concerned with natural disasters caused by water. On the other hand, water has a spiritual meaning in the Gospel of John, where 'living water' is a principle of the hidden spiritual life and symbolically it refers to the Spirit. Symbolism predominates in the Johannine passages concerning water. Symbolism, value, and meaning are especially pertinent in the religious aspect of life; in a symbolic reference, the material may represent the immaterial, since words are used with an associated meaning.

As David Sopher, a scholar working in the field of cultural geography, suggests, 'as society and economy become more complex, symbolisation and abstraction of ecological matter increase';[52] as specific landscape features are invested with meanings of a religious kind, an appreciation of the Gothic rendering of the natural and symbolic perspectives may give insights into the processes of translation, to arrive at a larger picture of biblical Gothic.

Regarding the linguistic ways in which the link between geographical imagery and religious symbolism came to be expressed in biblical Gothic, it is time to return again to the opening statements about translation, which constitutes an analytical operation involving interpretation and transformation. The Gothic translation was propelled by the need to disseminate the Christian message. In the time of Wulfila, the Gothic Christian language (the so-called biblical Gothic) was at an early stage of development; the translator could not rely on any previous literary tradition, since Gothic was not a written language before Wulfila, meaning that he had to face a challenge in translating. He coined a Christian terminology in Gothic (above all calques) in order to express the Greek Christian culture, since the range of words of Koine Greek, along with the prestige that this source language enjoyed, was much larger than that of Gothic. In this respect, the linguistic need emerged as a result of

52 Sopher, *Geography of Religions* 19.

the influence that Christian culture and the Greek language had over Gothic culture and language, which reacted to the impact of the new terms introduced by the new message by filling in the blanks with, above all, loan meanings (and loan translations as well). These were created by a method by which a word in the recipient language developed a wider semantic range to accommodate the Christian meaning of the word, extending the semantics of secular terms:[53] this is the case of *wato* 'water' as a symbol for Christ, and of *daupjan*, which developed the Christian meaning 'baptise'. *Daupjan* exemplifies how common words could be used to cover theologically complex meanings, since conceptual clarity is the overriding interest of the Gothic translation.[54] Translation procedures comprise, above all, calques: the choice of the semantic extension reveals the accuracy of the translation work, resorting to the borrowing of a foreign term only when concepts cannot be rendered with a native word.[55]

An outstanding characteristic of the Wulfilian technique of translation, although not so rigidly maintained, is the uniformity with which the same Greek word has been translated by the same Gothic word, whenever the sense permitted. In any case, a number (though not a high number) of Greek words are represented by dual and, apparently, synonymous renderings; it is open to question as to whether many are due to the later revisers and scribes. Examples of dual renderings are *saiws* and *marisaiws* translating Greek λίμνη 'lake' in the Gospel of Luke. Another case of multiple rendering of one Greek word is represented by the verb νίπτω, which is translated as *þwahan* and its prefixed forms *afþwahan*, *biþwahan* and *usþwahan*. On the other hand, the same Gothic term *wegs* translates three different Greek words, σεισμός 'quake' in Matthew, κλύδιον 'tempest, raging waves' in Luke and (in the plural) κῦμα 'wave' in Matthew and in Mark. We have seen that *usþwahan* translates two different Greek verbs: νίπτω/νίζω (John 13:12; and John 13:14) and πλύνω (Luke 5:2).

As we draw the results of this research together, it is important to emphasise the role played by the norms pertaining to the translation of the Sacred Text in late Antiquity (and the early Middle Ages), a constraint of a kind that affects the translation process and its final result. The key to the literalism of the Gothic version lies in its approach to translating the Sacred Text, an approach

53 Mîrșanu D., "For Want of a Better Word: Religious Terms in the Gothic Bible in Relation to Greek and Latin", *Biblicum Jassiense* 2 (2011) 30–45, 33–34.

54 Costa E., "Traduzione tra le remore dell'ufficialità e le libere (quasi) invenzioni", *Servitium*, III serie, 150 (2003) 636–642, 639.

55 Scardigli P., *Die Goten. Sprache und Kultur* (Munich: 1973) 124. See also Götti E., *Die gotischen Bewegungsverben* (Berlin – New York: 1974); Kotin M., "Das Weltbild und das Sprachbild in der gotischen Bibelübersetzung", in Sroka K.A. (ed.), *Kognitive Aspekte der Sprache. Akten des 30. Linguistischen Kolloquiums, Gdansk 1995* (Tübingen: 1996) 141–145.

consistent with late antique ideas concerning the translation of the Bible. As recognised by Friedrichsen, the nature of the translation technique employed by Wulfila is a lexeme-to-lexeme translation of the Greek text: 'The fundamental principle underlying the translation is the systematic correspondence of the Gothic text with the Greek, word for word, and in precisely the same order'.[56] The influence of Greek is pervasive: the Gothic version has the tendency to replicate the structure of the Greek source. The social conditions under which the translation is carried out must be taken into consideration, since the Gothic Bible is a functionally communicative translation within its cultural and religious context.

The adherence to Greek is evident in the Gothic ordering of the elements, as it was the case with early translations of the Bible, committed as they were to the replication of the linguistic norms of the Greek model.[57] The principle of structural equivalence observed in the Gothic version can be seen in all the early translations; literalness is, for example, a characteristic of the Latin versions as well. The Goths were committed to preserve the sacrality (the 'mystery') which, as stated by Jerome in his *Letter to Pammachius*, is manifest in the very order of those words; in the same letter, dated 395 and entitled *De optimo genere interpretandi* ('On the best method of translating'), the great Christian translator of the Vulgate advocates a sense-for-sense strategy, making however Bible translation an exception, by reason of its sacredness.[58]

Religion and language constitute the foundations of institutions that allow the structuring of societies. As we inquire into the cultural significance of the Gothic Bible, and into the dialectic between religion and language, the Gothic version should be considered as a complex and specific response of a Germanic language to cultural and religious development. Given the complex nature of sacredness, the incorporation of the Christian symbolic elements of nature into the fabric of the Gothic language is imbued with cultural meaning. The Gothic translator had a role in mediating ideas across linguistic and cultural boundaries, and, due to the frequent difficulties with the absence of equivalents (a vocabulary 'gap'), his task involved creating a new vocabulary to convey the message of the source text. In the terms of Berman's 'circular approach', a Gothic Christian identity was being constructed by means of translation.

Finally, it should be noted that a word of caution is called for when assessing the characteristics of the Gothic Bible, since there is no established method of

56 Friedrichsen, *The Gothic Version of the Gospels* 15.

57 Nida, *Towards a Science of Translating* 12.

58 Bartelink G.J.M., *Hieronymus, De Optimo Genere Interpretandi – Epistula 57: Ein Kommentar* (Leiden – Boston, MA: 1980).

unravelling the processes of translation and the complex literary relationships which result from them. A perusal of the Gothic passages concerning water and semantically related terms brings to light the problems associated with an examination of biblical Gothic, notably the fragmentary nature of the remains, since only portions have been preserved, and the possibility of scribal modifications during transmission; the uncertainty of the Greek type of text used as a source is another obstacle that carries a warning. Thus, it must be borne in mind that the reason behind the lexical choices is difficult to assess: a wise analysis will be cautious about sweeping generalisations.

Appendix: the Semantic Field of 'Water' in Gothic

According to ancient Greek philosophical categories, the four elements are air, water, earth and fire. The first three correspond to the three forms of matter and the fourth to energy. The following terms pertaining to the semantic field of water appear in the Gothic manuscripts, though we should note that, since a number of passages of the New Testament are lost from the Gothic version, not all the occurrences of Greek ὕδωρ are represented in the extant manuscripts. As will be seen, even though some derivations and semantic developments are uncertain and disputed, the majority of the words in this semantic field can be given satisfactory etymologies.

Aƕa is a feminine noun translating Greek ποταμός 'river, waters' (Matthew 7:25; 7:27; John 7:38; Luke 6:48; 6:49; Mark 1:5; 2 Cor. (B) 11:26). It goes back to Proto-Germanic **ahwō*, 'water, stream, river' from which Old Icelandic ø, Old English *ēa*, Old Saxon, Old High German *aha* are derived. The only clear cognate in an Indo-European language is the Latin *aqua*. There has been much speculation on further possibly related forms, especially in river names, such as Lithuanian Akelé and Illyrian Aquilo, and in the Venetic name of the city of Aquileia.[59]

Brunna (m.) is a *hapax* attested in the expression *brunna bloþis* 'flow of blood' and translates Greek πληγή 'fount, source' in Mark 5:29. It is attested in Crimean Gothic as *brunna* 'spring, well' and is a cognate of Old English *brunna, burna* (Scots *burn*), Old Saxon, Old High German *brunno*, Old Icelandic *brunnr* 'source, fount'. The Proto-Germanic root is **brunno* 'fount, spring', from Proto-Indo-European **bhrun* 'fountain, wellspring, source' (from **bh*(*e*)

59 Lehmann W.P.A., *A Gothic Etymological Dictionary, Based on the Third Edition of Vergleichendes Wörterbuch der Gotischen Sprache by Sigmund Feist* (Leiden – Boston, MA: 1986) 12.

r- 'move rapidly'), also in Old Irish *tipra* 'source, fount', Armenian *atbiur* 'fount', and Russian *bruja* 'current'.[60]

The weak verb *daupjan* translates βαπτίζειν,[61] while the prefixed form *ufdaupjan* translates βαπτίζαι 'baptise' (John 13:26; Luke 3:21; 7:29; Mark 16:16) and ἐμβάψαι 'dip in' (John 13:26). Related forms are the present participle (with the grammatical function of a noun) *Daupjands* (m.) 'Baptist'[62] and the abstract noun *daupeins* (f.) 'baptism' (βαπτίσμα).[63] Cognates in other Germanic languages are Old Norse *deypa* 'dip', Old English *diepan*, *dyppan* 'dip in' (more rarely 'baptise'), Old High German *touffan* 'baptise' and Old Saxon *dōpjan*. The Proto-Germanic root is **deup*- 'deep', from Proto-Indo-European **dheub*- 'deep' (also in Lithuanian *dubus*, Welsh *dwfn* 'deep').[64]

The *hapax flodus* (m.) translates Greek ποταμός 'stream, river' in Luke 6:49. Germanic cognates are Old English *flōd*, Old Norse *flóð*, Old High German *flut* 'a flowing of water, mass of water, river'. The Proto-Germanic form is **flōduz*, from the Proto-Indo-European root **pleu*- 'to flow' (Greek πλίειν 'swim', Latin *fluctus* 'wave', *flumen* 'river', Ancient Slavonic *plavati* 'swim').[65]

The feminine noun *garunjo* 'flood' translates πλημμύρης; a *hapax* attested only in the Gospel of Luke (6:48), it is a *-jon-* suffix derivative of the verb *ga-rinnan* 'to (over)flow', cognate with Old English, Old High German, Old Saxon *rinnan* 'to run, to flow'.

Marei (f.) translates θάλασσα 'sea'.[66] Germanic cognates are Old Norse *marr* 'sea'; Old English *mere*, 'sea, inland sea, pond'; Old High German *meri* 'sea' and, with lengthened grade, Old Norse *mœrr* 'swampland'; Old English *mōr* 'moor, heath'; Old High German *muor* 'moor'. The terms derive from Proto-Germanic **marīn*- (f.), **marī*- (n.), whose base meaning is 'standing body of water' (inland or coastal), from the same Proto-Indo-European root **mar-y*, **mor-y-n* as in Latin *mare*. *Mari*- is the first element in the Gothic compound *marisaiws*, whose second element is *saiws*. *Marisaiws* (m.) is attested only in chapter 8 of Luke (8:22; 8:23; 8:33) and translates λίμνη 'lake'. *Saiws* (m.) also translates λίμνη

60 Lehmann, *A Gothic Etymological Dictionary* 82.

61 John 10:40; Luke 3:4; 3:7; 3:12; 3:16; 3:21; 7:30; Mark 1:4; 1:5; 1:8; 1:9; 7:4; 10:38; 10:39; 1 Cor. (A) 1:13; 1:14; 1:15: 1:16; 1:17; 10:2; 12:13; 15:29; *Skereins* 2:4, 3:1; 3:4; 3:11; 4:1; Gal. (A) 3:27.

62 Matthew 11:11; 11: 12; Luke 7:20; 7:28; 7:33; 9:19; Mark 6:14; 6:24; 6:25; 8:28.

63 Luke 3:3; 7:29; 20:4; Mark 1:4; 7:4; 7:8; 10:38; 10:39; 11:30; Eph. (A) 4:5; Eph. (B) 4:5; *Skereins* 2: 3; 2:4; 3:1; 3:2; 3: 3; 3:4; Col. (B) 2:12.

64 Lehmann, *A Gothic Etymological Dictionary* 89.

65 Lehmann, *A Gothic Etymological Dictionary* 118–119.

66 Matthew 8:18; 8:24; 8:26; 8:27; 8:28; 8:32; John 6:1; 6:16; 6:17; 6:18; 6:19; 6:22; 6:25; Luke 6:17; 17:6; Mark 1:16; 2:13; 3:7; 4:38; 4:39; 4:41; 5:1; 5:13; 5:21; 7:31; 8:13; 9:42; 11:23; 1 Cor. (A) 10:1; 10:2; 2 Cor. (B) 11:25; 11:26; Rom. A 9:27. See Lehmann, *A Gothic Etymological Dictionary* 245.

in the Gospel according to Luke 5:12, while in the Deed of Naples (1:1; 2:34) its meaning is 'marshland'. Germanic cognates are Old English *sæ* 'sea, lake, pool', Old High German *seo* 'sea', Old Norse *sær* 'sea'. The Proto-Germanic root is **saiwa-*, of unknown origin.

Both *natjan* (attested only in Luke 7:38) and *ganatjan* (attested only in Luke 7:44) are weak verbs and translate βρέχειν 'moisten',[67] from Proto-Germanic **nata-* as in Old Saxon *nat*, Old High German *naz* 'wet, rain', *nezzen* 'to moisten'. Etymology is unclear, but it possibly derives from the Proto-Indo-European root **ned/t-* and is related to Sanskrit *nadī* 'river', to Greek νοτός 'southwest wind which brings moisture' and its derivation νότιος 'humid', and to Latin *natare* 'swim'.

The *hapax qrammiþa* (f.), attested only in Luke 8:6 (in the Parable of the Sower, referring to the seed fallen in the earth without moisture), translates ἰκμάς 'moisture'[68] and is related, among the Germanic languages, only to Old Norse *krammr* 'damp, sticky wet', and, said of snow, 'partly thawed'. The noun is derived from the same Proto-Indo-European root of Latin *gramiae*, 'viscous humour that collects in the corner of the eye', and of Lithuanian *grimsti* 'sink under (water)'.

The noun *rign* (n.) translates βροχή 'rain'[69] and the weak verb *rignjan* translates βρέχειν 'to rain'. The noun survived in Crimean Gothic *reghen* 'rain'.[70] Germanic cognates are Old Norse *regn*, Old English *regn*, Old High German *regan*, 'rain'. Etymology is uncertain: possibly from Proto-Indo-European **rēg/k-nó* as in Latin *rigare* 'moisten, irrigate'.

Swumfsl (n.) translates κολυμβήθρα 'pool' in John 9:11 and, in the form of the manuscript variant reading *swumslf* in the proximate context John 9:7. It is derived with the infix *-slo-* from Proto-Germanic **swimman* 'swim', whose etymology remains uncertain, possibly from Proto-Indo-European **swe-* 'move'.[71]

Þwahal translates λουτρόν 'laver, bath' (in Ephesians 5:26; *Skeireins* 2:8) in the expression τῷ λουτρῷ τοῦ ὕδατος, *þwahla watins* ('by the laver of the water'). Here we have a reference to baptism, and an allusion to the bath of the bride before the wedding day. The noun is cognate to Old Icelandic *þvál* 'soap'; Old

67 Lehmann, *A Gothic Etymological Dictionary* 264.
68 Lehmann, *A Gothic Etymological Dictionary* 279.
69 Matthew 7:25; 7:27.
70 The spelling may reflect Middle Dutch *reghen*, since it was the Flemish Ambassador Ogier de Busbecq who published in 1589 a list of Crimean Gothic words he had heard on a visit to Crimea; his transcriptions were influenced by his own language. See Lehmann, *A Gothic Etymological Dictionary* 284.
71 Lehmann, *A Gothic Etymological Dictionary* 337.

English *þweal* 'washing, bath, soap', and Old High German *dwahal* 'bath'.[72] The etymology is the same as *þwahan.*

The weak verb *þwahan*[73] and its prefixed forms *afþwahan* (John 9:7; 9:11; 9:15) *biþwahan* (John 9:11) and *usþwahan* (John 13:12; 13:14; Luke 5:2) translate νίπτειν 'to wash'; *usþwahan* also translates πλύνω ('to wash', said of fishing nets in Luke 5:2). The past participle *unþwahanaim* (formed with the privative prefix *un-*) renders ἀνίπτοις and κοιναῖς ('unwashed')[74]. Germanic cognates are Old Norse *þvá* 'to wash', Old English *þwean* 'to wash, anoint', and Old High German *thwahan* 'to wash'. No certain etymology has been suggested. Related to the verb is the noun þwahl (see above).[75]

Wato translates ὕδωρ 'water',[76] from Proto-Germanic **watr-*, source also of Old Norse *vatn*, Old English *wæter*, Old Saxon *water* and Old High German *wazzar*, all of them meaning 'water'. The Proto-Indo-European root is **wed-* 'moisten', the same as in the Greek word.[77]

Wegs translates σεισμός 'quake' in Matthew (8:24), κλύδιον 'raging waves' in Luke (8:24) and (in the plural) κῦμα 'wave' in Matthew (8:24) and Mark (4:37). The Gothic noun is cognate of Old Norse *vágr* 'sea', Old English *wæg*, Old Saxon and Old High German *wag* 'waves, rushing waters'. Related to *wegs* are the Gothic verbs **ga-wigan* 'to move, shake', and *wagjan* 'to move' (both translate the Greek verb σαλεύειν). The Proto-Germanic root is **wag-*, from Proto-Indo-European **wegh-* 'to move'.[78]

Wis (Matthew 8:26; Luke 8:24; Mark 4:39) translates γαλήνη 'calm (of the sea)'. The Latin versions translate the Greek term as *tranquillitas*. *Wis* is the only Gothic term pertaining to the semantic field of 'water' with no direct cognates in the other Germanic dialects. Etymology is disputed: possibly derived from the verb *wisan* 'to be'[79] and related to the Gothic term for 'nature', *wists*.

72 Lehmann, *A Gothic Etymological Dictionary* 369.

73 Mattthew 6:17; John 9:7; 13:14; Mark 7:3; 1 Tim. (A) 5:10; 1 Tim. (B) 5:10.

74 Luke 7:2 and Luke 7:5 respectively.

75 Lehmann, *A Gothic Etymological Dictionary* 369.

76 Matthew 8:32; 10:42; John 7:38; Luke 3:16; 7:44; 8:24; 8:25; Mark 1:8; 1:10; 9:22; 14:13; Eph. (A) 5: 26; 1 Tim. (A) 5:23; 1 Tim (B) 5:23; *Skeireins* 2:3; 2:4; 3:3; 3:4.

77 Lehmann, *A Gothic Etymological Dictionary* 395.

78 Lehmann, *A Gothic Etymological Dictionary* 397.

79 Feist, *Vergleichendes Wörterbuch der Gotischen Sprache* (Leiden: 1939) 567; Lehmann, *A Gothic Etymological Dictionary* 405.

Bibliography

Primary Sources

Bennett W.H. (ed.), *The Gothic Commentary on the Gospel of John. Skeireins aiwaggeljons þairh Iohannen: A Decipherment, Edition and Translation* (New York: 1960).

Bidez J. – Hansen G.C. (eds.), *Sozomenus, Kirchengeschichte*, 2. durchges (Berlin: 1995).

Bidez J. – Winkelmann F. (eds.), *Philostorgius, Kirchengeschichte*, 3. durchges (Berlin – New York: 1981).

Finazzi R.B. – Tornaghi P. (eds.), "*Gothica Bononiensia*. Analisi linguistica e filologica di un nuovo documento", *Aevum* 87 (2013) 113–153.

Hansen G.C. – Širinjan M. (eds.), *Sokrates. Kirchengeschichte* (Berlin: 1995).

Mommsen T. (ed.), *Iordanis Romana et Getica,* Monumenta Germania Historiae Auctores antiquissimi V, 1 (Berlin: 1882).

Streitberg W. (ed.), *Die gotische Bibel, I. Der gotische Text und seine griechische Vorlage. Mit Einleitung, Lesarten und Quellennachweisen sowie den kleineren Denkmälern als Anhang. 7. Aufl. Mit einem Nachtrag von Piergiuseppe Scardigli* (Heidelberg: 2000).

von Friesen O. – Grape A. (eds.), *Codex Argenteus Upsaliensis jussu Senatus Universitatis phototypice editus* (Uppsala: 1928).

Secondary Sources

Bartelink G.J.M., *Hieronymus, De Optimo Genere Interpretandi – Epistula 57: Ein Kommentar* (Leiden – Boston, MA: 1980).

Barton S.C., "Introduction", in Barton S.C. (ed.), *The Cambridge Companion to the Gospels* (Cambridge: 2006) 1–9.

Berman A., *The Experience of the Foreign: Culture and Translation in Romantic Germany* (Albany, NY: 1992).

Brown S., "Water Imagery and the Power and Presence of God in the Gospel of John", *Theology Today* 72 (2015) 289–298.

Bruce F.F., *The Gospel of John. Introduction, Exposition, and Notes* (Basingstoke: 1983).

Carlson E., "The Gothic Vocabulary of Fear", *Journal of English and Germanic Philology* 111 (2012) 285–303.

Cooper A., "New Directions in the Geography of Religion", *Area* 24 (1992) 123–129.

Costa E., "Traduzione tra le remore dell'ufficialità e le libere (quasi) invenzioni", *Servitium*, III serie, 150 (2003) 636–642.

Culpepper R.A., *Anatomy of the Fourth Gospel: A Study in Literary Design* (Philadelphia: 1983).

Dean-Drummond C., *Eco-Theology* (Winona, MN: 2008).

de Nie G., "The Spring, the Seed, and the Tree: Gregory of Tours on the Wonders of Nature", *Journal of Medieval Studies* 11 (1985) 89–135.

Dodd C.H., *The Interpretation of the Fourth Gospel* (Cambridge: 1953).

Drinka B., "The Sacral Stamp of Greek: Periphrastic Constructions in New Testament Translations of Latin, Gothic, and Old Church Slavonic", in Welo E. (ed.), *Indo-European Syntax and Pragmatics: Contrastive Approaches*, Oslo Studies in Languages 3 (Oslo: 2011) 41–73.

Falluomini C., *The Gothic Version of the Gospels and Pauline Epistles* (Berlin – Boston, MA: 2015).

Feiler P.F., "The Stilling of the Storm in Matthew: A Response to Gunther Borkamm", *Journal of Theological Studies* 26 (1983) 399–406.

Feist S., *Vergleichendes Wörterbuch der Gotischen Sprache* (Leiden: 3rd ed. 1939).

Fickler P., "Fundamental Questions in the Geography of Religion", in Wagner P.L. – Mikesell M.W. (eds.), *Readings in Cultural Geography* (Chicago: 1962) 121–144.

Francini M., "L'edizione della Bibbia gotica tra testo vivente e tradizione quiescente", in Ferrari F. – Bampi M. (eds.), *Storicità del testo, storicità dell'edizione* (Trento: 2009) 313–340.

Friedrichsen G.W.S., *The Gothic Version of the Gospels: A Study of its Style and Textual History* (London: 1926; rev. ed. 1939).

Friedrichsen G.W.S., *Gothic Studies* (Oxford: 1961).

Gareiou Z. – Zervas E., "Analysis of Environmental References in the Texts of the Four Gospels Using Descriptive Statistics", *Religions* 9 (2018) 1–14.

Goetting L., "Greek Textual Influence on Gothic Complex Verbs with Pleonastic Prepositions", *Journal of Germanic Linguistics* 19 (2007) 303–347.

Götti E., *Die gotischen Bewegungsverben* (Berlin – New York: 1974).

Green D.H., *Language and History in the Early Germanic World* (Cambridge: 1998).

Green D.H., "Linguistic and Literary Traces of the Ostrogoths", in Barnish S.J. – Marazzi F. (eds.), *The Ostrogoths. From the Migration Period to the Sixth Century. An Ethnographic Perspective* (Woodbridge: 2007) 387–416.

Harrington D.J., *The Gospel of Matthew* (Collegeville, MN: 1991).

Hultkrantz Å., "An Ecological Approach to Religion", *Ethnos* 31 (1966) 131–150.

Jellinek M.H., *Geschichte der gotischen Sprache*, Grundriss der Germanischen Philologie 1,1 (Berlin: 1926).

Jordan-Bychkov T.G. – Domosh M. – Rowntree L., *The Human Mosaic: A Thematic Introduction to Cultural Geography* (New York: 1994).

Joubert J., "Johannine Metaphors/Symbols Linked to the Paraclete-Spirit and their Theological Implications", *Acta Theologica* 27 (2007) 83–103.

Klostermann E., *Das Matthäusevangelium* (Tübingen: 1927).

Kluge F., "Gotische Lehnwörter in Althochdeutschen", *Beiträge zur Geschichte der deutschen Sprache und Literatur* 35 (1909) 124–160.

Koester C., *Symbolism in the Fourth Gospel: Meaning, Mystery, Community* (Minneapolis, MN: 2003).

Kong L., "Geography and Religion: Trends and Prospects", *Progress in Human Geography* 14 (1990) 355–371.

Kotin M., "Das Weltbild und das Sprachbild in der gotischen Bibelübersetzung", in Sroka K.A. (ed.), *Kognitive Aspekte der Sprache. Akten des 30. Linguistischen Kolloquiums, Gdansk 1995* (Tübingen: 1996) 141–145.

Lefevere A., *Translation, Rewriting, and the Manipulation of Literary Fame* (London: 1992).

Lehmann W.P., *A Gothic Etymological Dictionary, Based on the Third Edition of Vergleichendes Wörterbuch der Gotischen Sprache by Sigmund Feist* (Leiden – Boston, MA: 1986).

Metzger B.M., *The Early Versions of the New Testament: Their Origin, Transmission, and Limitations* (Oxford: 1977).

Mîrșanu D., "For Want of a Better Word: Religious Terms in the Gothic Bible in Relation to Greek and Latin", *Biblicum Jassiense* 2 (2011) 30–45.

Nida E., *Towards a Science of Translating, With Special Reference to Principles and Procedures Involved in Bible Translating* (Leiden: 1964).

Park C., *Sacred Worlds. An Introduction to Geography and Religion* (London – New York: 1994).

Psaltes S.B., *Grammatik der Byzantinischen Chroniken* (Göttingen: 1913).

Ratkus A., "Greek ἀρχιερεύς in Gothic Translation: Linguistics and Theology at a Crossroads", *NOWELE* 71 (2018) 3–34.

Reiss K. – Vermeer H.J., *Grundlegung einer allgemeinen Translationstheorie* (Tübingen: 1984).

Scardigli P., *Die Goten. Sprache und Kultur* (Munich: 1973).

Scherer P., "The Theory of the Function of the Gothic Preverb *Ga-*", *Word* 20 (1964) 222–245.

Schneiders S., *Written That You May Believe: Encountering Jesus in the Fourth Gospel* (New York: 2003).

Sedlak D., *Water 4.0: The Past, Present and Future of the World's Most Vital Resource* (New Haven, CT – London: 2014).

Semple E., *Influences of Geographic Environment: On the Basis of Ratzel's System of Anthropo-Geography* (New York – London: 1911).

Sopher D.E., *Geography of Religions* (Englewood Cliffs, NJ: 1967).

Squires J.T., "The Gospel According to Luke", in Barton S.C. (ed.), *The Cambridge Companion to the Gospels* (Cambridge: 2021) 158–188.

Stolzenburg H., *Zur Übersetzungstechnik des Wulfila* (Halle: 1905).

Stutz E., *Gotische Literaturdenkmäler* (Stuttgart: 1966).

Tjäder J.O., *Die Nichtliterarischen Papyri Italiens aus der Zeit 445–700. III. Tafeln* (Lund: 1954).

Toury G., "The Nature and Role of Norms in Translation", in Toury G. (ed.), *Descriptive Translation Studies and Beyond* (Amsterdam: 1995) 53–69.

Van der Watt J., *Family of the King. Dynamics of Metaphor in the Gospel According to John* (Leiden – Boston – Cologne: 2000).

Velten H.V., "Studies in Gothic Vocabulary with Especial Reference to Greek and Latin Models and Analogues", *Journal of English and Germanic Philology* 29 (1930) 489–509.

Venuti L., *The Translator's Invisibility: A History of Translation* (London – New York: 1995).

West J., "The Semantics of Preverbs in Gothic", *Indogermanische Forschungen* 87 (1982) 138–165.

Wolfe B.N., "The Relevance of Certain Semiticisms in the Gothic New Testament", *NOWELE* 71 (2018) 249–256.

Websites (last accessed: November 3, 2023) http://www.wulfila.be/

CHAPTER 9

Sweet and Salt Water in Medieval Law

Luca Loschiavo

1 A Preliminary Distinction

The Middle Ages inherited from Antiquity an absolute awareness of the primary importance of water. The *Etymologies* (*Etym.*) of Isidore of Seville – the encyclopaedia from which early medieval people continued to draw their knowledge for centuries – taught not only that the sea, together with the sky and the earth, makes up the universe (*Etym.* 13.I.1), but also that water, together with fire, governs every other element (*Etym.* 13.XII.3). Above all, Isidore insisted that, like air, water is an essential element for human life (*Etym.* 5.XXVII.38 and 13.XII.2).[1]

This awareness was of course very much present in the world of medieval law. When dealing with the great subject of water, the lawmakers of the Middle Ages (likewise those of today) started from the *summa divisio* between sweet waters (such as rivers and lakes, i.e. inland or terrestrial waters) and salt waters (those of the seas). The uses that people can make of them are obviously different, and so are the rights connected to these uses: rights that must both ensure the usefulness of the waters and prevent any sort of damage that might arise from using them.

In the Middle Ages, the sea was merely the surface on which people sailed and fished, facing great difficulties and equally great dangers (for medieval people, the underwater world belonged to the sphere of mystery and fantasy).[2] Sailing and fishing were certainly also possible on rivers and lakes. And yet, terrestrial waters – flowing and fresh – are of course also suitable for a number of other uses that are essential for the life and prosperity of medieval people.

We can therefore begin to consider rights concerning inland waters. The fundamental uses that only these waters (and not the sea) had can be divided into

1 For Isidore's *Etymologiae*, see Lindsay W.M. (ed.), *Isidori hispaliensis episcopi Etymologiarum sive Originum libri XX*, 2 vols. (Oxford: 1911).

2 See Karstens S., "Imagine Oceans: The Many Meanings of the High Seas in the Time of Early European Expansion" 196–224.

 | DOI:10.1163/9789004696501_011

three main categories: a) hygienic and potable use (for humans and animals);[3] b) use for agricultural irrigation; c) use as motive power. The law was (and still is) interested in these waters with regard to issues about ownership (dividing waters into public and private) and about the various uses they may have. In particular, the law tended (and still tends) to draw up disciplines for the different regimes of use, which can be collective (i.e., reserved only for the members of a particular community or, alternatively, free and common to all within certain limits), or individual and private (exclusive or shared with others) on the basis of a concession by the authority.

2 Roman Law

It is precisely in the field of water regulation that Roman Law, with its centuries-long reflection and experience, had reached a remarkable level of complexity and refinement, and it is useful to begin our survey with a short reminder of the principles which inspired it and of the solutions proposed by the jurists.[4] Even after the fall of the western Roman Empire, Roman law remained absorbed (somewhere more, somewhere less) in local customs. On the other hand, in the Byzantine East, the old norms continued to represent a precise and solid reference for centuries.

A preliminary question that Roman jurists asked themselves concerned the difference between public and private waters. In their reasoning this distinction was useful not only and not so much in order to ascertain the ownership of the domain, but rather to enable better regulation and discipline of the availability or use of resources. Classical jurisprudence – aware of not only the

3 Isidore (*Etym.* 20.III.1) explains that drinking water, unlike wine, must be fresh (*recens*): 'immediately available from the river or spring or well' since 'the old one does indeed stink'. In ancient sources, the concepts of *aqua viva*, water flowing from a spring, and *aqua profluens*, water naturally flowing downwards within a natural (a river, a stream) or artificial (canal or aqueduct) riverbed, are of course present. And running water, together with the air and the sea, was considered to belong to the *res communes* by natural right: see, for example, Dig. 1.8.2.1, 'And indeed by natural law the following belong in common to all men: air, flowing water, and the sea, and therewith the shores of the sea', trans. A. Watson, *The Digest of Justinian*, 4 vols. (Philadelphia: rev. ed. 2009) 1, 24. For abbreviations used in citing law codes, see References, below.

4 See especially Costa E., *Le acque nel diritto romano* (Bologna: 1919); Werner D. (ed), *Wasser für das Antike Rom* (Berlin: 1986); Bruun C., *The Water Supply of Ancient Rome. A Study of Roman Imperial Administration* (Helsinki: 1991); Hermon E. (ed.), *Vers une gestion intégrée de l'eau dans l'empire romain. Actes du colloque international: Université Laval – octobre 2006* (Rome: 2008).

economic but also the social value of water (as reflected, for example, in the importance of the baths in the daily life of every Roman city) – focused mainly on public waters. Considering public and collective interests as prevailing over purely individual and private ones, jurists widely recognised the public use of water resources.

Public waters were said to be *in usu populi*, i.e., generally available for free use, so that everyone could satisfy their basic needs (drawing water for drinking, washing, watering domestic animals, irrigating fields). It was the 'law of nature', i.e., the law concerning all living beings (*ius naturae*) or at least the law that all human beings have in common (*ius gentium*), that required that the use of public waters (and, if they were unavailable, also of private ones) should be free and open to all, even to those who were not Roman citizens. However, it was necessary to regulate this use in order to avoid waste, abuse and pollution. And it was the specific task of the jurists to design these regulations. It was only in the late imperial age that greater attention was paid to the fuller use of water with the intention of favouring both public and private economy (anticipating in many ways the solutions of modern legal systems).

But how can we distinguish between public and private waters? First of all – given the difficulty of accessing them and therefore of claiming possession of them – high mountain springs were classified as public, while springs coming from private land belonged to the owners (and jurists defined them as *pars o portio fundi*, 'part of the land').[5] As far as rivers and lakes were concerned, the criterion of perenniality (*perennitas*) was adopted: all perennial rivers and lakes were public, while streams that appeared dry in the summer months and canals with a low flow could be private.[6]

The public nature of rivers and lakes allowed everyone the free exercise of navigation and fishing,[7] as well as the possibility of watering animals and

5 See Dig. 43.24.11.pr.: 'Someone who pours something into a well so as to pollute the water with it is, Labeo says, liable under the interdict against force or stealth; for fresh water is *part of the land*, just as if someone had done a work connected with water', trans. Watson 4, 120); also Dig. 39.3.8; Dig. 39.3.21. See also Scott S.P. (trans.), Cod. Iust. (Cincinnati: 1932), 3.34.4 (a. 223): 'The Edict of the Praetor does not permit water, whose source is on the ground of another, to be conducted on the land of someone else, without the consent of him to whom the use of said water belongs'.

6 See Dig. 43.12.1.1–3: '1. A river is distinguished from a stream by size, or by the opinion of the surrounding inhabitants. 2. Some rivers are perennial, some torrential [...] 3. Some rivers are public, some not. Cassius defined a public river as a perennial one; this opinion of Cassius, which Celsus also approves, is held to be acceptable', trans. Watson 4, 92.

7 Navigation: Inst. 2.1.4 'The public use of the banks of rivers is also subject to the Law of Nations, just as the use of the river itself is; and hence anyone has a right to secure a vessel to them, to fasten ropes to trees growing there, or to deposit any cargo thereon, just as he has to

drawing water for irrigating fields (through the creation of canals or streams) on condition that the withdrawal was not so large as to make navigation impossible. More variable criteria – which often referred to the assessment of local populations – led to the qualification of ponds, marshes or other natural or artificial basins as public or private.[8]

The banks – the strips of sloping land from the level to the water surface – were in principle considered public.[9] Even in this case, however, the land could belong to private individuals, though its use had to be public[10] so that everyone could be allowed landing and mooring privileges for boats, as well as the temporary loading and unloading of goods (the rules were the same as those for the sea shore). However, the principle was flexible and allowed for exceptions: the state could reserve certain stretches of lake or river banks for exclusive use by means of a special concession. It was even accepted that, by means of the *longi temporis praescriptio* 'defence with long duration', the prolonged exercise over time of a certain use by a private individual gave rise to an exclusive right for that individual.[11]

To turn to private waters, Roman law granted the owner an unlimited right of use, including the widest powers of enjoyment and disposal (even allowing the possibility of completely removing the availability of water from a neighbour). Only with Justinian (527–565 CE) was the prohibition of immissions (i.e. actions principally aimed to cause harm to others) introduced. Nevertheless, private waters – especially those coming from perennial sources – could also

---

navigate the river itself; but the ownership of the same is in those whose lands are adjacent, and therefore the trees growing there belong to them'; fishing: Inst. 2.1.2 'All rivers and ports are also public, and therefore the right of fishing in a harbor or in streams is common to all', trans. Scott.

8 Dig. 43.14.1.3–6: '3. A lake is what has water perpetually. 4. A pool contains water for the time being which is stagnant. This kind of water usually collects in winter. 5. A canal is a receptacle for water dug by human hands. 6. These can all be public too', trans. Watson 4, 95; Dig. 8.3.23.1: 'If you grant me a watercourse across your land without designating which particular part of the land I am to channel the water across, your whole estate will be subject to the servitude', trans. Watson 1, 261.

9 Dig. 41.1.65.1: 'If some island in a river is your property, there is no public right. Paul: No; with islands of this kind, the river banks and seashore are public, no differently from that which applies to the mainland'; Dig. 43.12.3.pr.–1: 'Public rivers are those which are always flowing, and their banks are public. 1. A bank is held to be what contains a river at its fullest' (trans. Watson 4, 15 and 94 respectively).

10 See Dig. 41.1.30

11 Dig. 44.3.7: 'If for several years someone has fished by himself in a side stream of a public river, he may prevent another person from exercising the same right (trans. Watson 4, 143).

be the object of wide use by third parties.[12] Roman jurists therefore considerably developed the subject of the easements that the owner could create: the main one was the servitude of aqueduct (*servitus aquae ducti*), which included both the faculty of diverting water from another person's property and that of conducting it over another person's property.[13] Also very important was the *servitus aquae haustus* with which one could draw water by crossing another's land and thus exercise the right of way.[14] This easement could also be extended to animals (*servitus pecoris ad aquam adpellendi*).[15]

However, the most important issue, the one that gave rise to the most complex problems, was that of the channelling of public waters for private use (for domestic or recreational use, for agricultural irrigation, as motive power for mills or grinders, or for certain types of artisan production). Jurists endeavoured to reconcile the freedom of derivations (which stemmed from the aforementioned idea of water as a common good) and the necessary limitations that had to be imposed to ensure that individual use did not seriously harm the rights of other users in the present and future.

Derivations could above all concern public rivers and it seems that originally the rule was that of freedom of derivation.[16] Only in the late imperial age, and then with Justinian, did the opposite principle of the necessary concession apply, granted on a case by case approach by the public authority (see the text of Pomponius in Dig. 43.12.2, clearly adapted by the Justinian commissioners[17]).

---

12 See Dig. 8.3.9.

13 Definition in Dig. 8.3.1.pr.: 'Aquae ductus is the right to channel water across another's land' (trans., Watson 1, 258). See here Dig. 8.5.10.1: 'This action can be brought not only against the man on whose land the source of the water is situated or across whose land it is channeled, but, in fact, against anyone at all who prevents one from channeling the water, as is the case with servitudes in general. In short, by means of this action, I can proceed against anyone who tries to stop me from channeling the water' (trans. Watson 1, 270).

14 Dig. 8.3.3.3: 'A man who has a right to draw water is also presumed to have an iter, to enable him to exercise his right. As Neratius states […] if a man is granted the right to draw water and a right of access for the purpose, he will have them both; if he is granted only the right to draw water, a right of access is also included; if he is only granted access to a spring, a right to draw water is included. These rules apply to the drawing of water from a private spring. However, with regard to a public river, Neratius tells us […] that, although an *iter* to it must be granted, a right to draw water from it need not be, and if a man grants only the right to draw water from it, this grant achieves nothing' (trans. Watson 1, 259).

15 Dig. 8.3.4

16 Although with various limits: see for example Papirius Iustus in Dig. 8.3.17, where it is clarified that the derivation must be made *sine iniuria alterius* 'without wronging another'.

17 'Nothing prevents water from being drawn off from a public river (unless the emperor or the senate forbid it), provided that it will not be water in public use. But if the river is

A second type of derivation was through connections to public aqueducts (mainly intended for public use and free of charge) to bring water to homes or private estates. It would take too long to consider here the importance of Roman aqueducts, which, as everyone knows, are among the most obvious and concrete symbols of Romanisation and which constitute one of the main human modifications of the landscapes of the Mediterranean area. We must limit ourselves to pointing out that, in the first centuries of Roman history, such derivations were granted only exceptionally and that it was only from the imperial age that they became frequent and even generalised. Jurists were therefore concerned to specify the type of such derivations (linked to persons or funds, temporary, renewable or perpetual) and to regulate them: specific administrative concessions were required and a fee had to be paid.[18]

3 Water in the Transformation of the Ancient World

As one might expect, the desire to escape this model of *ad hoc* concessions favoured the spread of abusive derivations (a problem that afflicted the whole of the late antique period).[19] The imperial legislators intervened repeatedly, but it was clearly a losing battle. And it was not just a question of surveillance to prevent abuse and wrongdoing. It was also necessary to ensure the necessary and constant maintenance of the aqueducts, starting with the complex and costly periodic cleaning of the pipes.[20]

Observing the legislation with which the emperors tried to obviate these tasks allows us to recognise with particular clarity the progressive detachment of the *pars Orientis* from the West, which by then rather than being 'Roman' was becoming increasingly 'Romano-Barbarian'. Thus, if in the East the imperial administration continued to find the necessary funds for the care of the aqueducts and the construction of other waterworks (it is sufficient to think of what Justinian did for the water supply of Constantinople), the new masters of the West hardly ever bothered to repair the damage caused to those complex structures due to neglect or warfare (often, during sieges, the aqueducts were cut off near the cities).

---

navigable or another river derives its navigability from it, it is not permissible to do this' (trans. Watson 4, 94).

18 I.e. the *vectigal pro aquae forma* or 'water charge' (see, for example, Dig. 30.39.5).

19 Biavaschi P., *Avida cupiditas. Profili giuridici degli acquedotti romani pubblici nel tardo antico* (Milan: 2018).

20 It is worth noting that both Theodosius's code (CTh. 15.2) and Justinian's code (Cod. 11.43) devote a special title to the subject of aqueducts.

However, the same legislation also allows us to grasp a significant aspect of the transition from Antiquity to the Middle Ages. To this end, it is useful to recall what happened in Ostrogothic Italy. King Theoderic (493–525 CE) – whose political project aimed at the insertion of a Goth element into Roman *civilitas* – endeavoured to keep alive the Roman ideal of water as a source not only of life but also of wellbeing, of aggregation of citizens and homogenisation of the social fabric, an ideal that was realised in the many thermal complexes (Roman *thaermae*) and in the aqueduct system. Through the work of Cassiodorus we know how, under Theoderic's rule, despite a context of profound and general reorganisation, the office of the *comes formarum* (or *aquarum*) was kept alive, responsible for the maintenance and supervision of the aqueducts of Rome and also for the surveillance of the correct distribution of water in the *Urbs* (Cassiodorus, *Variae* 7.6, but also 3.31 for the sending of a special commissioner).[21] We also know that the Gothic king was not only interested in the care of the water system of important cities like Rome and Ravenna (*Var.* 5.38) but was also concerned with medium or less important centres such as Pavia, Verona, Vercelli, Parma or the small Spoleto (see, for example, *Var.* 2.37) or the spa town of Abano (*Var.* 2.39).[22]

After the Byzantine conquest, this ideal was irrevocably lost and, as in the whole of the West, water was now given different priorities. The emperors, noting the social, economic and political importance that the ecclesiastical hierarchies were assuming, began to delegate the care of the aqueducts to the bishops (Cod. 1.4.26 a. 530).[23] Such a move favoured the local powers against the central ones and aimed at attributing functions of vital importance for the population to figures (the bishops) generally considered reliable, honest, and capable of protecting the interests of the weakest.

However, this move also inevitably entailed a progressive change in the way public water was used. The use of water also had to reflect the new system of values that Christianity was imposing to the detriment of the previous cultural model (it is enough to remember how even in Rome, after the

---

21 *Cassiodorus: The Variae. The Complete Translation*, trans. M.S. Bjornlie (Oakland, CA: 2019) 144–145 and 283–284 respectively.

22 Biavaschi, *Avida cupiditas* 224–258.

23 The oldest evidence of such interventions dates back to the years immediately following the sack of Rome by Genseric and concerns the construction of two *balnea* (baths) promoted by Pope Hilarius (461–466) at the basilica of San Lorenzo; see Stasolla F., *Pro labandis curis. Il balneum tra tarda antichità e medioevo* (Rome: 2002) 49–51; eadem, "Tra igiene e piacere thermae e balnea nell'alto medioevo", in *L'acqua nei secoli altomedievali*, 2 vols. II (Spoleto: 2000) 873–926, 883–884.

Byzantine re-conquest, the baths were no longer receiving water and were no longer used, as they were now considered voluptuary and expendable). On the other hand, water was an essential element in the celebration of Christian ritual (from baptism to the preparation of the dead)[24] and its regular availability (and therefore the provision of appropriate derivations) took on a central role in the design and construction of religious buildings, in cities as well as in the countryside.

4 The Early Middle Ages

As we approach the early Middle Ages, the discussion of water and its legal regulation necessarily becomes more difficult and uncertain due to the scarcity of evidence and the imprecision of data. Certainly, there are undeniable new elements, and new protagonists appeared on the scene. During the fourth and fifth centuries, new peoples coming from the north and east – the 'barbarians' – took over the ancient Roman West. Compared to the peoples who inhabited the Mediterranean coasts (who had always been accustomed to considering water a precious commodity), the barbarians had a different relationship with water: they were used to the constant and all-too-abundant presence of this element and therefore did not feel the need to regulate it. Moreover, they were not familiar with anything like large urban agglomerations and the associated water needs). They practised subsistence agriculture, where artificial irrigation was ignored or rudimentary (at least until the barbarians learned the techniques through contact with the Romance populations). On the other hand, they certainly used the rivers to fish and move around in, but river trade was little developed and even less organised.

A panoramic reading of the normative sources (which are all later and certainly reworked with respect to the customary law that the new peoples brought with them when they arrived in the West) reveals a still rudimentary discipline of water use. The rules seem to presuppose a right of generalised and free use of the water present in the natural environment. However, apart from the cases in which they mention water used in some typical ordeals (for example, by hot or cold water), they mostly limit themselves to threatening sanctions (*compositiones*) against those who prevent others from navigating or fishing, who steal or damage other people's nets and boats, who try to forbid

24 Keefe S.A., *Water and the World. Baptism and the Education of the Clergy in the Carolingian Empire: A Study of Texts and Manuscripts* (Notre Dame, IN: 2002).

others free access to the rivers to draw water for drinking or watering animals, or who obstruct the use of water by others (*clausurae aquarum*).[25]

In the compilations of Romano-Barbaric laws there is an embryonic recognition of fishing and navigation rights, but there is also no precise distinction between public and private waters or between public and private waterways. No distinction is made between large and small rivers, springs, still and flowing waters. There is no legal framework regulating the different uses of these waters, nor is there any system of easements for the taking or derivation of water, although there is plenty of generic evidence of the basic principle, already expressed by Isidore of Seville,[26] that no one may be denied access to water for drinking (Lombard law exempts from liability the owner of the land on which there is a well someone has fallen into. This exemption was devised to prevent landowners from taking advantage of such an eventuality, so as to prevent wayfarers from accessing wells[27]).

The few rules inspired by less general and less rudimentary criteria that are sometimes encountered always derive from Roman law (as in LVis VIII.4.29–31 or Lrb 17.2). The principle according to which the sovereign was entitled to lordship over the waters, or at least over the main watercourses was instead inspired – so it is argued – by the patrimonial idea of the state proper to the Germans. Contrary to the principle of Roman law according to which the main watercourses are goods destined for public use (*bona publico usui destinata*), this sovereign power is mostly expressed – as with what happens with other public goods of a patrimonial or fiscal nature – through acts of concession for exclusive use (for example, when the use of river water is granted for the movement of mills or for irrigation),[28] but also through acts of actual disposition or alienation in terms of ownership, in favour of ecclesiastical bodies or private individuals. Sometimes, as for example in the Edict of Rothari, the same royal law is manifested by imposing special responsibilities on those who exercise their profession on rivers or lakes.[29]

With the birth of Charlemagne's empire, a new and more defined discipline of the *bona publica* was affirmed, in direct connection with the early medieval

25 For example: LSal XXVII and XLIV; LRib 46(42).1 and 79(76); LBur XCIV.1–2; LAl LXXXIII.1–2; LBai X.22–23 and XIX.9; LVis VIII.4.28–31; Roth. 299.

26 See once again Isidore, *Etym.* 5.XXVII.38. For the definition of a well, see also *Etym.* 13.XXI.5.

27 Roth. 306; Liut. 136.

28 There is no shortage of deeds emphasising that the river continues to belong to the *publicum* and imposing limits on the concessionaire's use so that the flowing water is limited in a reasonable way (*ad cuiuspiam hominis potestatem subtracta non fiat*).

29 Consider, for example, the peculiar liability that the Edict of Rothari provides for the boatman who, even unknowingly, ferries a fugitive or a thief (see Roth. 265, 266 and 268).

vision of sovereignty in which public protection, supreme jurisdiction and the sovereign's dominion over everything that was not private (*domanium*) coexisted. And it is precisely in the Carolingian age that the idea of the 'patrimonial state' takes shape, and with it that of the 'patrimoniality' of public goods and rights. From this idea, in the so-called feudal age, the category of *iura regalia* or *fiscalia* was born. In the case of the regime of water, too, certain attributes typical of state sovereignty became increasingly patrimonial[30] (for example, the imposition of taxes for navigation and landing on rivers or for the exploitation of river currents as motive power for mills, the construction of locks and fishponds) and could be granted or alienated by the sovereign to bodies or private individuals and become dominical rights (*regalia*). On the other hand, private property could (more or less legally) take on public functions (for example, the collection of the taxes mentioned above and the exercise of jurisdiction over certain types of disputes related to the use of water).

It is difficult to establish to what extent the principle of freedom of public use of river waters – although sporadically reaffirmed – remained effectively protected at this time. Certainly, the system of concessions (the cause of frequent disputes between ecclesiastical bodies, feudal lords and urban municipalities or *universitates*, for example for the control of ports on rivers and lakes and navigation rights) greatly weakened the idea of free navigation on large rivers. And if awareness of the public character of the rivers themselves persisted, this was now detached from the need for their free availability and indeed appeared to be perfectly compatible with the attribution of the waters to bodies or individuals in exclusive possession and enjoyment.[31] This does not mean that the use of the waters was forbidden to third parties. It indicates only that some entities (mainly ecclesiastics) or private subjects obtained (or usurped) the seigniorial right to impose tributes or, alternatively, a particular discipline on navigation, fishing, construction, and use of mills, etc.

---

30 See, for example, the *capitularies* of Pepin (Pipp. 94.9 – a. 787 in *Capitularia regum Francorum* I, 199) and Louis II of Italy (Hludov. 213.8 of the year 850 and 218.3 of the year 866, respectively in *Capitularia regum Francorum* II, 87–88 and 95).

31 The process that led to the full 'patrimonialisation' of the rights concerning public waters and their possible uses was completed when, around the year 1000, at the end of the feudal age, Otto III granted to certain subjects also the flowing waters (*aquarum decursus*) and, to the owners of coastal lands, the exclusive use of mills, fishponds, ports and any other possible use of the waters they contained (*Ottonis III. Diplomata*, nrs. 11, 407–408; 14, 411; 22, 421; 35 and 36, 435 and others). Similarly, Conrad the Salic in 1037 donated to the Bishop of Brescia both banks of two rivers (from their beginning up to where they meet the Po river), over which the sovereign had full jurisdiction (*Conradi II. Diplomata*, nr. 248, 342).

5 The Age of the Communes and the Renaissance

In his *constitutio de regalibus* of the year 1158, Emperor Frederick I (who had listened to the opinion of the Bolognese *Doctores*, the disciples of Irnerius) listed the assets subject to *iura regalia* ('imperial rights') and included navigable rivers and their tributaries, as well as public roads.[32] Rivers and roads were both equally declared part of the sovereign patrimony. The expression *flumina navigabilia et ex quibus fiunt navigabilia* ('navigable rivers and those, tributaries, which are made navigable by them') is clearly taken from the passage by the Roman jurist Pomponius referred to earlier.[33] This expression is used with a different meaning, however. Clearly, the character of navigability becomes here the discretional basis on which the river is recognised as public, against the principle of classical Roman law, which was based on the perennial (or not) nature of running waters.

This new vision, which perhaps originated in the ancient habit of understanding the navigable river as a communication route comparable to roads, spread throughout the European continent thanks to the inclusion of Frederick's constitution in the *Corpus Iuris Civilis*, on which, from 1100 onwards, the new and influential professional class of jurists was based. And it was precisely the new jurists, combining this norm with those they found in Justinian's books, who elaborated a new theoretical construction that considerably broadened the category of 'public waters': public waters were precisely not only navigable rivers (and the tributaries that fed them) but also perennial watercourses of smaller capacity.

It is well known that, in the middle of the twelfth century, Frederick Barbarossa attempted to restore, with imperial authority, the complex of rights and goods belonging to the empire – and therefore to the *publicum* – after the serious erosion suffered during the triumph of feudalism. If, however, the feudal system was far from being defeated, a new adversary was now coming forward with arrogance. This was the Communes, new political bodies which, beyond a merely formal recognition of the sovereign authority, intended to exercise the full range of public powers independently. If in the feudal states then existing in Europe the transfer of *iura regalia* tied to property belonging to the public domain took place on the basis of old titles of investiture or even of the open disavowal of the feudal superiority of the emperor, in the territories of the empire just twenty-five years after the *Constitutio de regalibus*, Frederick

32 Weiland L. (ed.), *Constitutiones et acta publica imperatorum et regum*, Monumenta Germaniae Historica, *Legum sectio IV*, I (Hanover: 1893) nr. 175, 244.

33 Dig. 43.12.2; see fn. 17 above.

I himself had to formally acknowledge to the Italian municipalities the rights that they had already exercised in a customary way, including those concerning water (*consuetudines quas ab antiquo exercuistis* [...] *in aquis et molendinis*, 'customs you have followed since ancient times [...] about (the use of) water and mills').[34]

The control of public waters was, in fact, the subject of a major conflict that pitted city governments against not only the empire but also feudal lords and large ecclesiastical bodies. An important role was also played by jurists. They were fully involved in the political life of the cities and, interpreting the Justinian sources to a very large extent, endorsed this process of taking over the *regalia* on the waters and therefore the attribution of public rivers to the state property and assets of the new political bodies.[35] By declaring illegitimate the old feudal titles of privilege, the Communes then claimed *iure proprio* (i.e. on the sole basis of their own authority) the possession and enjoyment of public waters in which they recognised a necessary attribute of their juridical subjectivity (i.e. their being independent and sovereign political bodies).[36]

The city statutes reflect the great attention that the new urban societies paid to the control and different uses of water. At a time of full economic expansion, not only did traditional uses such as navigation and irrigation receive a considerable incentive (due to the growth of trade and the intensification of agricultural production) but a whole series of artisanal or even proto-industrial activities now undergoing a strong upswing (processing of textiles, leather, metals, etc.) required a large supply of water and at the same time created frequent grounds for disputes.[37]

---

34 Chap. 1 of the *Pax Constantiae* (1183), in *Constitutiones et acta publica* (ed. Weiland) nr. 289, 400.

35 With a few authoritative exceptions, such that of Baldus de Ubaldis († 1400), who, unlike the common opinion, maintained that 'those navigable rivers do not belong to the various peoples, but to the empire. [...] we wonder if the emperor can grant those rivers to anyone. And I answer: yes, since he can grant public property' (in his commentary at Dig. 1.8: *In Primam Digesti Veteris Partem*, fol. 45va–b, nrr. 10 and 13).

36 On this topic see Asheri M., "Beyond the *Comune*: The Italian City-State and its Inheritance", in Linehan P. – Nelson J.L. – Costambeys M. (eds.), *The Medieval World*, 2nd ed. (London – New York: 2018) 530–548.

37 Just two examples: Ascheri, M. "L'acqua nella Repubblica di Siena: un monito dai nostri 'antichi'?", in Serino V. – Brogi A. – Paolucci G. (eds.), *Civiltà delle acque. Storie, miti, leggende in terre di Siena e di Maremma* (Sovicille: 2009) 1–10 and Borghero F. – Salvestrini F., "Fuoco e acqua negli ordinamenti normativi della Sardegna bassomedievale", in Mélanges de l'École française de Rome – Moyen Âge (MEFRM) 135/1 (Rome 2023) 137–170 (here esp. 149–158) < https://journals.openedition.org/mefrm/12150 >.

It is, of course, impossible to follow all the various local legislations that were added to the early medieval customs and the 'rediscovered' Roman laws in regulating these uses. The juridical particularism, so typical of this age, had, moreover, a necessary complement in the extensive production of jurists.[38] The latter once again had the task of drawing up regulations that, without restricting economic needs, would safeguard the rights of the community, starting with the healthiness and free availability of water. The basis of their reasoning was the Roman norms they read in the Justinian *Corpus*. Naturally, however, jurists were children of their time and did not simply reiterate the rules and solutions of Roman law. In particular, they were more original in their approach to the subject of private waters. On the one hand, the jurists, forcing the interpretation of Roman texts, came to admit the possibility of acquiring rights over public waters by prescription. On the other hand, it was in the statutory law that the new category of legal easements emerged. Among the latter, the concept of the compulsory access to water aqueducts stands out. At first, it was designed as a forced sales of the strips of land necessary for digging canals. Later, recalling the Roman category of *iura in re aliena* ('rights in the property of others'), it was described as a compulsory easement for the passage of water with an obligation to pay compensation.

With regard to the public waters, and in particular the regulation of their availability, the statute-makers chose case by case whether to favour the principle of freedom of derivation of these waters or that of individual concessions. We can thus note how the world of the Communes, while remaining within the medieval vision in which the concepts of 'property' and 'sovereignty' are blurred, represents a first step towards overcoming the concept of the patrimonial state. According to a theory which met with lasting success, and which looked at use rather than property, it was stated that 'all common goods are public'.[39] And it is no coincidence that in almost all cities special magistrates' courts were created to ensure compliance with the established water regulations.

38 The process by which the jurist class was formed and increased its authority until it became a fundamental component of the late medieval society is reconstructed by Cortese E., "Legisti, canonisti e feudisti: la formazione di un ceto medievale", in *Università e società nei secoli XII–XVI* (Pisa: 1982) 195–281 and by Brundage J.A., *The Medieval Origins of the Legal Profession. Canonists, Civilians, and Courts* (Chicago: 2008).

39 *Omnia communia sunt publica*: the thesis supported by the glossator Placentinus († c. 1180) and then repeated and spread throughout Europe by the *Magna Glossa* of Accursius (c. 1240) and later by Bartolus († 1357).

6 The Mediterranean and the Ancient Laws of the Sea

We can now move on to the second half of our topic, namely marine waters. Here again, the Middle Ages were the direct heir of Antiquity. The sea, as in the past, was synonymous with danger and mystery but also a source of rich earnings.

As far as law is concerned, however, there is a distinctive element that helps us separate the ancient from the medieval age with a certain clarity. The ancient world was almost exclusively centred on the Mediterranean basin, where a 'law of the sea' developed over many centuries, linking in a line of continuity the Phoenicians (who were, for example, the first to theorise the sharing of losses incurred in navigation), the Greeks and the Romans. In the Middle Ages, the tradition that could be defined as Mediterranean or Oriental (and which was not interrupted by the arrival of the Arabs in the seventh century, as a famous thesis of Henri Pirenne would have it[40]) was joined by a new one, which was forming in the waters west of the Strait of Gibraltar and which we can therefore call Atlantic or Western.

The main centres where maritime activity developed during the Middle Ages can be identified along a route that runs from east to west and from south to north. For the centuries up to the tenth, the main centres were Constantinople, Trebizond, and Antioch (the main bases of the Byzantine Empire's fleet). For the period from the tenth to the eleventh century, we need to look at the cities bordering the Adriatic Sea, such as Venice, Ancona, Trani and Bari, and at those of the southern Tyrrhenian Sea, such as Amalfi (often, in this period, the fleets of these cities were as strong or even stronger than those of Byzantium). Then, from the fourteenth century onwards, we see the emergence of the centres of the Bay of Biscay, such as Bilbao (ancient Vizcaya), Bayonne and La Rochelle. Finally, when the Middle Ages came to an end, the centres on the English Channel, such as Portsmouth and Le Havre, or on the northern seas, such as Bristol and Amsterdam, came to prevail over others.

Legal history, albeit with some delay, also follows this geographical transition. And here too, we can recognise several phases. In the first one, coinciding with the centuries of the early Middle Ages, the common 'law of the sea'

40 As is well known, H. Pirenne expressed his thesis as early as 1925 and then developed it in his *Mohammed and Charlemagne* published posthumously in 1937 (repr. London: 2007); on the historiographical debate that this thesis soon aroused, see Riising A., "The Fate of Henry Pirenne's Thesis on the Consequences of the Islamic Expansion", *Classica et Medievalia* 13 (1952) 87–130.

finds its expression in the *Nómos Rhodíon Nautikós*.[41] It is a collection of private origin, certainly compiled in the East. The author makes explicit reference to the much older *lex Rhodia de jactu* ('Rhodian Law of Jettison'), which was widely used in the Greco-Roman Mediterranean and to which Roman jurists had also referred in matters of maritime law.[42] The content is in any case broad[43] and reveals a certain systematic effort. Above all, although tributary in some respects to Roman-Justinian maritime law, the *Nómos* offers quite a few original elements deriving from Eastern-Greek customs. For instance, the figure of the ship's scribe[44] and the forms of association between ship-owners and merchants appear for the first time. Moreover, the principle of the sharing of profits and of risks between all participants in the enterprise (*koinonia*) is widely affirmed, going well beyond the Roman-law tradition: the rule – applied in all Mediterranean ports – according to which *quod omnes tangit, ab omnibus debet comprobari* ('what concerns all, must be accepted by all') undoubtedly originates from the Middle Ages.

The possible dating of this Byzantine collection is still the subject of debate among scholars, but it must in any case date back to the eighth century. The *Nómos* was the only text not coming from Justinian's *Corpus Iuris* to be included in the *Basilika*,[45] the great normative compilation commissioned by the Macedonian emperors Basil I († 886) and Leo VI, the Wise († 912).[46] The Arabs also made translations and incorporated its contents.[47]

41 The *Nómos* (also known as 'pseudo-Rhodian law') received its critical edition at the beginning of the twentieth century in Ashburner, *Νόμος Ῥοδίων Ναυτικός*; but since more than a hundred years have passed and twice as many witnesses are now available, a new critical edition would be desirable. On *Nómos*, see now Letsios D.G., *Νόμος Ῥοδίων Ναυτικός Das Seegesetz der Rhodier* (Rhodes: 1996).

42 See Dig. 14.2.

43 After the prologue and the chapters deriving from the ancient *lex Rhodia*, a variety of topics are dealt with, touching on criminal profiles but also the transport of passengers and their belongings, maritime loans (*foenus nauticum*), the chartering of the ship and its supplies, the stay in the port, the responsibilities of seamen and merchants, the collision and the recovery of goods on the high seas.

44 According to Ashburner (*Introduction* to *The Rhodian Sea-law* CXXXVIII), this figure was an early medieval creation. As we shall see, the presence of the scribe on board is already required in the first maritime statutes from the eleventh century onwards.

45 There are well over a hundred passages dealing with maritime law in the Justinian's compilation, scattered among the Digest, Code and Novels.

46 The *Nómos* is placed at the end of Book 53, which is all dedicated to maritime law. Many scholars believe that this book already constituted a unit before the *Basilika* was drafted (the composers would have reproduced it in its entirety).

47 Translations of the *Nómos Nautikós* into Arabic continued until the thirteenth century and beyond: see Burgmann L., "Mittelalterliche Übersetzungen byzantinischer Rechtstexte",

7 The 'Explosion' of Maritime Trade in the Mediterranean: (a) the Coasts of the Adriatic Sea

From Byzantine waters (including Black Sea), the *Nómos* spread to the Adriatic and Tyrrhenian seas, where it established itself and remained a point of reference until the thirteenth century and beyond. With the beginning of the eleventh century – in coincidence with important historical events such as the Crusades and the birth of the Communes – maritime trade experienced a new, impetuous development mainly centred on the Italian coastal centres. And certainly, when, in these centres, the need was felt to collect maritime customs in writing, the principles and rules of the *Nómos* were kept well in mind in these new compilations.

It was generally the fellowships that sprang up spontaneously to bring together seamen and merchants that provided the material for the statutes of maritime cities (in Italy and other Mediterranean countries). These rules responded to the twofold need felt strongly by those engaged in navigation and trade by sea: a) to escape the rigid formalism of land-based law; b) at the same time to provide themselves with rules that the existing systems did not know. The answers to these needs were found with the invaluable help of notaries and judges (the 'Consuls of the sea') specially elected and called upon to decide disputes arising from these activities. In particular, these judges were required firstly to have a real knowledge of the problems of navigation and trade by sea and, secondly, to make quick and simple decisions on disputes (one speaks of a procedure *sine strepitu et figura iudicii* 'without the fuss and formality of judgement').[48] These decisions had to be based on the statutes, on the good customs of the sea and on an extensive use of *aequitas* (which, in this case, mainly means 'elasticity'). New principles (such as the limited liability of the capitalist) and new (or partly new) legal figures (such as the limited partnership, the commission, the insurance, the naval *condominium*) emerged

---

in Thür G. (ed.), *Antike Rechtsgeschichte. Einheit und Vielfalt* (Vienna: 2005) esp. 50–51 and 63–66 (and footnotes).

48 The tendency to simplify and shorten the judicial procedure appears to be common to the practice of both ecclesiastical and municipal courts between the twelfth and thirteenth centuries. The *Liber Extravagantium* (1234) reports in this sense some Decretals of Popes Alexander III (X. 2.1.6), Innocent III (X. 2.1.13 and 4.1.27), Honorius III (X. 8.3.35.8 in the middle) and Gregory IX (X. 5.1.26). This tendency was further strengthened – for both ecclesiastical and lay judges – between 1312 and 1314, with two famous Decretals of Clement V, known as *Dispendiosam* and Saepe (Clem. II.1.2 and V.11.2); see Belda Iniesta J. – Coretti M., "Le Clementine *Dispendiosam* e *Saepe Contigit* come paradigma di sommarietà. Alcune note in chiave utroquistica", *Monitor ecclesiasticus* 131.2 (2016) 361–423.

from these rulings. The main innovation, however, lay in the consideration given in this statutory law to seamen and their rights. In time, even learned jurists (*doctores legum*) began to concern themselves with these new technical arrangements, which were then transposed into terrestrial law and became the subject of teaching in law schools.

The influence of the *Nómos* can still be seen very clearly on the coasts of the Adriatic Sea. This is the case, for example, with the statutes of Trani (*Ordinamenta et consuetudo maris edita per consules civitatis Trani*).[49] In the text that has come down to us through a Venetian edition of 1507,[50] these statutes are dated 1063, making them the oldest known maritime statutes. Nevertheless, there are several scholars who, without being able to bring decisive evidence,[51] cast doubt upon this dating and propose later dates (1183, 1231, 1363). In addition to the statutes of Trani, the influence of the *Nómos* can also be recognised in the rules dedicated to navigation in the statutes of Bari, probably drawn up around 1200 and traditionally attributed to judges Andrea and Sparano.

In other cases, however, remaining on the Adriatic coast, Roman-Byzantine maritime law soon had to deal with the customs of the emerging Venetian power. Many Adriatic cities, such as the Dalmatian Zadar, Split, Hvar, Ragusa or, on the Italian coast, Ancona and perhaps Trani itself, looked to the Venetian regulatory complex as a model to be placed side by side with the old Byzantine *Nómos*. In any case, the importance and originality of Venice's contribution to the formation of maritime law cannot be denied.[52]

First of all, it must be said that the city of the lagoon (due to its peculiar political history) felt *exempta ab imperio* ('out of the Empire') even before becoming a major world power. It certainly did not think it had to conform its behaviour to Roman law. Moreover, it is true that Venetian maritime customs were in the process of being formed at the very time when the *Nómos Nautikos* was being drawn up on the Bosphorus. It would therefore be improper to attribute the similarity of certain rules or legal solutions to a precise derivation

49 Marvulli N., "*Nómos Rhodíon Nautikòs* e *Ordinamenta et consuetudo maris edita per consules civitatis Trani*", *Archivio storico pugliese* 16 (1963) 42–64.

50 *Statutae Terrae Appignani*, Venetiis, auspiciis cura ac diligentia [...] viri Marci Marcelli, 1507 (cc. I–III) (the text of the old edition may be found in the appendix to Nisio S., "Degli 'Ordinamenta et consuetudo Maris' di Trani", *Archivio storico pugliese* 16 (1963) 5–38, 34–38).

51 The text that has come down to us is in the vernacular, but this is not a decisive argument, since it may well be a later translation of an older version. More convincing is the argument (see below) of the influence of the Venetian law on the *Ordinamenta* of the Apulian.

52 Lane F.C., *Venice and History* (Baltimore, MD: 1966) especially 193–252, and Cassandro G., *La formazione del diritto marittimo veneziano* (Milan: 1969) 321–353.

and not to a parallel elaboration of pre-existing elements. It is also worth noting that in Venice, where the entire population perceived maritime traffic as a 'common good' to be defended, primarily it was the state, from a very early stage, which took the initiative and strove to regulate activities on the sea with care and technical competence. In 1227 (under Doge Pietro Ziani) and then in 1229 (under Doge Tiepolo) many nautical regulations (*ordinamenta*) were published, but the true monument of Venetian maritime law dates to 1255: the *Statuta et ordinamenta super navibus et aliis lignis* ('Statutes and regulations on ships and other vessels') commissioned by Doge Rainero Zeno, completed about a century later (1346) by the *Additiones et corretiones super statutis navium et navigantium* on the initiative of Doges Francesco and Andrea Dandolo.

It is worth mentioning that at least the main rules of this complex order, built up over time, were read publicly to the Venetians once a year. These statutes mainly dealt with the 'technical fact' of transporting people and goods by water and therefore left to customs (and civil law) the regulation of those contracts that were already practised in ancient times such as maritime loans, limited partnership (called *collegantia* in Venice, *societas maris* elswhere), exchange and insurance. In those rules, the reference to customs is at times implicit or explicit (for instance, in the case of damage, the rules had to coincide with those provided for by the *Nómos*).[53] On the other hand, there is a clear concern for the safety of navigation and for safeguarding the conditions that guaranteed the security and prosperity of the Venetian Republic (starting with the strict control of the construction and sale of ships to foreigners). With respect to the principle that dominates the *Nómos*, namely the tendency to unify the risks and gains of all participants in the maritime enterprise (the ship's master, merchants and sailors), Venetian maritime law clearly distinguishes, and almost contrasts, the interests and responsibilities of the various actors (masters, merchants and sailors) and sets as its primary objective that of settling possible conflicts that may arise between them.

The rules concerning the crew are also relevant and deserve to be noted. To begin with, the sailor was also a man-at-arms and came on board carrying his own weapons. He was paid a wage (probably before departure) and therefore did not share in the charterer's or merchant's earnings. However, he cannot consider himself a mere employee (and may also take on the role of merchant, albeit limited to the goods he has taken on board with him). He could not leave the ship until the voyage was over, but the master could not substitute him

53 Cassandro, *La formazione* 334–336 and 344–345.

during the voyage except with the agreement of the embarked merchants. The scribe was necessarily part of the crew. The statutes consider his task (to record all goods loaded on board by merchants, passengers, and sailors) as an *officium* ('specific office') and require that he be entrusted to specially selected persons.

The regulation of the rental contract was also important. This escaped a precise legal framework (it escaped the Roman law categories of *locatio rei* or *locatio operarum*). The freight was paid before departure. The expenses for loading and unloading the cargo were paid by the shipmaster, who also bore the responsibility for it once loaded onboard (this responsibility ceased only in cases of violence, fire, bad weather and jettison, i.e. *jactus*).

8 The 'Explosion' of Maritime Trade in the Mediterranean: (b) the Coasts of the Tyrrhenian Sea

If we shift our gaze westwards to the Tyrrhenian Sea, the first maritime statutes seem to be those of the town of Amalfi, on the coast of Campania. The Amalfi navy was of great importance in the western Mediterranean between the ninth century and the first half of the twelfth century. As it has come down to us, the text of the *Capitula et ordinationes curiae maritimae nobilis civitatis Amalphae*, or also *Tabula de Amalpha*, comprises a series of twenty-two chapters in Latin to which are added forty-four chapters in the vernacular (but a couple of the latter are grafted between the Latin ones).[54] The Latin chapters certainly represent the earliest core and may date from before the conquest of the city by the Normans (1131). The chapters contain both technical rules and rules of individual conduct. Here too, a certain influence of Byzantine maritime law is undeniable.

It is evident that the *Capitula* makes implicit reference to the rules of the *Nómos* for the regulation of the dominical aspects (ownership of the ship and cargo, throwing of the goods into the sea), which are otherwise hardly mentioned in the text. This circumstance can be explained by considering that the territory of Amalfi did not offer a great amount of produce for export. Amalfi's navy aimed instead at entering external markets. It is therefore understandable that *Capitula* chapters mainly concern the economic use of the ship and the collaboration agreements between financiers, builders, owners and sailors

54 Mignone A., *Nuovi Studi sulla Tabula de Amalpha* (Verona: 2016). The original text is available in *Archivio storico italiano*, 1, *Appendice* (1842/44) 257 folios (on google books under 'Tabula de Amalpha'). A French translation can be read in Senigallia L.A., "La Table d'Amalfi (Tabula de Amalpha)", *Revue du droit maritime comparé* 32 (1935) 38–53.

aimed at maximising investments and offering the merchant a series of necessary services (warehouses and bases in Mediterranean ports with logistic-military support and appropriate alliances with foreign sovereigns).[55] In particular, the so-called contract of 'Colonna'[56] practiced in Amalfi proposes an associative model of great practical effectiveness, based as it is on an awareness of all participants of the common interest and thus on the principle of mutual solidarity to be preserved over time (which also means a clear predisposition to the amicable settlement of any disputes).

The decline of Amalfi coincided with the rapid rise of Pisa in Tuscany. The rules concerning navigation and trade are already present in the *Constitutum usus* (the oldest Pisan legislation, the first version of which dates back to 1161).[57] Here the jurisprudential imprint is strong while the influence of learned law (Roman law taught in schools) is less evident (by contrast, Roman law is clearly present in the city's other body of legislation, that of the *Curia legis*). Important rules of maritime law can be read in the *Breve curiae maris pisanae civitatis* (1298–1305), a corporate statute resulting from the work of the *Curia consulum maris* where maritime law disputes were decided. Pisan maritime law also initially influenced that of Marseilles (c. 1253).

Even further north, Genoa imposed its rule over the Tyrrhenian Sea somewhat later than Pisa and therefore the legislation is also later. Rules of maritime law already existed in Genoa between the end of the twelfth and the beginning of the thirteenth century. They appear mainly in the *Capitula communis Janue* (1306). Those rules were later superseded when the *Officium Gazariae* was established (1316). That magistracy was initially created for the needs of seafarers and merchants interested in the emporium of Caffa (today Feodosia) in Crimea but later extended its jurisdiction to all matters pertaining to the city's navy (including the repression of piracy and the regulation of privateer warfare). In 1336, the so-called Statutes of Pera were enacted, and its books 5 and 6 are devoted to mercantile law and colony law respectively. In the course

55 The Amalfi chapters deal mainly with the crew, the captain, the ship-owners, the contract of *comanda*, the contribution to damage and naval privileges.

56 The so-called 'column' (*colonna*) is a contract of an associative nature concluded between the ship's owners or charterers, the patron (often also co-owner), the crew members, who participated by providing their services, and finally the partners, i.e. the merchants who loaded the goods onto the ship, entrusting them to the *patronus* or following them personally.

57 Classen P., "Kodification in 12. Jahrhundert: Die Constituta usus et legis in Pisa", in Classen P. (ed.), *Studium und Gesellschaft im Mittelalter* (Stuttgart: 1983, originally published in 1977) 82–88. Storti C., *Intorno ai Costituti pisani della legge e dell'uso* (*secolo XII*) (Naples: 1998).

of time (1403; 1441), a series of statutes were drawn up to form an organic discipline for the whole matter.

It is worth highlighting that, just as was the case with Venice, Genoese maritime law was also marked by a natural tendency towards internationalism and cosmopolitanism. This is also why Genoese maritime law influenced the life of ports like Marseilles and Barcelona, which, at that time, were emerging even further west.

9 The Nautical Revolution of the Middle Ages and the North-Western Seas

In the fourteenth century, driven by a new and significant increase in maritime trade, there was what a consolidated historiographical tradition calls 'the nautical revolution of the Middle Ages'. This expression means above all that, thanks to the introduction or improvement of technical instruments and new methods of navigation (the compass, nautical charts, pilot books, new ship models), it became possible to navigate on the high seas and during the winter period. The construction of larger ships, suitable for much larger cargoes than in the past, made sea trade even more profitable. In general, this led to a clearer separation of interests between ship-owners and cargo owners. Ship-owners and merchants stopped boarding and entrusted the ship and goods to the captain who, at the same time, became more independent in making decisions. For greater security against attacks by pirates or enemy shipping, more and more ships were travelling together or were escorted by ships with soldiers. It should also be noted that the emergence of new political balances contributed to the increasing importance of ports on the Atlantic. Looking at the law in particular, at the end of that century, the *Nómos* – as the common law of the sea – was flanked and then definitively replaced as a reference model by other, new normative compilations.

The oldest of these is the one that in time became the *Libre de Consolat de mar* of Barcelona which was destined to establish itself throughout the Mediterranean basin during the sixteenth and seventeenth centuries. This result is largely due to the publishing success of the *Libre*, which was produced in 1494 by one Celelles. Divided into 334 chapters, the *Libre* is actually the result of the combination of a series of older elements, including rules of customary origin, such as the 252 chapters of the *Costums de la mar*, and jurisprudential rules, such as the forty-two chapters of the *Orde judiciari de València*. It is quite possible that the nucleus of maritime customs – customs that were already widespread in the ports of the western Mediterranean (between Italy, France and Spain) – was collected and written down in Valencia and in the Catalan language at the end of the thirteenth century (the customs of the consulates are recalled in a ruling by Peter of Aragon in 1282 and were certainly observed in maritime relations between Catalonia and Sicily following the Spanish settlement on the island in 1284). At that time, Aragon power was just beginning to expand decisively in the Mediterranean. In any case, the collection of various elements (by then included in the *Libre of the Consulate*) was certainly circulating in several versions by the middle of the fourteenth century. In the following century, the version used in Barcelona succeeded in establishing itself as a reference text in the Mediterranean and beyond, especially in view of the breadth of the topics covered and the completeness of the rules. Even before the collection was printed for the first time in Barcelona in 1484, it was already translated into the Italian vernacular in 1479, and subsequently in French, English and German.

In the fourteenth and fifteenth centuries, French seamanship also underwent considerable development. This development was concentrated not only on the Mediterranean but also extended to the Atlantic coast, including Britain, especially because of the close relations that transalpine nobles often had with those on the other side of the Channel. Thus, in the middle of the twelfth century, the Duchy of Aquitaine belonged to the Crown of England, which supplied it with wine. The island of Oléron, which belonged to that duchy, became an important commercial base. According to a legend, it was on the island (where she had come to settle legal disputes) that Eleanor of Aquitaine, the second wife of Henry Plantagenet, promulgated the famous *Rôles d'Oléron* (Rolls of Oléron) in 1160 (in Gascon dialect). More likely, the *Rôles* did not originate on Oléron but elsewhere and was then copied here.[58] The intention was to regulate traffic between France (Bordeaux, Brittany and Normandy) and England, Scotland and Flanders. In France the text soon became widespread (the first draft in French was made in 1286 or shortly before) and in 1364 it became the official maritime code. Even earlier, in England, Eleanor's son, Richard the Lion-Heart, extended its force to the whole of his kingdom. The *Rôles* were then promulgated again by Henry III (1266) and confirmed by Edward III (1329) before being republished under Henry VIII as *The Judgement of the Sea of Masters, of Mariners, and Merchants, and All their Doings*. The *Rôles* (originally consisting of twenty-four chapters which in time became thirty-eight)

58 Krieger K.F., *Ursprung und Wurzeln der Rôles d'Oléron* (Cologne – Vienna: 1970) 120.

regulated both the social discipline and safety of seafaring and commercial relations. More than once, it seems possible to identify the inspiration from the ancient rules of the Byzantine *Nómos* (it has been thought that Eleanor might have become acquainted with the ancient text during her stay in the Holy Land during the Second Crusade, where she had gone following her first husband Louis VII of France). The *Rôles d'Oléron* spread throughout Western Europe in the fourteenth century.

We know of two translations of the *Rôles*: one into Dutch dating from the end of the thirteenth or beginning of the fourteenth century and bearing the name *Vonnesse* (Judgments) *van Damme* (Damme is named the port of Bruges) and the other into Scots, entitled *Of lawis of scyppis* ('Of laws of ships') that can be dated to the second half of the fourteenth century.

Continuing northwards, it is now time to consider the situation in the northern seas. As far as we know, the oldest written maritime rules for these seas date back to the late-twelfth century. They reflect an organisation based on a joint venture between the skipper (at the same time owner of the ship or at least part of it) and the owners of the cargo of a particular vessel. All the participants to the venture appear to have had the same rights and duties aboard the ship. Decisions were taken by mutual agreement among the participants, and the skipper acted as a *primus inter pares*. This state of affairs changed profoundly when, during the thirteenth and fourteenth centuries, there was a considerable development of trading activities in these seas too. The use of new types of ships (with much larger holds) became widespread. The consequent need to create a specialised crew led to a progressive separation of interests between ship-owner, skipper, crew, and merchants.

New rules were needed. A new compilation of maritime law was then compiled in the Netherlands in the middle of the fourteenth century under the name of *Ordinancie* (Ordinance). It was intended to regulate navigation and trade from the Zwiderzee to other European regions. It was usually copied together with the Dutch translation of the *Rôles d'Oléron*. In the fifteenth century, mariners, and merchants from Flanders to Scandinavia and eastern Germany usually referred to both texts as *Waterrecht* ('water-law'). The compilation of *Vonnesse van Damme* and *Ordinancie* – reworked and enriched with new rules (especially from Lübeck) – became a new law-book, which took the name *Gotland Sea Law* better known as *Wisby* (or *Wisbuy*) *Sea Law*, although none of the laws in this compilation originated on the island of Gotland.

Scholars have long repeated that, from the fifteenth century onwards, the *Rôles d'Oléron* and the so-called *Wisby* (or *Wisbuy*) *Sea Law* (and later the 'water-law') represented a kind of common law of supra-territorial validity for all those who worked and invested on the sea. More recent studies[59] have shown that the reality of the northern seas – particularly those of Scandinavia – was rather dominated by the simultaneous existence of several local regulatory systems. In contrast to the *Rôles* and *Wisby Sea Law*, which originated in custom and dealt mainly with private individual relations, the local rules were expressions of the autonomy of the main towns (and ports) bordering those seas and also dealt with what for us today is public law. Their binding force was therefore not based on ancient customs, but on the city authorities and the efficiency of their courts.

10 Concluding Remarks

The medieval age certainly inherited from the ancient world a clear awareness of the importance of the liquid element: a great social importance (especially, of course, in relation to sweet water), but also a high economic value. Like the people of the ancient world, those of the Middle Ages also perceived this value, both recognising water as an irreplaceable commodity for primary use and understanding that it made it possible to carry out important productive activities (as a necessary driving force for mills and as a way for people and goods to circulate very advantageously over small, medium, and even very long distances).

However, the end of the ancient world also brought about considerable changes in this area. During the Middle Ages, the economic and then the political and cultural centre of gravity gradually shifted from the Mediterranean to the northern regions of the European continent. Here, the great abundance of water meant that the idea of water as a public good was perceived with less urgency than on the Mediterranean coasts (which were almost always plagued by freshwater shortages). And when the use of water had to be regulated, it was thought of as a good that it was 'natural' to imagine taking possession of and guaranteeing its exclusive use. The customs that dominated Northern Europe in the Middle Ages reflected this different approach. And when the 'legal renaissance', from the twelfth century onwards, re-proposed the principles and norms of Roman imperial law, the new medieval and polycentric reality rejected the attempt of contemporary sovereigns to re-impose a centralistic (state) control of fresh water.

---

59 Frankot E., "Medieval Maritime Law from Oléron to Wisby: Jurisdictions in the Law of the Sea", in Pedersen F.J.G – Pan-Montojo J. (eds.), *Communities in European History: Representations, Jurisdictions, Conflicts* (Pisa: 2007) 151–172.

On the sea, on the other hand, the Mediterranean legal tradition (a tradition that has a customary basis and shows common features long before the Mediterranean becomes a 'Roman lake') continued to dominate navigation and trade for a long time even during the medieval age. Here too, however, this tradition was profoundly renewed (it suffices to think of the significance of the disappearance of servile labour as the driving force behind ships) and it is possible to clearly recognise the direction this movement took. Progress proceeded by virtue of an almost ineluctable push from the east to the west and from the south ever further north. From the Byzantine imperial ports to the increasingly autonomous ports on the Italian, French and Spanish Mediterranean coasts, to those that opened up on the Atlantic and the North and Baltic Seas, capable of intercepting southern mercantile currents with those from the north. And even on the sea, the dominant approach was a pluralistic one, strongly opposed to a unitary and unifying one.

Abbreviations for Law Codes

Cod.	Justinian, *Codex*, ed. Krüger; trans. Frier.
Dig.	Justinian, *Digesta*, ed. Mommsen and Krüger; trans. Watson.
Inst.	Justinian, *Institutiones*, ed. Krüger.
LAl	*Lex Alamannorum*, ed. Eckhardt; trans. Rivers.
LBai	*Lex Baiuwariorum*, ed. von Schwind; trans. Rivers.
LBur	*Lex Burgundionum sive Liber Constitutionum*, ed. de Salis.
Lrb	*Lex Romana Burgundionum*, ed. Salis.
LRib	*Lex Ribuaria*, ed. Beyerle and Buchner.
LSal	*Lex Salica*, ed. Eckhardt.
Liut.	*Leges Liutprandi regis*, ed. Bluhme.
LVis	*Lex Visigothorum*, ed. Zeumer.
Nov.	Justinian, *Novellae*, ed. Schöll and Kroll; trans. Miller and Sarris.
Roth.	*Edictus Rothari*, ed. Bluhme.

Bibliography

Primary Sources

Ashburner W. (ed.), *Νόμος Ῥοδίων Ναυτικός. The Rhodian Sea-Law* (Oxford: 1909).

Baldus de Ubaldis, *In Primam Digesti Veteris Partem Commentaria* (Venice: 1616).

Beyerle F. – Buchner R. (eds.), *Lex Ribvaria*. Monumenta Germaniae Historica, *Leges nationum Germanicarium*, III.2 (Hanover: 1954).

Bjornlie M.S. (trans.), *Cassiodorus: The Variae. The Complete Translation* (Oakland, CA: 2019).

Bluhme F. (ed.), *Edictus Rothari*, Monumenta Germaniae Historica, *Leges* IV, 3–90 (Hanover: 1868).

Bluhme F. (ed.), *Leges Liutprandi regis*. Monumenta Germaniae Historica, *Leges* IV, 96–182 (Hanover: 1868).

Boretius A. – Krause V. (eds.), *Capitularia regum Francorum*. 2 vols. Monumenta Germaniae Historica, *Leges* II.1–2 (Hanover: 1883 and 1897).

Bresslau H. (ed.), *Conradi II. Diplomata*, Monumenta Germaniae Historica, *Diplomatum Regum et Imperatorum Germaniae*, IV (Hanover – Leipzig: 1909).

de Salis L.R. (ed.), *Lex Burgundionum sive Liber Constitutionum*, Monumenta Germaniae Historica, *Leges nationum Germanicarum*, II.1, 29–122 (Hanover: 1892).

de Salis L.R. (ed.), *Lex Romana Burgundionum*, Monumenta Germaniae Historica, *Leges Burgundionum*, Monumenta Germaniae Historica, *Leges nationum Germanicarum*, II.1, 123–163 (Hanover: 1892).

Eckhardt K.A. (ed.), *Leges Alamannorum*, Monumenta Germaniae Historica, *Leges nationum Germanicarum*, V, 1 (Hanover: 1966).

Eckhardt K.A. (ed.), *Lex Salica*, Monumenta Germaniae Historica, *Leges nationum Germanicarum*, IV.2 (Hanover: 1969).

Frier B.W. (trans.), *The Codex of Justinian. A New Annotated Translation*. 3 vols (Cambridge: 2016).

Giardina A. – Cecconi G.A. – Tantillo I. – Oppedisano F. (eds.), *Flavio Magno Aurelio Cassiodoro Senatore Varia*. 6 vols (Rome: 2016).

Krüger P. (ed.), *Iustitiani Institutiones*, in Mommsen T. – Krüger P. (eds.), *Corpus Iuris Civilis*, I (Berlin: 1892) 1–56.

Krüger P. (ed.), Codex Justinianus, in Krüger P. – Mommsen T. (eds.), *Corpus Iuris Civilis*, II (Berlin: 1895).

Lindsay W.M. (ed.), *Isidori hispaliensis episcopi Etymologiarum sive Originum libri XX*, 2 vols. (Oxford: 1911).

Miller D.J.D. – Sarris P. (trans.), *The Novels of Justinian. A Complete Annotated English Translation* (Cambridge: 2018).

Mommsen T. (ed.), *Cassiodorus Senatoris Variae*, Monumenta Germaniae Historica, *Scriptores, Auctores antiquissimi*, XII (Berlin: 1894).

Mommsen T. – Krüger P. (eds.), *Iustiniani Digesta*, in Mommsen T. – Krüger P. (eds.), *Corpus Iuris Civilis*, I (Berlin: 1892) 29–926.

Pardessus J.M. (ed.), *Collection des lois maritimes antérieures au XVIIIe siècle*, vols. I–VI (Paris: 1828–1845).

Rivers T.J. (trans.), *Laws of the Alamans and Bavarians* (Philadelphia: 1977).

Schöll R. – Kroll W. (eds.), *Corpus Iuris Civilis*, III, *Novellae* (Berlin: 1963).

Sickel T. (ed.), *Ottonis III. Diplomata*, Monumenta Germaniae Historica, *Diplomatum Regum et Imperatorum Germaniae*, II.2 (Hanover: 1893).

von Schwind E. (ed.), *Lex Baiuwariorum*, Monumenta Germaniae Historica, *Leges nationum Germanicarum*, V,2 (Hanover: 1926).

Watson A. (trans.), *The Digest of Justinian*. 4 vols. (Philadelphia: rev. ed. 2009).

Weiland L. (ed.), *Constitutiones et acta publica imperatorum et regum*, Monumenta Germaniae Historica, *Legum sectio IV*, I (Hanover: 1893).

Zeumer K. (ed.), *Lex Visigothorum*, Monumenta Germaniae Historca, *Leges nationum Germanicarum*, I (Hanover: 1902).

Secondary Sources

Asheri M., "Beyond the *Comune*: The Italian City-State and its Inheritance", in Linehan P. – Nelson J.L. – Costambeys M. (eds.), *The Medieval World*, 2nd ed. (London – New York: 2018) 530–548.

Ascheri M. "L'acqua nella Repubblica di Siena: un monito dai nostri 'antichi'?", in Serino V. – Brogi A. – Paolucci G. (eds.), *Civiltà delle acque. Storie, miti, leggende in terre di Siena e di Maremma* (Sovicille: 2009) 1–10.

Astuti G., "Ancora sulla natura giuridica e le origini del contratto di commenda marittima", in Astuti G. (ed.), *Tradizione romanistica e civiltà giuridica europea*, III (Naples: 1984, originally published in 1935) 1345–1371.

Belda Iniesta J. – Coretti M., "Le Clementine *Dispendiosam* e *Saepe Contigit* come paradigma di sommarietà. Alcune note in chiave utroquistica", *Monitor ecclesiasticus* 131.2 (2016) 361–423.

Benoît P.F.L. – Mattéoni O. (eds.), *Pêche et pisciculture en eau douce: la rivière et l'étang au Moyen Âge*, Actes des 1res Rencontres internationales de Liessies 27–29 avril 1998 (Lille: 2004).

Biavaschi P., *Avida cupiditas. Profili giuridici degli acquedotti romani pubblici nel tardo antico* (Milan: 2018).

Bognetti G., "La funzione di Amalfi nella formazione di un diritto comune del Medioevo", *Atti delle manifetazioni culturali pro Tabula Amalpha* I (Naples: 1934) 29–58.

Borghero F. – Salvestrini F., "Fuoco e acqua negli ordinamenti normativi della Sardegna bassomedievale", in *Mélanges de l'École française de Rome – Moyen Âge* (MEFRM) 135/1 (Rome 2023) 137–170 (here esp. 149–158) < https://journals.openedition.org/mefrm/12150 >.

Brundage J.A., *The Medieval Origins of the Legal Profession. Canonists, Civilians, and Courts* (Chicago: 2008).

Bruun C., *The Water Supply of Ancient Rome. A Study of Roman Imperial Administration* (Helsinki: 1991).

Buonocore V., "*La* Tabula de Amalpha*: chiose*", *Diritto dei trasporti* 3 (2002) 743–761.

Burgmann L., "Mittelalterliche Übersetzungen byzantinischer Rechtstexte", in Thür G. (ed.), *Antike Rechtsgeschichte. Einheit und Vielfalt* (Vienna: 2005) 43–66.

Burnouf J. – Leveau P., *Fleuve et marais, une histoire au croisement de la nature et de la culture* (Paris: 2004).

Classen P., "Kodification in 12. Jahrhundert: Die Constituta usus et legis in Pisa", in Classen P. (ed.), *Studium und Gesellschaft im Mittelalter* (Stuttgart: 1983, originally published in 1977) 82–88.

Cassandro G., *La formazione del diritto marittimo veneziano* (Milan: 1969).

Corrieri S., "Profili di storia del commercio marittimo del diritto della navigazione nel Mediterraneo: dal periodo statutario all'era delle scoperte geografiche", in Camarda G. – Corrieri S. – Scovazzi T. (eds.), *La formazione del diritto marittimo nella prospettiva storica* (Milan: 2010) 1–79.

Cortese E., "Legisti, canonisti e feudisti: la formazione di un ceto medievale", in *Università e società nei secoli XII–XVI* (Pisa: 1982) 195–281.

Costa E., *Le acque nel diritto romano* (Bologna: 1919).

Engelhardt E., *Histoire du droit fluvial conventionnel; précédée d'une Etude sur le régime de la navigation intérieure aux temps de aux temps de Rome et au Moyen âge* (Paris: 1889).

De Martino F., "Sul foenus nauticum", *Rivista di diritto della navigazione* (1935) 217–247.

De Martino F., "Ancora sul foenus nauticum", *Rivista di diritto della navigazione* 2 (1936) 433–445.

De Martino F., "Note di diritto romano. *Lex Rhodia*", *Rivista di diritto della navigazione* 3 (1937) 335–347, and 4 (1938) 3–38.

de Robertis F.M., "*Lex Rhodia.* Critica e anticritica su D. 14.2.9", in *Scritti vari di diritto romano* I (Bari: 1987) 309–327.

Fiori R., "L'allocazione del rischio nei contratti relativi al trasporto", in Lo Cascio E. – Mantovani D. (eds.), *Diritto romano e economia: due modi di pensare e organizzare il mondo (nei primi tre secoli dell'Impero)* (Pavia: 2018) 507–567.

Frankot E., "Medieval Maritime Law from Oléron to Wisby: Jurisdictions in the Law of the Sea", in Pedersen F.J.G – Pan-Montojo J. (eds.), *Communities in European History: Representations, Jurisdictions, Conflicts* (Pisa: 2007) 151–172.

García Sanz A., "Estudios sobre los origines del derecho marítimo hispano-mediterráneo", *Anuario de Historia del Derecho Español* 39 (1969) 213–316.

Goldschmidt L., *Handbuch des Handelsrechts. Dritte völlig umgearbeitete Auflage, Erste Abt.: Universalgeschichte des Handelrechts* (Stuttgart: 1891).

Guillerme A., *Les temps de l'eau: la cité, l'eau et les techniques: nord de la France: fin IIIe-début du XIX siècle* (Paris: 1983).

Hermon E. (ed.), *Vers une gestion intégrée de l'eau dans l'empire romain. Actes du colloque international: Université Laval – octobre 2006* (Rome: 2008).

Hilaire J., *Introduction historique au droit commercial* (Paris: 1986).

Iglesia Ferreirós A., "La formación de los Libros de Consulado ae Mar", *Initium: Revista catalana d'historia del dret* 2 (1997) 1–372.

Khalilieh H.S., *Admiralty and Maritime Law in the Mediterranian Sea (ca. 800–1050). The Kitāb Akriyat al-Sufun vis-a-vis the Nomos Rhodion Nautikos* (Leiden – Boston, MA: 2006).
Keefe S.A., *Water and the World. Baptism and the Education of the Clergy in the Carolingian Empire: A Study of Texts and Manuscripts* (Notre Dame, IN: 2002).
Krieger K.F., *Ursprung und Wurzeln der Rôles d'Oléron* (Cologne – Vienna: 1970).
Lane F.C., *Venice and History* (Baltimore, MD: 1966).
Lattes A., *Il diritto marittimo nelle carte liguri dei sec. XII e XIII* (Rome: 1939).
Lattes A., *Il diritto commerciale nella legislazione statutaria delle città italiane* (Milan: 1884).
Leguay J.P., *L'eau dans la ville au Moyen Âge* (Rennes: 2002).
Letsios D.G., *Νόμος Ῥοδίων Ναυτικός. Das Seegesetz der Rhodier* (Rhodes: 1996).
Marvulli N., "*Nómos Rhodíon Nautikòs* e *Ordinamenta et consuetudo maris edita per consules civitatis Trani*", *Archivio storico pugliese* 16 (1963) 42–64.
Mignone A., *Nuovi Studi sulla Tabula de Amalpha* (Verona: 2016).
Müller-Boysen C., *Kaufmannschutz und Handelsrecht in frühmittelalterlichen Nordeuropa* (Neumünster: 1990).
Nisio S., "Degli 'Ordinamenta et consuetudo Maris' di Trani", *Archivio storico pugliese* 16 (1963) 5–38.
Pirenne H., *Histoire économique et sociale du Moyen Âge* (Paris: 1963).
Riising A., "The Fate of Henry Pirenne's Thesis on the Consequences of the Islamic Expansion", *Classica et Medievalia* 13 (1952) 87–130.
Romanelli G., "Aspetti dello Statuto del mare di Ancona (XIV secolo)", *Archivio giuridico 'Filippo Serafini'* 160 (1981) 149–172.
Runyan T., "The Rolls of Oleron and the Admiralty Court in Fourteenth Century England", *American Journal of Legal History* 19 (1975) 95–111.
Schaube A., "Neue Aufschlüsse über die *Anfänge des Consulats des Meeres*", *Zeitschrift für Geschichtswissenschaft* 9 (1893) 223–258.
Scialoja A., "La commenda nel diritto comune del Mediterraneo dei secoli XI–XIII, I, Natura economica e giuridica della commenda", in Scialoja A. (ed.), *Saggi di storia del diritto marittimo* (Rome: 1946) 69–176.
Senigallia L.A., "La Table d'Amalfi (Tabula de Amalpha)", *Revue du droit maritime comparé* 32 (1935) 38–53.
Shepard J.W.,"The Rôles d'Oléron – A Lex Mercatoria of the Sea?" in Piergiovanni V. (ed.), *From Lex Mercatoria to Commercial Law* (Berlin: 2005) 207–254.
Stasolla F., *Pro labandis curis. Il balneum tra tarda antichità e medioevo* (Rome: 2002).
Stasolla F., "Tra igiene e piacere: thermae e balnea nell'alto medioevo", in *L'acqua nei secoli altomedievali*, 2 vols. II (Spoleto: 2000) 873–926.
Storti C., *Intorno ai Costituti pisani della legge e dell'uso (secolo XII)* (Naples: 1998).
Tangheroni M., *Commercio e navigazione nel Medioevo* (Rome – Bari: 1996).

Tetley W., "The General Maritime Law – The Lex Maritima", *European Transport Law* 31 (1996) 469–506.
Vignoli P. (ed.), *I Costituti pisani della legge e dell'uso (secolo XII)* (Rome: 2003).
Werner D. (ed.), *Wasser für das antike Rom* (Berlin: 1986).
Zeno R., *Storia del diritto marittimo italiano nel Medioevo* (Milan: 1946).
Zordan G., "Le leggi del mare", in Tenenti A. – Tucci U. (eds.), *Storia di Venezia. Temi. Il mare* (Rome: 1991) 621–662.

PART 4

Romance, Religion, Myth

CHAPTER 10

Water, Ubiquity and Multiplicity: Shapes of Water and their Narrative Power in a Byzantine Twelfth-Century Novel

Laura Borghetti

Introduction

Thales of Miletus (sixth century BCE), who – according to Aristotle's *Metaphysics* – was the originator (ἀρχηγὸς) of philosophy and mathematics, considered water as the principle of everything, since 'the seeds of all things have a moist nature, and water is the origin of the nature of moist things'.[1] In a Judeo-Christian context, the Septuagint depicts a similar scenario: in the very first lines of Genesis, the Holy Spirit is hovering over the face of the waters (ἐπάνω τοῦ ὕδατος) on a still primordial and not fully created earth.[2] Aristotle, whose physical doctrine constituted the foundation of the Byzantine view on natural philosophy, envisioned the earth as enclosed by a sphere of water.[3] Similarly, in a later *recensio* of the *Alexander Romance* (fifteenth century), the sea

1 Aristotle, *Metaphysics*, ed. and trans. W.D. Ross (Oxford: 1924) 983b26–27: διὰ τὸ πάντων τὰ σπέρματα τὴν φύσιν ὑγρὰν ἔχειν, τὸ δ' ὕδωρ ἀρχὴν τῆς φύσεως εἶναι τοῖς ὑγροῖ.

2 Gen. 1:1–2 in Rahlfs A. (ed.), *Septuaginta* (Stuttgart: 1935): 'Ἐν ἀρχῇ ἐποίησεν ὁ θεὸς τὸν οὐρανὸν καὶ τὴν γῆν. ἡ δὲ γῆ ἦν ἀόρατος καὶ ἀκατασκεύαστος, καὶ σκότος ἐπάνω τῆς ἀβύσσου, καὶ πνεῦμα θεοῦ ἐπεφέρετο ἐπάνω τοῦ ὕδατος' ('In the beginning God created the heavens and the earth. Now the earth was formless and empty, darkness was over the surface of the deep, and the Spirit of God was hovering over the waters', trans. New International Version).

3 Aristotle, *De Caelo*, in Moraux P. (ed.), *Aristote. Du ciel* (Paris: 1965) 287a32: ὕδωρ ἐστὶ περὶ τὴν γῆν ('earth is enclosed by water'). Aristotle was read continuously in Byzantium and the Byzantine interpretation of his works was mostly influenced by late antique commentators. See Trizio M., "Reading and Commenting on Aristotle in Byzantium", in Kaldellis A. – Siniossoglou N. (eds.), *The Cambridge Intellectual History of Byzantium* (Cambridge: 2017) 397–398. On the reception of Aristotle's view of the physical world in Middle-Byzantine time, see Telelis I.G., "Οι λόγιοι του 11ου αιώνα και ο Αριστοτελισμός: η περίπτωση των 'Μετεωρολογικών'", in Chrysos E. (ed.), Η Αυτοκρατορία σε Κρίση (;) Το Βυζάντιο τον 11ο Αιώνα (1025–1081) (Athens: 2003) 426–427; Telelis I.G., "Meteorology and Physics in Byzantium", in Lazaris S. (ed.), *A Companion to Byzantine Science* (Leiden – Boston, MA: 2020) 186–187, 191–197.

 | DOI:10.1163/9789004696501_012

is evocatively described as being like a snake surrounding the world.[4] *Ubiquity*, or the pervasive presence of an element in the natural environment, has been perceived as an essential feature of water ever since the most archaic phase of the Euro-Mediterranean cultures that also constitute the geographical environment and intellectual roots of Byzantine civilisation.

The same perception of water's ubiquity is present in Byzantine novels from the twelfth century, love stories set in a fictional world, in which the element of water is a constant, ubiquitous presence within the storyline, playing an active role as a driving force in the development of the different narrative sequences.[5] Moreover, we are going to observe that water comes on the scene in different forms and functions within diverse frameworks.[6] In this chapter, I aim to shed light on the different shapes and on the various contexts and modalities by means of which water reveals its literary agency, and its narratological and symbolic implications. In other words, I will investigate how the ubiquity of water, within the specific literary framework of a premodern novel, influences the narrative dimension of the text and how, in turn, it is entangled with the physical, philosophical and theological background of the textual source. By bringing to light the metaphorical meaning of the different shapes of water, it will be evident how the cultural and theological relevance of this element legitimises and motivates its leading role within the narratological apparatus of the text. Furthermore, I also aim to show how the same cognitive process

4 *Historia Alexandri Magni* (*Recensio β, e cod. Leidensi Vulc. 93*), ed. L. Bergson (Stockholm: 1965) 2.41.48: ὁ γὰρ ὄφις ἡ θάλασσά ἐστιν ἡ κυκλοῦσα τὴν γῆν ('The snake is the sea that surrounds the world').

5 The term 'novel' is adopted here following the convention that distinguishes between 'novels', namely the late antique and Komnenian fictional works, and 'romances', indicating the Byzantine vernacular works. See Agapitos P.A, *Narrative Structure in the Byzantine Vernacular Romances. A Textual and Literary Study of Kallimachos, Belthrandos and Libistros* (Munich: 1991) 4, fn. 5.

6 Besides his conception of water as an individual, ubiquitous element, Aristotle also argues that we never see water existing as a separate and distinct physical substance. Its different shapes, such as rivers and seas, consist of a mixture of water and other elements, such as earth. Therefore, on the one hand water exists in its pure and ubiquitous form, on the other hand water also appears in different, multiple shapes and with heterogeneous nature. Aristotle, *Meteorologica*, ed. F.H. Fobes, *Aristotelis meteorologicorum libri quattuor* (Cambridge, MA: 1919) 339 b9–13: ὕδατος δὲ φύσιν συνεστηκυῖαν καὶ ἀφωρισμένην οὔθ' ὁρῶμεν οὔτ' ἐνδέχεται κεχωρισμένην εἶναι τοῦ περὶ τὴν γῆν ἱδρυμένου σώματος, οἷον τῶν τε φανερῶν, θαλάττης καὶ ποταμῶν, κἂν εἴ τι κατὰ βάθους ἄδηλον ἡμῖν ἐστιν. ('While water we never see existing as a separate and distinct physical substance, nor it can so exist apart from the mass of it situated round the earth, by which I mean both that which we can see, for instance see and rivers, and any that may be hidden from us underground'). Lee H.D.P. (ed. and trans.), *Aristotle: Meteorologica*, Loeb Classical Library 397 (Cambridge, MA: 1952). See Althoff J., "hydôr / Wasser", in Höffe O. (ed.), *Aristoteles-Lexikon* (Stuttgart: 2005) 270.

that underlies the appearance of conceptual metaphors in the text also constitutes the conceptual scaffolding for the construction of the narrative.

1 Water and the Middle-Byzantine Fictional Narrative

My approach can be fruitfully applied to the genre of the twelfth-century Byzantine novel, which – I argue – displays a narrative blending of ancient Greek literature and Christian doctrine. The overarching presence of settings, figures and themes from the Hellenic past in the Constantinopolitan narrative, and their close dependence on ancient models – from which the whole apparatus of water imagery stems – does not imply the absence of an underlying pattern of originally Byzantine elements.[7] By reading more closely in between the lines, I aim to uncover the rich weave of symbols and references to the Christian present and bring to the fore the sophisticated hypotext of theological and philosophical erudition connected to the element of water. This textual de-codification is fundamental to the purpose of my analysis, due to the characteristic of Byzantine narrative sources of revealing their full meaning only after our complete understanding of the classical and/or biblical references that they contain.[8]

7 The two ancient novels, which constitute the main sources of Theodore Prodromos's novels, are *Leukippe and Kleitophon* by Achilles Tatius and *Aithiopika* by Heliodorus. On the reception of Heliodorus in Byzantium from the ninth to the twelfth century, see Agapitos, "Narrative, Rhetoric, and 'Drama' Rediscovered". According to Roderick Beaton, the novels take place in a hypothetical first-century BCE (Beaton R., *The Medieval Greek Romance* (Cambridge: 1996) 50). Beaton considers the novels 'attempts not just to revive past literature by allusion and imitation, but to recapture, in the fictional world they create, the world of the past in which that literature took shape' (Ibidem 54). Panagiotis Agapitos defines the literary allusion to the ancient time as 'an educationally ingrained concept of conservationist cultural unit' during the Komnenian time (Agapitos P.A. – Smith O.L., *The Study of Medieval Greek Romance: A Reassessment of Recent Work* (Copenhagen: 1992) 36). On the combination of ancient Greek heritage and Christian elements in the Byzantine novels, Joan Burton writes, 'in the process of reviving the ancient Greek novels, the Byzantine novelists also rewrote them, which included infusing them with Christian themes and motifs' (Burton J., "Reviving the Pagan Greek Novel in a Christian World", *Greek, Roman, and Byzantine Studies* 39 (1998) 180). On the presence of Christian references in the novels, see also Nilsson I., *Erotic Pathos, Rhetorical Pleasure: Narrative Technique and Mimesis in Eumathios Makrembolites' Hysmine and Hysminias* (Uppsala: 2001) 279–283.

8 See Messis C. – Nilsson I., "Byzantine Storytelling and Modern Narratology. An Introduction", in Messis C. – Mullet M. – Nilsson I. (eds.), *Storytelling in Byzantium. Narratological Approaches to Byzantine Texts and Images* (Uppsala: 2018) 4–5. See also Nilsson I., *Raconter Byzance la littérature au XIIe siècle*, Séminaires byzantins 3 (Paris: 2014) 29.

Out of four twelfth-century Byzantine novels, my analysis focuses on Theodoros Prodromos's *Rhodante and Dosikles* as a main source, due to its fine and rich repertoire of water imagery and its artful narrative.[9] *Rhodante and Dosikles*, composed probably during the early 1130s in twelve-syllable verses, is divided in nine books and consists of a love story in which the protagonists, Rhodanthe and Dosikles, are separated, travel by sea and eventually reunite.[10] Theodore Prodromos was a prolific and versatile poet active within the framework of the Komnenian clan, and received both Eirene Dukaina's and Anna Komnene's patronages. Together with other intellectuals, such as Michael Italikos, Theodore attended Eirene's *theatron*, perhaps also during the composition of *Rhodante and Dosikles*.[11] The Byzantine *theatra* – similarly to literary salons – served as meeting places where authors and patrons could interact, experiment and adapt works written by the ancient authors.[12] The refined entanglement of the ancient Greek heritage with thematic and figurative allusions to Byzantine culture contributed to produce fictional works, such as Prodromos's *Rhodante and Dosikles*, whose entertaining purpose is gracefully fulfilled thanks to the homogeneous blending of the novelistic discourse with the vibrant rhetorical-philosophical tradition.[13]

---

9 In the present study I use Markovich M. (ed.), *Theodori Prodromi de Rhodanthes et Dosiclis amoribus libri ix* (Stuttgart: 1992). Other Byzantine textual sources of various genres will be taken into account in the course of my investigation by way of example to further support my argumentations. Three more twelfth-century Byzantine novels have been transmitted: Eumathios Makrembolites's *Hysmine and Hysminias*, Constantine Manasses's *Aristandros and Kallithea* and Niketas Eugenianos's *Drosilla and Charikles*.

10 The date of composition is debated: see Jeffreys E. (trans.), *Four Byzantine Novels* (Liverpool: 2012) 7–10, and Nilsson, *Raconter Byzance* 59.

11 For further details concerning Theodore Prodromos's life and intellectual activity, see Hörandner W. (ed.), *Theodoros Prodromos. Historische Gedichte, Text und Kommentar*, Wiener Byzantinistische Studien 11 (Vienna: 1974) 21–35, Jeffreys, *Four Byzantine Novels* 3–6 and Agapitos P.A., "New Genres in the Twelfth Century: The Schedourgia of Theodore Prodromos", *Medioevo Greco* 15 (2015) 1–41.

12 On the intellectual background of the Komnenian novel, see Mullett M., "Aristocracy and Patronage in the Literary Circles of Comnenian Constantinople", in Angold M. (ed.), *The Byzantine Aristocracy IX to XIII Centuries* (Oxford: 1984) 173–201; Jeffreys, *Four Byzantine Novels* 4; Nilsson I., "Romantic Love in Rhetorical Guise: The Byzantine Revival of the Twelfth Century", in Cupane C. – Krönung B. (eds.), *Fictional Storytelling in the Medieval Eastern Mediterranean and Beyond* (Leiden – Boston, MA: 2016). 40–41. On the *theatra*, see Marciniak P., "Byzantine Theatron – A Place of Performance?" in Grünbart M. (ed.), *Theatron: Rhetorische Kultur in Spätantike und Mittelalter / Rhetorical Culture in Late Antiquity and the Middle Ages* (Berlin: 2007).

13 See also Nilsson, "Romantic Love in Rhetorical Guise" 54–55.

2 A Methodological Foreword

Before setting sail for our literary journey at the side of Rhodanthe and Dosikles's, a few lines regarding the methodological approach to the text are necessary.

To focus on water as a natural element, in its ubiquity and multiplicity, I aim to overcome the rather anthropocentric view that considers humans as the cornerstone of any narrative action and set water's agency as the focus of my analysis. Defining water agency as non-anthropocentric implies a reconsideration of the traditional assumption that agency is the exclusive property of the human individual.[14] In the course of my analysis I aim to retrace water's agentive faculty, suggesting that it does not show on a very physical / natural level. Physically water is not an active element per se. Byzantine authors were well aware that water is mostly moved by the wind, and that rivers do not spontaneously run, but that their flowing is instead connected to terrain inclination.[15] In light of this, the type of agency that I could identify in water is of a rather narratological nature. By elaborating on what Jones and Cloke claim about the trees – namely that, within the physical environment, they 'bring [agency] to the continuous unfolding of places and landscapes' – I may rephrase by saying that water's agency contributes to the 'continuous unfolding' of the various settings and events within the storyline.[16] From this perspective, water's agency can be classified as *transformative*, due to its capacity to 'make *new directions and formations*' on a narratological level.[17]

On the other hand, a literary analysis cannot overlook a fundamental human-centred aspect: the filter operated by the author's mind while translating their perception of the natural environment into text. This filtering

14 Jones O. – Cloke P., *Tree Cultures: The Place of Trees and Trees in their Place* (Oxford – New York: 2002) 8. See Knappett C. – Malafouris L., "Material and Nonhuman Agency: An Introduction", in Knappett C. – Malafouris L. (eds.), *Material Agency. Towards a Non-Anthropocentric Approach* (New York: 2008) x.

15 *Historia Alexandri Magni, Recensio α sive recensio vetusta* (ed. Kroll) 3.6.8.3: οὐ γὰρ κινεῖται θάλασσα, εἰ μὴ πνεύσῃ ἄνεμος, οὐδὲ σαλεύεται δένδρα, εἰ μὴ ῥιπίζῃ πνεῦμα, οὐκ ἐνεργεῖται ἄνθρωπος εἰ μὴ ἐκ τῆς ἄνωθεν προνοίας ('The sea does not move unless the wind blows it, and the trees do not tremble unless the breezes disturb them; and likewise, man does nothing except by the motions of divine Providence'). Trans. R. Stoneman, *The Greek Alexander Romance* (London: 1991). *Drosilla and Charikles*, ed. F. Conca, *Nicetas Eugenianus, De Drosillae et Chariclis amoribus* (Amsterdam: 1990) 4.304: Πηγαὶ ποταμῶν συγκινείσθωσαν ἄνω· θνήσκω γὰρ ὡς μόρσιμος, ἀλλὰ πρὸ χρόνου· ('Let river-water flow upwards; for I die as is my doom, but before my time'). Jeffreys (trans.), *Four Byzantine Novels*.

16 Owain – Cloke, *Tree Cultures* 6.

17 Ibidem 57.

mechanism plays a crucial role within the creation of metaphors and symbols since, according to cognitive linguistics, 'human intellectual processes are highly metaphorical'.[18] According to Lakoff and Johnson, in fact, metaphors consist of the understanding of 'one thing [target domain] in terms of another [source domain]'.[19] Let's take 'Life is a sea journey' as an example. In this case, the 'sea journey' is the source domain of the metaphor allowing the full representation of the target domain, which is 'life'. The natural environment and its elements are indeed ideal types of source domain, due to the constant interaction of our body and mind with the surrounding environment. The consequent outcome of this steady dialogue between the landscape and the human domain is the description of 'the mind in terms of the environment'.[20] The relevance and frequency of nature-related metaphors – in our case, water – depends on humans being 'physically embodied organisms in sensory contact with the physical environment through which we move'.[21] The human perception of nature, when expressed in words and texts, necessarily undergoes a cognitive process of interpretation and – often – of transformation, a process which in turn conveys a subjective and, therefore, human-centred point of view.

As we can see, a non-anthropocentric and a more human-centred perspective imply and even complement each other, allowing for a *multiperspective* analysis of the surrounding environment.

3 Ubiquitous Water within the Narrative Journey

As already noted, water is a ubiquitous presence within the storyline of *Rhodante and Dosikles*. Its literary agency is essential to the unfolding of the narrative events and to our understanding of them. An analogy that clearly depicts the relation between water's ubiquity and its narrative agency is the one used by Anne Buttimer regarding her concept of water as a *wholeness*. According to Buttimer, 'water is a *sine-qua-non* for terrestrial life' classifying

18 Lakoff G. – Johnson M., *Metaphors We Live By (with a New Afterword)* (Chicago – London: 2003, original ed. 1980) 5.

19 Ibidem 4. See also Kövecses Z., *Metaphor: A Practical Introduction* (New York – Oxford: 2nd ed. 2010) 4.

20 Cairns D., "Mind, Metaphor and Emotion in Euripides (Hippolytus) and Seneca (Phaedra)", *Maia: rivista di letterature classiche* 69.2 (2017) 267.

21 Ibidem. To further explore the ongoing interaction and connection between the human mind and its environment, refer to Chapter 7 in this volume, specifically Mike Bintley's "Fluid Dynamics: Aquatic Agency in the Poetic Edda of the Codex Regius", with a particular focus on 227–252.

itself thereby as 'whole', namely something that should be viewed in its entirety, not as a collection of parts (in our terms, ubiquity).[22] Buttimer looks at this 'wholeness as horizon, [...] which recedes as the journey through life unfolds'.[23] Our concept of water's ubiquity also acts similarly to a horizon, receding as the narrative journey(s) of our novel unfold(s).

The following is a summary of the events in the narrative of the novel that are linked to and influenced by water. The story begins *in medias res* during a long analepsis, after the barbarian Mistylos's pirates have attacked Rhodes by sea and kidnapped Rhodanthe and Dosikles. Kratandros, Dosikles's friend during their captivity tells him his own story (first analepsis): cleared after a trial, he had left his own island of Cyprus by ship. Dosikles responds to Kratandros with his story (second analepsis): he had fallen in love with Rhodanthe after first seeing her while she was taking a bath; Dosikles had kidnapped the consenting Rhodanthe and fled with her, after which their ship had docked at a safe harbour on Rhodes. Back in the present time, the satrap Artaxanes comes by sea and defeats Mistylos's fleet during a bloody naval battle. Rhodanthe and Dosikles are now Artaxanes's prisoners and separated into two different ships. Rhodanthe's ship sinks, and she is cast away and carried by the waves. Rhodanthe is saved by merchants, who host her on their ship and bring her to Cyprus, where Kratandros's family lives. Meanwhile, the pirates in Pissa are going to sacrifice Dosikles and Kratandros to the pagan gods, but a miraculous rain saves them by blowing out the pyre.[24] Released, Dosikles and Kratandros navigate back to Kratandros's family in Cyprus, where they find Rhodanthe. She is given a poisoned liquid by Kratandros's jealous sister, Myrilla, who is in love with Dosikles, but Rhodanthe survives. After getting married, Rhodanthe and Dosikles sail back home to Abydos.

In order to understand the role of the highlighted events and the relevance of water, it is necessary to clarify what 'event' actually means. According to Porter H. Abbott and Peter Hühn, an *event* is a constitutive part of narrativity and consists in a change of state within the storyline.[25] A first typology

22 Buttimer A., "Nature, Water Symbols, and the Human Quest for Wholeness", in Seamon D. – Mugerauer R. (eds.), *Dwelling, Place and Environment: Towards a Phenomenology of Person and World* (New York: 1985) 260.

23 Ibidem.

24 The exact location of the city of Pissa is unclear. Elizabeth Jeffreys notes that it is 'tempting to see an allusion here to the Italian city of Pisa', since some commercial agreements had been established between 1111 and 1136 (Jeffreys, *Four Byzantine Novels* 67, fn. 122).

25 Abbot P.H., "Narrativity", in Hühn P. (ed.), *The* Living *Handbook of Narratology* (Hamburg: 2014); Hühn P., "Event and Eventfulness", in Hühn P. (ed.), *The* Living *Handbook of Narratology* (Hamburg: 2014).

encompasses *general* events, which present no special features or functions. A second category of events, which I will define as *major*, is accredited with specific features such as 'relevance, unexpectedness, and unusualness'.[26] A *major* event is a non-trivial change of state whose relevance matches intradiegetic expectations.[27] In other words, the *major* event is brought into being by an intradiegetic entity and it has relevant, meaningful implications for the characters in the story. The water-related events in *Rhodante and Dosikles* are all second type events, namely *major* events. This means that water's literary agency, by virtue of its ubiquity, plays a major role throughout the whole novel in influencing and steering the development of most important events according to the main characters' perspective.

The substantial relevance of water-related events in our story lies in the correspondence between them and the fundamental structural phases of the plot. I shall analyse this correlation by relying upon J. M. Adam's synthesis of the sequential structure of an archetypical narrative as a methodological reference.[28] Adam's paradigm consists of a perhaps old-fashioned but still effective tool to individuate crucial narrative turning points and, in our specific case, their link to water. Even though the temporal sequentiality of the plot is interrupted by two major analepses, it is still possible to individuate and highlight its main phases and their correspondence to water:

- *Situation initiale* (beginning)
  - a in the plot: Rhodanthe and Dosikles are together in Rhodes
  - b in the storyline (first analepsis): Rhodanthe and Dosikles are still in Abydos before their meeting
  - c in Kratandros's story (second analepsis): Kratandros is still in Cyprus
- *Complication* (inciting incident)
  - a in the plot: pirates' arrival by sea
  - b in the storyline (first analepsis): Dosikles first sees and falls in love with Rhodanthe at the bath
- *Actions* (rising actions)
  - a in the storyline (first analepsis):
    - Dosikles kidnaps Rhodanthe at the bath
    - Rhodanthe and Dosikles's elopement from Abydos by sea

26 Hühn, "Event and Eventfulness".

27 Ibidem.

28 In the section of his essay dedicated to narratology, Adam collects and compares previous theories, such as those of Claude Bremond, Paul Ricoeur, Julien Greimas, Boris Tomachevski, Jean-Paul Sartre and Gerard Genette, in order to synthetise the criteria for a definition of narrative. See Adam J.M., *Les textes. Types et prototypes. Recit, description, argumentation, explication et dialogue* (Paris: 1992) 54.

  - b in the plot:
    - Rhodanthe and Dosikles reach Rhodes by sea
    - Artaxane's arrival by sea and naval battle
    - Rhodanthe and Dosikles are separated into two ships
  - c in Kratandros's story (second analepsis): Kratandros sets sail from Cyprus (life change)
- *Spannung* (climax)
  - a Rhodanthe:
    - Shipwrecked at the mercy of the waves
    - Poisoned by Myrilla's potion
  - b Dosikles: about to be burnt on the pyre, saved by the rain
- *Situation finale* (resolution and finale):
  - a rescue by the merchants
  - b Dosikles and Kratandros's navigation back to Cyprus
  - c Rhodanthe and Dosikles's navigation back to Abydos (happy ending)

By subdividing the storyline into its main stages, we are able to see that most *major* events in the plot are connected to or influenced by water. By examining the partition of plot and storyline, some initial observations are possible. The relevance and unexpectedness of the *major* events are reflected by the pirates' attacks and by Rhodanthe's shipwreck: aggressions from the sea and sinking ships were not uncommon on Byzantine coastal areas, yet occurred suddenly and were highly feared due to their devastating consequences.[29] Moreover, the transition from one narrative sequence to the other is frequently and smoothly accompanied by a sea trip: from this perspective, not only does the sea promote a change in the setting, but it also opens new scenarios to the protagonists' narrative development. Furthermore, it is also possible to track some narratological signals that pave the way for a change of narrative circumstances: for example, before Rhodanthe's ship sinks, the transitional stage in the storyline is reflected by the instability of the situation, namely by the raging of meteorological phenomena and by water's threatening movement.[30]

---

29 See Dennis G.T., "The Perils of the Deep", in Sode C. – Takács S.A. (eds.), *Novum Millennium: Studies on Byzantine History and Culture Dedicated to Paul Speck* (London – New York: 2001) 86; Veikou M. – Nilsson I., "Ports and Harbours as Heterotopic Entities in Byzantine Literary Texts", in von Carnap-Bornheim C. – Daim F. – Ettel P. – Warnke U. (eds.), *Harbours as Objects of Interdisciplinary Research – Archaeology + History + Geosciences* (Mainz: 2018) 269–270.

30 Prodromos, ed. Marcovich, *Rhodanthe and Dosikles* 6.207–216: Ἀπέπλεον μέν, ὡς ἔφην, αἱ φορτίδες ὑπὸ προπομποῖς ἀνέμοις εὐθυπνόοις. νὺξ δευτέρα μετῆλθε, καὶ θρασὺς Νότος δριμὺς προελθὼν ἐκ μέσης μεσημβρίας κυρτοῖ μὲν εὐθὺς τῆς θαλάσσης τὴν ῥάχιν, ὑψοῖ δὲ ταύτης κυμάτων πολλοὺς λόφους ἐκ χασμάτων κάτωθε μέχρι πυθμένος. ὁ μὲν Ποσειδῶν ἠρέθιζε τὸν

Almost following a cyclical tendency, water acts both as a trigger for narrative events and as a soothing element: on the one hand, in the beginning of the novel, Rhodanthe inflames Dosikles's passion while taking a bath; on the other hand, towards the conclusion, the miraculous rain saves Dosikles's life, allowing him to join Rhodanthe again. Finally, when the reader is just about to sigh with relief because Rhodanthe and Dosikles have been reunited, there comes Myrilla with her poisonous liquid, ready to threaten the two lovers one last time. Nevertheless, the happy ending is again connected to one last sea trip, the one from Cyprus to Abydos. Besides in the main characters' story, water also plays a most relevant role within parallel narrations. In fact, the second analepsis finds its intersection with the protagonists' events by means of Kratandros's sea trip from Cyprus to Rhodos: the time when the young man decides to change his own life's direction is also the time when he first meets Rhodanthe and Dosikles on the island.

However, the presence of water as a ubiquitous element and as a narratological thread-line within the plot does not only depend on water being a *sine-qua-non* element.[31] Earlier in this chapter, I argued for the importance of human embodiment within the physical environment as the source of a powerful cognitive link between the human mind and literary production, which has its outcome in metaphorical images. The exemplary metaphor 'life is a sea journey' does not serve the purpose of merely enriching and embellishing the verse or line where it is used. It illustrates instead the author's broader conception of both life (target domain) and sea (source domain).[32] By employing the more traditional view of structural narratology as a starting point, I argue that it is possible to individuate a similar shift also on the narratological dimension: the role of water within the storyline does not have a merely pragmatic

Νότον, τὸν πόντον ἀντήγειρεν ὁ θρασὺς Νότος, τὰς ναῦς ὁ πόντος, αἱ δὲ ναῦς τοὺς ἐν μέσῳ· κλύδωνα κοινὸν καὶ ζάλην εἶχον μίαν οἱ ναυλοχοῦντες, ἡ θάλασσα, τὰ σκάφη. ἐπεβρυχᾶτο τῆς θαλάσσης τὸ στόμα· ἀντεβρυχᾶτο τῶν πλεόντων τὸ στόμα. ('The freighters, as I said, sailed away with straight-blowing winds as their escorts. A second night intervened and a vigorous southerly wind driving up fiercely full from the south immediately mounded up the sea's ridges, it raised up on it many hills of waves from the gulfs below as far as the ocean floor. Poseidon provoked the southerly wind, and the vigorous southerly in turn roused the sea, and the sea the ships and the ships those within; the passengers, the sea, the vessels endured one and the same storm and tempest. The roaring of the sea's expanse found its counterpart in the anguished roars of the voyagers', trans. Jeffreys, *Four Byzantine Novels*).

31 Buttimer, "Nature, Water Symbols, and the Human Quest for Wholeness" 260.

32 Evola V., "La metafora come *carrefour* cognitivo del pensiero e del linguaggio", in Casadio C. (ed.), *Vie della metafora: linguistica, filosofia, psicologia* (Chieti: 2008) 55–56.

function, serving the simple purpose of a smooth shift of the different narrative sequences and accompanying *major* events. Water's narratological ubiquity corresponds instead to the author's broader metaphorical conception of water and life, a conception that consistently emerges all along the narration, in terms of, I argue, conceptual narratology.[33]

For example, in Book 7 of the novel, Rhodanthe and Dosikles are parted, and the girl is desperately wondering where her lover might be and in what conditions:

> εἰς Πίσσαν οἰκεῖς, εἰς θάλασσαν ἐμπλέεις;
> κλύδων, γαλήνη· κῦμα, λειότης πλόου;
> ἀντεμφυσήσεις ἀνέμων, ἡσυχίαι;
> χρηστή, σαπρὰ ναῦς· ὑγιές, σαθρὸν σκάφος; [34]
>
> Do you dwell in Pissa, do you sail the sea?
> Tempest or calm? Billows or a smooth passage?
> Opposing winds or unruffled waters?
> Is your ship sound or rotten, its hull whole or perished?[35]

Rhodanthe is very distressed and, of course, she worries about the conditions of Dosikles's voyage, since the last time she saw the young man, he was about to set sail on a different ship. However, the maritime context and the threatening features of the sea combined with stormy meteorological conditions also show a deeper, metaphorical significance. The trip by sea symbolises the course of life, and water presents multiple allegorical similarities with life. The course of fluids, as it happens during a sea trip, can have intradiegetic control on a character's life, that can variate and evolve according to the effect of liquids within the storyline (e.g., during a sea trip, through a meeting at the bath, due to the ingestion of poison, etc.). The 'life is a sea journey' motif is not unique to Prodromos. In one of his poems, John Geometres (tenth century) explicitly compares life and the sea: 'In no way does life differ from the sea, both unsteady,

33 The inspiration for my interpretation of 'conceptual narratology' stems from the field of cognitive narratology. See Herman D., *Cognitive Narratology*, in P. Hühn – Pier J. – Schmid W. – Schönert J. (eds.), *Handbook of Narratology* (Berlin: 2009). On the concepts of experientiality and embodiment, see Fludernik M., *Towards a Natural Narratology* (London: 1996) 20–22, 28, 37.

34 Prodromos, *Rhodanthe and Dosikles*, ed. Marcovich 7.32–35.

35 Jeffreys (trans.), *Four Byzantine Novels*.

bitter, feeding wild beasts, the(ir) harbours small'.[36] One century later, John Mauropos also refers to life and sea in terms of close analogy: 'I was leading a night battle in the sea of life […]', and then, addressing Christ as a guiding figure, Mauropus adds, 'you are my radiant beacon and a calm harbour'.[37] As we can observe from these examples, harbours too are crucial elements within the framework of the life-sea metaphor and, in the case of our novel, also within the storyline. In Book 2, during Dosikles's analepsis, the young man describes the harbour in Rhodes as a safe and peaceful place, protected by the dangerous movement of wavy seawaters. However, Rhodanthe and Dosikles's ship encountered serious difficulties in entering the harbour, due to its narrow and rocky access.[38] The harbour is indeed considered as a safe haven, but the path to reach it might not be easy. The peak of the danger, in fact, is reached right at its entrance, shortly before the easing of the narrative tension. Veikou and Nilsson, in their essay about Byzantine harbours as heterotopic entities, describe the link between the sea, the ship, and the harbour as a 'trialectic scheme'.[39] This is valid both in terms of narratological space and, in the case of our analysis, on the metaphorical and cognitive dimension: during one's life, each person is like a ship at the mercy of the waves and in constant search of a calm harbour as a shelter where to rest.[40]

As we have observed so far, life – as well as water – is characterised by constant fluency, instability, metamorphosis.[41] The overarching significance of water and liquids in the development of the human experience – both for each individual and for humankind as a whole – accompanies and triggers the evolution of the main characters and of some of their crucial experiences, such as the beginning and progress of their love, throughout the whole narrative plot. The metaphorical meaning of water as a ubiquitous presence within the novel

36 John Geometres, *Paradisus*, in Migne J.-P. (ed.), *Patrologia Graeca* 106, 873.29: Οὐδὲν ἁλὸς ῥ ὁ βίος διενήνοχεν· ἄστατοι ἄμφω, πικροί, θηροτρόφοι, οἱ λιμένες βραχέες. For discussion on the connection between *mare* ('sea') and *amarus* ('bitter'), see (in the present volume) Borysławski R., "Marie de France's Writing on Water: The Etymological Narrativity of the Sea in the *Lais*" 344–371.

37 Joannes Mauropus, *Canones in Sanctum Nicolaum*, ed. A.D. Panagiotou (Athens: 2008) 2.5.106–107, 113–114: Νυκτομαχῶν ἐν πελάγει τοῦ βίου […] σὺ δέ μου πυρσὸς φαεινὸς καὶ λιμὴν γαληνός.

38 Prodromos, *Rhodanthe and Dosikles*, ed. Marcovich 2.1–21.

39 Veikou – Nilsson, "Ports and Harbours as Heterotopic Entities in Byzantine Literary Texts" 267. See also Ibidem 265.

40 The liminal place of ports and harbours between danger and security constitutes the fifth heterotopic attribute of harbours (ibidem 269).

41 In this regard, see also Classen A., *Water in Medieval Literature. An Ecocritical Reading* (Lanham, MD – Boulder, CO – New York – London: 2017) XXXIII.

emerges within the narratological dimension by steering and influencing it. In other words, the way water is perceived and conceptualised in real life is also consistently translated into metaphors and employed as a thread line along the narratological development of the novel. The tight connection between one's own experience and its rendering in narratological and metaphorical terms is illustrated, within the field of cognitive narratology, as the transfer from experientiality to narrative through embodiment. As explained by Fludernik, 'experiencing can be used as building stones for the mimetic evocation of a fictional world'.[42] The embodiment, in turn, serves as a cognitive bridge between the experiencing and the creation of narrative, since the individual experience takes shape through the full immersion of the human being into the surrounding environment and its occurrences. Again, according to Fludernik, narrative is 'identified as the presentation of [this] experientiality'.[43]

Let me draw some preliminary conclusions. The ubiquitous presence of water within the storyline corresponds to the narratological rendering of conceptual metaphors (e.g., life / sea journey). In other words, it serves the purpose of building a conceptual bridge between the cognitive domain (human experience and embodiment) and the literary dimension (fictional narrative). Water as a whole – consisting of the totality of different shapes, such as sea, bath, rain, poison, etc. – is indeed similar to the horizon theorised by Anne Buttimer: it recedes – and hence becomes wider – as the journey through life unfolds and embraces the wholeness of the human practical and emotional life experience.

Having so far analysed water as a ubiquitous element and touched upon some of its different shapes, I am going to examine some of them more in depth. In spite of the overall similarity of plot and water imagery to the ancient models, I aim to show how the shapes of water play a fundamental role also within the more specifically Byzantine mind-set: their profound cultural and theological value, both from a narratological and from a metaphorical perspective, integrates the Byzantine worldview within the framework of the intersection of human life and nature.[44]

---

42 Fludernik, *Towards a Natural Narratology* 20.

43 Ibidem 36.

44 Messis – Nilsson, "Byzantine Storytelling and Modern Narratology: An Introduction" 4: 'This does not mean that narratology is not useful, but that we Byzantinists must combine our knowledge of the texts and their historical context with careful theoretical readings and sometimes also some scholarly imagination in order to shape a theory and a method for narrative analysis that is adapted to Byzantine literature and its interpretative needs'.

4 Harmful and Beneficial Shapes of Water

Throughout the novel's plot, it is possible to pinpoint how different shapes of water are differently connoted according to their positive or negative influence on and their connection to the main characters and their vicissitudes. Seawater, for example, can be both beneficial and harmful to the protagonists, depending on whether navigation is smooth, or if hostile meteorological conditions endanger their lives. Interestingly, calm weather and favourable winds do not inspire particularly evocative depictions of water and of the maritime environment. On the contrary, when storms rage and waves grow high, we encounter a manifold spectrum of references and metaphors. The most recurring and vivid images associate water to four different semantic fields: corporeality, downwards directionality, darkness and, ultimately, death. For example, when Gobryas, one of the pirates that first abducted Rhodanthe and Dosikles, falls in love with Rhodanthe and intends to marry her, Dosikles is desperate and wishes – very dramatically – that she were dead instead:

> Οἰκτρὸν μέν, οἰκτρὸν ταῖς κόραις Δοσικλέος
> ἰδεῖν Ῥοδάνθην εἰς βυθὸν κατηγμένην,
> εἰς τὴν ἀχανῆ τῆς θαλάσσης γαστέρα,
> λαχοῦσαν ὑγρὸν καὶ κατάρρυτον τάφον [...].[45]
>
> Pitiful, pitiful for Dosikles's eyes would it have been
> to see Rhodanthe swept into the deeps,
> into the sea's gaping belly,
> achieving a damp and watery grave [...].[46]

As we can observe in these verses, the deep seawater within an endangered context is very precisely connoted. The sea becomes an abyss (βυθός), into whose depths Rhodanthe is going to plummet (ἰδεῖν [...] εἰς) following a downwards trajectory.[47] This image evokes different scenarios, starting from the funerary context, which is reinforced by the depiction of the sea as a humid, watery grave (ὑγρὸν καὶ κατάρρυτον τάφον) two lines below. The correspondence

---

45 Prodromos, *Rhodanthe and Dosikles*, ed. Marcovich 3.444–447.

46 Jeffreys (trans.), *Four Byzantine Novels*.

47 The intertextual similarity to the ancient model and to a contemporary novel is striking: a similar storm and shipwreck occur both in Makrembolites's *Hysmine and Hysminias* (Book 7), contemporary to Prodromos' *Rhodanthe and Dosikles*, but also in a major ancient source, Achilles Tatius's *Leucippe and Clitophon* (Book 3). In this regard, see also Nilsson, *Erotic Pathos, Rhetorical Pleasure* 213–219.

between the concept of 'abyss' and the framework of death is very common in Byzantine literature. We meet it, for example, in one of Manuel Straboromanus's carmina (ninth century) dedicated to a mother's grief after her son's death. While expressing the excruciating desperation caused by her loss, the woman describes her crying as a sea of tears (θάλασσα δακρύων); having introduced the maritime context, she describes herself being held down by deep darkness (με [...] κατέσχε καὶ βαθύ σκότος), where the adjective βαθύ ('deep') shares a common root with βυθός ('abyss').[48] Prodromos uses the same expression as Straboromanus (εἰς βαθύ σκότος) in Book 5, when one of the pirates, Bryaxes, ruler of Pissa, describes throwing (ῥίπτω) some enemies to the bottom of the sea.[49] In the above quoted passage from Book 3, the sea's depth is also described as a 'gaping belly', as if it were an open mouth ready to devour the girl, and a voracious stomach, ready to digest her. The bodily connotations of the sea recur often in the novel. A couple of line afterwards, the sea has a mouth (στόμα) into which Rhodanthe might fall into (ἐκπίπτω).[50] Later on, in Book 5, when the pirates' fleet is threateningly approaching, the back of the sea is depicted as entirely (ὅλη) black (μέλαινα).[51] Once again, the sea is associated with both a dark colour range and the corporeal dimension: the word for 'back' (ῥάχις), in fact, is a precise anatomical term indicating the very lower part of the back, the chine.[52] One last interesting example of harmful shape of water, again in connection to the sea, occurs during Dosikles's lamentation, when the young man fears that Rhodanthe might be dead after the shipwreck. On that occasion, seeing the possibilities of marrying the girl fade away, Dosikles describes the sea in terms of a deathly bridal chamber:

> θάλαμος ὑγρὸς καὶ κατάρρυτον λέχος,
> ὑδαρὰ παστὰς καὶ θαλάσσιος γάμος,
> καὶ πόντος οἶκος καὶ μετ' ἰχθύων βίος;[53]

---

48 Manuel Straboromanus, *Carmina*, in Gautier P. (ed.), "Le dossier d'un haut fonctionnaire byzantin d'Alexis Ier Comnène, Manuel Straboromanos", *Revue des études byzantines* 23 (1965) 3.6, 8.

49 Prodromos, *Rhodanthe and Dosikles*, ed. Marcovich 5.234–235.

50 Prodromos, *Rhodanthe and Dosikles*, ed. Marcovich 3.455. The same expression describing the sea as having a mouth also appears in 5.240.

51 Prodromos, *Rhodanthe and Dosikles*, ed. Marcovich 5.100–101.

52 The Online Liddell-Scott-Jones Greek-English Lexicon, s.v. ῥάχις (http://stephanus.tlg.uci.edu/lsj/#context=lsj&eid=94854). Last accessed May, 2, 2024.

53 Prodromos, *Rhodanthe and Dosikles*, ed. Marcovich 6.268–270. A similar image occurs also in Achilles Tatius's *Leucippe and Clitophon*, ed. E. Vilborg (Stockholm: 1955) 3.5.4: εἰ δὲ καὶ θηρίων ἡμᾶς βορὰν πέπρωται γενέσθαι, εἷς ἡμᾶς ἰχθὺς ἀναλωσάτω, μία γαστὴρ χωρησάτω, ἵνα καὶ ἐν ἰχθύσι κοινῇ ταφῶμεν ('If we are doomed to feed the fish, let a single monster

A watery marriage chamber and a liquescent couch,
a watery bridal bower and a sea-girt wedding,
the ocean for your house and a life with the fishes?[54]

Such an unhappy depiction includes, nevertheless, a few interesting details. The bridal chamber is connoted by the very same adjectives that we met earlier, in Book 3, regarding the sea as a grave, namely 'humid' (ὑγρὸς) and 'watery' (κατάρρυτος). The wedding takes place in the sea and the sea is the new home, even though within a deathly framework. Clearly, the sea, along with the concepts of depth, darkness, and corporeality, constitutes one of the multiple faces of an overarching target domain of a metaphor, which is death.

Moreover, these various conceptual domains present an interesting theological background. Without going into too much detail regarding doctrinal matters, I am going to sketch some significant aspects relevant to our analysis. In Byzantine Christianity, body and soul were generally considered as two separate entities. Already in the biblical Book of Wisdom (third to second century BCE), the body was considered as potentially sinful, being the seat of sensory perception and pleasure.[55] Origen (third century) believed that the soul was linked with the body as a punishment, while Evagrius Ponticus (fourth century) saw the estrangement from the body and the material dimension as a way to salvation and reconnection with God.[56] These few theological ideas indicate a correspondence between the harmful role of water and the rather negative connotation of body within the Christian philosophical framework. From this perspective, the sensory aspect of the body matches very well with the negative, corporeal depiction of seawater in Book 3, when the pirate Gobryas intends to marry Rhodanthe. Gobryas's carnal and lascivious desire for the girl is reflected by Dosikles's words, since – according to him – the pirate's lust aims at the pure sensory pleasure that has its seat in the body.

Along with the corporeality, the downwards directionality, recurrent in the metaphorical depictions of harmful water, might conceal theological

engorge us together, a single stomach accommodate us as one, so that even among the fish we may share a sepulchre' Winkler J.J. (trans.), "Achilles Tatius: 'Leucippe and Clitophon'", in Reardon B.P. (ed.), *Collected Ancient Greek Novels* (Berkeley – Los Angeles, CA – London: 1989) with the only difference that the fishes are, in this case, the agents of death.

54 Jeffreys (trans.), *Four Byzantine Novels*.

55 Wisdom 1:4 (*Septuagint*, ed. Rahlfs): ὅτι εἰς κακότεχνον ψυχὴν οὐκ εἰσελεύσεται σοφία οὐδὲ κατοικήσει ἐν σώματι κατάχρεῳ ἁμαρτίας. ('Because into a soul that plots evil wisdom does not enter, nor does she dwell in a body under debt of sin', trans. New American Bible, revised edition).

56 Meyendorff J., *Byzantine Theology: Historical Trends and Doctrinal Themes* (New York: 1974) 26; Kapriev G., *Philosophie in Byzanz* (Würzburg: 2002) 32.

implications. The *topos* of the deep downfall, in fact, seems to recall the episode of Lucifer's banishment from living in heaven, described in the Book of Isaiah as follows: 'But you are brought down to the realm of the dead, to the depths of the pit'.[57] Furthermore, the concept of darkness, often recurring in connection with harmful seawater, clearly evokes the estrangement from the divine light. In John's Gospel, Jesus claims, 'I am the light of the world. Whoever follows me will never walk in darkness but will have the light of life'.[58] The identity of light and the Divine is a crucial theological cornerstone within the Christian philosophy and theology. According to Dionysius the Areopagite (fifth century), for example, the divine light invests nature as a whole and spreads through all created things as if reflected by means of mirrors.[59] If (divine) light equals life and radiates all over the Creation, it comes as no surprise that Prodromos, in Book 5, describes death with the simple adjective 'black' (τὴν μέλαιναν).[60]

In addition to its harmful, darker shapes and their Christian background, water also shows beneficial features and influence throughout the novel. A type of water shape that plays a crucial role, as we have seen, is bath water. Thanks to the bath, in fact, Dosikles gets to see Rhodanthe for the first time and, later on, he gains a chance to abduct her.[61] Already in the early Byzantine period, in continuity with late Antiquity, bath evokes a pagan erotic context, and, both in archaeology and literature, motifs with Venuses, *Gratiae* and *Amores* often recur in connections to baths.[62] While up until the sixth century the bath was

57 Isaiah 14:15 (*Septuagint*, ed. Rahlfs): νῦν δὲ εἰς ᾄδου καταβήσῃ καὶ εἰς τὰ θεμέλια τῆς γῆς. Early Christian writers had already developed a similar concept of the sea as a demonic space into a proper theological form. For a discussion of this, see Chapter 4 of this volume, Karl Whittington's "Opicinus de Canistris's Demonic Seas", especially 265–266.

58 John 8:21 (*Septuagint*, ed. Rahlfs): Ἐγώ εἰμι τὸ φῶς τοῦ κόσμου: ὁ ἀκολουθῶν ἐμοὶ οὐ μὴ περιπατήσῃ ἐν τῇ σκοτίᾳ, ἀλλ' ἕξει τὸ φῶς τῆς ζωῆς; trans. New International Version. Concerning the light and darkness antithesis in John's Gospel, see Trocmé E., "Light and Darkness in the Fourth Gospel", *Disdaskalia* 6.2 (1995) 3–13.

59 Ronchey S., *Lo Stato Bizantino* (Turin: 2002) 86; Maguire H., *Nectar and Illusion. Nature in Byzantine Art and Literature* (Oxford: 2012) 127.

60 Prodromos, *Rhodanthe and Dosikles*, ed. Marcovich 5.178. In both the English (Jeffreys, *Four Byzantine Novels* 87) and the Italian translations (Conca, *Il Romanzo Bizantino* 181) τὴν μέλαιναν is integrated with the substantive 'doom', by Jeffreys, and 'destiny' ('destino') by Conca, who points out that the adjective μέλαιναν implies the presence of the non-expressed substantive κῆρα, which literally means 'heart' and – again – constitutes a reference to the bodily dimension.

61 Likely, a typical Byzantine reader of the novel would have immediately thought of King David glimpsing Bathsheba while she bathed in the Old Testament: 'And it came to pass in an evening tide, that David arose from off his bed, and walked upon the roof of the king's house: and from the roof he saw a woman washing herself; and the woman was very beautiful to look upon' (2 Samuel 11:2, King James Version).

62 Berger A., *Das Bad in der byzantinischen Zeit* (Munich: 1982) 126.

presented to women as a vehicle to erotic allurement, and the women's tempting nudity could even be scary for men, from the seventh century on literary depictions rather focus on monastic baths or their religious meaning.[63] However, in the twelfth-century Byzantine novel – and later on also in the Palaiologan romances – the bath shows an interesting combination of classical and Christian aspects and, in comparison with the ancient models, takes on a new meaning in the Christian context.[64] Dosikles is not at all scared at the sight the Rhodanthe's nudity; on the contrary, he immediately falls in love with her:

ταύτην δὲ πατὴρ ὁ προλεχθεὶς ὁ Στράτων
ἔσωθεν ἐγκέκλεικε μικροῦ πυργίου,
ὡς ἂν δύσοπτος ἀρσένων κόραις μένῃ,
οὐδ' ἐκτὸς ἐλθεῖν τῆς φυλακῆς ἠξίου,
εἰ μὴ ῥυπανθὲν[65] τῆς κόρης τὸ σαρκίον
ἔχρῃζε λουτροῦ καὶ ῥοῆς καθαρσίου.
ταύτην μὲν οὕτως ἠσφαλίζετο Στράτων,
ὡς ἂν ἐραστοῦ λίχνον ὀφθαλμὸν φύγῃ·
οὐ μὴν φυγεῖν ἴσχυσε τὸν Δοσικλέος.[66]

Straton, her father whom I have just mentioned,
enclosed her in a little tower,
so that she should remain difficult for men to see,
and he did not think it right for her to go out from her prison
except when the girl's soiled flesh
had need of a bath and a cleansing stream.
So Straton protected her in this way,
so that she should escape an admirer's curious gaze,
but she could not escape that of Dosikles.[67]

---

63 Ibidem 127–128.

64 See Ingela Nilsson's interesting analysis of Eros and bath in "Ancient Water in Fictional Fountains: Waterworks in Byzantine Novels and Romances", in Shilling B. – Stephenson P. (eds.), *Fountains and Water Culture in Byzantium* (Cambridge: 2016) 287–291. See also Agapitos P.A., "The Erotic Bath in the Byzantine Vernacular Romance *Kallimachos and Crysorrhoe*", *Classica et Mediaevalia* 51 (1990) 257–273.

65 The use of ῥυπανθὲν is here ambivalent, meaning 'soiled' in terms of both material and spiritual impurity.

66 Prodromos, *Rhodanthe and Dosikles*, ed. Marcovich 2.175–183.

67 Jeffreys (trans.), *Four Byzantine Novels*. Jeffrey has chosen a rather mild translation ('curious') for the adjective λίχνον, while the original Greek term displays more intense meanings, such as 'luxurious' and 'gluttonous', both showing a more corporeal and erotic connotation that better suits the context of our text. Eustathios of Thessalonica uses the

These verses belong to Dosikles's analepsis, where he describes to Kratandros that Rhodanthe's father used to lock her up in a tower in order to protect her from men's insidious intentions.[68] The terms employed by Dosikles to illustrate Rhodanthe's bath are of relevance for the purpose of our analysis. Two verses are particularly worth noticing: '[…] except when the girl's soiled flesh had need of a bath and a cleansing stream'.[69] The conceptual core of these verses resides in the cleansing water and in the flesh.

Earlier on, I highlighted some relevant aspects of the theological debate concerning body and soul. I would now like to further stress and illustrate the concept of flesh (σάρξ), which is functional to our understanding of the theological significance of water in its beneficial shape. Flesh is an ambivalent subject within Christianity. According to Matthew's and Mark's Gospels, while the spirit is ready (πρόθυμος), flesh is weak (ἀσθενής).[70] In the Epistle to the Galatians, the desires of flesh are defined as opposite to the spirit.[71] Nevertheless, this lower status of flesh in comparison to the spirit is still susceptible to improvement and salvation through sanctified water. By virtue of Christ's incarnation, 'all people [σάρξ, namely flesh] will see God's salvation'.[72] After Christ's baptism in the Jordan's waters and the Theophany, the possibility of grace became visible and tangible for everyone and, since 'flesh and blood cannot inherit the kingdom of God', baptism was presented as the way *par*

---

same adjective λίχνον when describing his audience as having 'gluttonous ears', namely as being 'greedy for learning', Van der Valk M., *Eustathii archiepiscopi Thessalonicensis commentarii ad Homeri Iliadem pertinentes* (Leiden – Boston, MA: 1971–1987) 1:11.17–31; see also Pizzone A., "Audiences and Emotions in Eustathios of Thessalonike's Commentaries on Homer", *Dumbarton Oaks Papers* 70 (2016) 225–244, 227–230. Rather than simple curiosity, the right connotation for λίχνον in our context surely involves the emotions of desire and lust.

68 The image of a girl locked in a tower by her father, worried that her beauty could attract men's attention, is a common fairy-tale *topos*, but a hagiographic one too. The same happens to Saint Barbara, who manages to baptise herself in a bath right before being imprisoned. See Symeon Metaphrastes, *Passio sanctae Barbarae* in Migne J.J. (ed.), *Patrologia Greca* 116, 301–316.

69 Again, the reference to Bathsheba seems clear: 'And David sent messengers and took her; and she came in unto him, and he lay with her; for she was purified from her uncleanness: and she returned unto her house' (2 Sam. 11:4, King James Version). Similarly to Rhodanthe, Bathsheba had also just been purified by means of her bath. On the cleansing role of water in the biblical context, see Chapter 12 in this volume, Magennis H., "Water in Saints' Lives: A Study of the Old English Corpus", 372–406.

70 Matthew 26:41, Mark 14:38.

71 Galatians 5:17.

72 Luke 3:6. See also John 1:14: Καὶ ὁ λόγος σὰρξ ἐγένετο ('The Word became flesh').

excellence to the salvation of the soul.[73] According to Pseudo-Macarius (fourth century), for example, baptism is considered instrumental for the union with Christ and the 'deification' of the corporeal aspect of human existence.[74] If we look back at the two verses highlighted above, we can see how Rhodanthe's flesh is described as soiled (ῥυπανθὲν, meaning 'contaminated'), and only a bath and a purifying (καθάρσιος) torrent (ῥοή) of water will be able to cleanse her.[75]

However, Rhodanthe's flesh is far from being soiled. As a virgin, Rhodanthe represents an icon of purity, and the preservation of her virginity constitutes another crucial thread line of our novel, while also presenting a tight bond with the ubiquitous presence of water and its shapes.[76] Dosikles's intentions towards Rhodanthe are respectable, and divine providence (θεῶν πρόνοια) blesses the purity of their love from the very beginning, when in Book 2, right after abducting Rhodanthe, Dosikles is provided with a boat for their elopement.[77] According to Dosikles's respectable intentions and to Rhodanthe's consensus to him, the context of her bath recalls the typical pre-marriage bath, which was an important tradition and social occasion in late Antiquity.[78] The girl herself is very well aware and outspoken about the 'desire without passion (πόθος πάθους δίχα) and the pure bond (δεσμὸς ἁγνὸς) of bodies that

73 1 Cor 15:50: σάρξ καὶ αἷμα βασιλείαν θεοῦ κληρονομῆσαι οὐ δύναται. During the Feast of Theophany, the Orthodox Church celebrates the rite of the Great Blessing of Water, during which nature is sanctified and becomes a 'fountain of immortality [...] and a remission of sins' (Meyendorff, *Byzantine Theology* 170). On the role of water in the Byzantine euchology, see Ruggieri V., "Il pozzo e la cisterna: l'acqua nella vita quotidiana e nell'eucologio bizantino", *Orientalia Christiana Periodica* 83 (2017) 35–57.

74 Meyendorff, *Byzantine Theology* 68.

75 Prodromos, *Rhodanthe and Dosikles*, ed. Marcovich 2.179–180. It is important to underline that the Byzantine bath can acquire different meanings in different contexts. In the novel *Hysmine and Hysminias*, a passionate Hysminias dreams about bathing with Hysmine in a bathhouse, and the couple indulges with caresses and kisses. The dreamlike context allows them to reach a peak of eroticism that would otherwise be forbidden and that is also exceptionally free from any moral constraint or condemnation (Makrembolites, *Hysmine and Hysminias* 5.1.3–4, ed. Marcovich). On the variety of bath-related contexts in the Byzantine literature, see also Agapitos P.A., "Zwischen Grauen und Wonne. Das Bad in der byzantinischen Literatur", *Jahrbuch der österreichischen Byzantinistik* 54 (2004) 19–37, 30–32.

76 On the role of the virgin as an icon of purity, see Cooper K., "The Virgin as Social Icon: Perspectives from Late Antiquity", in Van Dijk M. – Nip R. (eds.), *Saints, Scholars, and Politicians. Gender as a Tool in Medieval Studies* (Turnhout: 2005) 9–24.

77 Prodromos, *Rhodanthe and Dosikles*, ed. Marcovich 2.455.

78 DeForest D., "Between Ideology and Social Practice: Baths and Bathing in Christian Communities in Late Antiquity", *Journal of Ancient History* 6.1 (2018) 157.

are not linked', until they will be married.[79] However, the two lovers' wedding intentions fade away as soon as they are imprisoned by the pirates. From that moment on, the preservation of Rhodanthe's virginity becomes an absolute priority in the plot, to the point that the adjective 'virgin' (παρθένος) is by far the most recurrent epithet attributed to Rodhanthe.[80] Already in Book 1, right after being imprisoned by the pirates, one of Dosikles's first worries is that one of those brutes could violate Rhodanthe's purity.[81] Moreover, when Dosikles believes Rhodanthe to be dead after the shipwreck, he torments himself with the thought that the ocean might have thrown out (ἐκρίπτω) her body on the shore, unveiled (ἀσυγκάλυπτος) and in the public eye.[82]

A recurrent shape of water in connection with purity and chastity is dew (δρόσος). Even though Rhodanthe had decided upon maintaining a chaste relationship with Dosikles until they are able to get married, their choice does not exclude the presence of corporeal desire.[83] While Rhodanthe appears firm and determined in her intentions, erotic temptation sometimes afflicts poor Dosikles. In Book 2, he wonders which dew will ever be able to quench the burning fire of his passion.[84] Later on, the divine (θεῖος) dew is essential in keeping their love warm, safe and chaste, not as a metaphor for the preservation of virginity but as a symbolic element that accompanies the legitimate and refreshing union of the protagonists' wedlock and extinguishes the flame of an unsuitable passion.[85] Dew in connection with virginity and chastity is a recurrent *topos*, especially in the Byzantine novel. For example,

---

79 Prodromos, *Rhodanthe and Dosikles*, ed. Marcovich 2.62–67. The substantive πάθος, translated by Jeffreys as 'passion', can also be intended with the meaning of 'incident, unfortunate accident' or even 'disaster', with a reference to the loss of virginity out of wedlock. The conflict between chastity and connubial life is a frequent topic of hagiography: see Kazhdan A., "Byzantine Hagiography and Sex in the Fifth to Twelfth Centuries", *Dumbarton Oaks Papers* 44 (1990) 44, 133, 199.

80 Ibidem 132: 'On the scale of Byzantine ethical values, virginity occupied the topmost place, much higher than legitimate marriage; it is, however, to be noted that virginity was considered a type of marriage, the virgin (or nun) being proclaimed Christ's marriage. A legitimate marriage, concluded with the approval of parents, was desirable'.

81 Prodromos, *Rhodanthe and Dosikles*, ed. Marcovich 1.104–105.

82 Prodromos, *Rhodanthe and Dosikles*, ed. Marcovich 6.481–486.

83 Prodromos, *Rhodanthe and Dosikles*, ed. Marcovich 3.66–65.

84 Prodromos, *Rhodanthe and Dosikles*, ed. Marcovich 2.282–283: Εἰ δ' οὐ θελήσει τὴν συναφὴν ὁ Στράτων, ποίαν ἐφεύρω τοῦ πυρὸς τούτου δρόσον ('But if Straton does not want this connection, what dew can I find to quench this fire?' trans. Jeffreys).

85 Prodromos, *Rhodanthe and Dosikles*, ed. Marcovich 2.463–465: Ἔρως δὲ θάλποι καὶ πόθου θεία δρόσος ὑμῶν δροσίζοι τὴν φίλην συζυγίαν, καὶ μηδὲν ὑμῖν ἐμποδὼν παρεμπέσοι ('May Eros warm you, and the divine dew of desire besprinkle your precious union, and may no obstacle come your way', trans. Jeffreys).

in Eumathios Makrembolites's *Hysmine and Hysminias*, whose plot follows a similar love story between a boy (Hysminias) and a girl (Hysmine), the girl's love is described by the boy as a high-branching cypress planted in the midst of his soul and anointed by her maidenly dew (παρθενική δρόσος).[86] The combination of dew and fire belongs to a deep-rooted biblical tradition that links dew – and, in the case of our novel, also Rhodanthe – to Mary, mother of Christ. In the Book of Daniel, an angel saves the three young Jews Shadrach, Meshach and Abednego, thrown by king Nabuchadnezzar in a fiery furnace, by blowing against the flame wind full of dew (πνεῦμα δρόσου).[87] Being completely opposite elements, dew and fire belong to a series of biblical paradoxes – such as the 'unwatered root' in Isaiah (53:2) – which are later understood to refer to Mary's immaculate conception and to the virgin birth. If the paradoxical juxtaposition of opposed elements is possible, the virgin birth is possible.[88] Along with Christ's birth, the celestial dew also recalls the fecundity of the soil. In the *Vita of Barlaam and Josaphat* the divine dew (οὐράνιος δρόσος) is considered to be sent by the Creator to irrigate the unfertile desert.[89]

Therefore, beneficial water – in the shapes of bath and dew – not only accompanies the positive development and major events of the protagonists throughout the narrative but also provides them with a theological legitimisation. Despite the misadventures and sometimes tragic experiences of the protagonists, the metaphorical connection of water with theological elements – such as baptism and the element of virginity – acts as guarantee for a positive finale. In the very last verse of the novel, Rhodanthe finally surrenders and has 'full knowledge of her bridegroom Dosikle's.[90] The corporeality of their carnal union has been refreshed, blessed and legitimated throughout the narrative by the metaphorical sanctification through water. Therefore, flesh is holy again and Rhodanthe and Dosikles 'are no longer two, but one flesh'.[91]

5 Concluding Remarks

Water as a multifaceted element is a ubiquitous presence within the natural environment. The recurring interaction between water and humans constantly

86 Eusthatius Macrembolites, *Hysmine and Hysminias*, ed. Marcovich 10.11.12–16.
87 Daniel 3:49–50.
88 Gador-Whyte S., *Theology and Poetry in Early Byzantium: The Kontakia of Romanos the Melodist* (Cambridge: 2017) 116.
89 *Vita of Barlaam and Josaphat*, ed. Volk 18.78–80.
90 Prodromos, *Rhodanthe and Dosikles*, ed. Marcovich 9.486: ἔγνω Δοσικλῆν ἡ Ῥοδάνθη νυμφίον.
91 Matthew 19:6: οὐκέτι εἰσὶν δύο ἀλλὰ σάρξ μία.

triggers human embodied cognition, leading to the literary transfer of human experientiality of the surrounding environment into the literary form. This cognitive translation results not only in conceptual metaphors having water as source domain but also in an overarching narrative structure that reflects the human perception of water's agency as a pervasive and leading presence in nature. Moreover, we have observed how, by analysing water's wholeness in its specific shapes, each of these shapes reflects individual theological meanings that, in many instances, attribute to the narratological structure the function of *exempla*.

It is also important to stress that water – especially seawater – played a crucial role for the citizens of the Byzantine Empire, especially someone like Theodore Prodromos, who lived in Constantinople within the sphere of the intellectual and political elite at the imperial court. As Nilsson and Veikou point out, 'maritime communication was essential for the function of the state and it was associated with financial investment in maritime trade as a central strategy for Byzantium's economic prosperity and growth'.[92] As we have seen, the literary transfer of water into conceptual metaphors and narrative does not merely reflect human experientiality in a broader sense, but – by virtue of its specific historical, cultural and theological value – it actually contributes to shaping the conceptualisation of Byzantine life experience.

From this perspective, the contradictions embedded in water's various shapes – harmful and beneficial character, body and soul, darkness and light, sin and salvation – express the full nature of human life experience, with its constant alternation of shadows and lights, within the framework of a Christian society. In Numbers 20:13 we read, 'This is the water of contradiction (ὕδωρ ἀντιλογίας), where the children of Israel strove with words against the Lord and where he was sanctified in them'.[93]

Contradictions, variegation, and intertextual echoes are key elements of the dialogue between the more typically Byzantine aspects of the novel and its ancient heritage. In the framework of such a literary and cultural *poikilia*, the ubiquitous element of water serves as a cognitive and thematic thread line throughout the whole novel. Theodore Prodromos indulges in playing with the combination of Christian and profane elements, allowing them to constantly intertwine and, often, to re-emerge – as if they were twisting and turning under the surface of a body of water – to carve the multifaceted structure of the novel.

92 Veikou – Nilsson, "Ports and Harbours as Heterotopic Entities in Byzantine Literary Texts" 267.

93 Num. 20:13 (*Septuagint*, ed. Rahlfs): τοῦτο ὕδωρ ἀντιλογίας, ὅτι ἐλοιδορήθησαν οἱ υἱοὶ Ισραηλ ἔναντι κυρίου καὶ ἡγιάσθη ἐν αὐτοῖς. See Classen, *Water in Medieval Literature* XIII.

Bibliography

Primary Sources

Bergson L. (ed.), *Historia Alexandri Magni (Recensio β, e cod. Leidensi Vulc. 93)* (Stockholm: 1965).

Conca F. (ed.), *Nicetas Eugenianus, De Drosillae et Chariclis amoribus* (Amsterdam: 1990).

Fobes F.H. (ed.), *Aristotelis meteorologicorum libri quattuor* (Cambridge, MA: 1919).

Gautier P. (ed.), "Le dossier d'un haut fonctionnaire byzantin d'Alexis Ier Comnène, Manuel Straboromanos", *Revue des études byzantines* 23 (1965) 201–204.

Hörandner W. (ed.), *Theodoros Prodromos. Historische Gedichte, Text und Kommentar*, Wiener byzantinistische Studien 11 (Vienna: 1974).

Jeffreys E. (trans.), *Four Byzantine Novels* (Liverpool: 2012).

Kroll W. (ed.), *Historia Alexandri Magni, Recensio α sive recensio vetusta* (Berlin: 1926).

Lee H.D.P. (ed. and trans.), *Aristotle: Meteorologica*, Loeb Classical Library 397 (Cambridge, MA: 1952).

Marcovich M. (ed.), *Theodori Prodromi de Rhodanthes et Dosiclis amoribus libri IX* (Stuttgart: 1992).

Marcovich M., *Eustathius Macrembolites De Hysmines et Hysminiae amoribus libri XI* (Munich – Leipzig: 2001).

Migne J.P. (ed.), *Patrologia Greca*, 162 vols. (Paris : 1857–1886).

Moraux P. (ed.), *Aristote. Du ciel* (Paris: 1965).

Panagioto A.D. (ed.), Ὁ Ἰωάννης Μαυρόπους Ὑμνογράφος τοῦ Ἁγίου Νικολάου (Athens: 2008).

Rahlfs A. (ed.), *Septuaginta* (Stuttgart: 1935).

Ross W.D. (ed. and trans.), *Aristotle: Metaphysics* (Oxford: 1924).

Stoneman R. (trans.), *The Greek Alexander Romance* (London: 1991).

Vilborg E. (ed.), *Achilles Tatius. Leucippe and Clitophon* (Stockholm: 1955).

Van der Valk M., *Eustathii archiepiscopi Thessalonicensis commentarii ad Homeri Iliadem pertinentes* (Leiden – Boston, MA: 1971–1987).

Volk R. (ed.), *Die Schriften des Johannes von Damaskos*, VI/2 (Berlin – New York: 2006).

Winkler J.J (trans.), "Achilles Tatius: 'Leucippe and Clitophon'", in Reardon B.P. (ed.), *Collected Ancient Greek Novels* (Berkeley – Los Angeles, CA – London: 1989) 170–284.

Secondary Sources

Abbot P.H., "Narrativity", *The* Living *Handbook of Narratology* (Hamburg: 2014), last updated January 20, 2014, https://www.lhn.uni-hamburg.de/node/27.html last accessed May, 2, 2024.

Adam J.M., *Les textes. Types et prototypes. Recit, description, argumentation, explication et dialogue* (Paris: 1992).

Agapitos P.A., "The Erotic Bath in the Byzantine Vernacular Romance *Kallimachos and Crysorrhoe*", *Classica et Mediaevalia* 51 (1990) 257–273.

Agapitos P.A., *Narrative Structure in the Byzantine Vernacular Romances. A Textual and Literary Study of Kallimachos, Belthrandos and Libistros* (Munich: 1991).

Agapitos P.A., "Narrative, Rhetoric, and 'Drama' Rediscovered: Scholars and Poets in Byzantium Reinterpret Heliodorus", in Hunter R. (ed.), *Studies in Heliodorus* (Cambridge: 1998) 25–156.

Agapitos P.A., "New Genres in the Twelfth Century: The Schedourgia of Theodore Prodromos", *Medioevo Greco* 15 (2015) 1–41.

Agapitos P.A., "Zwischen Grauen und Wonne. Das Bad in der byzantinischen Literatur", *Jahrbuch der österreichischen Byzantinistik* 54 (2004) 19–37.

Agapitos P.A. – Smith O.L., *The Study of Medieval Greek Romance: A Reassessment of Recent Work* (Copenhagen: 1992).

Althoff J., "hydôr / Wasser", in Höffe O. (ed.), *Aristoteles-Lexikon* (Stuttgart: 2005) 270.

Beaton R., *The Medieval Greek Romance* (Cambridge: 1996).

Berger A., *Das Bad in der byzantinischen Zeit* (Munich: 1982).

Burton J., "Reviving the Pagan Greek Novel in a Christian World", *Greek, Roman, and Byzantine Studies* 39 (1998) 179–216.

Buttimer A., "Nature, Water Symbols, and the Human Quest for Wholeness", in Seamon D. – Mugerauer R. (eds.), *Dwelling, Place and Environment: Towards a Phenomenology of Person and World* (New York: 1985) 259–280.

Cairns D., "Mind, Metaphor and Emotion in Euripides (Hippolytus) and Seneca (Phaedra)", *Maia: rivista di letterature classiche* 69.2 (2017) 247–267.

Classen A., *Water in Medieval Literature. An Ecocritical Reading* (Lanham, MD – Boulder, CO – New York – London: 2017).

Cooper K., "The Virgin as Social Icon: Perspectives from Late Antiquity", in Van Dijk M. – Nip R. (eds.), *Saints, Scholars, and Politicians. Gender as a Tool in Medieval Studies* (Turnhout: 2005) 9–24.

DeForest D., "Between Ideology and Social Practice: Baths and Bathing in Christian Communities in Late Antiquity", *Journal of Ancient History* 6.1 (2018) 136–165.

Dennis G.T., "The Perils of the Deep", in Sode C. – Takács S.A. (eds.), *Novum Millennium: Studies on Byzantine History and Culture Dedicated to Paul Speck* (London – New York: 2001) 81–88.

Evola V., "La metafora come *carrefour* cognitive del pensiero e del linguaggio", in Casadio C. (ed.), *Vie della metafora: linguistica, filosofia, psicologia* (Chieti: 2008) 55–80.

Fludernik M., *Towards a Natural Narratology* (London: 1996).

Gador-Whyte S., *Theology and Poetry in Early Byzantium: The Kontakia of Romanos the Melodist* (Cambridge: 2017).

Harvey D., *Justice, Nature, and the Geography of Difference* (Oxford: 1996).

Herman D., "Cognitive Narratology", *The* Living *Handbook of Narratology*, (revised version; uploaded 22 September 2013) last updated December 24, 2013, https://www.lhn.uni-hamburg.de/node/38.html; last accessed November, 3, 2023.

Hühn P., "Event and Eventfulness", *The* Living *Handbook of Narratology*, last updated September 13, 2013, https://www.lhn.uni-hamburg.de/node/39.html last accessed May, 2, 2024.

Jones O. – Cloke P., *Tree Cultures: The Place of Trees and Trees in their Place* (Oxford – New York: 2002).

Kapriev G., *Philosophie in Byzanz* (Würzburg: 2002).

Kazhdan A., "Byzantine Hagiography and Sex in the Fifth to Twelfth Centuries", *Dumbarton Oaks Papers* 44 (1990) 131–143.

Knappett C. – Malafouris L., "Material and Nonhuman Agency: An Introduction", in Knappett C. – Malafouris L. (eds.), *Material Agency. Towards a Non-Anthropocentric Approach* (New York: 2008) IX–XIX.

Kövecses Z., *Metaphor: A Practical Introduction* (New York – Oxford: 2nd ed. 2010).

Lakoff G. – Johnson M., *Metaphors We Live By* (*with a New Afterword*) (Chicago – London: 2003, original ed. 1980).

Maguire H., *Nectar and Illusion. Nature in Byzantine Art and Literature* (Oxford: 2012).

Marciniak P., "Byzantine Theatron – A Place of Performance?" in Grünbart M. (ed.), *Theatron: Rhetorische Kultur in Spätantike und Mittelalter / Rhetorical Culture in Late Antiquity and the Middle Ages* (Berlin: 2007) 277–286.

Messis C. – Nilsson I., "Byzantine Storytelling and Modern Narratology: An Introduction", in Messis C. – Mullet M. – Nilsson I.(eds.), *Storytelling in Byzantium. Narratological Approaches to Byzantine Texts and Images* (Uppsala: 2018) 1–11.

Meyendorff J., *Byzantine Theology: Historical Trends and Doctrinal Themes* (New York: 1974).

Mullett M., "Aristocracy and Patronage in the Literary Circles of Comnenian Constantinople", in Angold M. (ed.), *The Byzantine Aristocracy IX to XIII Centuries* (Oxford: 1984) 173–201.

Nilsson I.,"Ancient Water in Fictional Fountains: Waterworks in Byzantine Novels and Romances", in Shilling B. – Stephenson P. (eds.), *Fountains and Water Culture in Byzantium* (Cambridge: 2016) 281–298.

Nilsson I., *Erotic Pathos, Rhetorical Pleasure: Narrative Technique and Mimesis in Eumathios Makrembolites' Hysmine and Hysminias* (Uppsala: 2001).

Nilsson I., *Raconter Byzance: la littérature au XIIe siècle*, Séminaires byzantins 3 (Paris: 2014).

Nilsson I., "Romantic Love in Rhetorical Guise: The Byzantine Revival of the Twelfth Century", in Cupane C. – Krönung B. (eds.), *Fictional Storytelling in the Medieval Eastern Mediterranean and Beyond* (Leiden – Boston, MA: 2016) 39–66.

Pizzone A., "Audiences and Emotions in Eustathios of Thessalonike's Commentaries on Homer", *Dumbarton Oaks Papers* 70 (2016) 225–244.

Ronchey S., *Lo Stato Bizantino* (Turin: 2002).

Ruggieri V., "Il pozzo e la cisterna: l'acqua nella vita quotidiana e nell'eucologio bizantino", *Orientalia Christiana Periodica* 83 (2017) 35–57.

Telelis I.G., "Οι λόγιοι του 11ου αιώνα και ο Αριστοτελισμός: η περίπτωση των 'Μετεωρολογικών'", in Chrysos E. (ed.), Η Αυτοκρατορία σε Κρίση (;) Το Βυζάντιο τον 11ο Αιώνα (1025–1081) (Athens: 2003) 425–442.

Telelis I.G., "Meteorology and Physics in Byzantium", in Lazaris S. (ed.), *A Companion to Byzantine Science* (Leiden – Boston, MA: 2020) 177–201.

Thrift N., *Spatial Formations* (London: 1996).

Trizio M., "Reading and Commenting on Aristotle in Byzantium", in Kaldellis A. – Siniossoglou N. (eds.), *The Cambridge Intellectual History of Byzantium* (Cambridge: 2017) 397–412.

Trocmé E., "Light and Darkness in the Fourth Gospel", *Disdaskalia* 6.2 (1995) 3–13.

Veikou M. – Nilsson I., "Ports and Harbours as Heterotopic Entities in Byzantine Literary Texts", in von Carnap-Bornheim C. – Daim F. – Ettel P. – Warnke U. (eds.), *Harbours as Objects of Interdisciplinary Research – Archaeology + History + Geosciences* (Mainz: 2018) 265–277.

CHAPTER 11

Marie de France's Writing on Water: the Etymological Narrativity of the Sea in the *Lais*

Rafał Borysławski

Introduction

In his introduction to *The Anonymous Marie de France*, Robert Howard Bloch calls Marie a disruptor of prevailing cultural values and a founder of new ones, asserting that the *Lais* she authored[1] appear in and reflect on one of the liminal moments in the history of the medieval West.[2] Indeed, if we were to search for the narrative essence of Marie de France's *Lais*, for the narrative spaces they are located in and for their hermeneutic and moralising dimension, the likely answers would point to the idea of liminality. Liminal spaces and the qualities of in-betweenness permeate almost every aspect of the *Lais*, ranging from the way they function in between the linguistic boundaries of three languages – French, Breton, and, occasionally, English – to the in-betweenness of genders and gender relations they depict, to the transgressive spaces between value systems, morality, and immorality. The in-betweenness of the *Lais* appears as prominent also in their settings: from their imagined geographic locations spanning Brittany, Wales, and England, to the liminality of the human and the supernatural worlds, and, equally importantly, to the liminality of the human and the natural world.

In this last interaction, water and, in particular, the waters of the sea play a narratively crucial role. Culturally associated with the rites and moments of symbolic and literal passage, water itself is steeped in liminal connotations. Likewise, the waters of the sea, particularly in the context of the *Lais*, are abundant in the multivalent senses that they bring to the ideas of parting, distance, death, and to the concept of the voyage with its capacity to both sever lovers

1 A *lai* or, more properly, Breton *lai* is a relatively short narrative, octosyllabic poem that contains elements of adventure, magic, and romance, whose origins, in the words of Marie, are related to Breton orature.

2 Bloch R.H., *The Anonymous Marie de France* (Chicago – London: 2003) 19.

 | DOI:10.1163/9789004696501_013

and reunite those that have been separated.[3] The sea waters pose questions, lure with answers, and they are an instrument of transformation: despite the cursory nature of his overview of the motif of water in medieval literature, Albrecht Classen observes that in the *Lais,* water represents challenges that are not merely physical but enforces transformations of the protagonists and offers solutions to the issues they face.[4] We may perhaps speculate then that when in the epilogue to her fables, Marie provides the only personal detail about herself, *Marie ai num, si sui de France*, 'My name is Marie and I am of France',[5] it is also a testimony to a likely transformation in Marie's life connected with a sea voyage. Indirectly, the line evokes its presence as a transitional space which Marie must have at least once crossed, leaving France from where she tells us she came. Incidentally, phonetically and from the perspective of the medieval etymologising, the sea is also distantly echoed in Marie's name.[6] Whether she ever spared a thought to the correspondences between the sea and the first syllable of her own name or to its etymology, we can never know, but phonetic and semantic analogies such as this will be among the points explored in this chapter. Bloch also notes a specific linguistic and etymological sensitivity of Marie de France, describing her as 'obsessed with the opacity of words'.[7] In Bloch's view, Marie's preoccupation with language displays both clarity and ambiguity as it is 'expressed with unmistakable consciousness sometimes, and at other times seemingly without awareness'.[8] It is from the perspective of such linguistic and semantic sensitivity grounded in the medieval practice of etymologising that this chapter aims to address the presence and the liminal significance of the waters of the sea in Marie de France's *Lais*. The sea's semantic potential is both significantly coded and unravelled in them by means of intentional and, seemingly perhaps, unintentional uses of the etymologically inspired connotations, which, as will be shown, are also related to the semantic areas of love and death so fundamental for the *Lais*. Thus, the underlying

3 The paradoxical nature of the sea in a Byzantine novel roughly contemporary to Marie de France's *Lais*, along with its narrative and symbolic ubiquitousness, is explored by Laura Borghetti in the preceding chapter (317–343). In Chapter Six (196–224), Simon Karstens discusses later medieval contradictory perceptions of the ocean as the space of danger and mystery, while in Chapter Eight (253–283), Marusca Francini examines the symbolic meanings of water in the Gothic Gospels.

4 Classen A., *Water in Medieval Literature* (Lanham, MD – Boulder, CO – New York – London: 2018) 91.

5 Warnke K. (ed.), *Die Fabeln der Marie de France* (Geneva: 1974) 327.

6 As *mar*, the sea phonetically resounds in the name's first syllable and it is also present in the etymology of the name Mary (see notes 21 and 22 below).

7 Bloch, *The Anonymous Marie de France* note 24, 327.

8 Ibidem 35.

assumption of this chapter is that of the critical role that etymologies and etymologising held for medieval Christian hermeneutics. By grounding this examination in Isidore of Seville's *Etymologies* and Andreas Capellanus's *De amore*, etymologically instructed by Isidore's work, I intend to demonstrate that the possibilities that etymologising brought with itself heighten the sense of ambiguity and the liminality of the *Lais*. Their meanings escape uniform readings and are intentionally fluid, much like water running through one's fingers. Furthermore, the potential for ambiguity may also be related to how the *Lais* were experienced by their courtly audiences. Evelyn Birge Vitz has argued that the octosyllabic couplet of the *Lais* is strongly rooted in the oral culture of the Old French language[9] and that Marie de France was likely aware of the interconnectedness between the text written in such couplets and its oral rendition.[10] Therefore, the oral delivery and aural reception of the *Lais* may have contributed to the possibility of both stressing and hearing, by the performer and the audience respectively, the words and syllables that invited conjecturing connections between homophonically and etymologically correlated elements.

This chapter will argue that of all the four elements, water, and in particular the sea, is given unique agency by Marie de France in her *Lais* and that this agency is intensified by the logocentric perspective instructed by Isidorean etymologising. Even though it is impossible to prove its direct influence on Marie's oeuvre, etymologising was evidently among the most influential and widespread cognitive approaches to the linguistic dimension of reality throughout the Middle Ages and beyond. Thus, the chapter will discuss the significance of the aquatic motif in the *Lais* following Isidore's conceptual etymologies and their correlations of the Latin words for the sea, love and death, with some additional remarks on the expanded etymology of love provided by Andreas Capellanus's *De amore*. The chapter's primary argument will be focused on the ambivalent narrative roles of the aquatic motifs in *Equitan*, *Lanval*, *Yonec*, and *Milun*, paying closer attention to the functions related to the sea in the *lais* opening and closing the collection in the late thirteenth-century manuscript BL Harley 978,[11] *Guigemar* and *Eliduc*. Concluding with a closer study of the *lai* of *Guigemar*, the chapter will explore the sea as narratively meaningful through its linguistic multivalences, which serve the heuristic purposes of the *Lais*. From this perspective, I shall read passages from *Guigemar* as

9 Vitz E.B., *Orality and Performance in Early French Romance* (Cambridge: 1999) 4–25.

10 Ibidem 38–45.

11 For a comprehensive discussion of Harley 978 MS, see Taylor A., *Textual Situations: Three Medieval Manuscripts and Their Readers* (Philadelphia: 2002) 76–136.

revealing, concealing, and constructing its narrative essence presented there as concerned with the eschatology of love.

1 The Sea, Death, and Love in the Tradition of Medieval Etymologising

The roots of the belief that etymons or alleged etymons of words hold keys to the origins of the concepts behind them and constitute a reflection of the divine order of creation are immemorial. In medieval and later Europe, the principal source of such a logocentric approach to reality was Isidore of Seville's *Etymologiarium libri XX sive Origines*. The range of Isidore's primary material is vast[12] and the work's unparalleled influence on and fundamentality to medieval science and thought have been widely discussed.[13] The cognitive mode that the *Etymologies* have been built upon is present, even if circuitously, whenever we encounter instances of linguistic self-referentiality, toponymy, and onomastics, but also of the allegorical ways of thinking so characteristic of medieval hermeneutics.[14] The *Etymologies* have been, therefore, at the centre of scholarly attention with regard to their role on linguistic, rhetorical, and theological conceptualisations of medieval forms of thought.[15] Their logocentric understanding of the nature of reality and of how its elements are conceptualised into language are of significant interest in the present attempt to hypothesise about the effect of the medieval etymological knowledge exerted on the interpretations of the *Lais*. The influence of Isidore's *Etymologies* on

12 A survey of the sources of the *Etymologies* is provided in S.A. – Lewis W. J. J. – Beach A. – Berghof O. (trans.), *The Etymologies of Isidore of Seville* (Cambridge: 2006) 10–17.

13 Davide Del Bello provides a selection of contemporary scholars' voices who assert the large magnitude of the *Etymologies* on the European Middle Ages: Del Bello D., *Forgotten Paths. Etymology and the Allegorical Mindset* (Washington, DC: 2007) 95. In the introduction to their translation, Barney, Lewis, Beach, and Berghof outline the chief directions of the influences of the *Etymologies* 24–26.

14 See Del Bello's discussion of allegorical etymology as a form of thought, *Forgotten Paths* 95–115.

15 See, for instance, the discussion of the heuristic and cognitive specificity of the classical and medieval practice of etymologising in Ansler M.E., *Etymology and Grammatical Discourse in Late Antiquity and the Early Middle Ages* (Amsterdam: 1989); Del Bello presents medieval etymologising as a cognitive tool that contributed to the allegorical ways of thinking characteristic of medieval hermeneutics in *Forgotten Paths*; most recently, Mercedes Salvador-Bello grounds the entire tradition of early medieval riddling in Isidorean etymological categories by extension claiming that the fundaments of early medieval learning and intellectual mindset are to a large extent indebted to his *Etymologies*: Salvador-Bello M., *Isidorean Perceptions of Order* (Morgantown, WV: 2015).

the *Lais* may be compared to a hypertext that invites one to see connections between concepts existing across semantic categories. Those among the audiences that were familiar with Isidorean etymologising would hear and discern such connections. Their presence would also sensitise the learned recipients of the *Lais* to their ambivalent potential, which is, as will be shown, the effect that Marie claims she wishes to achieve.

Before this chapter proceeds to explore the ambivalence and multivalence of the meanings related to the presence of the sea in the *Lais*, let us consider the semantic parallels which become evident once Isidorean etymology of *mare*, the sea, is juxtaposed with his etymologies for *mors*, death, and *amor*, love. Although they appear in different books of the *Etymologies*, all of them are related to the experience of bitterness and entrapment. Book XIII on the cosmos and its parts describes the sea as a general gathering of water: 'Mare est aquarum generalis collectio. Omnis enim congregatio aquarum, sive salsae sint sive dulces, abusive maria nuncupantur, iuxta illud' (Genesis 1:10): 'Et congregationes aquarum vocavit maria'. Proprie autem mare appellatum eo quod aquae eius amarae sint'.[16] 'A sea is a general gathering of waters, for every confluence of waters, whether they are salt or fresh, is, loosely, a "sea", according to this (Genesis 1:10): 'And the gathering together of waters, he called seas (*maria*)'. But strictly speaking, something is called sea (*mare*) because its waters are bitter (*amarus*)'.[17] Of additional importance for the allegorical reading of the sea are Isidore's naturalistic comments on its mercurial appearance. The sea, he claims, 'autem certum non esse colorem, sed pro qualitate ventorum mutari; nam modo flavum est, modo lutulentum, modo atrum',[18] 'has no specific colour; it changes with the quality of the winds. Sometimes it is golden, sometimes muddy, and sometimes black'.[19] However, what Isidore distinguishes as the chief difference between the seawater and other waters is not the changeability of its colour and its expanse, but its taste, described not as salty but as bitter. Bitterness is then the key concept to the etymon for the sea. At the same time, saltiness, and salt, the most important of medieval food condiments, mainly held positive connotations, and Isidore speaks highly of it, seeing its etymology as cognate to *salus*, 'health' in Book XVI.[20] Given the fact that salt was obtained from seawater, it may seem odd that the taste of the sea was described as *amarus*, 'bitter'. However, it appears that the notion of the

16 Isidore, *Etymologiarium sive originum* libri xx, ed. W.M. Lindsay (Oxford: 1911) XIII.XIV.1.

17 *The Etymologies of Isidore of Seville* 277.

18 Isidore, *Etymologiarum* XVI.II.3–6.

19 *The Etymologies of Isidore of Seville* 277.

20 Isidore, *Etymologiarum* XVI.II.6.

bitterness of the sea may have its distant roots in the etymological connection that in Book VII Isidore claims exists between the name of Mary and the sea itself as *stella maris*, 'the star of the sea',[21] which, as Maryam, ultimately may derive from the Hebrew word מר *mar* signifying 'bitterness' and ים *yam*, 'sea'.[22]

As for the etymology of *mors* 'death', Isidore provides three approaches to it in Book XI devoted to the human being and portents. Its etymon, he claims, may be relatable to Mars, the god of war from the pre-Christian pantheon. Alternatively, it may originate in the Judeo-Christian narrative of the Fall as the bite of the satanic serpent. However, the very first explanation he proposes, as in the case of the sea, involves the reference to bitterness. As in the case of the sea, Isidore finds the etymon for death in the adjective *amarus*: 'Mors dicta, quod sit amara, vel a Marte, qui est effector mortium [sive mors a morsu hominis primi, quod vetitae arboris pomum mordens mortem incurrit]'.[23] 'Death (*mors*) is so called, because it is bitter (*amarus*), or by derivation from Mars, who is the author of death; [or else, death is derived from the bite (*morsus*) of the first human, because when he bit the fruit of the forbidden tree, he incurred death]'.[24] And, as was the case with the bitterness of the sea, 'bitter' had functioned as a collocation with 'death' long before it appeared in Isidore's *Etymologies*. Perhaps one of the most famous associations of death with bitterness that clearly was familiar to Isidore is presented by Virgil in the *Aeneid* when Aeneas descends into Hades and hears the weeping souls of the infants, whose premature death is described there as *acerbo*, 'bitter'.[25] Associating untimely death with bitterness was therefore common – *bitera deaþ*, 'bitter death' remains one of the most typical collocations in Old English poetry.[26] For instance, Ælfric lists *mors acerba*, in the sense of a sudden death, as one of the

21 Ibidem VII.X.1.

22 Klein E., *A Comprehensive Etymological Dictionary of the Hebrew Language for Readers of English* (Tel Aviv: 1987) s.v, "ים," and s.v. "381 ,259 ",מרד. Bitterness as the attribute of the sea is also well-attested in pre-Isidorean sources, from the Old Testament to the works of ancient Greeks, such as Herodotus's *Histories*. Sebastian I. Sobecki lists a number of authors and texts where the sea is qualified as 'bitter', seeing the roots of such qualification in the associations of the sea with demonic abyss and only later with it as a patristic allegory of *mare vitae*. See Sobecki S.I., *The Sea and Medieval English Literature* (Cambridge: 2008) 35–41.

23 Isidore, *Etymologiarum* XI.II.31.

24 *The Etymologies of Isidore of Seville* 243.

25 In his seminal study of the afterlife in pre-Christian Roman culture, Franz Cumont discusses it in the chapter on "Untimely Death", Cumont F., *After-Life in Roman Paganism* (New Haven, CT: 1922) 128–147.

26 For the discussion of this motif, see Thompson V., *Dying and Death in Later Anglo-Saxon England* (Woodbridge: 2004) esp. 9–11, and 103–107.

three forms of death in his homily XI, along with *mors inmatura*, 'untimely death', and *mors naturalis*, 'natural death'.[27]

Lastly, the etymology of the word *amor*, 'love', is presented by Isidore in Book X *De vocabulis* on the vocabulary of the terms intended to characterise people and human types. *Amor*, we are told, is cognate with the word *amicus*, 'friend', both of which are derived from *hamus*, 'hook', signifying the bond between two people: 'Amicus, per derivationem, quasi animi custos. Dictus autem proprie: amator turpitudinis, quia amore torquetur libidinis: amicus ab hamo, id est, a catena caritatis; unde et hami quod teneant. Amabilis autem, quod sit amore dignus'.[28] 'And *amicus* is appropriately derived; the term for someone tormented by carnal desire is *amator turpitudinis* ('lover of wickedness'), but *amicus* is from 'hook' (*hamus*), that is, from the chain of charity, whence also hooks are things that hold. Lovable (*amabilis*), too, because one is worthy of love (*amor*)'.[29] Though it may not be immediately evident, the etymological elucidation appears to bear a somewhat tortuous association with water through the reference to the hook as its etymon. Elsewhere in the *Etymologies*, Isidore associates the word *hamus* with fishing, and thus with the aquatic environment: in Book XIII on animals, he speaks about various types of fish and the methods of catching them, and in Book XIX on ships, building, and clothing, when he discusses shipbuilding. This association seems to have been sensed by Andreas Capellanus, likely a contemporary figure to Marie de France, whose *De amore* provides an extended explanation of the origins of the word love, which is clearly inspired by Isidore's reference to the hook, but which hinges upon the angling metaphor and is thus related to water:

> Dicitur autem amor ab amo verbo, quod significat capere vel capi. Nam qui amat, captus est cupidinis vinculis aliumque desiderat suo capere hamo. Sicut enim piscator astutus suis conatur cibiculis attrahere pisces et ipsos sui hami capere unco, ita vero captus amore suis nititur alium attrahere blandimentis totisque nisibus instat duo diversa quodam incorporali vinculo corda unire vel unita semper coniuncta servare.[30]
>
> *Amor* is derived from the verb *amo*, meaning catch or be caught, for the lover is caught in bonds of desire and longs to catch another on his hook

27 Pope J.C. (ed.), *Homilies of Ælfric. A Supplementary Collection* (London – New York – Toronto: 1967) vol. I, 420, line 113.

28 Isidore, *Etymologiarum* X.A.5.

29 *The Etymologies of Isidore of Seville* 213.

30 Andreas Capellanus, *On Love*, trans. and ed. P.G. Walsh (London: 1982) 36.

> (*hamo*). Just as a clever angler tries to entice the fish with his morsels and to catch them on his bent hook, so the man ensnared by love tries to attract another by his charms. He strives with every effort to join two separated hearts with an imperceptible bond, or to keep them once joined always together.[31]

Capellanus's interpretation of the nature of the lover's behaviour through the etymology of love also evokes the impression of torment, and perhaps even death with the image of a lover caught like a fish on a hook. Some two centuries later, it is the likely inspiration for Chaucer's vision of unfortunate lovers in *The Parliament of Fowls*, caught like fish and then abandoned on a bank of a 'sorewulf were' (line 138) and left as 'the fish in prysoun is al drye' (line 139).[32]

The three Isidorean etymological explanations carry within themselves either a sense of ambivalence or a sense of elusiveness: from *mare*, the sea, whose colour and appearance are changeable, to *mors*, death, the etymologies of which are derived by Isidore from three alternatives corresponding to the three types of death, to *amor*, 'love', etymologically explained by Isidore as being both positive in how it attaches people and destructive when the attachment is carnal. Isidore's explanations of these concepts carry, then, a sense of an interrelatedness that exists between their etymologies. Between them, however, they also carry a phonetic similarity that resounds in the **mr* consonantal cluster. That this similarity must have been audible to medieval authors and audiences is likely. The correspondence between the 'bitterness' of the sea and the 'bitterness' of the experience of love is substantiated by the French and Latin puns based on the aural similarities between *mare* and *amor*. One such significant example is observed by Sebastian I. Sobecki in the twelfth-century French romance of *Tristan*. While little can be said about its author, one Thomas of Britain, he appears to have flourished around the same time as Marie de France. Towards the beginning of the poem (lines 40–52), Ysolt speaks at some length about the bitterness of the sea and the bitterness of human sorrows. As he is listening to her speech, Tristan becomes confused as to whether Ysolt speaks about *la mer*, 'the sea', *l'amur*, 'love', or *amer*, 'bitterness'.[33] All three evidently not only aurally blend and morph into one another, but they also mutually enhance each other's significance as key terms for the narrative of love where the sea and a potion play crucial roles. We shall later

31 Ibidem 37.

32 Benson L.D. (ed.), *The Riverside Chaucer* (Oxford: 1988) 387.

33 Sobecki, *The Sea and Medieval English Literature* 66–67.

see a similarly constructed example of a play on meanings between the sea, love and evil in Marie de France's *Guigemar*.

Altogether, an intriguing and challenging range of meanings appears to emerge from the awareness of the correspondences between the etymologies of the sea, love, and death. Their ambivalently transformative powers are defined by the bitterness that is either directly present as the etymon for the 'sea' and 'death' or that is inherent in the conflicted experience of love in Marie's *Lais* and is likened to the torment of being caught on the etymological hook. From the etymological perspective, the sea is immensely crucial for several of the *Lais*. The etymologies related to it encompass the issues of changeability, emotional relations, death, and the experience of uncertainty. Thus, as I intend to show below, it is not impossible to see the presence of the sea in the *Lais* as an invitation to decode it in instances of punning, but also as a contribution to ambivalent understandings of the *Lais* in general and of their individual motifs. The etymological dimension of the sea may also be present as an element onomastically resounding in the names of three of the *Lais* protagonists who are either related to the sea voyage or who reside by the sea, Guigemar, Meriaduc, and Muldumarec. Lastly, none of these ideas excludes any other; on the contrary, with their etymological fluidity and their meshing into one another, they all add their senses to the understanding of the ambivalence of the *Lais* as literally and figuratively located in-between the sea, love, and the threat of either physical or spiritual death.

If we are to believe the words of her Prologue, Marie de France herself intended her *Lais* to invite individual interpretations, discussions, and debates about their implications. As she opens her *Lais*, she points to the writing and reading practices of 'the ancients' as grounded in the principle of writing enigmatically which is meant to lead to a more refined and subtle understanding of a text. She is thus clearly aware that what appears to be outwardly obscure in ancient texts served heuristic purposes. Writing obscurely, *oscurement*, appears to be the purpose of her compositions as well:

Custume fu as anciëns,
Ceo testimoine Preciëns,
Es livres que jadis faiseient
Assez oscurement diseient
Pur ceus ki a venit esteient
E ki aprendre les deveient
K'i peüssent gloser la letter
E de lur sen le surplus metre.
Li philesophe Ie saveient,

Par els meïsmes l'entendeient,
Cum plus trespassereit li tens,
Plus serreient sutil de sense
E plus se savreient guarder
De ceo qu'i ert, a trespasser.[34] (Prologue, 9–22)

It was customary for the ancients, in the books which they wrote (Priscian testifies to this), to express themselves very obscurely so that those in later generations, who had to learn them, could provide a gloss for the text and put the finishing touches to their meaning. Men of learning were aware of this and their experience had taught them that the more time they spent studying texts the more subtle would be their understanding of them and they would be better able to avoid future mistakes.[35]

Notably, Priscian is the only rhetorical authority whom Marie mentions in her *Lais*. At the very least, the allusion testifies to her familiarity with the author of *Institutiones grammaticae* and *Praeexercitamina*, but, as Mary-Louise Zanoni observes, it may also testify to her familiarity with his principles.[36] And thus, it is likely that Marie may have been conscious of the role Priscian attributed to etymologies in his work.[37] Yet, although clearly present in her oeuvre, Marie's reliance on etymology as a way of indicating a broader spectrum of meanings does not need to be explicit, and it may at times be overlooked when approaching the *Lais*. For one thing, her Prologue rather clearly implies that any directness in interpretation is precisely what Marie de France would see as unsubtle. Secondly, so fundamental was the role of etymology in any consideration of language, rhetoric, and hermeneutics that applying an etymological

34 Rychner J. (ed.), *Les Lais de Marie de France* (Paris: 1983). All subsequent citations from the *lais* are from this edition and will henceforth by indicated by the *lai*'s title and line numbers in parentheses.

35 Marie de France, *The Lais of Marie de France* 41. All subsequent translations are from this work and will be henceforth indicated by page numbers in parentheses.

36 For a discussion of Priscian as an inspiration behind the *Prologue* to the *Lais* see, Zanoni M.L., "'Ceo testimoine Preciëns:' Priscian and the Prologue to the 'Lais' of Marie de France", *Traditio* 36 (1980) 407–415.

37 Mark Amsler emphasises the impact of etymologising on Priscian's grammatical analysis describing it as 'governed by the explanatory procedures of *derivatio* and *compositio* [...], Priscian redeploys these technical elements by elaborating them within a different metalanguage, based on etymological strategies', Amsler, *Etymology and Grammatical Discourse* 73.

perspective would have been implicit in any deliberation over meanings of texts.[38]

2 The Aquatic Motifs in the *Lais*

As a narrative component, water is significantly present in six out of the twelve *Lais*. Three of them, *Guigemar*, *Milun*, and *Eliduc*, albeit to a different degree, are founded on the motif of the sea voyage across the Narrow Seas between Brittany (*Bretaigne mineur*) and England (*Logres*), South Wales (*Suhtwales*), or, in the case of *Guigemar*, an unspecified land bearing some resemblance to the Otherworld. In the other three where water is present, that is in *Equitan*, *Lanval*, and *Yonec*, it is narratively important but without any reference to the sea. I shall briefly view the roles of water in the three *lais* which make no mention of the sea, before discussing the different implications carried by the sea in the other three *lais*.

In *Equitan*, water is crucially present in the *lai*'s central motif, that of the scalding bath. The lethal bath of boiling water is prepared by an adulterous wife and her lover, Equitan, the king of Nantes, with the intention of eliminating the wife's husband, the king's seneschal, by boiling him to death. In the heat of the moment, as the adulterous lovers are discovered by the husband, it is the king, who, trying to hide from the seneschal, jumps into the tub and dies there. He is followed by his lover, tossed into the scalding bath by her enraged husband. Thus, the purifying association that water typically bears with itself is ironically combined here with the chastisement of fire needed to prepare the deadly, boiling bath, originally intended by the scheming wife as the method of disposing of her husband. The climax to the *lai* is therefore essentially formed by the effect of the fusion of the presence of two opposing elements, water, and fire, much as adulterous love is a fusion of desire with sinfulness.

The water appearing in *Lanval* takes the form of a stream, with a plethora of possible symbolic meanings related to rebirth, transformation, and the boundary of the Otherworld. It is beside a stream that Lanval, one of King Arthur's knights, shunned and ostracised by many at the court, meets a mysterious fairy

38 Logan E. Whalen, for instance, examines one such example of how Marie de France may have manifested her indebtedness to Priscian's rhetoric of grammar and its correlation to etymologising in the Prologue to the *Lais*. In Marie's opening motif of poetry likened to blossoming flowers, Whalen traces the implications of the etymology and interplay of *florii*, 'flowers', *florilegium*, and *folii*, 'leaves' as her implicit comment on memory and the body of writing: Whalen L.E., *Marie de France and the Poetics of Memory* (Washington, DC: 2008) 39–41.

lady who becomes his beloved and whom he may continue visiting on condition that he keeps her identity secret. The fairy lady's demand to keep her identity secret jeopardises Lanval's honour and his position at the court when he is confronted by Queen Guinevere, who desires him. As Lanval rejects her advances, she accuses him of homosexuality, and he is soon put on trial by the king for offending the queen, only to be saved by and reunited with his beloved at the last moment. The water of the stream in the *lai* seems then to bear the sense of the element facilitating the transformation and the coming of age in the knight. Yet, as this may only happen after a test of Lanval's honesty to the fairy lady, the stream seems also to imply a connection with the frequent Indo-European motif of the river of forgetfulness,[39] as the hero is expected to remain silent as to the true identity of his beloved.

The fairy world in *Yonec* is also associated with flowing water, and, even more conspicuously than in *Lanval*, with the theme of forgetting. The *lai*'s nameless heroine follows her mortally wounded, shapeshifting fairy lover Muldumarec to the Otherworld, where she learns that he is the lord of a port city beside a river. Meanwhile, her cruel husband is the lord of Caerwent in southern Wales, the city which also lies on a river that is named as Duelas in the *lai*. Both rivers appear to be reflecting on one another: the presence of Duelas is mirrored by the presence of the river in the *lai*'s Otherworld, which also constitutes a reflection of the earthly world of the *lai*. By virtue of broader associations with the flow of time and the boundary between life and death,[40] both rivers seem to be also related to the transition from life to death, and both are associated with the flow of time that is one of the *lai*'s themes. As such, they may also contribute to invoking the theme of forgetting, which constitutes one of the pivotal narrative motifs of the *lai*. In this way, the lady's transitions between the two river cities may be distantly reminiscent of the associations with the Lethe, the river of forgetfulness. Forgetfulness and remembering are crucial to the resolution of *Yonec*. While the lady's memories remain intact, her return from the Otherworld marks the transformation in her cruel husband: thanks

39 In his seminal essay "Waters of Memory, Waters of Forgetfulness", Bruce Lincoln attempts at reconstructing the proto-Indo-European myth of the river of forgetfulness by comparing its numerous occurrences in Classical, Vedic, Germanic, and Celtic sources. In his work, Lincoln focuses on the motifs of transformation and reversible forgetfulness associated with mythical representations of flowing water. Lincoln B., "Waters of Memory, Waters of Forgetfulness", *Fabula* 23 (1982) 19–34.

40 See Carol Zaleski's discussion of the associations between the river topos, death, and transformations in medieval narratives in her *Otherworld Journeys. Accounts of Near-Death Experience in Medieval and Modern Times* (New York – Oxford: 1987) 28–31, 62–65, 121–123.

to the magical ring of forgetfulness given to the lady by dying Muldumarec in his Otherworld city by the river, her husband miraculously forgets about his wife's infidelity. After the birth of Yonec, Muldumarec's son, whom the husband assumes to be his, the husband ceases to be his former callous self and is courteous and kind to his wife. She only reveals that it was Muldumarec who was the true father of Yonec years later, when together with her husband they unexpectedly visit Muldumarec's tomb. Dying, she discloses to Yonec that it was her husband who killed Muldumarec, Yonec's real father. On hearing this, Yonec kills his purported father, avenging both his parents. Thus, the narrative prominence of water in *Yonec* appears to be similar to its place in *Lanval*, since it largely seems to be associated with the domain of the Otherworld in the *lai*.[41]

Of the three *lais* whose protagonists embark on the voyages between Brittany and the shores of Britain, or otherwise unknown lands, it is in *Guigemar* that the sea acquires the status of an almost autonomous agent, and it is on *Guigemar* that I intend to focus in discussing etymological connections and their consequences for the *lai* in the closing part of the chapter. However, as a connecting and separating factor, an embodiment of what spatially and metaphorically lies in-between the locations of the *Lais*, and as the space of transition and transformation, the sea is a crucial setting not only in *Guigemar*, but also in *Eliduc*, and – although to a much lesser degree – in *Milun*, with which I shall begin the examination of the sea motif.

In this *lai*, Milun and his lover's love child grows to be a young man eager to rise to the reputation of his father, whom he never met. To earn fame, he crosses the sea from England, where he was raised by his mother's sister, to Brittany. His achievements are such that he becomes known as *Sanz Per*, the 'Peerless One'. Meanwhile, Milun, his father, hears of his fame and follows him, voyaging across the sea from Britain to Normandy and, subsequently, to Brittany. After some time there, Milun is confronted by Sanz Per and is unhorsed by him in a jousting confrontation. It is only then that the father introduces himself to his son who lovingly recognises him in the unhorsed opponent. They subsequently cross the sea together and return to South Wales, where Milun is finally able to marry his beloved, the mother of his son. Although the three sea crossings – the ones of Sanz Per and of Milun individually and of the two of them together – and the sea itself are mentioned only briefly in individual lines of the *lai*, they are fundamental to its plot, in the same way sea crossings are central to *Guigemar* and *Eliduc*. In *Milun*, they are synonymous with the identity quest, challenge, transformation, and revelation. Crossing the

41 It must be noted that, nonetheless, the water motif in *Yonec* is not as central to the *lai*'s plot as in *Lanval* and *Equitan*.

sea is thus the *sine qua non* of the *lai*; without it neither the son nor the father, nor, eventually, the mother are able to be reunited and find themselves. Crossing the sea is imperative to Milun's son who wishes to make a name for himself in order to match what he has been told of his father:

> Mut se deit hum preisier petit,
> Quant il issi fu engendrez
> E sis pere est si alosez,
> S'il ne se met en greinur pris
> Fors de la tere e del païs. (*Milun*, 306–310)

> A man must have a very low opinion of himself, if, when he has been born of such a famous father, he does not establish greater fame beyond his own region and country. (101)

The sea is thus the threshold to his transformation into a knight and, to some extent at least, his words also describe Eliduc's motivation.

The second crossing of the sea present in *Milun* is equally fundamental for its eponymous hero. Only after arriving in Brittany is Milun eventually able to meet his son and reveal himself to him. The sea thus joins the father with the son, and then, in a rare instance of a fortunate ending among the *Lais*, it eventually unites the entire family: Milun and Sanz Per cross the waters for the last time, *bon oré eurent e fort vent* (506), with the wind being favourable and strong. The sea of *Milun*, even when it separates and tries its protagonists, does so only to bring them back to one another.

Significantly, one other element of the *lai* obliquely related to water functions in it as the factor uniting its protagonists and conjoining elemental forces. Earlier in *Milun*, Marie presents a very peculiar, secret way which the lovers devised to communicate. For twenty years they pass messages to each other that are sent hidden in the plumage of a swan who was trained to fly between the lovers' locations. Along with other creatures featuring in the *Lais*, the swan of *Milun* has drawn considerable critical attention. Most critics support the idea that, despite the fact that swans carry a wealth of symbolic connotations, the particular bird in the *lai* essentially lacks any symbolic qualities and is a purely pragmatic detail of the plot.[42] However, the figurative dimension of

---

42 This was first proposed by Constance Bullock-Davies, who went as far as proposing the exact location where the swan originated from and understood 'the swan in its capacity as a practical love-messenger'. See Bullock-Davies C., "The Love-Messenger in 'Milun'", *Nottingham Mediaeval Studies* 16 (1972) 20–27. Likewise, despite acknowledging the

the swan motif is considered by R. Howard Bloch, who sees the *lai*'s bird in its richly figurative dimension reflecting the concept of writing as a deferral: from the nearly homophonic connection between the swan (*cisne*) and the sign (*signe*), to the swan's feathers, or quills, that should be associated with writing, to the shape of the swan's neck that itself resembles the letter S, phonetically connecting the initial sound in 'swan' with that in the word 'sign' and in *scripture* 'writing'.[43] Still, one other aspect of the swan that supports the *lai*'s perspective on the sea as a connecting force should be mentioned in connection to water. It is the bird's straddling of three elements: air, water, and earth. Its capacity to soar in the air connects the lovers in their locations, while the ability to land and float on water seems to indicate not just different locations of the lovers on land, but also the sea that later in the *lai* takes such a defining and character-shaping role.

Perhaps we should treat it as no coincidence that in the order they appear in Harley MS 978, both the opening *Guigemar* and the closing *Eliduc* are the *lais* in which the sea performs the most formative narrative roles. In *Guigemar*, admittedly the most symbolic and the most infused with magical objects of Marie's *Lais*, the eponymous hero is an accomplished knight suffering from the inability to feel love towards women.[44] On one of his hunts, he is wounded by his own arrow, which rebounds from a magical hind with stag's antlers. The androgynous beast, also wounded by the knight, curses Guigemar that his injury never heals unless he finds a woman who will suffer love for him as much as he does. Serendipitously, Guigemar then finds an empty, luxuriously fitted ship, *tute d'ebenus* (157), all of ebony, which appears to be waiting for him, and which he boards. The unmanned ship miraculously transports him to an unnamed land where he is met by a lady who bravely enters the ship and finds him there. She is unhappily married and jealously guarded by her old and cruel husband. In due course, Guigemar falls in love with the lady, experiencing the healing wound of love, which, agonising, he reveals to her. Thus, his involuntary sea voyage has led to breaking the spell of his inability and

symbolic connotations of swans, June Hall McCash rejects the possibility that any of their symbolic interpretations is relevant in *Milun*. See Hall McCash, "The Swan and the Nightingale. Natural Unity in a Hostile World in the *Lais* of Marie de France", *French Studies* 49.4 (1995) 391–392. For a discussion of *Milun*'s swan as non-symbolic element of the plot, see also Kinoshita S. – McCracken P., "Bodies and Embodiment", in Kinoshita S. – McCracken P. (eds.), *Marie de France. A Critical Companion* (Cambridge: 2012) 152–158.

43 Bloch, *The Anonymous Marie de France* 98–102.

44 This inability has been discussed as one of the homoerotic elements among Marie's *lais*, most recently by Liendo E., "The Wound that Bleeds: Violence and Feminization in the *Lais* of Marie de France", *Neophilologus* 104 (2020) 25–28.

unhealed wound, although also injuring him anew. After some time, their love is discovered and Guigemar is eventually forced to return to Brittany on the same ebony ship, which, once again, marvellously appears by the shore. The lovers part, having earlier exchanged knots on Guigemar's shirt and the lady's belt which only they can untie. The lady cannot bear her existence without Guigemar, and, just as she wishes to drown herself, she spots the ebony ship for the third time and decides to step aboard. What follows is the third sea crossing in the *lai*, when the magic vessel miraculously takes her to Brittany. There, she is captured by Lord Meriaduc, who is instantly consumed with desire towards her, but the knotted belt protects her from his attack. Eventually, both the knot on Guigemar's shirt and the knot on the lady's belt allow the lady and Guigemar to recognise one another, Meriaduc is defeated, and Guigemar is united with his beloved and finally able to find peace. It is notable, furthermore, that, as in *Milun*, the final transformation in *Guigemar* takes place after the third sea crossing. The number of crossings may bear some numerological importance because of the value of number three whose symbolic roots are to be found in the Platonic and Pythagorean traditions combined with exegetical approaches to the Holy Trinity.[45] However, when they do appear in the *Lais*, the numbers of sea crossings are not consistently triadic: in *Eliduc*, the last of the *lais*, there are four of them.

Of all the characters of Marie de France's *Lais*, it is Guigemar whose fate is, seemingly unwittingly, defined by the sea, which he involuntarily traverses to the mysterious land where he meets his beloved and, with her, his salvation from the stag's curse. Guigemar's destination has long been noted as bearing some semblance to the Celtic Otherworld, often represented as lying across the sea or as enterable through the sea.[46] In the *lai*, it is the sea, bearing the ebony vessel with the powerless and dying Guigemar, that quite literally facilitates the *aventure*, in the sense of spatial and narrative reaching of the places,

45 For instance, born a generation after Marie de France, Albertus Magnus in his *Liber de causis proprietatum elementorum* frequently refers to the presence of triads in the divinely ordained nature. For a relatively recent overview of medieval symbolism associated with numbers, see Morrison T., "The Art of Early Medieval Number Symbolism", *Journal of the Australian Early Medieval Association* 2 (2006) 169–181. For a more comprehensive, even if a little dated study of numerological symbolism in the Middle Ages, see Hopper V.F., *Medieval Number Symbolism* (New York: 1938). Hopper discusses Albertus Magnus's findings in the chapter devoted to medieval number philosophy, esp. 89–94.

46 For a survey of the motifs of the sea voyage to the Otherworld in *Guigemar* and its parallels in Celtic tradition, see Illingworth R.N., "Celtic Tradition of the Lai of Guigemar", *Medium Ævum* 31.3 (1962) 176–187 and Hodgson F., "Alienation and the Otherworld in *Lanval, Yonec*, and *Guigemar*", *Comitatus: A Journal of Medieval and Renaissance Studies* 5.1 (1974) 19–31.

people and experiences to which Guigemar is taken across the sea. There is nothing over which he appears to have any control once he steps aboard the vessel and once it sets sail across the sea:

> De ceo s'esteit il merveilliez.
> Il s'est sur le lit apuiez;
> Repose sei, sa plaie duelt.
> Puis est levez, aler s'en voelt;
> Il ne pout mie returner:
> La nefs est ja en halte mer!
> Od lui s'en va delivrement,
> Bon oret ot e suëf vent:
> N'i a nïent de sun repaire!
> Mult est dolenz, ne set que faire!
> N'est merveille se il s'esmaie,
> Kar grant dolur a en sa plaie.
> Suffrir li estuet l'aventure;
> A Deu prie qu'en prenge cure,
> K'a sun poeir l'amemt a port
> E sil defende de la mort. (*Guigemar*, 187–202)

> Guigemar marvelled at all this and in great pain from his wound reclined on the bed to rest. Then he rose intending to leave the ship, but he could not go back, as the ship was already on the high seas, speeding quickly away with him, the wind favourable and blowing gently. There was no question of his returning to land and he was grief-stricken, not knowing what to do. No wonder he was dismayed, for his wound was causing him great suffering. But he had to accept his fate and he prayed to God to take care of him, to bring him, if at all possible, to a safe harbour and protect him from death. (45–46)

The sea of *Guigemar* is thus a state of limbo, of in-betweenness, a purgatorial domain where the only hope lies in the prayer for divine mercy. With its three occurrences in the *lai*, the sea also lies in-between not just lands, but parts of the narrative, and, more importantly, the identities of its protagonists. In this respect the sea's role is similar to that in *Milun*. In *Milun*, however, the characters know precisely where they are heading on their voyages, whereas the sea of *Guigemar*, together with the ebony ship steered by some unknown force, have agencies of their own. It is not that the *lai*'s protagonists guide themselves across the sea, but that they are guided by it. The sea is thus a suspended and undefined space of transition between lands, worlds, and conditions of the protagonists, which acts as the space of potentiality. It is its interaction with the *lai*'s protagonists that defines their later predicaments.

As in *Milun*, the sea transforms and leads both lovers into their respective exiles, but, in contrast to *Milun*, it also offers guidance through chance, challenge, and access to the Otherworld, leading to escape, freedom, and fulfilment. The sea thus plays a crucial role in *Guigemar*, contributing to the profound sense of inconclusiveness of the *lai* noted by critics. Roberta L. Krueger sees it as a narrative which

> leaves several threads untied. The jealous husband remains a shadowy, if distant and unmentioned, menace; there is no word of marriage or children for Guigemar and his lady; and finally, there is no account of reunion with Guigemar's loving family from the *lai*'s opening lines. Marie's own discursive knot is both open and closed.[47]

Likewise, the sea, which plays such a pivotal role in the *lai*, is not devoid of ambivalence: is its role to be read as that of the curse cast on the wounded Guigemar and, subsequently, is the black vessel in which Guigemar falls asleep some kind of a funerary boat transporting the knight to the Otherworld? Comparably, does the ebony ship perform the same function for the lady who embarks on it instead of flinging herself into the sea to drown? Or, on the contrary, does the sea and the vessel it carries serve to save Guigemar from his wound and from his initial inability to love, described as the wrong (*mespris*, 57) done by nature to him? Finally, does the sea perform a similar role for the lady in the *lai*, who, instead of drowning herself, is carried away from the misery of her marriage? In both perspectives, however, the purgatorial agencies of the sea and of the vessel operate as renewal, hope, and, for Guigemar, as redemption. What seems essential for the purport of the *lai* is that none of them are the effect of the choices made by its protagonists.

Lastly, Eliduc, the eponymous protagonist of the final *lai* in Harley 978, and, simultaneously, one of the most morally ambiguous of Marie's heroes, crosses the sea between Brittany and Logres four times. First, when, banished by his king for no apparent reason, he leaves his faithful wife Guildeluëc in charge of his lands. While in Logres, he proves to be so worthy and valiant to the local king, that the king's daughter Guilliadun falls in love with him. Her beauty and

47 Kruegger R.L., "The Wound, the Knot, and the Book: Marie de France and Literary Traditions of Love in the *Lais*", in Whalen L. (ed.), *A Companion to Marie de France,* Brill's Companion to the Christian Tradition 27 (Leiden – Boston, MA: 2011) 55–88, 63.

virtues incline Eliduc to a deep platonic affection towards her, marred by his reluctance to tell Guilliadun of his wife in Brittany. The lovers must part, however, when Eliduc is called back to Brittany by his Breton king who requires Eliduc's help against enemies. After promising his return to Guilliadun, Eliduc crosses the sea for the second time in the *lai*. Having served his king and keeping his secret from his wife, Eliduc, distraught by his love for Guilliadun, leaves Brittany and crosses the sea for the third time to be reunited with his lover in Logres. Then, together, Eliduc and his beloved furtively set off for what is to be the *lai*'s fourth and last sea crossing back to Brittany. The moment of exposure comes when, during a fierce sea storm, a sailor accuses Eliduc of being responsible for the cataclysmic weather because of Eliduc's immoral conduct, at the same time revealing the truth to Guilliadun about her lover's marriage. Upon hearing this, she falls into a prolonged faint, while Eliduc, in terrible fury, kills the sailor. Once in Brittany, Eliduc entrusts a hermit with the care for unconscious Guilliadun, while his wife eventually learns about her husband's love for the maiden. Guildeluëc, perhaps the most charitable and unselfish of Marie's characters,[48] is not only able to revive Guilliadun, but she also decides to become an abbess in a convent that Eliduc is to found for her as a form of atonement, thus leaving her husband free to marry Guilliadun. Eventually, after many years of marriage, Guilliadun follows Guildeluëc into convent, while Eliduc joins a monastery. With its aura of penitence, the end of the *lai*, therefore, appears to imply a sense of guilt for the double marriage and for the duplicitous love that were facilitated by Eliduc's sea crossings.

Thus, although crossing the sea in both *lais* involves the changes brought by love, the treatment of the sea in *Eliduc* is notably different from that of *Guigemar*. If *Guigemar*'s characters are guided by unknown forces steering them towards the unknown across the sea, the protagonist of *Eliduc* knows full well where he is heading. On the final crossing of the *lai*, he even assumes the position at the ship's helm and steers the vessel himself. Instead of the consequences of chance, the sea in the *lai* poses questions about the consequences of loyalties and disloyalties, testing Eliduc's faithfulness towards his king and, predominantly, towards Guildeluëc, his wife, to whom he swore *Qu'il li porterat bone fei* (84), that he would keep good faith. Later, however, the sea crossing

48 Guildeluëc's remarkable charity and generosity continue to invite critical attention. Most recently Kimberly Tate Anderson and Danielle Gilbert have proposed reading her character in the capacities related to the cult of the Holy Virgin as a representation of an ideal wife, widow, and virgin-abbess. See Tate A.K. – Gilbert D., "Guildeluëc and the Virgin Mary: Legitimizing the Abbess as Holy Mother", *Neophilologus* (2021) (accessed November 2, 2023).

is also a challenge to Eliduc's promise to return to Guilliadun, while on the last crossing in the *lai* it becomes the test of his own truthfulness to himself and to Guilliadun, who only then learns about Eliduc's wife. As already noted, the raging storm that accompanies this last crossing seems to reflect a moral judgement passed on Eliduc, which is also explicit in the words of one of his sailors. It is notable then that Marie de France provides the only more detailed representation of the hazards and the dread of crossing the Narrow Seas in violently changing weather. On previous occasions in the *lai*, the wind was good, and the crossings were merely mentioned in passing. This time, Marie paints the scenes which in her time must count as among the most dramatic and most realistic depiction of the perils of seafaring and the terrors they involved:

> Bon vent ourent e bon oré
> E tut le tens aseüré.
> Mes quant il durent ariver,
> Une turmente ourent en mer,
> E uns venz devant els leva
> Ki luin del hafne les geta;
> Lur verge bruisa e fendi
> E tut lur sigle desrumpi. (*Eliduc*, 811–820)

> They had good wind and settled weather, but as they were about to arrive, they encountered a storm at sea and a wind arose before them that drove them far from the harbour. Their mast broke and split, and the sail was completely torn. (121)

Eliduc, Guilliadun, and Eliduc's men on board begin praying and imploring God, St. Nicholas, St. Clement, and the Virgin Mary to save them and let them reach the harbour, but the ferocity of the sea intensifies:

> Une hure ariere, une altre avant
> Issi alouent costeiant.
> Mult esteient pres de turment.
> Uns des eschipres haltement
> S'est escriëz: « Que faimes nus?
> Sire, ça enz avez od vus
> Cele par qui nus perissuns.
> Ja mes a terre ne vendruns!
> Femme leial espuse avez
> E sur celi altre en menez

Cuntre Deu e cuntre la lei,
Cuntre dreiture e cuntre fei.
Laissiez la nus geter en mer,
Si poüm sempres ariver. (*Eliduc*, 827–840)

They sailed back and forth along the coast and came extremely close to being shipwrecked. Then one of the sailors cried aloud: 'What are you doing? Lord, you have with you the woman who will cause us to perish. We shall never make land! You have a loyal wife and now with this other woman you offend God and his law, righteousness, and the faith. Let us cast her into the sea and we shall soon arrive safely.' (121)

Eliduc erupts into an impassioned rage and foul language towards the sailor, simultaneously trying to comfort the maiden who suffers from both seasickness and the effects of hearing the shocking news:

Mes entre ses braz la teneit
E cunfortout ceo qu'il poeit
Del mal que ele aveit en mer
E de ceo qu'ele oi numer
Eue femme espuse ot sis amis,
Altre que li, en sun pais. (*Eliduc*, 847–852)

But he held her in his arms and comforted her as best as he could against her seasickness and because she had heard that he had a wife in a foreign country. (122)

The sea in *Eliduc* is thus the space of a violent revelation and confrontation, an elemental force which removes deceitful trappings and lays bare the bullying and boorish character of the knight. Thinking that his beloved has died, he kills the shipman by striking him with an oar, kicking his body into the sea, and taking the helm himself:

Par le pié l'en a jeté fors;
Les undes en portent le corso
Puis qu'il l'ot lancié en la mer,
A l'estiere vait governer;
Tant guverna la nef e tint,
Le hafne prist, a terre vint. (*Eliduc*, 863–868)

> Then with his foot he pushed him overboard and the waves bore the body away. When he had cast him into the sea, he went to take charge of the helm, steering the boat and holding it on course so that he reached the harbour and land. (122)

Remarkably, the connection with the sea is also present after the party eventually lands in Brittany: Eliduc leaves the unconscious Guilliadun in what the *lai* describes as *recez fu pres de la mer* (887), a shelter close to the sea, perhaps a form of closure echoing the significance of the sea in the *lai*.

The ambiguity and fluidity of the sea's roles and significance are, therefore, crucial for the *lai*'s plot: in *Eliduc*, not only does the sea signify transition and constitute the space which leads to transformations, but it is also the way both in and out of political, emotional, and moral turmoil. It tests loyalties and being torn between allegiances. It also hooks the lovers to one another since, perversely, it facilitates a new life for Eliduc and Guilliadun, and yet it is also closely related to death and torment, not just that of the sailor, but of Guilliadun too, in her near-death loss of consciousness. Likewise, the suffering of the rejected Guildeluëc is related to Eliduc's sea voyages. The multiple sea crossings from Brittany to Logres and back, like the tidal flows, both hide his infidelity and untruthfulness, and ultimately reveal it. Therefore, among the *lai*'s audiences, the consideration of the roles of the sea may have contributed to debates on moral ordeal, purification, perdition, and condemnation that were likely envisaged and hinted at by Marie her Prologue. Lastly, the several etymological connections between the sea, death and love presented above in this chapter appear to be particularly strongly present in *Eliduc* too, if not in terms of how they resound in individual words, then in terms of the way their semantic areas mutually reinforce one another's ambivalences.

3 Towards a Conclusion: *Guigemar* and the Etymological Implications of the Sea

Let us finally return to the question of how the Isidorean etymology of *mare*, 'sea', and its relations to the etymologies of *amor*, 'love', and *mors*, 'death', may supplement the ambiguity of the narrative elements of Marie de France's *Lais*. To be sure, the intention of this chapter is not to present an examination of the entire body of the *Lais* from this perspective, rather, as stated above, it is to signal that not only was interpreting senses through etymologies familiar to and expected among the more learned audiences of the *Lais*, but that it may have contributed to and invited their multiple, not infrequently contradictory, interpretations. The aural reception of the recitation of the *Lais* was even more likely to lead to the understandings that were the effect of the connotations that the *mar/mer/mor* component carried with themselves. Central to them are the audible associations between the concepts of the sea, love, and death emphasised in the opening part of this chapter. As a way of probing into the consequences of such aural and semantic correlations, we shall take a closer look at an instance from *Guigemar* where the *mar/mer/mor* component appears to open possibilities of ambivalence.

Guigemar and Meriaduc, the only two names that appear in the *lai* both contain the phonetic element of **mr*, incidentally present also in Isidorean etymologising of the sea, love and death. Like most of the names appearing in the *Lais*, they are of Breton origin, and, as such, the *mar* and *mer* present in them bear no semantic likeness to the sea. In his detailed study of the proper names in Marie de France's *Lais*, Ernst Brugger presents an extensive overview of their etymons, the dazzling multiplicity of the variants of their spelling, and their roots and connections to other Celtic languages, particularly Welsh and Cornish.[49] The etymons of Guigemar connect the name with the multiple spellings of the Breton words for 'nobility' and 'horse',[50] while Meriaduc is toponymic in nature. Brugger derives its multiple forms from the north Welsh place name associated with a fourth-century Saint Meriadog.[51] Regardless of the popularity of the two names in Brittany, for Anglo-Norman audiences, their pronunciations resound with the syllables that resemble those of *mar*, the sea. As a secondary character, Meriaduc constitutes the final *aventure* and challenge to which the ship carries Guigmar and, later, the lady back across the sea to Brittany. But it is the sound of the syllables of Guigemar's name that may suggest several terms that delineate the scope of this opening *lai*: from the sense of the marvel at the *lai*'s supernatural elements, to the guidance that is offered both to Guigemar and the audience by the love that is meant to save him from death, to the bitterness of the suffering involved, all of which are the narrative effects of the presence of the sea in the *lai*. Of these, as we have

---

49 Brugger E., "Eigennamen in den Lais der Marie de France", *Zeitschrift für französische Sprache und Literatur* 49.4/6 (1927) 201–252.

50 Ibidem 203.

51 Ibidem 237–239. Intriguingly, the saint, whose name is also spelt 'Meriasek' and who, as Brugger asserts, is cognate with S. Cynan (Conan) Meriadog, crossed the channel to found a church in Cornwall. An interesting parallel ensues whereby Meriaduc is named after a place and saint, and in both the saint's and Meriaduc's lives the sea crossing is important. See also "S. Cynan (Conan) Meriadog, Prince, Confessor", in Baring-Gould S. – Fisher J. (eds.), *The Lives of British Saints* (London: 1908) 220–224.

seen, love, death, bitterness, and the sea are etymologically interrelated. And so, Guigemar, *merveillez*, marvelling at the mystery of the ship, is guided, *guieé*, towards the *amur*, love, which is found in the Otherworld beyond the sea, *mer*. As promised by the androgynous stag, love saves Guigemar from *mort*, death, precisely because it is tinged with *amer*, a bitter aftertaste, as it is marked by the necessary suffering and by its adulterous nature.

Clearly, the sea carrying the crewless vessel first with Guigemar and then with the lady between their worlds epitomises the space between inexperience and experience, death and life, desperation and hope. Like the elemental power that it is, and like the emotional and spiritual transitions it reflects, it is a force that confuses those subjected to it. Notably, it is only by the sea that Guigemar can reach the lady, and the same is true for her in reverse – the only unguarded point of entry to and exit from the keep where the lady's husband holds her imprisoned is by the sea. It is, therefore, a causative power in itself, simultaneously remaining beyond human control since the ship appears to be directed by the sea currents and the sea winds. All this is felt in the very first words that, on arriving upon her shore, Guigemar utters to the lady, who, having braved the sea waters, stepped onto the ship, and found him there. Frightened confusion resounds loudly in his request for guidance and explanation:

> Bele dame, pur Deu vus pri,
> Cunseilliez mei, vostre merci!
> Kar jeo ne sai queil part aler,
> Ne la neif ne puis governer. (*Guigemar*, 333–336)
>
> Fair lady, I beg you in God's name, please help me, for I do not know where to go, or how to steer the ship. (47)

But, begging for an explanation and instruction and imploring her to be merciful, Guigemar indirectly also implies that it was the mercy of the *mer*, *mer-ci*, that brought him to her. As may be expected, he shall be provided with the transformative guidance of love, but, again because of the role and the influence of the sea, it will also be mirrored in the transformational experience of the lady.

If, in *Guigemar*, the sea, with all of its etymological connotations, serves the purpose of the looking glass in which Guigemar can reflect himself and on himself, it is important to note that it serves a similar purpose for the unnamed heroine of the *lai*. Even though she may be interpretable as a fairy lady, which, among other things, is indicated by her beauty compared to that of a fairy (lines

703–704),[52] she nonetheless seems as confused and as vulnerable as Guigemar. Her behaviour and position mirror his earlier actions and status. Again, the sea waters are instrumental in directing her towards her own *aventure*, that is, towards the events in Brittany culminating in the discovery and reunion with Guigemar. Although not physically, but invisibly wounded like him (for 'amur est plaie dedenz cors', line 483, 'love is an invisible wound in the body'), after his forced departure from her, she is close to dying and threatens to drown herself. And, just as Guigemar spotted the ship on the waves in the first part of the *lai*, the lady also sights the very same ebony vessel at the last moment, boards it, and, in the same crewless manner as Guigemar, is ferried across the sea to Brittany, a land unknown to her. Immediately before she finds the ship awaiting her in the harbour, she curses ever meeting Guigemar and the distress that encounter has caused her. It is then that we witness perhaps the most distinct example of punning involving the sea that Marie de France employs. The paronomasia, which extends from Guigemar's name to the mention of the sea, seems predominantly aural in the assonance of vowels. However, it is also reinforced by the etymological connections with the sea familiar to us from Isidore's *Etymologies*:

> Guigemar, sire, mar vus vi!
> Mieuz voil hastivement murir
> Que lungement cest mal sufrir.
> Amis, si jeo puis eschaper,
> La u vus fustes mis en mer
> Me neierai! (*Guigemar*, 668–673)
>
> Guigemar, lord, how sad [bad] that I met you! I prefer to die a speedy death rather than suffer this misfortune too long. Beloved, if I could escape, I should drown myself just where you were put to sea! (52)

Incidentally, the *-emar* in Guigemar's name displays an aural similarity to the Anglo-Norman noun *amarre*, 'mooring rope', related to the verb *amarrer*, 'to moor', as well as some audible likeness to *amer*, 'to love' (with its related form of *amare*), both indicating the sense of attachment. And yet, in the same line, the name is also almost instantly echoed in the adverb *mar*, 'disastrously', denoting an unfortunate or calamitous experience. That very *mar* resonates also in the assonances in the line below as **-mur* of *murir*, a variant spelling

52 For a discussion of the ambiguity in the presentation of *Guigemar*'s heroine as a fairy-like mistress, see Wade J., *Fairies in Medieval Romance* (New York: 2011) 25–26.

of *morir*, 'to die', which, in the following line (line 670), is echoed in the *mal* of *mal sufrir*, 'severe suffering'. Unsurprisingly, the verb *eschaper*, 'to escape', which is what the lady wishes to do, rhymes with where it is that she wishes to escape, that is with the noun *mer*, 'the sea' (line 672). The final assonance and paronomasia in the passage are present between *en mer / Me neierai* (lines 672–673). Not only, as William Rothwell proves, is *neierai* a curious verb, which may be both the future first person form of the verb *neer*, 'to drown', as well as the future form of *noer*, 'to swim',[53] thus encapsulating the paradox of the sea as capable of both bringing death to the lady and capable of reuniting her with her lover. But the relation between these two capacities of the sea is also audibly bolstered in the assonance between *en mer / Me neierai*. And, since nasalisation in Old French begins to gradually take place only about a century or century and a half before Marie flourished,[54] the pronunciation of *en mer* was likely to be inversely echoed in *me neierai*.

Concluding, the example of the paronomasia found in *Guigemar* and grounded in the alterations and variants of the *mar/mer/mor* component is ultimately phonetically and semantically traceable to *mer*, 'the sea'. A statement may be proposed that it is so not only because of the central role that the sea plays in creating the *aventures* of the *lai*: it is by the sea that Guigemar arrives to meet the lady, and it is by crossing the sea that she is finally reunited with him. Importantly, however, it may also be caused by the curious etymological connections that are traceable to Isidore and other sources. They invite the connotations spanning the bitterness and changeability of the physical expanse of the sea with the bitterness of death and the ambivalence of the bonds of love. All these components are present in the fragment from *Guigemar* and in the entire *lai* itself. Crucially, however, they are equally present in all of Marie's *lais* that follow, closing with the moral and emotional ambivalences of *Eliduc*, in which the sea yet again plays a critical role in facilitating the *aventures* of love and the conflicts that they incur. Between the inexperience of Guigemar, who is yet to learn the suffering of love at the beginning of the *Lais*, and the experience of Eliduc, torn between conflicting loyalties of love, lie the tales that demand debates, answers, and taking sides. Perhaps then, the most disturbing message that the audience is to learn from the presence of the sea at the beginning and the end of Marie de France's *Lais* is the

53 For a discussion of the confusing relation between the two verbs, see Rothwell W., "Sink or Swim?" A Homonymic Dilemma in Medieval French", *Zeitschrift für romanische Philologie* 92.3–4 (1976) 386–393.

54 On the development of nasalised consonants and vowels in Old French, see Kibler W.W., *An Introduction to Old French* (New York: 1984) 95–96, 124–125.

sheer uncontrollability of the emotion it brings and symbolises. Like the vastness and depth of the sea, the vastness and depth of love amaze and terrify, engendering ambivalent conditions and reactions. And thus, if we are to read so many of the elements of the *Lais* through the phonetic and semantic presence of *mer* in them, then it is through the fluidity and changeability of the sea and love, and through their bitterness that the *Lais* open their meanings to their audiences. Much as the sea liberates Marie's protagonists, it also circumscribes them within its power, both literally and figuratively as *la mer amar de l'amur*, the bitter sea of love.

Bibliography

Primary Sources

Barney S.A. – Lewis W.J.J. – Beach A. – Berghof O. (trans.), *The Etymologies of Isidore of Seville* (Cambridge: 2006).

Benson L.D. (ed.), *The Riverside Chaucer* (Oxford: 1988).

Capellanus A., *On Love*, trans. and ed. P.G. Walsh (London: 1982).

Isidore of Seville, *Etymologiarium sive originum* libri XX, ed. W.M. Lindsay (Oxford: 1911).

Marie de France, *The Lais of Marie de France*, trans. G.S. Burgess – K. Busby (London: 1999).

Pope J.C. (ed.), *Homilies of Ælfric. A Supplementary Collection*. Vol. I (London – New York – Toronto: 1967).

Rychner J., (ed.), *Les Lais de Marie de France* (Paris: 1983).

Secondary Sources

Amsler M., *Grammatical Discourse in Late Antiquity and the Early Middle Ages* (Amsterdam – Philadelphia: 1989).

Baring-Gould S. – Fisher J., (eds.), *The Lives of British Saints*, Vol. II (London: 1908).

Bloch R.H., *The Anonymous Marie de France* (Chicago – London: 2003).

Brugger E., "Eigennamen in den Lais der Marie de France", *Zeitschrift für französische Sprache und Literatur* 49.4/6 (1927) 201–252.

Bullock-Davies C., "The Love-Messenger in 'Milun'", *Nottingham Mediaeval Studies* 16 (1972) 20–27.

Classen A., *Water in Medieval Literature* (Lanham, MD – Boulder, CO – New York – London: 2018).

Cumont F., *After-Life in Roman Paganism* (New Haven, CT: 1922).

Del Bello D., *Forgotten Paths. Etymology and the Allegorical Mindset* (Washington, DC: 2007).

Deshayes A., *Dictionnaire etymologique du breton* (Douarnenez: 2003).

Hodgson F., "Alienation and the Otherworld in *Lanval*, *Yonec*, and *Guigemar*", *Comitatus: A Journal of Medieval and Renaissance Studies* 5.1 (1974) 19–31.

Hopper V.F., *Medieval Number Symbolism* (New York: 1938).

Illingworth R.N., "Celtic Tradition of the Lai of Guigemar", *Medium Ævum* 31.3 (1962) 176–187.

Kibler W.W., *An Introduction to Old French* (New York: 1984).

Kinoshita S. – McCracken P. (eds.), *Marie de France. A Critical Companion* (Cambridge: 2012).

Klein E., *A Comprehensive Etymological Dictionary of the Hebrew Language for Readers of English* (Tel Aviv: 1987).

Krueger R.L., "The Wound, the Knot, and the Book: Marie de France and Literary Traditions of Love in the *Lais*", in Whalen G.E., *A Companion to Marie de France* (Leiden – Boston, MA: 2011) 55–87.

Liendo E., "The Wound that Bleeds: Violence and Feminization in the *Lais* of Marie de France", *Neophilologus* 104 (2020) 19–32.

Lincoln B., "Waters of Memory, Waters of Forgetfulness", *Fabula* 23 (1982) 19–34.

Morrison T., "The Art of Medieval Number Symbolism", *Journal of the Australian Early Medieval Association* 2 (2006) 169–181.

McCash J.H., "The Swan and the Nightingale. Natural Unity in a Hostile World in the *Lais* of Marie de France", *French Studies* 49.4 (1995) 385–396.

Rothwell W., "Sink or Swim?: A Homonymic Dilemma in Medieval French", *Zeitschrift für romanische Philologie* 92.3–4 (1976) 386–93.

Salvador-Bello M., *Isidorean Perceptions of Order* (Morgantown, WV: 2015).

Sobecki S.I., *The Sea and Medieval English Literature* (Cambridge: 2008).

Tate A.K. – Gilbert D., "Guildeluëc and the Virgin Mary: Legitimizing the Abbess as Holy Mother", *Neophilologus* (2021) https://doi.org/10.1007/s11061-021-09701-w (accessed 21 November 2021).

Taylor A., *Textual Situations: Three Medieval Manuscripts and Their Readers* (Philadelphia: 2002).

Thompson V., *Dying and Death in Later Anglo-Saxon England* (Woodbridge: 2004).

Vitz E.B., *Orality and Performance in Early French Romance* (Cambridge: 1999).

Wade J., *Fairies in Medieval Romance* (New York: 2011).

Warnke K. (ed.), *Die Fabeln der Marie de France*, 1898, reprint (Geneva: 1974).

Whalen L.E., *Marie de France and the Poetics of Memory* (Washington, DC: 2008).

Zaleski C., *Otherworld Journeys. Accounts of Near-Death Experience in Medieval and Modern Times* (New York – Oxford: 1987).

Zanoni M.L., "'Ceo testimoine Preciëns:' Priscian and the Prologue to the 'Lais' of Marie de France", *Traditio* 36 (1980) 407–415.

CHAPTER 12

Water in Saints' Lives: a Study of the Old English Corpus

Hugh Magennis

Water, though hardly ubiquitous in hagiographical narratives, has a significant role in many of them. Most commonly, water plays a part in miracle scenes; water also provides a medium of travel (often with miraculous occurrences incorporated here too); in some saints' lives water is presented as a means of cleansing and healing, particularly in baptism; and water can feature in reports of abstinence and self-mortification. In deploying these narrative water themes hagiographers were influenced primarily by the Bible, adapting and recasting biblical models in their work, though in some cases the influence of ancient romance is also discernible.

1 The Bible

Water figures widely in the Hebrew Bible and the New Testament. In the Hebrew Bible, as in some saints' lives discussed below, water is a powerful element under the control of God, who uses it as an instrument of his divine judgement, bringing danger and death through flood and storm, as fearfully acknowledged in the exclamations and prayers of the psalmist (for example, at Psalms 17(18):16, 31(32):6, 45(46):3 and 92(93):3–4)) and realised most spectacularly in Noah's Flood (Genesis 6–9) and in the destruction of the Egyptian army in the Red Sea (Exodus 13–14). Also, as again taken up in hagiography, in the Hebrew Bible water enables cleansing, in the ritual ablutions of priests, for example (as in Exodus 29:4, Leviticus 16:4 and Numbers 19:7–9), and in the removal of defilement (as at Leviticus 11:40, Leviticus 15:5–12 and Deuteronomy 23:11).

Above all, water is essential to life. Living in an arid region, nothing is more serious to the people of the Bible than the absence of water (see, for example, Jeremiah 14:3, Joel 1:20, Haggai 1:10). Water – most notably, flowing water, wells, rain and dew – is symbolic of God's blessing and of spiritual refreshment (as at Psalms 22(23):2, Isaiah 32:2 and Isaiah 35:6); God himself is 'the fountain of living waters' (Jeremiah 2:13). This theme of the preciousness of water is taken

 | DOI:10.1163/9789004696501_014

up in hagiography in episodes where saints bring forth water miraculously or experience or make use of its soothing powers, and, inversely, in the stories of self-mortification in which ascetics exist with little of it.

These strands of significance are also apparent in the New Testament. The danger of water is exemplified strikingly in Revelation's prophecy that there will be no sea in heaven (21:1) and in miraculous episode of Jesus calming the storm on the Sea of Galilee (Mark 4:35–41) to the relief of his terrified disciples, and again in the miracle of Jesus walking upon the waters (Matthew 14: 22–23) (while Peter is too afraid of the danger to attempt the same). The emphasis in these gospel episodes is on miracle, presenting a lead for writers of saints' lives, whose protagonists perform similar wonders. The Old Testament theme of cleansing appears in the New Testament as baptism, the 'water of repentance' (Matthew 3:11), in the ministry of John the Baptist, who baptises Jesus himself. The idea of water as refreshment is reflected in Jesus's image of giving cold water to drink to the least of those who are needy (Matthew 10:42, Mark 9:41) and in the parable of Lazarus and the rich man, where the rich man in hell begs to receive a drop of water from the tip of Lazarus's finger (Luke 16:24). Water as representing life and God's blessing is powerfully taken up in spiritual terms in Jesus's account, in the episode with the Samaritan woman at the well, of the salvation he brings: 'The man who drinks the water I give him will not know thirst anymore. The water I give him will be a spring of water within him, that flows continually to bring him everlasting life' (John 4:13–14). Water also figures in another miracle of Jesus, that of the changing of water into wine at the marriage feast at Cana.

A separate strand of significance for water is found especially in the Acts of the Apostles, where water, specifically the sea, becomes a medium of travel, as the apostles, obedient to Christ's instruction to go forth and teach all nations (Matthew 28:19), carry out their missionary activities voyaging across the Mediterranean. Such travels include the motif of storms at sea, as experienced by Paul when being taken to Rome as a prisoner (Acts 27:9–44); his ship is tossed by the waves and ends up being blown ashore and wrecked on an island that 'was strange to them' (it turns out to be Malta) (Acts 27:39).[1]

Travel across the Mediterranean, a sea of economic and cultural 'connectivity',[2] is also a staple theme in ancient romance and indeed the writer

---

1 References to travel by boat appear occasionally elsewhere in the Bible, in the story of Jonah, for example (Jonah 1–3), and in Jesus's crossing of the Sea of Galilee (Mark 4:35–6), but such travel becomes a recurrent theme in Acts.

2 On the 'connectivity' of the ancient (and medieval) Mediterranean, see Horden P. – Purcell N., *The Corrupting Sea: A Study of Mediterranean History* (Oxford: 2000) 123–172.

of Acts may have been guided by the example of romance in incorporating it.[3] Ancient romance also influenced some hagiographical texts directly in the deployment of plots involving sea travel[4] but the essential lead followed by hagiographers in their treatment of voyages of missionaries is that of Acts. Such voyages may be seen as instances of *imitatio apostolorum*, just as the performance of miracles (including those involving water) are instances of *imitatio Christi*.

The theme of travel is a common one in hagiography but is particularly relevant to the early insular missionary saints and the ascetics who were engaging in the practice of *peregrinatio pro amore Dei*, journeying away from their homeland to bring the gospel to the European mainland or withdrawing from worldly society altogether. The most spectacular journey of an insular saint is the one described in *The Voyage of St Brendan*,[5] in which the sixth-century abbot sails across the ocean to the Promised Land of the Saints, but plenty of less fanciful saints' lives (and other writings) relate stories of voluntary exile undertaken by insular monastics, including some of the most celebrated of them.[6]

Water in the Bible is extensively discussed in patristic and medieval exegesis, where it has a range of non-literal significations, often denoting the Holy Spirit, for example, or Christian teaching;[7] the sea, in particular, is a powerful symbol in Christian thought.[8] And, following biblical custom, water has an important role in the liturgy.[9] Such traditions of thought and practice have only tangential relevance for most hagiographers, however, who generally in

3 See Pervo R.I., "Early Christian Fiction", in Morgan J.R. – Stoneman R. (eds.), *Greek Fiction: The Greek Novel in Context* (London: 1994) 240.

4 As in 'Pseudo-Clementine' literature, discussed, for example, by Pervo in his survey of early Christian fiction: see Pervo, "Early Christian Fiction" 250–251; see also Heffernan T.J., *Sacred Biography: Saints and Their Biographers in the Middle Ages* (New York – Oxford: 1988) 142–144.

5 Selmer C. (ed.), *Navigatio Sancti Brendani* (Notre Dame, IN: 1959); O'Meara J.J. (trans.), *The Voyage of Saint Brendan: Journey to the Promised Land* (Dublin: 1976; repr. Gerrards Cross, Buckinghamshire: 1991).

6 See Mullins J., "*Herimum in mari*: Anglo-Saxon Attitudes towards *Peregrinatio* and the Ideal of a Desert in the Sea", in Klein S. – Schipper W. – Lewis-Simpson S. (eds.), *The Maritime World of the Anglo-Saxons* (Tempe, AZ: 2014) 239–254.

7 With particular reference to Old English literature, see DiNapoli R., *An Index of Theme and Image to the Homilies of the Anglo-Saxon Church* (Hockwold cum Wilton, Norfolk: 1994) *s.v. river, sea, water*.

8 Christian interpretations of the sea are well summarised by Sobecki S.I., *The Sea in Medieval English Literature* (Cambridge: 2008) 34–41.

9 For the liturgy, see Gaillard J., "Eau", in Viller M. – Cavallera F. – de Guibert J. (eds.), *Dictionnaire de Spiritualité chretien*, IV (Paris: 1960); see also "Introduction" above 16–17.

their narratives focus on water as a physical entity.[10] Certainly as far as prose saints' lives are concerned – and the vast majority of saints' lives are in prose – only rarely does water have an intended non-literal significance in narrative episodes. Spiritual cleansing represents one example of such significance, and river or sea crossings can symbolically signal moments of transition or change, but prose hagiography is for the most part a literal-minded genre and this is reflected not least in its deployment of water themes. Water almost invariably functions instrumentally rather than having any kind of non-utilitarian interest of itself.[11] As demonstrated below, it is a different story with saints' lives in verse, in which poetic imagination and the rich resources of poetic tradition can be drawn upon.

2 Saints' Lives in Old English

In exploring the significance of water in hagiography this chapter will concentrate on writings in Old English, since Old English saints' lives provide a manageable corpus of material for a chapter-length study and have the attraction of both being firmly representative of the genre of hagiography as a whole, but also, especially in the poetry,[12] presenting distinctive cultural features.[13] Saints' lives in Old English prose and verse, which range in length from a single sentence (as in some entries in the compilation known as the *Old English Martyrology*) to more than 10,000 words, are almost invariably based on Latin

10 On symbolic understandings of the sea in Old English literature, see Mullins, "*Herimum in mari*".

11 Non-utilitarian approaches to water (and other elements of nature) are explored in the burgeoning field of ecocriticism: see, with particular reference to Old English, Estes H., *Anglo-Saxon Literary Landscapes: Ecotheory and the Environmental Imagination* (Amsterdam: 2017) esp. 35–59; see also Siewers A.K., *Strange Beauty: Ecocritical Approaches to Early Medieval Landscape* (New York: 2009).

12 Below, all references to and quotations of Old English poetry, with the exception of *Beowulf*, are taken from Krapp G.P. and Dobbie E.V.K. (eds.), *The Anglo-Saxon Poetic Records*, 6 vols (New York: 1931–1953), abbreviated below as ASPR; for *Beowulf*, references are to Fulk R.D. – Bjork R.E. – Niles J.D. (eds.), *Klaeber's Beowulf and the Fight at Finnsburg* (Toronto: 4th ed. 2008). Translations follow Bradley S.A.J., *Anglo-Saxon Poetry: An Anthology of Old English Poems in Prose Translation with Introduction and Headnotes* (London: 1982) (with occasional modification, as indicated). Unattributed translations below are my own.

13 On saints' lives in early England, see, for example, Lapidge M., "The Saintly Life in Anglo-Saxon England", in Godden M. – Lapidge M. (eds.), *The Cambridge Companion to Old English Literature* (Cambridge: 2nd ed. 2013) 251–272; Magennis H., "Approaches to Saints' Lives", in Cavill P. (ed.), *The Christian Tradition in Anglo-Saxon England: Approaches to Current Scholarship and Teaching* (Cambridge: 2004) 163–183.

originals, often closely translated but often too reworked in creative ways. They comprise material on the saints of the universal church and also treatments of native saints, all adhering to the conventions and archetypes of hagiographical tradition as developed in the early Christian centuries.[14] In the present survey only writings on the Virgin Mary are excluded from consideration, as these do not follow the hagiographical pattern found in other saints' lives.[15]

The cultural distinctiveness of Old English verse saints' lives in their treatment of water derives from the fact that the very medium of Old English traditional poetry brings with it a rich repertoire of imagery, formulas and themes concerning water, particularly to do with voyaging over the (often dangerous) sea, which imparts a strong emotional dimension to narrative and description. In Old English poetry water enables travel and communication; it offers defence against enemies but also the possibility of sudden attack; and it works as an agent of separation.[16] Such ideas doubtless reflect the real-life experience of the Anglo-Saxons and other northern European peoples in the early Middle Ages,[17] but they find particularly imaginative expression in traditional poetry, in which water also figures as an object of wonder.[18]

---

14 The most influential early lives are collected and translated by White C., *Early Christian Lives* (Harmondsworth: 1998). For a comprehensive introduction to hagiography and the veneration of saints, see Bartlett R., *Why Can the Dead Do Such Great Things? Saints and Worshippers from the Martyrs to the Reformation* (Princeton, NJ: 2013).

15 On the Virgin Mary in Old English literature, see Clayton M., *The Cult of the Virgin Mary in Anglo-Saxon England* (Cambridge: 1990) and *Apocryphal Gospels of Mary in Anglo-Saxon England* (Cambridge: 2008).

16 Old English poetry brings the dangers and benefits of travel by water compellingly to life, as in the accounts of voyages in *Beowulf* and descriptions of the cruelty of the sea in *The Wanderer* and *The Seafarer* (ed. ASPR III, 134–37, 143–147). The possibility of attack from the sea is recognised by the warriors in *The Battle of Maldon* (ed. ASPR VI, 7–16), who (mistakenly) allow the Vikings to come ashore at a place of the defenders' choosing, while in *Exodus*, the sea is a *sæfæsten*, 'sea-stronghold' (line 127), stopping the progress of the Israelite tribes. The theme of separation by water is an especially potent one for Old English poets, secular and religious, who draw upon the expressive language of the theme of exile in treating it. Water entails separation and exile in *Wulf and Eadwacer* and *The Wife's Lament* (ed. ASPR III, 179–80, 210–211), and it brings hope of coming together – but only when good weather comes in spring – in *The Husband's Message* (ed. ASPR III, 225–227).

17 For an overview of the sea in 'real-life' in Anglo-Saxon England, see Carver M., "Travels on the Sea and in the Mind", in Klein S.S. – Schipper W.V. – Lewis-Simpson S. (eds.), *The Maritime World* (Tempe, AZ: 2014) 21–36.

18 On water in general in Old English poetry, see Discenza N.G., *Inhabited Spaces: Anglo-Saxon Constructions of Place* (Toronto: 2017) 164–178; Mullins, "*Herimum in mari*"; Frederick J.A., "From Whale's Road to Water under the Earth: Water in Anglo-Saxon Poetry", in Hyer

Though culturally distinctive and highly interesting in literary terms, verse lives represent a small proportion of the surviving corpus of Old English saints' lives, amounting to just over 5,000 lines of poetry altogether, across five narrative poems celebrating different types of saint: *Guthlac A* and *Guthlac B* (on an ascetic saint), *Juliana* (on a virgin martyr), *Andreas* (on a missionary expedition by an apostle) and *Elene* (on the finding of the true cross by St Helena).[19] There are also two catalogue-type poems, *The Fates of the Apostles* and *The Menologium*,[20] which are brief in the extreme in treating their saints, though *The Menologium* does have one mention of water, when it relates that Saint Augustine brought the Christian faith to Britain *ofer sealtne mere*, 'over the salty sea' (line 103).

The vast majority of Old English lives are written in prose and include those by the prolific Ælfric of Eynsham, writing at the end of the tenth and beginning of the eleventh century, who produced more than fifty separate items,[21] nearly all in his two volumes of *Catholic Homilies* and his collection *Lives of Saints*.[22] There also survive about thirty anonymous prose lives, mostly contemporary with Ælfric, as well as the 230 or so short entries that make up the ninth-century *Old English Martyrology*[23] and the 115 miracle stories associated with St Benedict and other Italian saints found in Wærferth's late ninth-century

M.C. – Hooke D. (eds.), *Water and the Environment in the Anglo-Saxon World* (Liverpool: 2018) 15–32; Sobecki, *The Sea in Medieval English Literature* 41–44: Sobecki, who writes of the 'deafening presence of the sea in pre-Conquest writings in English' (45), lists the relevant studies of Old English literature (43, n. 80). Water as an object of wonder is notably evident in the Exeter Book riddles: see *Riddles* 3 (storm at sea), 41 (water), 84 (water) (*Riddles*, ed. ASPR III, 180–210, 224–43); on water in the riddles, see Frederick, "From Whale's Road" 25–27.

19 *Guthlac A*, ed. ASPR III, 49–72; *Guthlac B*, ed. ASPR III, 72–88; *Juliana*, ed. ASPR III, 113–133; *Andreas*, ed. ASPR II, 3–51; *Elene*, ed. ASPR II, 66–102.

20 *Fates of the Apostles*, ed. ASPR II, 51–54; *Menologium*, ed. ASPR VI, 49–55.

21 Numbers are given as approximate since works may be identified and demarcated according to differing criteria.

22 *Ælfric's Catholic Homilies: The First Series*, ed. P. Clemoes (Oxford: 1997); *Ælfric's Catholic Homilies: The Second Series*, ed. M. Godden (Oxford: 1979); *Ælfric's Lives of Saints*, ed. and trans. W.W. Skeat (London: 1881–1900; reprinted as 2 vols, 1966); Skeat also includes at the end of vol. II of this edition a separate life by Ælfric, that of Saint Vincent. These collections are abbreviated in references below as *CH I*, *CH II* and *LS*, respectively, and are cited below by item number. Ælfric also has a few additional pieces here and there, such as the story about Saint John in the *Letter to Sigeweard*, ed. R. Marsden, *The Old English Heptateuch and Ælfric's Libellus de Veteri Testamento et Novo: Vol. I* (Oxford: 2008) 224–227.

23 *The Old English Martyrology: Edition, Translation and Commentary*, ed. and trans. C. Rauer (Cambridge: 2013); abbreviated in references below as *OEM*: referred to by page number.

translation of the *Dialogues* of Gregory the Great (there are eighty-five stories if we exclude the accounts of the deaths of holy people in Book IV, the final book of Gregory's work).[24] The *Dialogues* and *Old English Martyrology* are particularly useful for our purposes in that they tend to concentrate on the most noteworthy things about the saint(s) they are commemorating.

Taken together these individual hagiographical items in Old English make up a sizeable corpus, which can conveniently form the basis for my discussion in the rest of this chapter. The chapter will identify all of the water themes found in the corpus[25] and will go on to focus on one life in particular, the anonymous (prose) *Life of St Mary of Egypt*,[26] in which water may be seen as having a symbolic significance of a kind not discernible elsewhere in the material surveyed; finally, it will highlight the distinctive treatment of water in the verse lives.

3 Water Themes

3.1 *Miracles*

The most frequently-occurring miracles involving water in our sample are those in which the holy person is thrown into water (most commonly the sea) to die, on the orders of a persecutor. Often the saint is tied to a heavy weight, only for the bonds to break, however, with the saint either surviving unharmed or having their body washed ashore or otherwise preserved for veneration. The former outcome is exemplified in the cases of Saint Eugenia (*OEM*, 36; *LS* II, lines 389–95: the river Tiber), Saint Christina (*OEM*, 140: the sea) and Saint Pantaleon (*Life of Saint Pantaleon*,[27] lines 274–90: the sea). The latter outcome is exemplified in the cases of Saints Ananias, Petrus and companions (*OEM*, 52: the sea), the 'Four Crowned Ones' (*OEM*, 212: the sea), Saint Vincent (Ælfric, *LS* XXXVII, lines 255–275: the sea) and Saint Clement (*OEM*, 218; *CH I* XXXVII, lines 100–6: the sea), whose miracle is particularly spectacular: Clement (Pope

24 *Bischofs Wærferth von Worcester Übersetzung des Dialoge Gregors des Grossen*, ed. H. Hecht (Leipzig: 1900); abbreviated in references below as *Dial.*

25 Instances noted below reference only 'actual' H2O; thus mentions of tears or sweat are not included.

26 Ed. and trans. Skeat, *Ælfric's Lives of Saints* XXIIIB. This anonymous *Life*, along with another mentioned below, the *Life of Saint Eustace*, is one of four non-Ælfrician items copied into the manuscript of *Lives of Saints*; for *Eustace*, see *LS* XXX.

27 Pulsiano P. (ed.), "The Old English Life of St Pantaleon", in Hall T.N. (ed.), *Via Crucis; Essays on Early Medieval Sources and Ideas in Memory of J.E. Cross* (Morgantown, WV: 2002) 61–103.

Clement I) is tied to an anchor but his body is found to be preserved, along with the anchor, 'in a stone house made ready by the Lord' (*OEM*, 218; trans. Rauer, 219) in the middle of the sea, where the surrounding waters have temporarily dried up for thirty miles around and dry up again every year on Clement's feast day for seven days, to allow people to visit the church.

In other variants of this type of miracle, the saint is thrown into water without a heavy weight: the Forty Soldiers are thrown into an icy lake, where the waters warm up as they die (*OEM*, 62–64; *LS* XI, lines 142–199); Saints Rufinus and Secunda (*OEM*, 136) are thrown into the Tiber but they float; Saints Nazarus and Celsus are thrown into the sea but they walk on the water (*OEM*, 146); the head and body of Saint Chrysogonus are thrown into the sea but are later washed ashore together to be buried (*OEM*, 218–220); similarly the body of Saint Antoninus is cut up and thrown into a river (or canal) but discovered later (*OEM*, 208); the body of Saint Quentin is thrown into the Somme but discovered fifty-two years later (*OEM*, 208); Saint Caesarius is thrown to his death into a torrential river but his persecutor is killed the same day (*OEM*, 210). Saints Ananias and Petrus, as well as being thrown into the sea, had previously also been put into a boiling bath, only to emerge unharmed (*OEM*, 52), a miracle also associated with Saint Cecilia (*OEM*, 216; *LS* XXXIV, lines 342–51) and Saint Margaret (Tib, Chap. 18, 128–130; CC, Chap. 18, 168–170).[28]

In these miracles God intervenes to glorify his saints rather than the saints themselves performing wonders, though one example of the latter is the rescue by Saint Machutus of his servant who has been thrown into the sea with bound hands and feet (*Life of Machutus*,[29] 59, lines 8–13). God also acts when he causes persecutors of saints to come to a watery end after their evil deeds: Heliseus., the persecutor of Saint Juliana is drowned at sea in a shipwreck (*Juliana*, lines 671–682), while the (unnamed) persecutor of Saints Ananias and Petrus is devoured by wild beasts in the sea (*OEM*, 52).

Other instances of divine favour towards saints involving water are less gruesome. In an episode about Saint Maurus, the iron head of a billhook, which has fallen into deep water, miraculously ascends through the water to join the handle held by Maurus (*Dial.*, II.6, 113–114); in the story of Saint Adrian and his wife Saint Natalia, Natalia, blown off course on her voyage, is guided over the sea by the appearance of her dead husband to the place where his

28 The *Life of St Margaret* in British Library MS Cotton Tiberius A. III, Clayton M. – Magennis H. (eds. and trans.), *The Old English Lives of St Margaret* (Cambridge: 1994) 109–139 (Tib); the *Life of St Margaret* in Corpus Christi College, Cambridge MS 303 (eds. and trans.) Clayton – Magennis, *The Old English Lives* 149–171 (CC).

29 Yerkes D. (ed.), *The Old English Life of Machutus* (Toronto: 1984).

body is buried (*OEM*, 60); the church of Saint Zeno is preserved from a flood that engulfs everything else, even though its door is open (*Dial.* III.19); Saint Fulgentius is made to stand in a small circle in the open and when a sudden rain storm occurs, frightening his enemies, not a drop falls in Fulgentius's circle (*Dial.* III.12); in the *Life of Saint Guthlac,*[30] when a raven makes off with the document of a guest of the saint, dropping it into a lake, Guthlac assures his guest that all will be well with the help of God, and the document is found hanging undamaged on a reed (Chap. 9, lines 1–22); the miracle of Saint Clement being cast into the sea and then discovered in a shrine in the middle of the sea, mentioned above, is followed by another miracle, in which a woman visiting the shrine one year forgets her sleeping child and goes home without it, leaving it to be engulfed by the returning waters. 'Then after a year had passed, when the people arrived there on Clement's feast day, they found the child alive and sleeping in the church, and it went away with its mother' (*OEM*, 218; trans. Rauer, 219; compare *CH I* XXXVII, lines 128–147).

Instances of objects being divinely transported across the sea occur in two lives in the Old English corpus: two doors given to Saint Giles by the pope in Rome appear at his monastery, having somehow crossed the sea (*Life of Saint Giles,*[31] lines 501–529), and a psalter is miraculously transported by sea to Saint Machutus (*Life of Saint Machutus*, 9, lines 7–14).

Miracles in which the saint is the agent rather than the beneficiary are exemplified in *imitatio Christi* episodes, such as Saint Cuthbert turning water into wine (*OEM*, 68) and holy people walking on water: Saint Mary of Egypt walks across the Jordan (*LS* XXIIIB, lines 684–685); Saints Nazarus and Celsus walk on the sea, having been cast into the deep in a death sentence (*OEM*, 146); Saint Maurus walks on the sea to rescue a drowning monk (*Dial.* II.7); and Saint Martin of Tours tells fishermen where to cast their nets, whereupon they catch an enormous salmon (*LS* XXXI, lines 1268–1276; cf. Luke 5:1–11, John 21:1–14). Saint Martin also makes an evil river-snake depart, it having swum threateningly towards his monks (*LS* XXXI, lines 1259–1266). As mentioned in the next section, saints also intervene miraculously to help travellers caught up in storms at sea, emulating Christ on the Sea of Galilee.

Among other miracles performed specifically by the saint (though some of these might also be seen as divine interventions), a popular one is the bringing

30 Gonser P. (ed.), *Das angelsächsische Prosa-Leben des hl. Guthlac* (Amsterdam: 1909); translations below are taken from Swanton M. (trans.), *Anglo-Saxon Prose* (revised ed. London: 1993) 88–113.

31 Treharne E.M. (ed. and trans.), *The Old English Life of St Nicholas with the Old English Life of St Giles* (Leeds: 1997) 131–47, with translation 148–162.

forth or discovery of water in a waterless place. The biblical parallel here is that of Moses producing water from the rock at Horeb (Numbers 20:11), though it has also been argued that Anglo-Saxon occurrences were also influenced by pagan practice.[32] Saint Clement causes water to well up where there had been no spring (*OEM*, 218; see also *CH I* XXXVII, lines 78–80); Saint Cuthbert finds a spring of pleasant water in a waterless island (*CH II* X, lines 171–176); Saint Alban prays for water at a pleasant hill and a well-spring appears at his feet (*LS* XIX, lines 107–115); in response to a prayer of Saint Benedict, a fountain springs forth at the top of a mountain (*Dial.* II.5); and in the cave of Saint Martinus water springs up, giving him just as much for his needs (*Dial.* III.16). In a somewhat analogous miracle Saint Machutus brings it about that a single drop of water serves a whole group without decreasing (*Life of Saint Machutus*, 13, lines 12–16); elsewhere the same saint finds a miraculous spring (25, line 3–17, line 1). A final spectacular variation on the motif of bringing forth water is to be seen in the episode of Saint Andrew drowning the wicked Mermedonians when he calls forth water from a pillar in their city, causing a great flood (*Andreas*, lines 1492–1582; prose *Acts*,[33] lines 261–291).

A motif reflecting God's miraculous protection of his saints is the lifting of oppression from fire as though it were refreshing dew. Thus, Saint Martin emerges from a conflagration feeling 'as if he were in a pleasant dew' (*on wynsumum deaw*, *LS* XXXI, line 888); Saint Eustace and his family, undergoing the ordeal of martyrdom by fire, pray that the fire's heat may be turned to liquid dew (*LS* XXX, line 441), whereupon the heat cools as they give up their souls; Saint Agatha taunts her persecutor that if he tortures her by fire God will send a healing dew from heaven (*LS* VIII, line 88).

3.2 *Travel*

Travel by sea as a theme in Christian story goes back as far as the Acts of the Apostles and it figures widely in hagiography. Saints themselves travel and they help those who travel, while God watches over the voyages of the righteous and unrighteous as they cross the unpredictable sea, intervening as appropriate to protect the good and punish the ungodly (the latter exemplified in the deaths of the persecutors of Saint Juliana and of Saints Ananias and Petrus, mentioned above). Mention of river crossings, on the other hand, is very infrequent,

32 See Hooke D., "Rivers, Wells and Springs in Anglo-Saxon England : Water in Sacred and Mystical Contexts", in Hyer M.C. – Hooke D. (eds.), *Water and the Environment in the Anglo-Saxon World* (Liverpool: 2018) 107–135.

33 *The Acts of Matthew and Andrew in the City of the Cannibals*, ed. Cassidy F.G. – Ringler R.N., *Bright's Old English Grammar and Reader* (New York: 3rd ed. 1971) 205–219.

though there are occasional examples in the Old English material: Saint Eustace crosses a river with his two sons but loses them to wild animals (*LS* XXX, lines 173–174), and later he crosses another river on a military expedition (lines 308–9); Saint Giles crosses the Rhone (*Life of Saint Giles*, lines 176–177); in the story of Saint Mary of Egypt, both Mary and the monk Zosimus, and indeed his brethren, cross the River Jordan, as discussed more fully below. Travel by river is exemplified in the episode of Saint Martin's body being taken from Poitiers to Tours (*LS* XXX, lines 1472–1480; *CH II* XXXIV, lines 323–327 [though in the *CH II* version Ælfric does not explicitly mention that the ship journey is by river]).

Travel by sea has a range of purposes and reasons, from the noble to the mundane. Like the apostles, saints travel as missionaries, as (in our sample) Saint Germanus sailing to Britain (*OEM*, 150), 'where he performed many miracles both on sea and on dry land' (trans. Rauer 151), Saint Andrew voyaging to Mermedonia (*Andreas*, lines 235–836; prose *Acts*, lines 72–101), Saint Fursa, who travels from Ireland, first to Britain and then on to King Clovis of the Franks (*OEM*, 48), Saint Giles travelling in stages from Greece until he reaches Marseilles (*Life of Saint Giles*, lines 101–132) and Saint Machutus, making a circuitous journey to preach in Brittany (*Life*, 35, line 2–43, line 2).

Some travellers in hagiography embark on journeys that we might view more specifically as quests. Before his time in Brittany, Machutus had been a companion of Saint Brendan in his search for the Promised Land of the Saints, encountering many wonders on the way (*Life of Saint Machutus* 7, lines 17–31, line 3); after the martyrdom of her husband Adrian, Saint Natalia boards a ship from Nicomedia to travel to Byzantium 'where Christians had taken the body of Adrian' (*OEM*, 60; trans. Rauer 61); Saint Helena sails to the Holy Land to find the True Cross (*Elene*, lines 225–275); in the story of Saint Guthlac, Guthlac's servant goes by boat across the fens on a quest to bring back the saint's sister (*Guthlac B*, lines 1326–1343; prose *Guthlac*, Chap. 20, lines 119–121); Guthlac himself had travelled by boat to reach his fenland retreat (prose *Guthlac*, Chap. 3, lines 23–25). Even the voyage of Mary of Egypt from Alexandria to Jerusalem can be seen as a kind of perverted quest as Mary goes in search of sexual partners (*LS* XXIIIB, lines 376–392).

Sea travel can also be for reasons of necessity, where the saint needs to escape or is forced to depart: Saint Natalia escapes an unwelcome suitor (*OEM*, 60), Saint Erasmus escapes the persecution of Maximianus (*OEM*, 112), Saints Vitus and Modestus also escape persecution (*OEM*, 116), Saint Eustace and his family flee their homeland after conversion (*LS* XXX, lines 163–172), Saint Clement is compelled by Trajan to go into exile by ship (*CH I* XXXVII, lines 53–62).

Hagiography also has instances of travel to and fro for official, commercial and personal reasons, particularly again travel across the Mediterranean Sea.

Such voyages are usually of interest to hagiographers, however, only when something unexpected has happened on them, most commonly a storm. Storms at sea affect saints on their travels, as is the case with Saint Andrew on his way to Mermedonia (*Andreas*, lines 360–468; prose *Acts*, lines 75–83), Saint Natalia on her voyage to Byzantium (*OEM*, 60), Saint Agazatha travelling from Pannonia to Rome (*Dial.* IV.59) and the devout Nathan, who, in the quasi-hagiographical *Legend of Saint Veronica*[34] is providentially blown off course by high winds on his way to Rome, leading to the cure of the cancer-stricken lord of the city of Lybia (lines 17–20).

More typically, however, it is other people going about their business that are afflicted by storms, from which they are then miraculously saved by a saint. Saint Agazatha, having safely made land in his storm, prays for a mariner who has become separated from the ship and thus saves him (*Dial.* IV.59) and Saint Martin of Tours is invoked by a sailor during a storm, which suddenly abates (*LS* II, 290, lines 1135–1142; *OEM*, 214). Storms threatening ships are also calmed by Saint Giles (*Life of Saint Giles*, lines 64–85), Saint Machutus (*Life of Saint Machutus* (97, line 21–99, line 10) and Saint Nicholas (*Life of Saint Nicholas*,[35] lines 186–200), while Nicholas also prays that a fleet driven into his harbour by a storm may sail safely to its destination (lines 342–345, 418–419). Bishop Maximianus's ship, returning to Italy from Constantinople, is filled with water in a storm but makes its way safely to port over the course of eight days, sinking only when the holy man has disembarked (*Dial.*, II. 36): God has granted this miracle 'through Maximianus, his servant'.

3.3 *Cleansing, Healing, Abstinence, Self-Mortification*

Mentions of baptism, referred to in Ælfric's *Life of Saint Eugenia* as a 'cleansing of all filth' (*LS* II, line 337), are included in many saints' lives, though usually writers feel no need to say anything concerning the administering of the sacrament, not even mentioning water. Exceptions to this rule occur only when there is something remarkable about the baptism, as in the account of the young Saint Athanasius baptising children in the waves of the sea at Alexandria (*OEM*, 92) and that of Saint Christina being immersed three times in the sea by Christ, who has appeared to her walking over the waves (*OEM*, 140). Saint James the Greater asks for water at his execution and baptises the pharisee Josias (*CH II* XXVII, lines 171–179; *Concerning James, the Brother of John*,[36]

34 Assmann B. (ed.), *Angelsächsische Homilien und Heiligenleben* (Kassel: 1889) 181–192.

35 Treharne (ed.), *The Old English Life of St Nicholas* 83–100, with translation 101–117.

36 Warner R.D.N. (ed.), *Early English Homilies of the Twelfth Century MS Vesp. D. XIV* (London: 1917) 21–25.

24, lines 30–36); similarly, Saint Laurence blesses a jug of water and baptises a believing Roman soldier in the midst of his own tortures (*CH I* XXIX, lines 175–178). A perversion of baptism appears in Ælfric's *Life of Saint Stephen* when a widow dips her hair in a font vessel and curses her children (*CH II* II, lines 114–117).

Water is also used to cleanse or heal people from affliction. Saint Machutus washes the eyes of a blind king and thus gives him his sight (*Life of Saint Machutus*, 65, lines 2–8); Saint Guthlac cures an insane man by bathing him in consecrated water after others have failed (prose *Guthlac*, Chap. 12, lines 17–46); Saint Cuthbert heals a woman with holy water (*CH II* X, lines 273–275); healing water also occurs in a legend of Saint Michael retold in one of the Blickling homilies (*Blickling* XVII, 209).[37] The sinner Mary of Egypt washes her hands and face in the waters of the Jordan (*LS* XXIIIB, line 502). The miracle of fire turning to pleasant dew, mentioned earlier, may also be seen as an instance of healing; indeed in one of the examples cited (*LS* VIII, line 88) Saint Agatha refers to dew as *halwendlic*, 'healing'.

The consumption of water is referred to in hagiography in the lives of ascetic saints, the emphasis being on how little water these saints drink. Saint Paul the Hermit drinks only a little water 'from his hollow hand' (*OEM*, 46; trans. Rauer, 47) while Saint Antony goes without food or water for two or three days in a row and when he does break his fast it is to take some bread and salt and 'just a little water' (*OEM*, 50; trans. Rauer, 51). Saint Guthlac lives on barley-bread and water, breaking his fast only after sunset (prose *Guthlac*, Chap. 4, lines 24–32). Saint Mary of Egypt is very thirsty 'on account of the want of water in this wilderness' (*LS* XXIIIB, lines 537–538; trans. Skeat 37), living only on plants (line 569); when the monk Zosimus brings her a few lentils soaked in water, she touches the lentils with her fingers and puts only three in her mouth (lines 714–716). In the monastery of Zosimus he and the other monks subsist on bread and water alone (lines 93–94). Most extreme is Saint Mary Magdalen, who during her thirty years in the desert 'never ate human food, or drank' (*OEM*, 142; trans. Rauer 143). There is one example in Old English of the deprivation of food and water as part of the affliction imposed on a saint by a persecutor: Saint Anastasia, deprived of bread and water in her prison cell, is fed with heavenly food by a spirit (*OEM*, 34).

37 Morris R. (ed. and trans.), *The Blickling Homilies* (Oxford: 1874, 1876, 1880; reprinted as one vol., 1967): according to this legend, in a grotto-church dedicated to Saint Michael there is a stream issuing from the roof; many people suffering from illness and affliction are healed by drinking this 'heavenly liquid'.

The other ways that water figures in the lives of ascetic saints are as an instrument of self-mortification and by providing a location for withdrawal from the world. In self-mortification Saint Cuthbert stands in the sea at night (*CH II* X, lines 78–81), Saint Neot prays standing in a pond (*Life of Saint Neot*,[38] 130, lines 31–32) and Saint Æthelthryth denies herself the indulgence of bathing 'save at high festivals' (*LS* XX, line 45; trans. Skeat 435). With respect to location, saints retreat to isolated places surrounded by or beside water. Saint Eadberht of Lindisfarne spends each Lent and Advent in a secret place 'surrounded by sea currents' (*OEM*, 98; trans. Rauer 99), Saint Cuthbert retreats to Farne Island, *on flowendre yðe*, 'amid the flowing waves' (*CH II* X, lines 163–164), and Saint Guthlac withdraws to a tiny island in the watery Fens, 'where there are immense swamps, sometimes dark stagnant water, sometimes foul rivulets running; and also many islands and reeds and tummocks and thickets' (prose *Life*, ed. Gonser, Chap. 3, lines 3–6; trans. Swanton 91). It is notable that these holy people in our corpus who retreat to watery places are insular saints engaging in their version of withdrawal to the desert as they emulate in a new environment the ascetics of the earlier church who dwelt in arid places.

4 Saint Mary of Egypt

One such ascetic from the earlier church is Saint Mary of Egypt, as celebrated in the anonymous Old English *Life* that tells her story. The Old English *Life of Saint Mary of Egypt* is a generally faithful translation, though with frequent rhetorical augmentation, of a Latin *vita*, in which Mary withdraws to the Syrian desert after a life of debauchery in Alexandria and after many years is encountered there by the monk Zosimus, who is following the practice of his monastery in spending the season of Lent alone in the desert; Mary tells Zosimus her story and brings spiritual enlightenment to him. The *Life of Saint Mary of Egypt* is of particular interest with regard to water, uniquely so indeed among Old English hagiographical writings: here is a saint's life that is far from literal-minded in its use of water themes.

The *Life* does have the familiar motifs of travel by sea – Mary making her way to Jerusalem – and the miracle of walking on the water – Mary crossing the river to meet the monk Zosimus – but otherwise water operates symbolically in this life. The focus of this symbolism is the River Jordan, 'holiest of all waters' (lines 63–64; trans. Skeat 7), the river that separates the desert, a place

38 Warner, *Early English Homilies* 129–134.

of isolation but also spiritual enlightenment, from the 'world' on the other side. In the desert Mary asks Zosimus about life in the world – 'tell me how nowadays Christ's people are governed in the world and how the emperor's [matters stand]; or how the flock of Christ's right-believing congregation is now being looked after' (lines 252–225; trans. Skeat 17–19, modified) – and later (lines 684–685) she walks across the river to 'the side of the Jordan pertaining to the world' (lines 624–625; trans. Skeat 43). The 'world' is a place of activity, crowds and consumption, the desert on the other side a place of solitariness and contemplation, without the most basic necessities for life. Mary suffers 'on account of the want of water in this wilderness, scarcely enduring the terrible necessity' (lines 537–538; trans. Skeat 37). When Zosimus comes across her it is at the site of a dried-up stream (lines 196–197). Water must have been present in it in the past, but presumably only very rarely; the stream is still dried-up when Zosimus returns there two years later (line 740).

Zosimus's monastery is on the side of the Jordan that belongs to the world but it is appropriately *near* the holy river and its monks go across to the other side each Lent:

> When they had gone over the river Jordan, then each one sundered himself far from the others, and none of them joined himself again to his companions; but if any one of them saw another afar [coming] towards him, immediately he turned out of this [chosen] direction, and went another way, and lived by himself, and continued in perpetual prayers and fasting. (lines 133–138; trans. Skeat 11)

Crossing the river is an act of spiritual resolution.

Mary's moment of conversion comes when she is in Jerusalem (where she has gone for immoral purposes) and is mysteriously prevented from entering the great church on the feast of the Exaltation of the Cross. Suddenly 'knowledge of salvation (*hælo andgit*) touched her mind' (lines 424–425; trans. Skeat, 29) and she repents her former life. Soon afterward she hears a voice calling from far away, saying, 'If you cross the waters of the Jordan, there you will experience and find good repose' (lines 483–484; trans. Skeat 33, modified). Mary makes her way to the Jordan, where, spiritually cleansing herself, she first visits the church of John the Baptist and then washes her hands and face in the holy water of the river, before returning to the church to receive the sacrament of the Eucharist (lines 501–506).

The next morning Mary crosses the Jordan and begins her life of self-mortification in the desert, alone, until she is encountered by Zosimus forty-seven years into her sojourn (line 515). After telling Zosimus her story, she instructs him to wait for her one year later 'on the side of the Jordan that pertains to the world' (lines 624–625; trans. Skeat 43) so that she may receive the sacrament from him, and the monk duly waits for her there, though not understanding how she will get across to him, 'now that there is no ship wherein she may come to me' (lines 674–675; trans. Skeat 45). It is then that Mary walks upon the water, both to get to him and then to return to the other side: 'She went towards him, walking on the soft waves (*uppan þa hnescan yða*) as if on dry land' (lines 684–685; trans. Skeat 47). At this meeting she asks Zosimus to cross the river again the following year and return to the dried-up stream where they first met. He does so and finds her body there laid out properly for burial (line 742).

In the *Life of Saint Mary of Egypt* the River Jordan functions as a boundary, both in a literal sense and symbolically. The Jordan and its waters represent a key narrative fulcrum in the story of Mary. The Jordan is a physical reality but also a powerful symbol of the theme of advancing to spiritual perfection and also that of cleansing from sin, that are at the heart of the message that the *Life* conveys. It is in this anonymous *Life* that the theme of water finds its fullest development among Old English hagiographical writings. It should be emphasised, however, that there is nothing distinctively Anglo-Saxon in the treatment of water in the *Life*. The Old English text is a thoughtful rendering that frequently contributes rhetorical heightening to emphasise the meaning of its Latin original, but it is notable that such heightening is not very evident where water is concerned.[39] With respect to water, the Old English version faithfully transmits rather than elaborates the content of its original, and it is examined here as an example of hagiography in general, rather than as a specifically Old English work. For a distinctively Anglo-Saxon approach to hagiography we need to turn to lives in Old English verse.

5 Verse Saints' Lives

The surviving verse saints' lives display most of the same range of water themes as found in Old English hagiographical writings as a whole: water figures in miracles, is a medium of travel and represents cleansing; the one category identified above that is not represented in the poetry is that of water as contributing

39 For example, the descriptive phrase *uppan þa hnescan yða*, 'on the soft waves', quoted above, reflects (slightly simplifying, indeed) the Latin *super liquidum equoris fluctum*, 'on the limpid waves of the surface' (see *Vita Sanctae Mariae Egiptiacae*, ed. and trans. Magennis, *The Old English Life of St Mary of Egypt* 140–209), quoted phrase at line 761.

to ascetic practice, water as a means of self-mortification. The themes are the same as those found in prose but can be developed in the verse saints' lives with considerable imagination, exploiting the resources of Old English poetic tradition. And, though water is *used* by people in the verse lives, it can also take on an interest in and of itself. As Heide Estes, suggests, for example, 'The expression of joy in journeying on the sea [...] might be read as celebrating, in its own terms, a landscape in which humans cannot build or live, and unstructured by human intervention'.[40]

5.1 Guthlac A *and* Guthlac B

The theme of water as a means of self-mortification is entirely absent from the (small) corpus of verse lives, even though we have two poems on an ascetic saint, Saint Guthlac, who retreated to a watery wilderness accessible only by boat. We might especially expect this theme to play a part in *Guthlac A*, which tells of Guthlac settling in his place of solitary retreat, where he fights against demons for ownership of it. The Latin *Vita Sancti Guthlaci* by Felix of Crowland[41] and its Old English prose translation[42] describe the saint's island abode in atmospheric terms. According to Felix, to get there Guthlac makes his way through 'a most dismal fen of immense size [...] now consisting of marshes, now of bogs, sometimes of black waters overhung by fog, sometimes studded with wooded islands and traversed by the windings of tortuous streams' (trans. Colgrave 87).[43] Travelling in a fisherman's skiff, the saint arrives at an island in the middle of the marsh, 'which on account of the wildness of this very remote desert had hitherto remained untilled and known to a very few' (trans, Colgrave 89).[44] On the island Guthlac finds 'a sort of cistern' (*velut cisterna*), which supplies him with 'muddy water' (*lutulentae aquae*), though he restricts himself to drinking this water and eating a small amount of barley-bread along

40 Estes, *Anglo-Saxon Literary Landscapes* 38.

41 Colgrave B. (ed. and trans.), *Felix's Life of Saint Guthlac* (Cambridge: 1956).

42 See n. 30, above.

43 Chap. 24, 'inmensae magnitudinis aterrima palus ... nunc stagnis, nunc flactris, interdum nigris fusi vaporis laticibus, necnon et crebris insularum nemorumque intervenientibus flexuosis rivigarum anfractibus' (Colgrave 86). On Guthlac's life in the fens, see Wickham-Crowley K.M., 'Living on the *Ecg*: The Mutable Boundaries of Land and Water in Anglo-Saxon Contexts", in Lees C.L. – Overing G. (eds.), *A Place to Believe In: Locating Medieval Landscapes* (University Park, PA: 2006) 85–110; see also Wickham-Crowley's chapter "Fens and Frontiers" in Hyer M.C. – Hooke D. (eds.), *Water and the Environment in the Anglo-Saxon World* (Liverpool: 2018) 68–88.

44 Chap. 25, 'quae ante paucis propter remotioris heremi sloitudinem inculta vix nota habebatur' (ed. Colgrave 88).

with it only after sunset, 'for the needs of this mortal life' (*mortalis vitae*) (ed. and trans. Colgrave, Chap. 28 94–95).

These details are faithfully reflected in the Old English prose translation of the *Vita* (ed. Gonser, Chap. 3, lines 1–31; Chap. 4, lines 1–32) but are conspicuously absent from the account of Guthlac's place of retreat in *Guthlac A*. In *Guthlac A* there is no mention of the fens, the island, or water, muddy or otherwise. Instead, the poem re-presents Guthlac's territory as a pleasant green landscape, though one infested by evil spirits when he first arrives, thus both attractive and foreboding, 'a place of contradictions', as Nicole Guenter Discenza refers to it.[45] The territory is dominated by a hill (*beorg*), a key spiritual symbol in the poem that has been much discussed by critics.[46] Here Guthlac has 'the most cherished abode on earth' (*þam leofestan eard on eorðan*, lines 427–428). Having established ownership of the land where he wishes to live,[47] he becomes the *bytla*, 'builder' (lines 148, 723), of a home, a place of joy that anticipates the joy of the heavenly home to come.[48] The poet of *Guthlac A* has strikingly rejected all references to water in line with the poem's emphasis on Guthlac's abode as a desirable piece of land which the saint wrests from evil hands. As Catherine Clarke has explained, in describing this desirable place, the *Guthlac A* poet draws upon traditions of the *locus amoenus* (describing an idealised landscape), familiar particularly in Latin literature.[49] I have suggested elsewhere that the absence of water in *Guthlac A* may also serve to strengthen correspondences between Guthlac, with his *beorg*, and the archetypal desert saint Antony, whose dwelling place was mountainous (and who fought demons for possession of it), and Paul, who lived at a 'rocky mountain' (*saxeum montem*).[50]

---

45 Discenza, *Inhabited Spaces* 154; see also Michelet F.L., *Creation, Migration, and Conquest: Imaginary Geography and Sense of Space in Old English Literature* (Oxford: 2006) 172–173, on the 'ambiguous' nature of the place.

46 Most notably, Guthlac's *beorg* has been interpreted as a barrow (Shook L.K., "The Burial Mound in *Guthlac A*", *Modern Philology* 58 (1960) 1–10), a heathen tumulus (Wentersdorf K.P., "*Guthlac A*: The Battle for the *Beorg*", *Neophilologus* 62 (1978) 135–142) and a mountain (Reichardt P.F., "*Guthlac A* and the Landscape of Spiritual Perfection", *Neophilologus* 58 (1974) 331–338).

47 See Clark S., "A More Permanent Homeland: Land Tenure in *Guthlac A*", *Anglo-Saxon England* 40 (2012) 75–102.

48 On building and home, see Discenza, *Inhabited Spaces* 151–156; also Jones C.A., "Envisioning the *coenobium* in the Old English *Guthlac A*", *Mediaeval Studies* 57 (1995) 259–291.

49 Clarke C.A.M., *Literary Landscapes and the Idea of England* (Cambridge 2006) 45–66: Clarke focuses particularly on lines 732–748 of the poem.

50 Magennis H., *Images of Community in Old English Poetry* (Cambridge: 1996) 183. For the *saxeum montem* reference, see *Vita Sancti Pauli primi eremitae*, Chap. 5, ed. J.-P. Migne *Patrologia Latina* 23 col. 21B.

Reference to water as a means of self-mortification is also absent from *Guthlac B*, a poem on the death of Guthlac. *Guthlac B* describes the saint's retreat as being 'in a wilderness' (*on westenne*, line 935) but mentions that it was on an island only once, towards the end of the poem, when it is declared that at the death of Guthlac, as his place of retreat is filled with sweet odours and angelic singing, 'the island trembled, the flat ground convulsed' (lines 1325–1326, trans. Bradley 282; *Beofode þæt ealond, / foldwong onþrong*).

Guthlac B lacks the theme of water as a means of self-mortification but it does have a description of travel by water, in the closing episode of the poem, in which Guthlac's servant embarks on a journey by boat to inform the saint's sister of his death. Concerning this boat trip Felix's *Vita*, followed by the Old English prose version, simply recounts that the attendant 'took a boat and, leaving the landing-place, then began the journey which the man of God had commanded him' (trans. Colgrave 159).[51] *Guthlac B* develops this detail at some length, turning what would have been a paddle through marshy shallows into a speeding voyage across the waves. Steeped in the formulaic language of traditional poetry, the lofty description of the voyage in *Guthlac A* would have resonated familiarly with Anglo-Saxon audiences:

gewat þa ofestlice
beorn unhyðig, þæt he bat gestag,
wæghengest wræc, wæterþisa for,
snel under sorgum. Swegle hate scan
blac ofer burgsalo. Brimwudu scynde,
leoht, lade fus. Lagumearg snyrede,
gehlæsted to hyðe, þæt se hærnflota
æfter sundplegan sondlond gespearn,
grond wið greote. (lines 1327–1335)

Then, unhappy, he went hastily on until he boarded a boat and launched the wave-steed – the ship moved off – brisk beneath his sorrows. The sun was shining hotly, shimmering above the dwellings. The light timber vessel sped along, striving forward on its way. Into the landing-place the laden water-steed hurried so that after tossing on the open water the

51 Chap. l, 'arrepta navicula portum reliquit ac deinde, quo vir Dei praeceperat, coepto itinere perexit' (ed. Colgrave 158); the Old English prose version has 'eode þa on scip, and þa ferde to þære stowe, þe se godes wer ær bebead' (ed. Gonser Chap. 20, lines 120–121), 'he went onto a boat and travelled to the place that the man of God had instructed' (trans. Swanton 110).

buoyant boat trod upon the sandy shore and grated on the gravel. (trans. Bradley 282, modified)

In these lines the *Guthlac B* poet daringly presents a version of what may be seen as a recurrent motif or type scene of Old English poetry, that of the journey of the hero by ship. This motif occurs in *Beowulf*, twice (lines 205–228 and 1896–1913), and also in the verse saints' lives *Elene*, *Juliana* and *Andreas*. It is characterised by the mention of a number of definitive descriptive details, expressed in formulaic terms.

The definitive details mentioned both in the above passage from *Guthlac B* and in *Beowulf* (parallels in *Elene*, *Juliana* and *Andreas*, respectively, are discussed separately below) may be listed as follows:

Embarkation, often with mention of the shore (though the latter detail does not occur in *Guthlac B*): *he bat gestag* (line 132), 'he boarded a boat'. *Beowulf* parallels: *on stefn stigon* (line 212), 'boarded onto the stern', *Gewat him on naca* (line 1903), 'departed on the ship'.

Urgency and (wind-driven) speed: *ofestlice* (line 1327), 'speedily', *snel* (line 1330), 'quick', *snyrede* (line 1332), 'hastened'. *Beowulf* parallels: *winde gefysed* (line 217), 'sent forth with the wind', *wind ofer yðum* (line 1907), 'wind above the waves', *lyftgeswenced* (line 1913), 'impelled by the wind'.

Sea and/or waves: *wæghengest* (line 1329), 'wave-steed', *sundplegan* (line 1334), 'play of the sea'. *Beowulf* parallels: *flota wæs on yðum* (line 210), 'the boat was on the waves', *Gewat ofer wægholm* (line 217), 'departed over the billowy sea', *fleat famigheals forð ofer yðe* (line 1909), 'floated foamy-necked forth over the waves'.

Cargo: *gehlæsted* (line 1333), 'loaded'. *Beowulf* parallels: *on bearm nacan beorhte frætwe* (line 214), 'bright treasures into the hold of the ship', *hladen herewædum* (line 1897), 'loaded with battle-armour'.

The hero/leader, specifically highlighted: *beorn unhyðig* (line 1328), 'unhappy man'. *Beowulf* parallel: *se goda ... sundwude sohte* (lines 205–208), 'the good man sought the sea-wood'. In *Guthlac B*, however, this feature is deployed ironically, as the traveller is not a leader but a servant, who before the voyage is described as lacking in courage – *ða afyrht wearð / ar elnes beloren* (lines 1326–1327), 'then the messenger, dispossessed of courage, grew frightened' – and is portrayed elegiacally as *unhyðig* (line 1138), 'unhappy', rather than displaying heroic resolution.

Most insistently, the ship on which the voyage is made: the ship is referred to in descriptive alliterating compound words, including kennings. There are five such compounds in the *Guthlac B* passage: *wæghengest*, 'wave-steed', *wæterþisa*, 'water-rusher' (both line 1329), *brimwudu* (line 1331), 'sea-wood', *lagumearg* (line 1332), 'sea-steed', *hærnflota* (line 1333), 'wave-floater'. *Beowulf* parallels: *wegflota* (line 1907), 'wave-floater', *sæwudu* (line 226), 'sea-wood', *sundwudu* (lines 208 and 1906), 'sea-wood'.[52]

Destination harbour or landing place: *to hyðe* (line 1333), 'to the landing-place', *sondlond* (line 1334), 'sea-shore', *wið greote* (line 1335), 'on the gravel'. *Beowulf* parallels: *on wang stigon* (line 225), 'went ashore', *ceol up geþrang* (line 1912), 'the ship drove ashore', *on lande stod* (line 1913), 'stood on the land'.[53]

Structurally the motif moves towards this reference to arrival by means of an *until/so that* pattern: in *Guthlac B*, 'so that ... the buoyant boat stood upon the sandy shore'.[54]

Other discernible components of the motif of the journey of the hero by ship, not included in the *Guthlac B* version, are:

Sight of land at the end of the voyage: *land gesawon* (*Beowulf*, line 221), 'saw the land', *þæt hi Geata clifu ongitan meahton* (*Beowulf*, line 1911), 'that they could perceive the cliffs of the Geats'.

Reference to the admirable crew/warriors travelling with their leader: *weras on wilsið*, (*Beowulf*, line 216), 'men on their desired journey'.

These latter details are perhaps not considered relevant by the *Guthlac B* poet, since the boat is never far from land and the servant is travelling alone, without a crew. The curious detail of the sun shining over the dwelling places (in what seems to be an uninhabited wilderness) may also reflect the fact that the journey is not really a sea journey.

Edward B. Irving refers to the spirit of heroic endeavour evident in the *Beowulf* sea journeys, observing that '[the] voyages in *Beowulf*, in actuality, become resonant renderings of heroic power'.[55] Such a spirit is missing in

---

52 Of the compounds found in *Guthlac B*, *lagumearg* and *hærnflota* are *hapax legomena*, though formulaic parallels are *sæmearh* (*Andreas* line 267, *Elene* line 245), *ægflota* (*Andreas* line 258), *sæflota* (*Andreas* line 381), *scipflota* (*The Battle of Brunanburh* line 11) and (as referred to in the main text, above) *wægflota* (*Beowulf* line 1909, *Andreas* line 487, and *Elene* line 246). *Wæghengest* is also found in *Elene* (line 236) and has formulaic parallels in *merehengest* (*Meters of Boethius* 26.25), *brimhengest* (*Rune Poem* lines 47 and 66, *Andreas* line 513), *sæhengest* (*Andreas* line 488) and *fearoðhengest* (*Elene* line 226). *Wæterþisa* appears in *The Whale* (line 50), while *brimwudu* appears in *Elene* (line 244) and is paralleled in *flodwudu* (*Christ* line 853), *sæwudu* (*Beowulf* line 226) and *sundwudu* (*Beowulf* lines 208 and 1906, *Christ* line 667) (the last two *Beowulf* occurrences are also mentioned in the main text, above).

53 A further detail, that of the comparison of the ship to a bird, is confined to *Beowulf* and *Andreas* and is best taken as a direct borrowing by the *Andreas* poet (see n. 61, below): *fugle gelicost* (*Beowulf* line 218); *fugole gelicost* (*Andreas* line 497).

54 This point is made by Ogilvie J.D.A. – Baker D.C., *Reading Beowulf: An Introduction to the Poem, Its Background, and Its Style* (Norman, OK: 1983) 150.

55 Irving E.B., *A Reading of Beowulf* (New Haven, CT: 1968) 51.

Guthlac B. Instead, as suggested above, in including the phrases *beorn unhyðig* (line 1328), 'unhappy man', and also *snel under sorgum* (line 1330), 'brisk beneath his sorrows' (an ironic variant is the upbeat *snel under segle*, as found in *Andreas*, line 505, 'brisk beneath the sail'), the *Guthlac B* poet has injected an element of elegiac colouring, appropriate to the servant's mournful mission. The incorporation of the voyage motif has allowed the poet to emphasise the urgency and speed of the journey and through its elevated register to suggest its importance, while also maintaining an elegiac tone. In including this passage the poet is attempting something ambitious, doing so (to quote the editor of the standard edition of the poem) in 'a high style, with ornateness its most distinctive feature',[56] and thereby presenting the hagiographical theme of travel by water in a distinctively Anglo-Saxon manner.

5.2 *Elene and Juliana*

Voyages by sea also feature in *Elene* and *Juliana*, two poems on female saints by the named author Cynewulf. In *Elene* the sea-voyage passage is even more elaborate than that in *Guthlac B*, extending over thirty-three lines (lines 225–258), and whereas the *Guthlac B* passage was prompted by a brief mention of a boat journey in the Latin original, there is no reference at all to a sea voyage in the *Acta Quiriaci*, a version of which is thought to have been the source for Cynewulf in *Elene*.[57] Elene and her 'mighty army' (Chap. 3, trans. Allen and Calder, 61) must have travelled by ship in order to get to Jerusalem but this is not thought worthy of mention in the Latin work. In the Old English poem, however, a fully-fledged journey-by-sea motif is developed, executed in the same elevated style and with the same identifying components as we saw in *Guthlac B*, except that the detail of the accompanying admirable warriors (understandably, not in *Guthlac B*) is also included, and, as in *Beowulf*, the mood of the passage is cheerful and optimistic.

The *Elene* passage begins,

> Ongan þa ofstlice   eorla mengu
> to flote fysan.   Fearoðhengestas
> ymb geofenes stæð   gearwe stodon,
> sælde sæmearas,   sunde getenge. (lines 225–228)

56 Roberts J. (ed.), *The Guthlac Poems of the Exeter Book* (Oxford: 1979) 57.

57 Henschenius G. – Papebrochius D., *Acta Sanctorum, Maius, Tom. I*, 445–448 (Antwerp: 1680); trans. M.J.B. Allen – D.G. Calder, *Sources and Analogues of Old English Poetry: The Major Latin Texts in Translation* (Cambridge: 1976) 59–69.

> A throng of men, then, quickly hastened down to the sea. Ships, the horses of the ocean, lay ready along the seashore, sea-steeds moored afloat upon the sound. (trans. Bradley, 171)

These lines signal the appearance of the sea-journey motif and begin to highlight its constituent features, identification of which builds up further as the description develops. Thus, mention is made of embarkation at the shore (*gehlodon ... werum ond wifum*, lines 234–236, 'loaded with men and women'); urgency and speed (*ofer fifelwæg famige scriðan*, line 237, 'slip spuming over the ocean wave'; *brecan ofer bæðwæg, brimwudu snyrgan, / under swellingum, sæmearh plegean, / wadan wægflotan*, lines 244–246, 'scudding over the waterway, the timbered vessel sweeping along under swelling sails, the sea-steed racing, the wave-skimming ship forging onwards'); sea and/or waves (*ofer fifelwæg*, line 237, 'over the ocean wave'; *ofer earhgeblond yða swengas*, line 239, 'the buffeting of the waves over the surging of the sea'; etc.); cargo (*þa gehlodon hildesecuum, / bordum ond ordum*, lines 234–235, 'loaded with battle-coats, with shields and with spears'); the hero/leader (*idese siðfæt*, line 229, 'the woman's expedition'; *sio guðcwen*, line 254, 'the warlike queen'); the ship(s) (*fearoðhengestas*, line 226, 'horses of the sea'; *sæmearas*, line 228, 'sea-steeds'; *brimwudu*, line 244, 'timbered ships'; *hringedstefnan*, line 248, 'ring-prowed ships'; etc.); destination harbour and landing place (*to hyðe*, line 248, 'to the harbour'; *æt sæfearoðe, sande bewrecene*, line 251, 'at the sea-shore whipped with sand'); admirable warriors (*wlanc manig*, line 231, 'many a high-mettled man'; *mægen fægere*, line 242, '[no] finer-looking force').

In *Elene* the resources of traditional poetry are deployed here to bring a sense of urgency and resoluteness to the saint's military expedition to Jerusalem and to magnify Elene as a heroic figure. The expedition is recreated in Germanic terms as the voyage of the hero, who, accompanied by a retinue of noble warriors, confidently travels with the aim of carrying out a mighty exploit. The exploit here is the glorious one of discovering the true cross, and the enemies to be faced are not powerful armies but the spiritually obstinate Jews, who must be cowed into revealing the location of the cross. Through this description the heroic stature of Elene is emphasised. As she sets out on her mission to the Holy Land she is established for Anglo-Saxon audiences as a formidable leader, admirably purposeful and determined to succeed.

Elene also has two references to baptism: Elene's chief opponent, Judas, ends up converting to Christianity and is baptised (*gefulwad*, line 1043), receiving 'the bath of baptism' (*fulwihtes bæð*, line 1033). In an earlier episode in the poem, however, in the account of Judas's opposition to Elene, it is the absence of water that is highlighted, when Judas is terrorised by the implacable Elene by being confined for seven days in a dry well or pit – *in drygne seað* (line 693). The *seað* may be seen as symbolic of the spiritual aridity of Judas before his conversion.

The motif of the journey of the hero by sea also makes a brief, and highly ironic, appearance in *Juliana*, a poem on the martyrdom of the eponymous saint at the hands of the evil Heliseus. In his account of the final voyage of Heliseus after he has had Juliana executed, Cynewulf subverts the motif by associating it with an unheroic persecutor and including elegiac imagery where a cheerful mood is normally to be expected. The Latin original has the simple statement that when the persecutor was sailing to his estate 'a mighty tempest came and swamped his boat' (Chap. 22, trans. Allen and Calder 132), killing all on board.[58] The hagiographical/romance theme of the storm at sea is deployed, and there is no elegiac imagery.

In the Old English we encounter the familiar sea-journey tropes of the hero/leader (who is, however, *sceohmod*, line 672, 'craven-hearted') going to sea (*ehstream sohte*, line 673, 'sought the ocean stream'), with his band of warriors (*sceaþena þreate*, line 672, who are, however, *heane mid hlaford*, line 681, 'abject with their lord'), speeding across the sea (*leolc ofer laguflod*, line 674, 'bounded over the water-flood'), the swan's way (*on swonrade*, line 675), in their ship (*to scipe*, line 672), seeking the harbour or landing place (*ærþon hi to lande geliden hæfdon*, line 677, 'before they had voyaged to land'), but, in contrast to other journeys of the hero by sea, they never arrive there. And instead of the optimistic outlook of such purposeful voyagers as Beowulf and Elene, Heliseus and his followers are unhappy in the 'welter of the waves' (*wæges wylm*, line 680) as they go to their deaths: they are *heane* (line 681), 'abject, suffering', *hroþra bidæled / hyhta leas* (lines 681–682), 'deprived of comforts, destitute of hopes'.

Though it is not explicitly stated in the Latin original, Cynewulf portrays the deaths of Heliseus and his men as a punishment inflicted upon them: *þurh þearlic þrea* (line 678), 'by way of severe punishment'. Another layer of irony is apparent here, since earlier in the poem, in his argument with Juliana in her

58 The source of the Old English poem is best represented by the variant of the *Passio Sanctae Iulianae* in Lapidge M. (ed.), "Cynewulf and the *Passio S. Iulianae*", in Amodio M.C. – O'Brien O'Keeffe K. (eds.), *Unlocking the Wordhord: Anglo-Saxon Studies in Memory of Edward B. Irving, Jr* (Toronto: 2003), edited text at 156–171. For translation (though not specifically of the Lapidge variant), see Allen and Calder, *Sources and Analogues* 121–132. The Old English here presents a literal translation of 'venit tempestas valida, et mersit navem ipsius' (ed. Lapidge Chap. 22, 165).

prison cell, the devil had boasted of visiting destruction on sea travellers as one of his stratagems in his onslaught against humanity:

> Sume on yðfare
> wurdon on wege wætrum bisencte,
> on mereflode, minum cræftum
> under reone stream. (lines 478–481)

> Some on a sea journey were by my powers engulfed in the waters on their way to the ocean, beneath the dismal tide. (trans. Bradley, 313–314)

After being drowned in the sea Heliseus and his men join that devil in hell: *helle sohton* (line 682), 'they sought hell'. It is remarked sardonically that Heliseus's followers share none of the joys of fellowship with their lord *in þam neolan scræfe* (line 684), 'in that deep pit'.

5.3 *Andreas*

Michael Bintley observes that '*Andreas* is unusually wet, even for an Old English poem'.[59] It is certainly the hagiographical poem that makes the most use of water. *Andreas*, which tells of the apostle Andrew's mission to rescue Saint Matthew from the cannibalistic Mermedonians and then to convert them, has an extended account of the saint's sea journey to Mermedonia; it also has an episode in which Andreas miraculously calls forth a destructive flood from a marble pillar in the city and a short report of his departure by sea, after he has converted the city's inhabitants. Taken together these passages amount to over seven hundred lines in a poem of 1722 lines (though it should be said that a large portion of the account of the voyage to Mermedonia is taken up by speeches).

These story elements have been inherited from the poem's Latin source.[60] In *Andreas*, however, the story elements involving water are presented with vastly

59 Bintley M., "*Aquas ab Aquis*: Aqueous Creation in *Andreas*", in Twomey C. – Anlezark D. (eds.), *Meanings of Water in Early Medieval England*, Studies in the Early Middle Ages 47 (Turnhout: 2021) 159.

60 The immediate source is no longer extant but it must have been a version of the *Acts of Saint Andrew and Matthew* closely related to that found in the 'Casanatensis' recension and similar too to that used for the separate translation into Old English prose: the Casanatensis version is edited by F. Blatt, *Die lateinischen Bearbeitungen der Acta Andreae et Matthiae apud Anthropophagos* (Giessen – Copenhagen: 1930); for a translation, see Allen and Calder, *Sources and Analogues* 4–34. For the Old English prose version, see n. 33, above. The original Greek *Praxeis Andreou kai Mattheia* is also regarded as having

greater elaboration, as the poet creatively develops the imagery and symbolism of water, exploiting the rich water-related language of traditional Old English poetry in general, and of *Beowulf* in particular, to do so.[61] The *Beowulfian* overtones may be seen as imparting a level of heroic grandeur to the narrative, though it has recently been suggested that in *Andreas* the language of *Beowulf* is also used with irony to undercut the pagan world of *Beowulf*.[62]

The voyage to Mermedonia is treated much more descriptively in *Andreas* than in the analogues. This voyage becomes an extended version of the sea journey of the hero/leader, as seen above in other poems, with the 'small boat' of the analogues being rewritten as a *widfæðme scip* (line 240), 'broad-beamed ship', and Andrew's disciples as noble retainers, *collenfyrðe, / ellenrofe* (lines 349–350), 'bold in spirit and strong in courage', who will not countenance the ignominy of ever deserting their lord: if they do so, they exclaim, *We bioð laðe on lande gehwam, / folcum fracoðe* (lines 408–409), 'In every land we shall be despicable and contemptible to people'.

The constituent features of the motif of the sea journey of the hero/leader are all present in the *Andreas* instance, some of them several times, indeed, over the course of a long passage (lines 235–828) interspersed with an extended dialogue between Andreas and the captain of the ship, who, unknown to Andreas, is really Jesus. Thus there is mention of the shore and embarkation (*gangan on greote*, line 238, 'to walk on the shingle'; *in ceol stigon*, line 349, 'climbed into the ship'); urgency and speed over the sea/waves (*færeð famigheals, fugole gelicost / glideð on geofone*, lines 497–498, 'foamy-prowed it forges onwards, glides over the flood most like a bird'); cargo (literally, the passengers) (*ceol gehladenne / heahgestreonum*, lines 361–362, 'a ship laden with noble treasures'); the hero or

significant correspondences to the hypothetical source of *Andreas*; for a translation of this, see Elliott J.K. (trans.), *The Apocryphal New Testament: A Collection of Apocryphal Christian Literature in an English Translation* (Oxford: 1993) 283–299. For a succinct discussion of *Andreas*'s source and analogues, see North R. – Bintley M.D.J. (eds.), *Andreas: An Edition* (Liverpool: 2016) 4–6.

61 The debt of the *Andreas* poet to *Beowulf* is now well established in scholarship: see Riedinger A., "The Formulaic Relationship between *Beowulf* and *Andreas*", in Damico H. – Leyerle J. (eds.), *Heroic Poetry in the Anglo-Saxon Period: Studies in Honor of Jess B. Bessinger, Jr.* (Kalamazoo, MI: 1993) 283–312; Powell A.M., "Verbal Parallels in *Andreas* and its Relationship to *Beowulf* and Cynewulf" (Unpublished PhD dissertation, University of Cambridge: 2002); Orchard A., "The Originality of *Andreas*", in Neidorf L. – Pascual R.J. – Shippey T. (eds.), *Old English Philology: Studies in Honour of R.D. Fulk* (Cambridge: 2016) 331, 352; North – Bintley, *Andreas* 62–81.

62 See North R., "Meet the Pagans: On the Misuse of *Beowulf* in *Andreas*", in Cesario M. – Magennis H. (eds.), *Aspects of Knowledge: Preserving and Reinventing Traditions of Learning in the Middle Ages* (Manchester: 2018) 185–209.

leader specifically highlighted (*þriste on geþance*, line 237, 'bold of mind'); ship compounds, including kennings (*sæhengeste*, line 488, 'sea-horse'); destination or landing place (*on land becwom*, line 827, 'came to land'); sight of land at the end of the voyage (*wang sceawode*, line 239, 'surveyed the landscape'); admirable crew/warriors travelling with their lord (*beornas beadurofe*, line 848, 'men brave in battle').

There is even a short voyage description within this account of a voyage, as the sea captain tells Andreas about his journey to the place where they meet:

> Us mid flode bær
> on hranrade heahstefn naca,
> snellic sæmearh, snude bewunden,
> oðþæt we þissa leoda land gesohton,
> wære bewrecene, swa us wind fordraf. (lines 265–269)

> The high-stemmed ship, the speedy sea-horse, lapped in swiftness, carried us with the tide upon the whale-road until, swept on by the sea, we reached the land of this people according as the wind drove us. (trans. Bradley 118)

This passage is unparalleled in the analogues, as is the mention of a sea journey at the end of the poem as Andreas departs from Mermedonia (lines 1706–1714). In *Andreas* the people bring the saint to a ship at a headland (*æt brimes næsse / on wægþele*, lines 1710–1711), watching as he sails away *on yðum, ofer seolpaðu* (lines 1713, 1714), 'among the waves', 'along the seal's paths'. Francis Leneghan identifies this passage as an instance of what he refers to as the motif of the departure of the hero by ship, which may be viewed as a sub-type of the motif of the journey of the hero by sea.[63]

In *Andreas*, however, the main description of the sea journey of the hero is not without its incongruities. For one thing, Andreas is initially distinctly unheroic about going to Mermedonia at all: later he is described in heroic terms but at first he makes excuses to avoid the mission and even complains that the *herestræta ofer cald wæter* (line 200–201), 'highways across the cold water', are unfamiliar to him; for another, although Andreas is in command of his men, it is the captain of the ship (who is really Jesus) who is the leader on board, a youth but clearly the superior figure on the voyage; and, though loyal, Andreas's men prove to be fearful and in need of reassurance when they

63 See Leneghan F., "The Departure of the Hero in a Ship: The Intertextuality of *Beowulf*, Cynewulf and *Andreas*", *SELIM* 24 (2019) 105–132.

experience a storm at sea: *Nu synt geþreade þegnas mine / geonge guðrincas* (lines 391–392), 'now the young soldiers, my thanes, are dismayed', laments Andreas. Such incongruities arise, in my view, from the unwieldiness of the poem's source when approached in the terms of traditional Old English poetry. The Old English poet evidently wishes to remain faithful to the spirit of the source but some of the details of that source cannot be accommodated easily in the course of 'Germanicising' it.

The journey by sea in *Andreas* is also complicated by the occurrence of a storm during it, vividly described. Storms at sea are a familiar theme in hagiography but do not feature in other instances of the sea journey of the hero in Old English poetry; storms at sea are associated in Old English poetry with suffering, as in the poems referred to as elegies. In fact there are two storms at sea in *Andreas*, the one on the voyage to Mermedonia (lines 369–468) and the biblical storm on the Sea of Galilee, as described by Andreas in his dialogue with the sea-captain (lines 438–460); there is also a short passage (lines 511–525) in a speech by the captain describing storms at sea and comparing God's control of the sea (*se ðe brimu bindeð, brune yða / ðyð ond þreatað*, lines 519–520, 'he who binds the waters and rebukes and checks the brown waves') with his control of men, a comparison that by implication evokes the metaphor of the voyage of life.[64] The words are spoken, of course, by one who is a literal helmsman on the voyage to Mermedonia but also the divine controller of all things.

The storm on the Sea of Galilee is miraculously calmed by Jesus, and the captain informs Andreas that it is because the sea has recognised the grace of the Holy Spirit in him (*gife ... haliges gastes*, lines 530–531) that the storm on the way to Mermedonia has subsided: the miracle has occurred through God's protection of his saint (lines 533–536), whom the captain praises as *cyninges ... / þegen geþungen* (lines 527–528), 'the excellent thane of the King'.

The episode in which Andreas calls forth a flood from a pillar in his prison cell in waterless[65] Mermedonia may be seen as a kind of fulfilment of this pronouncement of the sea-captain. Critics of *Andreas* have rightly viewed the flood as bringing cleansing to Mermedonia and thus as a figure of baptism,[66]

64 On the idea of the voyage of life in the poem, see North – Bintley, *Andreas* 89–90; Bintley, "*Aquas ab aquis*" 162–316; McBrine P., "The Journey Motif in the Poems of the Vercelli Book", in Zacher S. – Orchard A. (eds.), *New Readings in the Vercelli Book* (Toronto: 2009) 298–317.

65 On Mermedonia as a waterless place, see Bintley, "*Aquas ab aquis*" 163–164: Bintley sees Mermedonia as a place of blood.

66 On the flood in *Andreas* as a figure of baptism, see Hill T.D., "Figural Narrative in *Andreas*", *Neuphilologische Mitteilungen* 70 (1969) 261–273; Hieatt C.B., "The Harrowing of Mermedonia: Typological Patterns in the Old English *Andreas*", *Neuphilologische Mitteilungen* 77

but the flood also links with the earlier storm at sea. The analogues do not explicitly develop this link but in *Andreas* the flood becomes a raging sea (*beatende brim*, line 1543), terrifying the people with its storms (*stormas*, line 1576). Andreas has miraculously precipitated this storm and miraculously he stills it, demonstrating power over water. Through this miracle he emulates Jesus on the Sea of Galilee and bears out the observation of the sea-captain that he has special favour with God as his *þegen geþungen*.

Michael Bintley goes further in a recent essay, seeing water in *Andreas* as having an overarching symbolic function in which it signifies divine creation but can also be associated with the role of the author transmitting Christian teaching. In accordance with John 7:37–39, Christ is understood to be the source of living water, and the poet, as Bintley puts it, 'having drunk at Christ's wellspring, like the stone in Andrew's cell, becomes a conduit for living water, that is the Holy Spirit'.[67] This is an attractive reading, suggestive of the artistic possibilities of water symbolism in hagiography. I would also stress, however, the insistent association of water with miracles in *Andreas*: water as controlled by God and water as controlled by Andreas.

6 Conclusion

Andreas is the Old English poem in which water themes are explored most richly. In *Andreas* water provides a means of travel, is instrumental in miracles (at sea and on land) and represents the spiritual cleansing of baptism. Water may also be seen as marking the boundary of the waterless land of Mermedonia, an *igland*/*ealand* (lines 15, 28, 'island'),[68] and, as argued by Bintley, *Andreas* may be read as drawing biblical ideas of the water of life. The traditional poetic language cultivated by the *Andreas* poet enables the creative expression of such themes, as is the case with travel in *Guthlac B*, *Elene* and *Juliana*; only *Guthlac A* (surprisingly?) eschews water themes altogether.

(1976) 49–62; Walsh M.M., "The Baptismal Flood in the Old English *Andreas*: Liturgical and Typological Depths", *Traditio* 33 (1977) 137–158 ; Anlezark D., *Water and Fire: The Myth of the Flood in Anglo-Saxon England* (Manchester: 2006) 174–240.

67 Bintley, "*Aquas ab aquis*" 171.

68 For discussion of *ealand*/*iglond*, which has also been taken to mean 'land beyond the waters'," see Grosz O.J.H., "The Island of Exiles: A Note on *Andreas* 15", *English Language Notes* 7 (1970) 240; Bolintineanu A., "The Land of Mermedonia in the Old English *Andreas*", *Neophilologus* 93 (2009) 153–154; North – Bintley, *Andreas* 218–220.

Images of travel, miracle and cleansing are developed in the Old English prose *Life of Saint Mary of Egypt* as well, a work that is also concerned with water, and with the lack of it, in ascetic practice. The Old English text is a close translation of a Latin original (which is itself a translation from the Greek) and is therefore, like most of the several hundred items trawled earlier in the chapter, representative of hagiography as a whole rather than being distinctively Anglo-Saxon. In general, the saints' lives and shorter pieces surveyed here are remarkably consistent in the ways in which water features in them: travel (including storms), miracles, cleansing and self-denial provide an inventory of themes in hagiographical writings from the simple to the complex. Water is life-giving and powerful but always under the power of God; in hagiography it can also be under the power of the saint.

Bibliography

Primary Sources

Allen M.J.B. – Calder D.G., *Sources and Analogues of Old English Poetry: The Major Latin Texts in Translation* (Cambridge: 1976).

Assmann B. (ed.), *Angelsächsische Homilien und Heiligenleben*, Bibliothek der Angelsächsischen Prosa 3 (Kassel: 1889).

Blatt F. (ed.), *Die lateinischen Bearbeitungen der Acta Andreae et Matthiae apud Anthropophagos*, Beihefte zur Zeitschrift für neutestamentliche Wissenschafte 12 (Giessen – Copenhagen: 1930).

Bradley S.A.J. (trans.), *Anglo-Saxon Poetry: An Anthology of Old English Poems in Prose Translation with Introduction and Headnotes* (London: 1982).

Cassidy F.G. – Ringler R.N., *Bright's Old English Grammar and Reader* (New York: 3rd ed. 1971).

Clayton M. (ed. and trans.), *Apocryphal Gospels of Mary in Anglo-Saxon England*, Cambridge Studies in Anglo-Saxon England 26 (Cambridge: 2008).

Clayton M. – Magennis H. (eds. and trans.), *The Old English Lives of St Margaret*, Cambridge Studies in Anglo-Saxon England 9 (Cambridge: 1994).

Clemoes P. (ed.), *Ælfric's Catholic Homilies: The First Series. Text*, Early English Text Society, Supplementary Series 17 (Oxford: 1997).

Colgrave B. (ed. and trans.), *Felix's Life of Saint Guthlac* (Cambridge: 1956).

Elliott J.K. (trans.), *The Apocryphal New Testament: A Collection of Apocryphal Christian Literature in an English Translation* (Oxford: 1993).

Fulk R.D. – Bjork R.E. – Niles J.D. (eds.), *Klaeber's Beowulf and the Fight at Finnsburg* (Toronto: 4th ed. 2008).

Godden M. (ed.), *Ælfric's Catholic Homilies: The Second Series. Text*, Early English Text Society, Supplementary Series 5 (Oxford: 1979).

Gonser P. (ed.), *Das angelsächsische Prosa-Leben des hl. Guthlac*, Anglistische Forschungen 27 (Amsterdam: 1909).

Hecht H. (ed.), *Bischofs Wærferth von Worcester Übersetzung des Dialoge Gregors des Grossen*, Bibliothek der Angelsächsischen Prosa 5 (Leipzig: 1900).

Henschenius G. – Papebrochius D., *Acta Sanctorum, Maius, Tom. I*, 445–8 (Antwerp: 1680).

Krapp G.P. – Van Kirk Dobbie E. (eds.), *The Anglo-Saxon Poetic Records*, 6 vols (New York: 1931–1953).

Lapidge M. (ed.), "Cynewulf and the *Passio S. Iulianae*", in Amodio M.C. – O'Brien O'Keeffe K. (eds.), *Unlocking the Wordhord: Anglo-Saxon Studies in Memory of Edward B. Irving, Jr* (Toronto: 2003) 147–171.

Magennis H. (ed. and trans.), *The Old English Life of St Mary of Egypt* (Exeter: 2002).

Marsden R. (ed.), *The Old English Heptateuch and Ælfric's Libellus de Veteri Testamento et Novo: Vol. I*, Early English Text Society, Original Series 330 (Oxford: 2008).

Migne J.-P. (ed.), *Vita Sancti Pauli primi eremitae, Patrologia Latina* 23, 17–30.

Morris R. (ed. and trans.), *The Blickling Homilies*, Early English Text Society, Original Series 58, 63, 73 (Oxford: 1874, 1876, 1880; reprinted as one vol., 1967).

North R. – Bintley M.D.J. (eds.), *Andreas: An Edition* (Liverpool: 2016).

O'Meara J.J. (trans.), *The Voyage of Saint Brendan: Journey to the Promised Land* (Dublin: 1976; reprinted Gerrards Cross, Buckinghmashire: 1991).

Pulsiano P. (ed.), "The Old English Life of St Pantaleon", in Hall T.N. (ed.), *Via Crucis: Essays on Early Medieval Sources and Ideas in Memory of J.E. Cross*, Medieval European Studies 1 (Morgantown, WV: 2002) 61–103.

Rauer C. (ed. and trans.), *The Old English Martyrology: Edition, Translation and Commentary*, Anglo-Saxon Texts 10 (Cambridge: 2013).

Roberts J. (ed.), *The Guthlac Poems of the Exeter Book* (Oxford: 1979).

Selmer C. (ed.), *Navigatio Sancti Brendani Abbatis* (Notre Dame, IN: 1959).

Skeat W.W. (ed. and trans.), *Ælfric's Lives of Saints*, Early English Text Society, Original Series 76, 82, 94 and 114 (London: 1881–1900; reprinted as 2 vols, 1966).

Swanton M. (trans.), *Anglo-Saxon Prose* (revised ed. London: 1993).

Treharne E.M. (ed. and trans.), *The Old English Life of St Nicholas with the Old English Life of St Giles*, Leeds Texts and Monographs, new series 15 (Leeds: 1997).

Warner R.D.N (ed.), *Early English Homilies of the Twelfth Century MS Vesp. D. XIV*, Early English Text Society, Original Series 152 (London: 1917).

White C. (trans.), *Early Christian Lives* (Harmondsworth: 1998).

Yerkes D. (ed.), *The Old English Life of Machutus*, Toronto Old English Series 9 (Toronto: 1984).

Secondary Sources

Anlezark D., *Water and Fire: The Myth of the Flood in Anglo-Saxon England* (Manchester: 2006).

Bartlett R., *Why Can the Dead Do Such Great Things? Saints and Worshippers from the Martyrs to the Reformation* (Princeton, NJ: 2013).

Bintley M., "*Aquas ab Aquis*: Aqueous Creation in *Andreas*", in Twomey C. – Anlezark D. (eds.), *Meanings of Water in Early Medieval England*, Studies in the Early Middle Ages 47 (Turnhout: 2021) 159–175.

Bolintineanu A., "The Land of Mermedonia in the Old English *Andreas*", *Neophilologus* 93 (2009) 149–164.

Carver M., "Travels on the Sea and in the Mind", in Klein S.S. – Schipper W.V. – Lewis-Simpson S. (eds.), *The Maritime World of the Anglo-Saxons*, Medieval and Renaissance Texts and Studies 448, Essays in Anglo-Saxon Studies 5 (Tempe, AZ: 2014) 21–36.

Clark S., "A More Permanent Homeland: Land Tenure in *Guthlac A*", *Anglo-Saxon England* 40 (2112) 75–102.

Clarke C.A.M., *Literary Landscapes and the Idea of England, 700–1400* (Cambridge: 2006).

Clayton M., *The Cult of the Virgin Mary in Anglo-Saxon England*, Cambridge Studies in Anglo-Saxon England 2 (Cambridge: 1990).

DiNapoli R., *An Index of Theme and Image to the Homilies of the Anglo-Saxon Church* (Hockwold cum Wilton, Norfolk: 1994).

Discenza N.G., *Inhabited Spaces: Anglo-Saxon Constructions of Place* (Toronto: 2017).

Estes H., *Anglo-Saxon Literary Landscapes: Ecotheory and the Environmental Imagination* (Amsterdam: 2017).

Frederick J.A., "From Whale's Road to Water under the Earth: Water in Anglo-Saxon Poetry", in Hyer M.C. – Hooke D. (eds.), *Water and the Environment in the Anglo-Saxon World* (Liverpool: 2018) 15–32.

Gaillard J., "Eau", in Viller M. – Cavallera F. – de Guibert J. (eds.), *Dictionnaire de spiritualité chretien*, IV (Paris: 1960) 8–29.

Grosz O.J.H., "The Island of Exiles: A Note on *Andreas* 15", *English Language Notes* 7 (1970) 240.

Heffernan T.J., *Sacred Biography: Saints and Their Biographers in the Middle Ages* (New York – Oxford: 1988).

Hieatt C.B., "The Harrowing of Mermedonia: Typological Patterns in the Old English *Andreas*", *Neuphilologische Mitteilungen* 77 (1976) 49–62.

Hill T.D., "Figural Narrative in *Andreas*", *Neuphilologische Mitteilungen* 70 (1969) 261–273.

Hooke D., "Rivers, Wells and Springs in Anglo-Saxon England: Water in Sacred and Mystical Contexts", in Hyer M.C. – Hooke D. (eds.), *Water and the Environment in the Anglo-Saxon World* (Liverpool: 2018) 107–135.

Horden P. – Purcell N., *The Corrupting Sea: A Study of Mediterranean History* (Oxford: 2000).

Hyer M.C. – Hooke D. (eds.), *Water and the Environment in the Anglo-Saxon World* (Liverpool: 2018).

Irving E.B., *A Reading of Beowulf* (New Haven, CT: 1968).

Jones C.A., "Envisioning the *coenobium* in the Old English *Guthlac A*", *Mediaeval Studies* 57 (1995) 259–291.

Klein S. – Schipper W. – Lewis-Simpson S. (eds.), *The Maritime World of the Anglo-Saxons*, Medieval and Renaissance Texts and Studies 448, Essays in Anglo-Saxon Studies 5 (Tempe, AZ: 2014).

Lapidge M., "The Saintly Life in Anglo-Saxon England", in Godden M. – Lapidge M. (eds.), *The Cambridge Companion to Old English Literature* (Cambridge: 2nd ed. 2013) 251–272.

Leneghan F., "The Departure of the Hero in a Ship: The Intertextuality of *Beowulf*, Cynewulf and *Andreas*", *SELIM* 24 (2019) 105–132.

Magennis H., *Images of Community in Old English Poetry*, Cambridge Studies in Anglo-Saxon England 18 (Cambridge: 1996).

Magennis H., "Approaches to Saints' Lives", in Cavill P. (ed.), *The Christian Tradition in Anglo-Saxon England: Approaches to Current Scholarship and Teaching* (Cambridge: 2004) 163–183.

McBrine P., "The Journey Motif in the Poems of the Vercelli Book", in Zacher S. – Orchard A. (eds.), *New Readings in the Vercelli Book* (Toronto: 2009) 298–317.

Michelet F.L., *Creation, Migration, and Conquest: Imaginary Geography and Sense of Space in Old English Literature* (Oxford: 2006).

Mullins J., "*Herimum in mari*: Anglo-Saxon Attitudes towards *Peregrinatio* and the Ideal of a Desert in the Sea", in Klein S. – Schipper W. – Lewis-Simpson S. (eds.), *The Maritime World of the Anglo-Saxons*, Medieval and Renaissance Texts and Studies 448, Essays in Anglo-Saxon Studies 5 (Tempe, AZ: 2014) 239–254.

North R., "Meet the Pagans: On the Misuse of *Beowulf* in *Andreas*", in Cesario M. – Magennis H. (eds.), *Aspects of Knowledge: Preserving and Reinventing Traditions of Learning in the Middle Ages* (Manchester: 2018) 185–209.

Ogilvie J.D.A. – Baker D.C., *Reading Beowulf: An Introduction to the Poem, Its Background, and Its Style* (Norman, OK: 1983).

Orchard A., "The Originality of *Andreas*", in Neidorf L. – Pascual R.J. – Shippey T. (eds.), *Old English Philology: Studies in Honour of R.D. Fulk* (Cambridge: 2016) 331–352.

Pervo R.I., "Early Christian Fiction", in Morgan J.R. – Stoneman R. (eds.), *Greek Fiction: The Greek Novel in Context* (London: 1994) 239–254.

Powell A.M., "Verbal Parallels in *Andreas* and its Relationship to *Beowulf* and Cynewulf" (Unpublished PhD dissertation, University of Cambridge: 2002).

Reichardt P.F., "*Guthlac A* and the Landscape of Spiritual Perfection", *Neophilologus* 58 (1974) 331–338.

Riedinger A., "The Formulaic Relationship between *Beowulf* and *Andreas*", in Damico H. – Leyerle J. (eds.), *Heroic Poetry in the Anglo-Saxon Period: Studies in Honor of Jess B. Bessinger, Jr.*, Studies in Medieval Culture 32 (Kalamazoo, MI: 1993) 283–312.

Shook L.K., "The Burial Mound in *Guthlac A*", *Modern Philology* 58 (1960) 1–10.

Siewers A.K., *Strange Beauty: Ecocritical Approaches to Early Medieval Landscape* (New York: 2009).

Sobecki S.I., *The Sea in Medieval English Literature* (Cambridge: 2008).

Walsh M.M., "The Baptismal Flood in the Old English *Andreas*: Liturgical and Typological Depths", *Traditio* 33 (1977) 137–158.

Wentersdorf K.P., "*Guthlac A*: The Battle for the *Beorg*", *Neophilologus* 62 (1978) 135–142.

Wickham-Crowley K., "Fens and Frontiers", in Hyer M.C. – Hooke D. (eds.), *Water and the Environment in the Anglo-Saxon World* (Liverpool: 2018) 68–88.

Wickham-Crowley K., "Living on the *Ecg*: The Mutable Boundaries of Land and Water in Anglo-Saxon Contexts", in Lees C.L. – Overing G. (eds.), *A Place to Believe In: Locating Medieval Landscapes* (University Park, PA: 2006) 85–110.

CHAPTER 13

'Where Mighty Rivers Sprayed Poison': Aspects of Water in Old Norse Myth

Tom Birkett

Of the four elements thought to constitute the basis of all matter, water is remarkable for the varieties of its isolated forms – ice, liquid, vapour – and its ability to carry, divide and transform matter. Rather than setting out to evaluate the role of water in these different forms throughout the many overlapping narratives that constitute Norse mythology, this short survey addresses several important aspects of the element at a structural level within the Norse cosmos and as expressed through its anthropomorphic representatives.

Firstly, it will sketch out the key roles of water in the creation, destruction and structuring of the cosmos as outlined in the surviving sources, focusing on the issue of whether this presentation accords with the prevailing medieval understanding of the four elements. Secondly, it will consider the deities Njǫrðr, Ægir and Rán, and what these beings connected with the sea can tell us about the Norse relationship with water and its coding within the cosmology. Finally, it will consider the Norse god Thor's antagonistic relationship with water in its most destructive forms, arguing that whilst in certain cases we are clearly dealing with a confrontation between a feminised natural force and an attempt at control by a hypermasculine agent of the gods, in other cases we see an expression of water's primacy outside the dynamic of chaos and civilisation through which much of the surviving mythology can be parsed.

1 Creation and the Cosmos

Water is the element that heralds the beginning of the Norse cosmos. As the thirteenth-century Icelandic mythographer Snorri Sturluson presents it in the *Prose Edda*, life emerges in association with a primordial river, or system of rivers, known as Élivágar, which arise from the icy realm of Niflheimr and flow into the void Ginnungagap. We are told that this ice realm was primary: 'Fyrr

 | DOI:10.1163/9789004696501_015

var þat mǫrgum ǫldum en jörð[1] var skǫpuð er Niflheimr var gǫrr, ok í honum miðjum liggr bruðr sá er Hvergelmir heitir' ('Niflheimr was formed many ages before earth was created, and at its middle lies a spring which is called Hvergelmir', 9).[2] In Snorri's retelling of this creation narrative, it is the meeting of the frozen runoff from the icy realm of Niflheimr with the heat from the fiery realm of Muspell that leads to the creation of a temperate space in the middle of the void. This in turn creates the necessary conditions for the first life to emerge from the spitting poison and melting frost.

Snorri's account of creation through the meeting of fire and ice relies heavily on what Carolyne Larrington refers to as 'a synthesis of oppositions', an idea which has clearly been influenced by 'medieval scientific theory'.[3] Snorri, whose importance as a prolific recorder of Norse mythology, poetry and Scandinavian history cannot be overstated, served as lawspeaker for two separate terms, spent several years in Norway by royal invitation, and was a politician involved in European as well as Icelandic affairs. He was certainly not writing in a vacuum and his systematised prose retelling of diffuse mythical narratives is written under the influence of both Christian dualism and conceptions of the cosmos deriving from the Classical philosophers. The idea of life arising from the meeting of elements follows a learned tradition that can be traced back to the fifth-century BCE philosopher Empedocles, who theorised that 'all matter consists of four elements in shifting combination'.[4] Empedocles's work *On Nature* also discusses the elements in relation to cosmogony, suggesting that they combine under the oppositional forces of 'love' and 'strife', the two constantly vying between the creation of the cosmos, and its destruction.[5] Aristotle's conception of matter borrows much from Empedocles: he characterises the four elements as different combinations of wet, dry, hot and cold

1 Here the primary referent of *jǫrð* is 'the earth' (personified elsewhere as the giantess Jǫrð), though the term is also used to refer to the element 'earth', as in the collocation *jǫrð eða stein*, 'earth or stone': in this respect it is very similar to the modern English terms 'earth/Earth'.

2 All quotations of *Gylfaginning* are from Faulkes A. (ed.), *Edda: Prologue and Gylfaginning* (London: 2nd ed. 2005); all translations are my own unless otherwise stated.

3 Larrington C., *The Norse Myths: A Guide to the Gods and Heroes* (Oxford: 2017) 57.

4 Cohen J.J. – Duckert L., "Introduction: Eleven Principles of the Elements", in Cohen J.J. – Duckert L. (eds.), *Elemental Ecocriticism: Thinking with Earth, Air, Water, and Fire* (Minneapolis, MN: 2015) 2.

5 Kingsley K.S. – Parry R., "Empedocles", in Zalta E.N. (ed.), *The Stanford Encyclopedia of Philosophy* (Online, Summer 2020 edition) https://plato.stanford.edu/archives/sum2020/entries/empedocles/. Last accessed November 3, 2023.

(water being cold and wet, and thus diametrically opposed to hot and dry fire). The four elements are also key to Plato's *Timaeus*, as Guðrún Nordal notes, along with the Neoplatonic conception of the 'world-body' – or the structural parallel between the microcosm of the body and macrocosm of the cosmos.[6] Snorri almost certainly had knowledge of the microcosm-macrocosm theory, most probably via a Latin copy of the Elucidarius or vernacular translation of this work, as Clunies Ross argues.[7]

The details of Snorri's creation narrative seem to accord with the concept of elements as the basis of all matter, as well as with the idea of dualistic forces creating balance. We are told that *úr*, or drizzle, blew out of the cold regions and was 'cleared up' by the dry heat from Muspell, leaving the centre of Ginnungagap 'svá hlætt sem lopt vindlaust' ('as mild as windless air', 10). Snorri's choice of words here – including referencing 'still winds' – suggests that he may have conceived this locus of creation in Ginnungagap as being carefully balanced between the Empedoclean elements. Indeed, Muspell, elsewhere referring to a being associated with the destruction of the world, may have been reconceived as a realm of fire by Snorri, as Faulkes suggests, perhaps to further his elemental presentation of creation.[8] In his version of the creation narrative, Snorri certainly stresses the 'dialectic interaction of these two contraries [...] yield[ing] a synthesis of cosmic importance',[9] and he also locates the diametrically opposed realms of cold Niflheimr and burning Muspell in geographical space, to the north and south respectively, with the temperate space between them perhaps related to 'a geographical system of climate zones reminiscent of that found in the learned culture of medieval Europe', as Langeslag suggests.[10] Clunies Ross has further argued that Snorri favours this elemental conception of creation – along with the notion that the giants were personifications of natural forces – as it served 'to further his general theme that the "earthly understanding" of the Pre-Christian Scandinavians

6 Nordal G., *Tools of Literacy: The Role of Skaldic Verse in Icelandic Textual Culture of the Twelfth and Thirteenth Centuries* (Toronto: 2001) 273. See also Dronke U. – Dronke P., "The Prologue of the Prose Edda: Explorations of a Latin Background", in Pétursson E.G. – Kristjánsson J. (eds.), *Sjötíu Ritgerdir helgadar Jakobi Benediktssyni 20. júlí 1977* (Reykjavik: 1977).

7 Clunies Ross M., "Snorri Sturluson's Use of the Norse Origin-Legend of the Sons of Fornjótr in his *Edda*", *Arkiv för Nordisk Filologi* 98 (1983) 51–52.

8 See the discussion of Muspell in the Index of Names in Faulkes, *Edda: Prologue and Gylfaginning* 173.

9 Lincoln B., "In Praise of the Chaotic", in Dill U. – Walde C. (eds.), *Antike Mythen: Medien, Transformationen und Konstruktionen* (Berlin: 2009) 378.

10 Langeslag P.S., *Seasons in the Literatures of the Medieval North* (Cambridge: 2019) 105.

[…] showed in this case a correct understanding of the elemental composition of the world'.[11] By moving away from the consistent representation of giants as hostile forces (often associated with the earth) in skaldic poetry, and also avoiding too close an association with the biblical Nephilim, Snorri could show the pagan Norse ancestors to be followers of natural law: ultimately, it was this 'power to observe the material world which led [these ancestors] to an understanding of God'.[12]

Whilst Snorri clearly saw an opportunity to overlay the prevailing medieval understanding of the elements onto his pre-Christian sources, the synthesis is far from complete, and there are some details in the creation narrative which do not fit comfortably into this superficially neat dialectic.[13] It is in these contradictions – and the differences in accounts of creation in the poetry – that we can perhaps glimpse an older conception of the role of water in the cosmos.

First of all, though Snorri clearly attempts to balance water and fire, not all elements are equi-present at the moment of creation. Whilst we might locate air (or wind) in the gusts of heat and driving drizzle, earth is constituted only later, in the killing of the being Ymir and use of his body to make the world: it does not seem to be present prior to this violent act. As the dismembering of Ymir is clearly stated to be the origin of the terrestrial world in *Vafþrúðnismál* (st. 21) and *Grímnismál* (st. 40), it was perhaps too difficult for Snorri to weave earth into the meeting of elements at the moment of creation whilst still incorporating this subsequent myth of world building.[14] Snorri's suggestion that layers of frost and rime accumulate in the void to provide a space for Ymir to inhabit is, perhaps, something of a fudge – it is terrain of a sort, though comprised of frozen water rather than earth. This clever, if slightly awkward, co-option of ice by Snorri to stand in for earth raises the possibility that water may have been understood to have primacy in a conception of creation that did not rely on the learned theory of the meeting and balancing of elements.

---

11 Clunies Ross, "Snorri Sturluson's Use of the Norse Origin-Legend" 53. In support of her argument that Snorri favoured an understanding of giants as anthropomorphised natural forces, Clunies Ross discusses Snorri's casting of the three sons of Fornjótr known from Norwegian genealogies as personifications of three of the elements: Ægir (sea), Logi (fire) and Kári (a poetic name for wind).

12 Clunies Ross, "Snorri Sturluson's Use of the Norse Origin-Legend" 52.

13 Snorri's sources include eddic poetry, which he quotes extensively, as well as references in skaldic verse which he expands upon in the later sections of his *Prose Edda*: other details 'obviously came directly from oral tradition', as Heimir Pálsson argues, "Reflections on the Creation of Snorri Sturluson's Prose Edda", *Scripta Islandica* 68 (2017) 210.

14 All references to eddic poems are to Kristjánsson J. – Ólason V. (eds.), *Eddukvæði*, 2 vols. (Reykjavík: 2014). Translations are my own unless otherwise stated.

Secondly, whilst the dichotomy of ice and fire is unmistakeable in Snorri's telling, water is not encountered as a pure element that mingles with others to create life: in Snorri's account it is the *eitr-kvikja*, 'poisonous fermentation' ('poisonous flow' in Faulkes's translation), and the frost-forming vapours from this same poisonous brew which ultimately form the first being, whilst in the eddic poem *Vafþrúðnismál*, it is the spray of poison-drops from the river that more directly gives rise to the first giants. *Eitr* is a term that most often means poison, and *kvikja* is usually translated as 'yeast' or 'fermentation', so the matter of life arises from a dynamic process happening in rather than *as* water: either a kind of primordial fermentation, or an ejection of *eitr*. Water here is clearly the conduit for life, but not clearly the matter *of* life. Later in *Gylfaginning*, Snorri provides a source for the poisonous sediment and associated effluvia of Élivágar by mentioning the innumerable serpents which writhe in the waters of the originating spring. As strange as this image of life arising from a poisonous and yeasty liquid might be, the image of 'fermentation' occurring in a river does more or less map onto observed nature: in the frothy churning of brown waters of a river in spate, or indeed the production of *haf-lauðr* or 'sea-foam' in the surf. However, this poisonous river also undergoes what Heather O'Donoghue characterises as 'a bafflingly complex series of chemical and physical reactions'.[15] Firstly, the water slows and turns to ice as it flows away from the freezing Niflheimr and into Ginungagap. Then, poisonous vapours arise from the frozen river to form layers of *hrím*, rime or frost, which accumulate in the void. Finally, the poisonous rime melts under the 'quickening' influence of warm air from Muspell: 'þá er mœttisk hrímin ok blær hitans svá at bráðnaði ok draup, ok af þeim kvikudropum kviknaði […]' ('when the rime and the warm breeze met, so it melted and dripped, and there was a quickening from those quickened drops […]' 10). The first two of these transformations rely on extreme cold, yet they are said to happen the further the river flows from its source in Niflheimr. Snorri resorts to a strange image to describe this process, telling us that ice formed in the poisonous water 'svá sem sindr þat er renn ór eldinum' ('like slag that runs from a [forge] fire' 10).

There is an obvious contradiction in the idea of ice and rime forming from water as it approaches the elemental heat emanating from Múspellsheimr, and an obvious disjunct between this image of smelting and its location in ice cold water. O'Donoghue speculates that this peculiar conception may be related to observing a lava flow in Iceland, 'which would first flow and then harden into

15 O'Donoghue H., *From Asgard to Valhalla: The Remarkable History of the Norse Myths* (London: 2007) 15.

fantastic shapes'.[16] Mathias Nordvig, taking his cue from Anne Holtsmark's discussion of these and other inconsistencies in Snorri's creation narrative,[17] has argued more forcibly that the image of a gushing liquid hardening in layers makes sense as a reference to a 'low discharge effusive eruption', reflecting both the geothermal realties of Iceland and a lack of native Norse vocabulary to describe this activity.[18] Following this reading, we might see the serpents writhing in the spring of Hvergelmir as dragons associated more with the element of fire than with water, and the poisonous droplets as gas and steam attending an eruption: Snorri's account of creation through the meeting of fire and ice could thus be read as overlaying a contemporary quasi-scientific understanding of opposing elements – and geographies – onto an inherited tradition based more on the observed realities of the volcanic Icelandic environment.

Nordvig's reading productively complicates the elemental dialectic that Snorri may have imposed on the narrative, but a volcanic conception of creation does not sit well with the centrality of water to other versions of the creation narrative. Firstly, the primal river system Élivágar is clearly understood as a body of water. Halvorsen reads the name as a compound of *él*, 'bad weather, storm' and *vágr*, 'sea',[19] a reading that Rudolf Simek accepts.[20] Snorri himself lists eleven rivers that arise from the spring Hvergelmir, a catalogue borrowed from the poem *Grímnismál*.[21] Even if the idea of eleven rivers is 'obviously an invention of Snorri's', it makes it clear that he understood the referent to be one of water and was drawing evidence for this from the surviving poems.[22] More compellingly, *Vǫluspá* presents an alternative creation myth in which water is primary: in this poem the sons of Burr are said to have 'biǫðom um ypðo'

16 Ibidem.

17 Holtsmark A., *Studier i Snorres Mytologi* (Oslo: 1964).

18 Nordvig M., "Creation from Fire in Snorri's Edda: The Tenets of a Vernacular Theory of Geothermal Activity in Old Norse Myth", in Hermann P. – Mitchell S.A. – Schjødt J.P. – Rose A. (eds.), *Old Norse Mythology in Comparative Perspectives* (Cambridge, MA: 2017) 284. Nordvig also queries Snorri's portrayal of Niflheimr as a cold region, in support of his argument that Élivágar represents a lava flow rather than a freezing river, and understands the build-up of *hrím* ('rime') as a reference to ash rather than frost; applying the vocabulary of water to a new phenomenon.

19 Halvorsen E.F., "Élivágar", *Kulturhistorik Leksikon for nordisk Medeltid* 3 (1958) 597–598.

20 Simek R., *Dictionary of Northern Mythology* (Cambridge: 1993) 73. Ursula Dronke has 'Snow Storm Waves', Dronke U. (ed. and trans.), *The Poetic Edda III* 78.

21 Henning Kure points out that in the context of *Grimnismál* 'Hvergelmir might just as well (or perhaps even better) be a variant name for Aurgelmir or Ymir, from whose body fluids all water originated', Kure H., "Wading Heavy Currents: Snorri's Use of *Vǫluspá* 39", in Gunnell T. – Lassen A. (eds.), *The Norse Apocalypse: Approaches to* Vǫluspá *and Nordic Days of Judgement* (Turnhout: 2013) 84.

22 Simek, *Dictionary* 73.

('lifted up seashores', st. 4) from the gaping void.[23] Dronke makes a strong case that the poetic term for land (*biǫð*) used here is consistently associated with the margins of water, so the implication is that these 'cosmic weightlifters' are drawing the world up from the primordial ocean, with no mention of the heat of Muspell.[24]

As Dronke points out, the poet of *Vǫluspá* clearly regards water as 'a primordial and seemingly vacuous element from which matter may emerge', a green land also rising from the ocean 'ǫðro sinni' ('for a second time', st. 56), after the destruction of Ragnarǫk.[25] The poem *Vafþrúðnismál*, quoted by Snorri, makes no reference to the cold of the river Élivágar, or to the formation of ice and rime, but is more direct in linking the poisonous emanations from the river with the moment life comes into being. The strophe preserved in the *Poetic Edda* differs in a few places from that quoted by Snorri but is substantially the same: 'Ór Élivágum / stukku eitrdropar, / svá óx, unz varð ór jǫtunn' ('from Élivágar sprayed drops of poison which increased until from it arose a giant', st. 31).[26] *Grímnismál* does not refer directly to the creation but does give us an account of the waters that feed Hvergelmir and from which 'eigu vǫtn ǫll vega' ('all streams have their course', st. 26). Water is also bound up with acts of creation on a smaller scale to that of world formation. Snorri tells us that the gods themselves are licked from the frost by the cow Auðhumla: 'fjórar mjólkár runnu ór spenum hennar' ('four rivers of milk ran from her teats', 11), these 'rivers' feeding the proto-giant Ymir. Later, all the giants bar two are drowned in the deluge from Ymir's blood when he is killed by Odin and his brothers, in what may be an echo of the biblical flood narrative. Even humans wash up as driftwood on the beach before they are given life by the gods, suggesting once again the importance of water as a conduit of life.

Water, present before the creation of the world, also outlasts, and certainly contributes to, the world's destruction. Ragnarǫk, the series of events that leads to the destruction of the world and defeat of the gods and their human allies, proceeds through a sequence of apocalyptic imagery including a winter lasting three years, a trembling World-Tree, the breaking loose of the wolf Fenrir, and the advance of the fire giant Surt. Clunies Ross suggests that Snorri is 'polarising the giant race into representatives of *hrím* ["frost"] and *hiti* ["fire"]', with water associated with creation, and fire, through Surt, associated with

23 This is the translation given in Dronke, *The Poetic Edda*, II 7.
24 Dronke, *The Poetic Edda*, II 115.
25 Ibidem 33.
26 'Strophe' is the preferred term for referring to stanzas of Old Norse poetry, which often vary in line length rather than consisting of more regular units.

apocalypse.[27] However, fire is by no means the only elemental force present at Ragnarǫk: water is very much implicated in the apocalypse as well, the sea churned up by the Midgard Serpent flooding the land, and the gods themselves destroyed, as Judy Quinn points out, 'not by the engulfing sea but in direct combat with the animated forces who also control, or in some sense are, the sea'.[28]

Snorri quotes nine strophes from *Vǫluspá* following his paraphrase of the events of Ragnarǫk, and it is clear that whilst his account is based closely on the poem, it expands on the depiction in places. The poem tells us that 'ormr knýr unnir' ('the serpent thrashes the waves', st. 47); Snorri, perhaps mindful of the central place of the elements in his creation narrative, tells us more emphatically that 'þá geysisk hafit á lǫndin' ('then the sea will gush onto the land', 50) as the serpent emerges. Both accounts refer to the sons of Muspell travelling across the ocean in a ship piloted by Loki: this should probably be understood as analogous to the giants riding over the rainbow bridge Bifröst, in that barriers meant to exclude and protect now become the giants' means of access to the gods. However, the fact that the sons of Muspell – the realm Snorri associates with fire in his story of creation – travel by ship on the water suggests that Snorri is not concerned to present Ragnarǫk through the paradigm of elemental disaggregation or as a reversal of creation. The strophe describing the final collapse of the world in *Vǫluspá* is notable, however, in that it touches on all the elements:

Sól tér sortna,	The sun begins to blacken,
sígr fold í mar,	the earth sinks into the sea,
hverfa af himni	from heaven disappear
heiðar stiǫrnor.	the bright stars.
Geisar eimi	Steam surges
við aldrnara,	against the life-nourisher [fire]
leikr hár hiti	the heat billows high
við himin siálfan.	against heaven itself.

Here we have earth, water, and fire, with wind implied by the surging of steam and the gusting of heat into the heavens. We could even read Aristotle's fifth element, aether, into the repeated reference to *himin*, 'heaven, sky', which raises the question of why Snorri, who is so clearly influenced by the microcosm/

27 Clunies Ross, "Snorri Sturluson's Use of the Norse Origin-Legend" 51.

28 Quinn J., "Mythologizing the Sea: The Nordic Sea-Deity Rán", in Tangherlini T. (ed.), *Nordic Mythologies: Interpretations, Institutions, Intersections* (Berkeley – Los Angeles, CA: 2014) 73.

macrocosm theory of the elements in his retelling of the creation narrative, did not make more of this elemental collapse at Ragnarǫk. The fact that he does not paraphrase this strophe, and that no mention is made of *Vǫluspá*'s reference to the earth arising from the ocean *iðiagrœna* ('green once more', st. 56) is perhaps due to the cyclical nature of time that this 'second emergence' suggests: a step too far, perhaps, for a scholar versed in the linear progress of Christian salvation history. What we can say is that in the poem *Vǫluspá* that Snorri is drawing on, water is a constant, whether it be the 'positive presence of the primordial waters that await their first event', or the cataclysmic floods that bring the old world to an end.[29]

2 Water as Enclosure and Conduit

Between the violent events of creation and environmental collapse, water plays a central role in the structuring of the cosmos: it circumscribes and divides the world, creating barriers and conduits between realms and also encompassing all land through the concept of the encircling sea. The sea that surrounds the world, referred to as *umsjá* by Snorri and elsewhere as *úthaf*, represents the ultimate circumscribing by untamed nature. This is a concept that seems to be paralleled by the Old English *garsecg*, a poetic reference to the sea as a 'spear warrior' further specified as *gársecges begang*, 'the circuit of ocean', in the Old English poem *Andreas* (line 530a).[30] The idea of the encircling sea overlaps somewhat with the primal river system, Simek going so far as to suggest that Élivágar 'can be understood to be a name for the proto sea surrounding the world'.[31] There is contradiction in the representation of this enveloping body of water: sometimes the crossing to giant land seems to involve crossing the sea, whilst other sources make it clear that this primal ocean encompasses all worlds: after all, Thor sets out from Hymir's home in giant land in his quest to fish the Midgard Serpent from the depths, passing into seas that even the giant Hymir is scared to traverse. Snorri also tells us that it is Ymir's blood that is used to form the lakes and seas, and is clear that the encircling ocean is also created from this deluge in Snorri's account: indeed, he tells us 'þar af gerðu þeir sjá þann er þeir gerðu ok festu saman jörðina, ok lǫgðu þann sjá í hring útan um hana' ('out of this [blood] they made the sea with which they encompassed

29 Dronke, *The Poetic Edda*, II 34.

30 Krapp G.P. (ed.), *The Vercelli Book*, Anglo-Saxon Poetic Records 2 (New York: 1932) 3–51. *Garsecg* is used four times for the sea in this single poem.

31 Simek, *Dictionary* 73.

and fastened the world, and they placed this sea in a ring around the outside of it', 12). Snorri goes on to add that 'most men think it impossible to cross'.[32]

The idea of encirclement by water is also attested in kennings, which Quinn points out have predominantly negative connotations: the sea is 'sometimes glittering like a necklace but more frequently described as a constraining force, isolating land masses and inhibiting travel between them'.[33] It is this *úthaf* – beyond control of even the gods – that we can perhaps think of as coterminous with the primordial sea from which the earth is dragged at creation, and into which *Vǫluspá* tells us the world will sink. Of course, fear of this space did not stop Norse settlers sailing out into the expanse of the North Atlantic, but it is interesting that Ginnungagap was later associated with the seas between Greenland and Vinland, 'the wildest of real oceans known to sailors'.[34] The fact that this sea is home to the Midgard Serpent is a clear indication that it was viewed as a 'chthonic force on the outside of society' and a space beyond the effective control of the gods.[35]

Bodies of water, as well as circumscribing the world, also serve as divisions within it. *Vafþrúðnismál* (st. 16) refers to the river Ífingr as the border between the realm of the gods and that of the giants, stressing that the river will not freeze and will flow freely for all time, ensuring that it remains an effective barrier against the incursion of the Jǫtnar. In other contexts, including *Hymiskviða*, it seems to be the primordial river(s) that Thor must cross to access Jǫtunheimr: Hymir's home is located somewhere 'east of Élivágar' (st. 5). Several other rivers are mentioned as barriers, including the river Vimur whose ferocity almost sweeps the god away when crossing to the hall of the giant Geirrǫðr. *Grímnismál* (st. 29) also refers to four rivers that Thor must cross 'every day' when he goes to pass judgements at the foot of the world tree: what these rivers, referred to as 'Kǫrmt ok Ǫrmt / ok Kerlaugar tvær' (st. 29), are dividing is unclear. However, it does add to the impression that Thor's crossing

32 It is tempting to make a further distinction between Ymir's blood comprising the saline sea, and fresh-water rivers which flow from the spring Hvergelmir, but there is another myth accounting for the saltiness of the sea. The giant girls Fenia and Menia grind out salt from a mill at the bottom of the ocean creating a whirlpool above them, according to Snorri, who relates the myth in *Skáldskaparmál*; the Old Norse poem *Gróttasǫngr* also references the giant girls and the mill, but not the myth of the grinding of salt in the sea.

33 Quinn, "Mythologizing the Sea" 80.

34 Dronke, *The Poetic Edda, II* 34. Simek points out that the north-west passage is labelled Ginnungagap in Bishop Guðbrandur Þorlaksson's map of 1066, *Dictionary* 110.

35 Clunies Ross M., *Prolonged Echoes: Old Norse Myths in Medieval Northern Society, Part I* (Odense: 1994) 268.

of watery boundaries is really fundamental to his divine role – wading rivers is an everyday activity rather than an occasional adventure.[36]

Another river that seems to serve as a boundary is Gjǫll (or 'Clamour'): Snorri tells us that Hermod must cross this river using a bridge, the Gjallarbrú, in order to gain entrance to the underworld, and Simek suggests on this evidence that 'it clearly forms the border to Hell'.[37] Whether this idea predates Snorri or not is difficult to judge, but it does fit with the function of rivers as boundaries elsewhere in the mythology. Several other rivers are connected to the underworld: the river Slíðr ('Sharp-edged') is mentioned as a river of Hel in *Vǫluspá* (st. 36) and the list of rivers in *Grímnismál* (st. 28), and Snorri tells us – probably drawing on a trope of Christian visionary literature as Simek suggests – that it flows from a poisonous source and is full of knives and swords.[38] Finally, we have reference to adulterers, murderers and perjured men wading 'þunga strauma' (*Vǫluspá*, st. 38) in the build-up to Ragnarǫk. In a particularly grizzly image, the dragon Níðhǫggr is said to suck corpses there, and whether or not this is the 'Corpse Shore' mentioned in the previous strophe, it is clear that we are dealing with the underworld and the wading of a river to 'cross over' into Hel. A similar image appears in *Reginsmál* st. 4, where the river is named as Vaðgelmir and those who lie and slander are forced to cross it. Snorri relocates this image of wading the poisonous river to a time after Ragnarǫk and understands it as a punishment, but, as Kure argues, in doing so Snorri may have been 'deliberately seeking to make the strophes of *Vǫluspá* more similar to Revelation than they actually are'.[39] Kure goes on to suggest that 'wading heavy currents', rather than a punishment, could instead be read as a 'a passing across water from the world of the living to the realm of the dead': all must pass through water, in other words, but there are different destinations.[40] *Grímnismál* certainly associates wading a river with entrance to a more positive afterlife in Valhalla: the strong current of the river Þund 'þikkir ofmikill / valglaumi at vaða' ('seems to the joyful dead warriors too great to wade') (st. 21). There is also a longstanding association between ships and passage to the afterlife in the Norse tradition, including through the famous ship burials and stone ship settings that are both probably related to the idea of a crossing of water

36 John Lindow points out that 'Thor's crossing of rivers may have to do with the fact that he does his business mostly in the realm of the giants, who live on the other side of boundaries', *Norse Mythology: A Guide to the Gods, Heroes, Rituals and Beliefs* (Oxford: 2001) 290.

37 Simek, *Dictionary* 111.

38 Ibidem 294.

39 Kure, "Wading Heavy Currents" 84.

40 Ibidem 86.

to reach the afterlife, or at the very least attest to 'a strong connection between death and the domain of the sea'.[41]

Water, then, particularly in the form of treacherous rivers, seems to play a key role as a boundary in Norse myth, dividing the realm of the gods from that of the giants, delineating the Norse underworld and the passage from life to death, and circumscribing the world through the concept of the encircling ocean. This may reflect the reality that certain bodies of water restrict passage through the landscape: indeed, the crossing of dangerous rivers must have been a lived experience for most early medieval Scandinavians, and rivers also constituted boundaries between various polities, including the Göta River where 'the three kingdoms of Scandinavia converged'.[42] However, there is also something in the properties of water that makes it particularly effective as a symbol of the imperfect division of worlds: a river in spate, or the noisy river to the underworld *can* be crossed, even if it is dangerous for humans to do so; it is a fluid barrier that allows for both effective separation and the possibility to move (with difficulty) between realms. Clunies Ross also points out that 'in keeping with its physical properties as a liquid, water in Old Norse myth belongs to both the vertical and horizontal dimensions, for it flows both across and below',[43] whilst Quinn draws attention to the fact that bodies of water have 'both surface and depth': humans can traverse the surface but also access this depth, even if this underwater realm is associated almost exclusively with death.[44] This ability of water to move across different planes – to fall from the sky and to drain into the underworld as well as forming a surface to traverse – undoubtedly contributes to this element's role as an effective conduit within the cosmos, as well as representing at times an effective barrier to human access. This property is of course once more a function of observed reality: just as the tree is a suitable symbol of the link between different realms (with the tree branches reaching into the sky, and the roots running into the earth beneath our feet), water falls from the heavens, flows across, and evaporates from the land, and both emerges from and disappears into the ground.

The presentation of water in the myth may well be based in its real world properties, but it can also be said to amount to a structuring principle in the Norse cosmos, constituting a complex hydrological process which Quinn,

---

41 Quinn, "Mythologizing the Sea" 75.

42 Helle K., "Norway", in Pulsiano P. – Wolf K. (eds.), *Medieval Scandinavia: An Encyclopedia*, (Abingdon: 2016) 436.

43 Clunies Ross, *Prolonged Echoes* 53.

44 Quinn, "Mythologizing the Sea" 76.

in her discussion of the concept of 'Liquid Knowledge' in Old Norse poetry, identifies as a 'global liquid recycling system'.[45] The well Hvergelmir, itself the source of all rivers, is fed by dew dripping from the horns of the stag Eikþyrnir who is said to stand on the roof of Valhalla and feed on the foliage of Yggdrasill *Grímnismál* (st. 26). Yggdrasill itself is fed by the three wells under its roots, themselves representing access to the underworld, and the dews that form in its leaves in turn feed the rivers. A goat who eats the foliage of Yggdrasill also provides an endless supply of mead to the warriors in Valhalla, and the tree itself is doused in milky water from the well known as *Urðarbrunnr* in order to protect it from decay. With all these liquid exchanges taking place – the tree raising water from the wells which feed the rivers which are themselves fed by the dews from the tree – the cosmos might well be represented as a 'hydroponic cycle', in which wells and rivers, Norns and mythical animals, and even the world tree itself, are conduits.[46]

This hydroponic cycle also seems to transcend worlds and the difficulty of crossing between them: rivers rise in Niflheimr (the primordial world); and according to Snorri's *Gylfaginning* the sacred wells are located in Midgard, Jotunheim and Niflheimr – perhaps synonymous here with Hel, the underworld which even the gods cannot gain access to except through death.[47] It is thus a medium that finds its way between realms far more effectively than the gods. The movement of water between realms is important to stress because it has a bearing on the essential connectivity of the different worlds in the Norse cosmos: the flow of water demonstrates the *possibility* of access, but also frustrates and represents the difficulty of this same access. We can see water bubbling up from a spring, but it is not easy to trace where it comes from; we see rain falling from the sky but cannot access its source. It also serves to highlight the complex symbolism of water: in 'being capable of movement and therefore difficult to define' it is 'particularly associated' with uncontrollable natural forces, as Clunies Ross suggests:[48] yet it also serves to bound and protect the world of gods and humans, and it is a source of creation, riches, wisdom, and sustenance. These somewhat contradictory qualities are in turn reflected in the various personifications of water within the myth cycle.

45 Quinn, "Liquid Knowledge" 204.
46 Ibidem 205.
47 Or by borrowing the horse Sleipnir, as Hermóðr does in an ultimately unsuccessful attempt to parley with Hel and have Baldr released from her domain.
48 Clunies Ross, *Prolonged Echoes* 54.

3 Water Personified

T. S. Eliot's disarmingly simple observation that 'The sea has many voices, / Many gods and many voices' in his maritime poem 'The Dry Salvages' could well have been written about Norse conceptions of the sea.[49] The deities and supernatural beings associated with water – and particularly with the sea – reflect this element's particular qualities as well as its essential duality as both a productive and destructive force and its positioning beyond the direct control of the gods. There is, it should be pointed out, nothing quite like the layered anthropomorphism of water afforded to the other elements: wind, for example, is associated with the giant eagle Hræsvelgr – a figure only alluded to briefly in *Vafþrúðnismál* (st. 37) and in Snorri's *Gylfaginning* – or with the figure of Kári, about whom even less is known.[50] Fire, personified through the fire giant Surt at Ragnarǫk, does not seem to have a positive counterweight to express its utility, though Loki at one point does contend with a giant named Logi (wildfire personified) in a contest to devour a trough of food.[51] The element of earth, anthropomorphised through Thor's mother Jǫrð, is a more complex case, as it is also strongly associated with the jǫtnar as a whole, whose domain is more often than not mountain, rock, and lava field. Water, however, is afforded special treatment in the surviving myth, and its functions seem to have been divided between several deities.

Njǫrðr, patriarch of the Vanir gods, is the deity associated most closely with the sea. Snorri tells us that he is to be prayed to before sea-voyages and fishing, and his unsuccessful marriage to the mountain giantess Skaði expresses the irreconcilable differences between life on land and in the sea, or perhaps more precisely between coast and interior.[52] In Snorri's account of the god, Njǫrðr is not associated exclusively with the sea, however, but with those same three elements that are represented most often as destructive forces: Snorri tells us in *Gylfaginning* that 'Hann ræðr fyrir gǫngu vinds ok stillir sjá ok eld' ('He rules over the course of the winds, and stills the sea and fire', 23), suggesting a broader protective function than that of seafaring alone. His dwelling – Nóatún ('Ship Enclosure') does further associate him with the sea: his feet are

49 In Eliot T.S., *Four Quartets* (New York: 1943).

50 On the figure of Kári, see Clunies Ross, "Snorri Sturluson's Use of the Norse Origin-Legend" 57, where she also makes the important observation that 'the earliest known representations in West Norse of the four primal elements tend to be more concrete in their expression than the Latin nouns they translate'; these include *sæ* ('sea') for 'water' in an early Norwegian Book of Homilies.

51 *Gylfaginning* 40.

52 *Gylfaginning* 23–24.

also exceptionally beautiful – a fact that wins him his bride Skáði – and this may also be because they are 'cleanest' from his association with the sea, as Judith Jesch intimates.[53] Furthermore, the insult that Loki directs at Njǫrðr in *Lokasenna* – that giantesses urinated into his mouth (st. 34) – is an image that makes most sense if his domain is considered to be the sea into which anthropomorphised rivers drain. All of these images seem to associate him more with coastal waters than the deep sea, and particularly with the shoreline: the fact that he is said to be able to grant wealth to anyone who asks for his aid may reflect not only the riches of the sea at large, but particularly the importance of the wealth deposited on the shore.[54] These 'gifts' of beached wood and stranded sea mammals would have been of particular importance in Snorri's Iceland: after the rapid deforestation of the island in the settlement age, driftwood became the primary source of timber, and the sagas record many conflicts over rights of access both to this precious commodity and to the carcasses of beached whales.[55]

Whilst Njǫrðr is often referred to as the god of the sea, the dialectic encompassed by the ocean, and by water more generally, is represented most prominently by the sea giant Ægir and his wife Rán. Both these figures are clearly personifications of the sea, but their roles in the myth cycle are quite distinct. Ægir – whose name may be related to Germanic **ahwo*, Lat. *aqua* ('water') as Simek points out[56] – is a well attested figure who is usually portrayed as having a more positive relationship to the gods than is typical for giants. Snorri uses Ægir's visit to Asgard and the questions he asks of the god of poetry, Bragi, as the main structuring conceit of his *Skáldskaparmál*.[57] Ægir also hosts a feast to which the gods and elves are invited, an episode with some prominence in the myth cycle, it being the setting for Loki's slandering of the gods in the poem

53 Jesch J., *Women in the Viking Age* (Woodbridge: 1991) 138.

54 *Gylfaginning* 23–24.

55 See, for example, the decidedly unheroic battle between men of Kaldbakr and Vík over the disputed carcass of a beached whale in *Grettis saga*. Earlier in the same saga we are told that there was so much driftwood that it 'um rekann var ekki skilit, því at þeir váru svá nógir þá, at hverr hafði þat, er vildi' ('about driftwood rights nothing was decided, because there was so much there that whoever wanted it could have it'), Guðni J. (ed.), *Grettis Saga Ásmundarsonar* (Reykjavík: 1936) 23.

56 *Dictionary* 2.

57 In this account the giant Ægir is also referred to as Hlér, and elsewhere he is called Gymir: these are names which are also used to reference the sea in Old Norse poetry, blood referred to as *sárgymir* ('Gymir/sea of the wound') in *Hákonarmál* st. 7, for example. Snorri states of these names for Ægir that 'alt er eitt' ('all are one'), Snorri Sturluson, *Edda: Skáldskaparmál*, ed. A. Faulkes (London: 1998) 37. All references to *Skáldskaparmál* refer to this edition.

Lokasenna; the quest to find a receptacle for Ægir to brew drink for the gods also underlies Thor's encounter with Hymir and famous fishing trip, related in *Hymiskviða*. Although this feast does not have a happy outcome, Ægir is represented as a generous host, and it is notable that the light at the banquet is supplied by gold piled at the centre of the hall (the origin of the kenning 'Ægir's fire' for gold, according to Snorri).[58] Ægir with his positive associations, conspicuous wealth, and amicable relationship with the gods, seems to primarily represent the riches that sea provides, amongst which we might count fish and seafood, cetaceans and seals, sea-birds, seaweed and salt – all important resources for communities along the North Atlantic littoral.

Rán, on the other hand, referred to in the *Prose Edda* as 'Gymir's spray-cold spæ-wife',[59] is more consistently associated with the danger and unpredictability of seafaring. As well as associating Rán with 'the physicality of the sea – its iridescence and power', kennings seem to suggest a close association between this mythological figure and death at sea, as Quinn points out.[60] Indeed, drowning is 'more or less idiomatically equated with "falling into Rán's hands"', and with her net used to entrap sailors she 'embodies the sinister side of the sea, at least in the eyes of the late Viking Age Icelandic seafarers'.[61] The coding of the male personification of the sea as positive and 'civilised' (seen in the roles of interlocutor and attentive host) and the female deity as a wild, dangerous force who drags sailors to their death of course maps on to the binary of the civilising patriarchy of the Æsir and the often feminised natural world. There is even some basis for considering Ægir to be a god rather than a giant because of his unusually positive relationship with the Æsir.[62] Clunies Ross points out that Ægir is 'never said to be a giant in the *Edda*', whilst Lindow suggests that 'his inclusion in the thulur as a giant seems questionable' precisely because of this positive relationship with the gods.[63] However, this duality expressed through a (giant) husband and wife, one unusually generous to the gods and another overtly hostile to seafarers, does not need to be explained by positioning Ægir

58 *Skáldskaparmál* 40–41.

59 Faulkes (trans.), *Snorri Sturluson: Edda* 91. 'spæ-wife' is Faulkes' translation of ON *vǫlva*, meaning seeress or prophetess.

60 Quinn, "Mythologizing the Sea" 84.

61 Simek, *Dictionary* 260.

62 Andy Orchard refers to him as 'the Norse god of the sea', though later stating that 'it may well be that Ægir's adoption into lists of the Æsir is a late development', *Cassell's Dictionary of Norse Myth and Legend* (London: 1996) 21 and 23. Rán is also sometimes referred to as a deity.

63 Lindow, *Norse Mythology* 47. Clunies Ross, "Snorri Sturluson's Use of the Norse Origin-Legend" 56.

as a god: it may also be a function of the particular relationship of a seafaring community to the constant presence of the ocean. The sea is a source of sustenance and wealth, and a means of rapid transit between communities, but also a space inherently inhospitable to humans and wholly resistant to control.[64] This dichotomy is appropriately expressed by a giant couple, the husband disarmingly generous and amenable – almost god-like – and wife wild and deadly. That this dichotomy is also clearly gendered does not mean that Ægir belongs with the Æsir, even if Snorri pushes him in this direction: he is still an outsider, and a representative of an occasionally benevolent force that remains beyond the effective control of the gods.

The blending of these positive and negative characteristics of the sea can clearly be seen in the nine giant daughters of Ægir and Rán known as the waves. Their names (related by Snorri) reference the diverse states of the sea – from the relatively neutral Hefring ('Raising') and Bylgja ('Billowing') to the more polarised Himinglæva ('Heaven-Bright') and Blóðughadda ('Bloody-Haired') – and thus take on characteristics of both mother and father. Clunies Ross further suggests that the duality in the personification of the sea can be linked to observed nature, arguing that 'the reason why the sea was conceptualised as an entity where male and female principles met and mingled was that the foam that results from the action of the waves was likened to semen',[65] the reference to the god Heimdall being birthed by these same nine daughters of the waves possibly related to 'the foam, scum and other flotsam that is produced in the turmoil and accumulates upon beaches', this being 'a place at the fertile margin of earth and sea'.[66] If this reading is accepted, then the sea – capable of procreation with or within itself – seems to be somewhat independent of the more common union between procreatively powerful giantesses and pacifying gods.

4 Thor as Water Queller

Aside from the sea, rivers and other bodies of water also seem to have received anthropomorphic treatment, and in some cases, they are clearly associated with female reproductive potential as well as with the dangers of a destructive, and feminised, natural world. The clearest example of this is the myth

64 It is notable that this dual personification is not replicated in respect to other elemental forces: fire, for example, personified through the fire giant Surt, does not seem to have a positive counterweight to express its utility.

65 Clunies Ross, *Prolonged Echoes* 175.

66 Ibidem 176.

of Thor's wading of the river Vimur to the home of the giant Geirrǫðr.[67] In Snorri's retelling the river, conceived as one of the bodies of water dividing the realm of the gods from that of the giants, is swollen by a giantess named Gjálp straddling the water upstream and 'causing it to rise': either urinating or filling the river with her menstrual blood. After struggling against the swollen waters with the aid of a staff, Thor uses a well-aimed rock to stem the flow, Snorri having Thor exclaim, 'At ósi skal á stemma' ('at its source a river must be stemmed!').[68] Clunies Ross, in a sustained analysis of this episode as related in both *Skaldskaparmál* and the skaldic poem *Þórsdrápa*, has pointed out that in the poem Gjálp is not clearly identified as the source of the rising waters, even if the river Thor contends with is certainly personified as female. As she argues, the river Vimur may in fact be linked to the god's giant mother Jǫrð (or Earth), and the emergence of rivers in the mountains may have been associated in the Norse imagination with menstruation or urination of female chthonic beings.[69] In this psychosexual reading of the narrative of *Þórsdrápa*, Thor's summoning of his Æsir strength and use of his stabilising staff against a river being swollen by a female giantess is an act of male assertion when faced with his mother's attempt to return him under her power. Supporting this reading is the reference in *Þórsdrápa* to the god declining to take 'Njǫrðr's option', this most likely being an oblique reference to the myth alluded to earlier in this essay: the insult levelled at the (costal) sea god Njǫrðr that the giant daughters of Hymir urinated into his mouth.[70] So, Thor will not swallow the river water / urine as Njǫrðr does (and indeed that all personified seas must do if the rivers that drain into them are imagined as female waters) and instead uses his strength, and his phallic staff Gríðarvǫlr, to brace against the current, as well as using a rock to figuratively and literally stem the flow emanating from Gjálp. Expanding the imagery from rivers to antagonistic waters more generally, he and

67 The most extensive elaboration of this myth is in the skaldic poem *Þórsdrápa* ('Eulogy on Thor') by the tenth-century skáld Eilífr Goðrúnarson (composed, perhaps, for his patron in Trøndelag, the redoubtable pagan Earl Hákon of Hlaðir) which Snorri quotes in the *Skáldskaparmál* section of his *Prose Edda* and which was probably the main source for the account that he provides. A related narrative is recorded by Saxo, and the fact that Snorri quotes 'a fragment of a poem in ljóðaháttr [...] implies that there was an eddic poem that related this story too' (Faulkes, *Skáldskaparmál* XXIII).

68 *Skáldskaparmál* 25.

69 Clunies Ross M., "An Interpretation of the Myth of Þórr's Encounter with Geirrøðr and His Daughters", in Dronke U. – Helgadottir G.P. – Weber G.W. – Bekker-Nielsen H. (eds.), *Speculum Norroenum: Studies in Memory of Gabriel Turville-Petre* (Odense: 1981) 372–374.

70 *Lokasenna* st. 34.

Thjalfi's hearts are described, in the words of Eilífr Goðrúnarson, as 'djúpakǫrn [...] dolgs' ('sea-acorns of enmity').[71]

Thor is referred to by Adam of Bremen, in his account of the cultic centre of Uppsala, as the god who 'presides over the air, which governs the thunder and lightning, the winds and rains, fair weather and crops'.[72] He is often depicted as riding a chariot driven by goats, and the entrenched notion of Thor as a god of thunder and lightning (challenged recently by Declan Taggart in his monograph *How Thor Lost His Thunder*) has tended to reinforce the impression that his primary association is with the upper atmosphere, like the god Jupiter with whom he became associated. Yet when Thor and his companions undertake a series of contests against elemental forces at the home of Utgarða-Loki, it is notable that whilst Loki competes with wildfire, and Thjalfi attempts to race thoughts through the air, Thor combats water, and his attempt (and failure) to drain a horn whose end actually extends into the ocean, is said to be the origin of the tides.[73] There is thus some basis for understanding Thor as a god with a particularly important relationship to water. This does not amount to Thor personifying the sea like the figures just discussed, or 'resembling a giant of the ocean' as Lotte Motz argues, though he does take on 'aspects' of this force: rather, the son of earth's association with water is almost exclusively antagonistic.[74]

Thor's most renowned exploit is his fishing trip in search of the World-Serpent, recounted in the poems *Hymiskviða* and *Ragnarsdrápa*, retold in Snorri's *Prose Edda*, and represented in several stone carvings from the Viking Age. This episode is perhaps so prominent because it encapsulates the danger posed by the deep ocean for a seafaring people, and the god's doomed conflict with the world-encircling serpent Jǫrmungandr – a monster itself personifying 'the uncouth power of the sea' – represents Thor's greatest challenge as a bulwark against the hostile elements.[75] Many of Thor's other exploits centre on river crossings: Clunies Ross notes that Thor 'is often obliged to enter the element of water' in the course of his duties which she equates to engaging with representatives of nature and particularly 'chthonic forces'.[76] In addition to his difficult crossing of the river Vimur, recounted in *Þórsdrápa*, Thor

71 *Þórsdrápa* st. 11. 'Sea acorns' refers to stones, with the image implying that they are strongly rooted to the sea-floor.

72 *The History of the Archbishops of Hamburg-Bremen*, ed. and trans. F.J. Tschan (New York: 2002) 207.

73 *Gylfaginning* 40–41.

74 Motz L., "Þórr's River Crossing", *Saga-Book* 23 (1990–1993) 481.

75 Ibidem.

76 Clunies Ross, *Prolonged Echoes* 267.

carries Aurvandill in a basket through the frozen waters of Élivágar and is said to wade across rivers on a daily basis to reach the well of Urðr where the gods hold council: his ability to cross the treacherous river (or rivers) that separate Asgard and Jǫtunheimr is also essential to his role in containing the giants and the natural forces they embody. In one telling episode, Thor is prevented from crossing a body of water by a ferryman who insults him from a safe distance (the ferryman is his father Odin in disguise). This myth, recounted in the Eddic poem *Hárbarðsljóð*, draws attention to how much Thor relies on being able to cross water to fulfil his function as a protective deity: being unable to cross water renders him as impotent as he is in the more famous episode in which his hammer is stolen.

Thor is often situated as the paradigmatic representative of the civilising gods contending with forces of nature, 'representative[s] of chaos',[77] or more specifically with 'ordering of chaos of a particular kind: unbridled female sexuality'.[78] His struggle with water is therefore situated as a variation on this wider scheme as protective deity. However, Thor's association with water is so extensive that it is tempting to understand his primary role to be a queller of this particular elemental force. Tarrin Wills has speculated on Thor's invocation before crossing bodies of water, for which he gathers evidence from runestones and placenames as well as Old Norse literature including *Þórsdrápa*,[79] and it may be that the god was particularly associated with protection against water in all its manifestations. Even when he battles with other elements, such as catching the fiery metal ingot hurled at him in Geirrǫðr's hall, it is notable that the poet of *Þórsdrápa* refers to this assault in terms of Thor's hands swallowing a molten drink, with the iron described as 'the heavy red mouthful of the seaweed of tongs' (trans. Marold, st. 17).[80] Thor may be born of the giantess Jǫrð (or Earth) and operate across many planes of existence, including sky and earth, but in the surviving sources he is more frequently associated with the

77 Motz, "Þórr's River Crossing" 482.

78 Clunies Ross, "An Interpretation of the Myth of Þórr's Encounter with Geirrøðr and His Daughters" 390.

79 See Wills T., "Þórr and Wading", in Krüger J. – Busch V. – Seidel K. – Zimmermann C. – Zimmermann U. (eds.), *Die Faszination des Verborgenen und seine Entschlüsselung–Rāđi saʀ kunni: Beiträge zur Runologie, skandinavistischen Mediävistik und germanischen Sprachwissenschaft Christiane* (Berlin: 2017) 411–428 and "The Landscape of Thor Worship in Sweden", in García Losquiño E. – Sundqvist O. – Taggart D. (eds.), *Making the Profane Sacred in the Viking Age: Essays in Honour of Stefan Brink* (Turnhout: 2020) 107–209.

80 Trans. E. Marold, "Eilífr Goðrúnarson, Þórsdrápa 17", in *Poetry from Treatises on Poetics*, trans. K.E. Gade – E. Marold (Turnhout: 2017) 113.

water he contends with – in its most primal and destructive forms – than with any other elemental force.

5 Conclusions

Water, in its many guises, is so ubiquitous that it is bound to feature prominently in any mythology, let alone that of a people for whom seafaring was such an essential activity, and this short survey has only ruffled the surface of the many encounters with and processes involving water in the surviving narratives. On a macrocosmic level, water seems to have been understood as a primordial element involved with but also preceding creation and implicated in the destruction and renewal attending the Norse apocalypse. Whilst Snorri's account of the cosmos is clearly influenced by the prevailing medieval understanding of the elements and their combination to form matter, as well as the theory of the world as body, by paying close attention to the poetic sources he was drawing on, and contradictions in his own presentation, a slightly different paradigm begins to emerge in which water can be understood to represent a primordial medium for life rather than one of its elemental building blocks. Water's ability to carry other substances – whether poison, sap, sea foam, or salt – is a characteristic that makes it so useful as a symbol of conduction and connection in the myth cycle, whilst its different states – ice, raging torrent, dew, vapour – means that it is often associated with transformative, including life-giving, processes. On a structural level, water performs a useful function of effectively dividing worlds, but also allowing for access – sometimes fraught with danger – not only to geographical space but to different realms on a vertical plane, most notably the underworld. Whilst the World-Tree might be thought of as the main structuring entity in the cosmos, certainly on the vertical axis, the complex hydrological cycle expresses the connection and interdependence of mythical space on all planes.

The many different attributes, and changeability, of water, are also expressed in Old Norse mythology through the various personifications of the sea, and particularly the polarisation of wealth/prosperity and dangerous unpredictability through the spouses Ægir and Rán and their daughters the waves. Thor, whilst associated with 'earth' through his mother Jǫrð, contends most often with the element of water, whether that be wading treacherous rivers, fishing for the Midgard-Serpent, or attempting to drain the sea in a drinking competition. It is Thor's relationship with water that is perhaps most revealing of the element's positioning within the cosmic struggle between gods and giants. Thor's wading of the river Vimur is clearly an example of male strength prevailing over a feminised element, but the god's relationship with the sea is less clear cut: here, rather than representing a productive energy which the male gods seek to control, it seems rather to signify a primordial force that is beyond containment, associating it with the third class of monstrous beings who 'represent uncontrollable natural forces that cannot be spatially confined', as Clunies Ross argues.[81] Water, from which the cosmos arose, will engulf the earth at Ragnarǫk, and will allow for the emergence of a new world. In that sense, water is both the beginning and the end, the element that precedes, encircles, infuses and outlasts the world. As for whether water in its different forms is more frequently coded positive or negative in the myth cycle, a much more extensive study is needed: but we might also consider whether the element exists outside these boundaries, and even beyond both defined space and mythic time.

Bibliography

Primary Sources

Dronke U. (ed. and trans.), *The Poetic Edda: Vol. II, Mythological Poems I* (Oxford: 1997).

Dronke U. (ed. and trans.), *The Poetic Edda: Vol. III, Mythological Poems II* (Oxford: 2011).

Eliot T.S., *Four Quartets* (New York: 1943).

Faulkes A. (ed.), *Edda: Prologue and Gylfaginning* (London: 2nd ed. 2005).

Faulkes A. (ed.), *Snorri Sturluson, Edda: Skáldskaparmál* (London: 1998).

Faulkes A. (trans.), *Snorri Sturluson: Edda* (London: 1998).

Gade K.E. – Marold E. (trans.), *Poetry from Treatises on Poetics*, Skaldic Poetry of the Scandinavian Middle Ages 3 (Turnhout: 2017).

Guðni J. (ed.), *Grettis Saga Ásmundarsonar*, Íslenzk Fornrit 7 (Reykjavík: 1936).

Krapp G.P. (ed.), *The Vercelli Book*, Anglo-Saxon Poetic Records 2 (New York: 1932).

Kristjánsson J. – Ólason V. (eds.), *Eddukvæði*, 2 vols. (Reykjavík: 2014).

Tschan F.J. (ed. and trans.), *The History of the Archbishops of Hamburg-Bremen* (New York: 2002).

Secondary Sources

Clunies Ross M., "An Interpretation of the Myth of Þórr's Encounter with Geirrøðr and His Daughters", in Dronke U. – Helgadottir G.P. – Weber G.W. – Bekker-Nielsen H. (eds.), *Speculum Norroenum: Studies in Memory of Gabriel Turville-Petre* (Odense: 1981) 370–391.

Clunies Ross M., *Prolonged Echoes: Old Norse Myths in Medieval Northern Society, Part I* (Odense: 1994).

81 Clunies Ross, *Prolonged Echoes* 54.

Clunies Ross M., "Snorri Sturluson's Use of the Norse Origin-Legend of the Sons of Fornjótr in his *Edda*", *Arkiv för Nordisk Filologi* 98 (1983) 47–66.

Cohen J.J. – Duckert L., "Introduction: Eleven Principles of the Elements", in Cohen J.J. – Duckert L. (eds.), *Elemental Ecocriticism: Thinking with Earth, Air, Water, and Fire* (Minneapolis, MN: 2015) 1–26.

Dronke U. – Dronke P., "The Prologue of the Prose Edda: Explorations of a Latin Background", in Pétursson E.G. – Kristjánsson J. (eds.), *Sjötíu Ritgerdir helgadar Jakobi Benediktssyni 20. júlí 1977* (Reykjavik: 1977) 153–176.

Halvorsen E.F., "Élivágar", *Kulturhistorik Leksikon for nordisk Medeltid* 3 (1958) 597–598.

Helle K., "Norway", in Pulsiano P. – Wolf K. (eds.), *Medieval Scandinavia: An Encyclopedia*, (Abingdon: 2016) 436–440.

Holtsmark A., *Studier i Snorres Mytologi* (Oslo: 1964).

Jesch J., *Women in the Viking Age* (Woodbridge: 1991).

Kingsley K.S. – Parry R., "Empedocles", in Zalta E.N. (ed.), *The Stanford Encyclopedia of Philosophy* (Online, Summer 2020 edition) https://plato.stanford.edu/archives/sum2020/entries/empedocles/. Last accessed November 3, 2023.

Kure H., "Wading Heavy Currents: Snorri's Use of *Vǫluspá* 39", in Gunnell T. – Lassen A. (eds.), *The Norse Apocalypse: Approaches to Vǫluspá and Nordic Days of Judgement* (Turnhout: 2013) 79–92.

Langeslag P.S., *Seasons in the Literatures of the Medieval North* (Cambridge: 2019).

Larrington C., *The Norse Myths: A Guide to the Gods and Heroes* (Oxford: 2017).

Lincoln B., "In Praise of the Chaotic", in Dill U. – Walde C. (eds.), *Antike Mythen: Medien, Transformationen und Konstruktionen* (Berlin: 2009) 372–390.

Lindow J., *Norse Mythology: A Guide to the Gods, Heroes, Rituals and Beliefs* (Oxford: 2001).

Motz L., "Þórr's River Crossing", *Saga-Book* 23 (1990–93) 469–488.

Nordal G., *Tools of Literacy: The Role of Skaldic Verse in Icelandic Textual Culture of the Twelfth and Thirteenth Centuries* (Toronto: 2001).

Nordvig M., "Creation from Fire in Snorri's Edda: The Tenets of a Vernacular Theory of Geothermal Activity in Old Norse Myth", in Hermann P. – Mitchell S.A. – Schjødt J.P. – Rose A. (eds.), *Old Norse Mythology in Comparative Perspectives* (Cambridge, MA: 2017) 269–289.

O'Donoghue H., *From Asgard to Valhalla: The Remarkable History of the Norse Myths* (London: 2007).

Orchard A., *Cassell's Dictionary of Norse Myth and Legend* (London: 1996).

Pálsson H., "Reflections on the Creation of Snorri Sturluson's Prose Edda", *Scripta Islandica* 68 (2017) 189–232.

Quinn J., "Liquid Knowledge: Traditional Conceptualisations of Learning in Eddic Poetry", in Rankovic S. – Melve L. – Mundal E. (eds.), *Along the Oral-Written Continuum: Types of Texts, Relations and their Implications* (Turnhout: 2010) 183–226.

Quinn J., "Mythologizing the Sea: The Nordic Sea-Deity Rán", in Tangherlini T. (ed.), *Nordic Mythologies: Interpretations, Institutions, Intersections* (Berkeley – Los Angeles, CA: 2014) 71–91.

Simek R., *Dictionary of Northern Mythology* (Cambridge: 1993).

Taggart D., *How Thor Lost his Thunder* (Abingdon: 2017).

Wills T., "Þórr and Wading", in Krüger J. – Busch V. – Seidel K. – Zimmermann C. – Zimmermann U. (eds.), *Die Faszination des Verborgenen und seine Entschlüsselung–Rāđi saʀ kunni: Beiträge zur Runologie, skandinavistischen Mediävistik und germanischen Sprachwissenschaft Christiane* (Berlin: 2017) 411–428.

Wills T., "The Landscape of Thor Worship in Sweden", in García Losquiño E. – Sundqvist O. – Taggart D. (eds.), *Making the Profane Sacred in the Viking Age: Essays in Honour of Stefan Brink* (Turnhout: 2020) 107–209.

FIGURE 14.1 The Sailing of Saint Cuthbert, 12th century. Life of Saint Cuthbert, London, British Library, Yates Thompson MS 26, fol. 26r.: https://commons.wikimedia.org/wiki/File:BLYates Thompson26LifeCuthbertFol26rCuthSails.jpg. Last accessed 14 December 2023.

Effusion on Affusion

Kelly Grovier

I take a basin of water and place a mirror in it
and put it under the wheel of the sun.
– MEISTER ECKHART

Off my skin, the world is nothing
but glint and shimmer – painted
light echoing from the spill
as I am ladled, situla to font,
in the vaulted narthex; nothing

but polished slosh as I slip
from the snarl of copper
dragons. Whether you map
my static waves in the still
pulse of veined vellum –

the cross and sphere
of Isidore's unnavigable mind –
or grind me down to the alchemy
of monks conjuring the surge
from pulverised lazurite

in Lindisfarne and Durrow –
I shape the shift from spirit
to engine, shoulder the weightless
glister of mill wheels as they spin
revolution to revolution.

Index

www.ingramcontent.com/pod-product-compliance
Lightning Source LLC
Chambersburg PA
CBHW071328230926
51849CB00002B/338
9789004548428